HANDBOOK OF URBAN MOBILITIES

This book offers the reader a comprehensive understanding and the multitude of methods utilized in the research of urban mobilities with cities and 'the urban' as its pivotal axis. It covers theories and concepts for scholars and researchers to understand, observe and analyse the world of urban mobilities.

The *Handbook of Urban Mobilities* facilitates the understanding of urban mobilities within a historic conscience of societal transformation. It explores key concepts and theories within the 'mobilities turn' with a particular urban framework, as well as the methods and tools at play when empirical, urban mobilities research is undertaken. This book also explores the urban mobilities practices related to commutes; particular modes of moving; the exploration of everyday life and embodied practices as they manifest themselves within urban mobilities; and the themes of power, conflict and social exclusion. A discussion of urban planning, public control, and governance is also undertaken in the book, wherein the themes of infrastructures, technologies and design are duly considered.

With chapters written in an accessible style, this handbook carries timely contributions within the contemporary state of the art of urban mobilities research. It will thus be useful for academics and students of graduate programmes and post-graduate studies within disciplines such as urban geography, political science, sociology, anthropology, urban planning, traffic and transportation planning, and architecture and urban design.

Ole B. Jensen is Professor of Urban Theory at the Department of Architecture, Design and Media Technology, Aalborg University (Denmark).

Claus Lassen is Associate Professor and Director of the Centre of Mobility and Urban Studies (C-MUS) at Aalborg University (Denmark).

Vincent Kaufmann is Associate Professor of Urban Sociology and Mobility at Ecole Polytechnique Fédérale de Lausanne (EPFL, Switzerland).

Malene Freudendal-Pedersen is Professor in Urban Planning at Aalborg University (Denmark).

Ida Sofie Gøtzsche Lange is Assistant Professor at the Department of Architecture, Design and Media Technology, Aalborg University (Denmark).

HANDBOOK OF URBAN MOBILITIES

Edited by Ole B. Jensen, Claus Lassen, Vincent Kaufmann, Malene Freudendal-Pedersen and Ida Sofie Gøtzsche Lange

Routledge
Taylor & Francis Group
LONDON AND NEW YORK

First published 2020
by Routledge
2 Park Square, Milton Park, Abingdon, Oxon OX14 4RN

and by Routledge
52 Vanderbilt Avenue, New York, NY 10017

Routledge is an imprint of the Taylor & Francis Group, an informa business

British Library Cataloguing-in-Publication Data
A catalogue record for this book is available from the British Library

Library of Congress Cataloging-in-Publication Data
A catalog record has been requested for this book

ISBN: 978-1-138-48219-7 (hbk)
ISBN: 978-1-351-05875-9 (ebk)

Typeset in Bembo
by Swales & Willis, Exeter, Devon, UK

CONTENTS

CONTRIBUTORS

Ransford A. Acheampong, PhD, is Presidential Academic Fellow in Future Cities at the Manchester Urban Institute and the Department of Planning and Environmental Management, University of Manchester. His research interests include understanding user adoption behavior, diffusion trends and socio-spatial implications of emerging mobility solutions such as shared-mobility, Mobility-as-a-Service (MaaS) and autonomous transport.

Ander Audikana is Postdoctoral Research Fellow at the Centre for Applied Ethics at the University of Deusto. Ander received a PhD in Sociology from University of Paris-Est. He was Fulbright-Schuman scholar at George Mason University and University of California, Berkeley, and visiting fellow at the École Polytechnique Fédérale de Lausanne. He holds an MA in City and Regional Planning from University Paris 12 and Sociology degrees from École normale supérieure and University of Deusto.

Michel Bierlaire is Full Professor at EPFL and directs the Transport and Mobility laboratory. He is also the Director of the EPFL Transportation Center. His main expertise is in the development and applications of models and algorithms for the design, analysis and management of transportation systems. Namely, he has been active in demand modelling, operations research and Dynamic Traffic Management Systems. He holds a PhD in Mathematical Sciences from the University of Namur, Belgium.

Thomas Birtchnell is Senior Lecturer in the School of Geography and Sustainable Communities, the University of Wollongong, Australia. His books are *Indovation* (Palgrave Macmillan, 2013), *3D Printing for Development in the Global South* (Palgrave Macmillan, 2014) co-authored with William Hoyle and *A New Industrial Future?* (Routledge, 2016), co-authored with John Urry.

Thomas Buhler is Assistant Professor in Urban Planning and Geography at the University of Burgundy-Franche-Comté (UBFC). His research focuses mainly on the role of habits in urban travel practices and domestic energy use. He is also interested in the process of shaping public policies aimed at changing behaviour in these two areas.

Weert Canzler, MA in Political Science at the Free University of Berlin, doctorate in Sociology at Institute of Sociology at Technical University Berlin, habilitation in Social Science Based Mobility Research at Technical University Dresden. Since 1994 he has been Research Fellow at the Social Science Center Berlin (Wissenschaftszentrum Berlin für Sozialforschung, WZB) and, together with Andreas Knie, cofounder of the 'Project Group on Mobility'. Since 2013 he has been Speaker of Leibniz Research Alliance Energy Transition.

Julie Cidell is Professor of Geography and GIS at the University of Illinois at Urbana-Champaign. Outside of the academy, she has worked as a transportation engineer. She studies how local governments and individual actors matter in struggles over large-scale infrastructure and policy development and the corresponding urban environments that are produced, including airports, railroads and logistics hubs as well as green buildings and urban sustainability policies. Her current research is on university student mobility.

Matteo Colleoni is Professor of Urban Sociology and Director of the Research Centre on Mobility, Tourism and Territory at the University of Milan-Bicocca (Italy). His research interests lie in the area of urban mobility, accessibility and transport exclusion, metropolitan areas and urban change and time–space analysis. Among his recent publications is *Understanding Mobilities for Designing Contemporary Cities*, with Paola Pucci (Springer, 2016).

Federico Cugurullo is Assistant Professor in Smart and Sustainable Urbanism at Trinity College Dublin. He has done extensive empirical research in the Middle East and Southeast Asia, exploring smart and eco-city projects. His most recent work examines the impact that artificial intelligence is having on urban design, planning and governance.

Guillaume Drevon is a Geographer. He is currently developing his research at the Urban Sociology Laboratory of the Swiss Federal Institute of Technology Lausanne. His research focuses mainly on rhythms of life. As part of his work, Guillaume Drevon studies strategies developed by individuals to deal with the temporal pressures of everyday life by putting the notion of temporal vulnerability into perspective. Today, Guillaume Drevon develops the theme of rhythms of life and city in the context of several researches whose objective is to better understand the evolution of time perception in contemporary societies. Guillaume Drevon is an Associate Researcher at the Luxembourg Institute of Socio-economic Research. He also participates in scientific editorial activity by being a member of the editorial board of the journal *espacestemps.net*.

Aurore Flipo is Sociologist, Associate Researcher at the University of Grenoble-Alpes (PACTE) and Post-doctoral Researcher at the Ecole Nationale des Travaux Publics de l'État (LAET). Her doctoral dissertation was dealing with the patterns of social and spatial mobility of young intra-European migrant workers. Her work focuses on the interactions between processes of social stratification and spatial mobility at different scales (local, national, international) and individuals' socio-spatial arrangements.

Malene Freudendal-Pedersen is Professor in Urban Planning at Aalborg University and has an interdisciplinary background linking Sociology, Geography, Urban Planning and the Sociology of Technology. Her research focuses on mobilities practices, the interrelation between spatial and digital mobilities and its impacts on everyday life, cities and societies.

She is co-organizing the International Cosmobilities Network, co-founder and co-editor of the journal *Applied Mobilities* and the book-series Networked Urban Mobilities, both at Routledge.

Alexis Gumy is a PhD student at the Urban Sociology Laboratory of the École Polytechnique Fédérale de Lausanne (EPFL). Graduated from the same school in Civil Engineering, his research is based on quantitative methods applied to the analysis of lifestyles and forms of mobility. Alexis seeks to transcend classical, functionalist and normative statistical approaches, considering experience or opinions as a determining factor in the study of behaviour. He is currently interested in the relationship between practices and representations in European cross-border agglomerations (SNF). Finally, Alexis is also a Lecturer in several courses at EPFL.

Luc Gwiazdzinski is a Geographer (Université Grenoble Alpes). A teacher-researcher in planning and urban development, he has also created and directed a time and mobility agency, a development agency and an urban planning and sustainable development agency. He has conducted numerous research projects, international conferences and published numerous articles and some 15 books on time and mobility issues, including: *La nuit dernière frontière de la ville*, l'Aube; *La nuit en questions*, l'Aube; *Nuits d'Europe, Pour des villes accessibles et hospitalières*, Ministère des transports, UTBM Éditions; *Si la route m'était contée, Un autre regard sur la route et les mobilités durables*, Eyrolles; *L'hybridation des mondes*, Elya; Chronotopies, *Lire et écrire les mondes en mouvement*, Elya; *Tourismes et adaptations*, Elya.

Katrine Hartmann-Petersen is Associate Professor in planning and mobilities, Roskilde University, Denmark. She has a transdisciplinary background investigating the interconnectedness between everyday life, urban planning and mobilities. She is former special advisor in municipal planning departments dealing with mobilities planning in practice. She is a member of the Cosmobilities Network taskforce.

Edward H. Huijbens, born in 1976, is a Geographer and graduate of Durham University, England. He chairs Wageningen University's research group in cultural geography. Edward works on tourism theory, issues of regional development, landscape perceptions, the role of transport in tourism and polar tourism. Edward has authored over 35 articles in several scholarly journals such as *Annals of Tourism Research*, *Journal of Sustainable Tourism*, *Tourism Geographies*, published three monographs in both Iceland and internationally and co-edited four books.

Hanne Louise Jensen is Associate Professor at the Department of Sociology and Social Work, Aalborg University. Her research interests are within the fields of mobility, everyday life and the social production of place. Currently she is engaged in research on the relation between belonging and spatial nostalgia and on bridges and leisure mobility on the Limfjord.

Martin Trandberg Jensen is Associate Professor in Tourism and Mobilities at the Department of Culture and Global Studies, Aalborg University. His current research focuses on nonrepresentational theories, tourism mobilities and sensuous scholarship. Epistemological questions and methodical innovations remain central to his work.

Ole B. Jensen is Professor of Urban Theory at the Department of Architecture, Design and Media Technology, Aalborg University (Denmark). He is Deputy Director and Co-founder of the Centre for Mobilities and Urban Studies (C-MUS). He is the author of *Staging Mobilities* (Routledge, 2013) and *Designing Mobilities* (Aalborg University Press, 2014), the Editor of the four-volume collection *Mobilities* (Routledge, 2015) and author (with Ditte Bendix Lanng) of *Mobilities Design: Urban Designs for Mobile Situations* (Routledge, 2017).

Gunnar Thór Jóhannesson, born in 1976, is Professor at the Department of Geography and Tourism at the University of Iceland. His recent research has been on destination dynamics and place making with a focus on the entanglement of nature and culture. He has published his research in various books and journals. Most recently, he co-edited a volume titled: *Co-creating Tourism Research: Towards Collaborative Ways of Knowing* (Routledge, 2018).

Vincent Kaufmann is Associate Professor of Urban Sociology and Mobility at Ecole Polytechnique Fédérale de Lausanne (EPFL). Since 2011, he is also Scientific Director of the Mobile Lives Forum in Paris. He has been Invited Lecturer at Lancaster University (2000–2001), Ecole des Ponts et Chaussées, Paris (2001–2002), Laval University, Québec (2008), Nimegen University (2010), Université de Toulouse Le Mirail (2011), Université Catholique de Louvain (2004–2018), Politecnico Milan (2016) and Tongji University in Shanghai (2018). His fields of research are: motility, mobility and urban life styles, links between social and spatial mobility, public policies of land planning and transportation.

Robin Kellermann is a Cultural Historian from Berlin working in the fields of transport and mobility. His current research is dedicated to the historical evolution of waiting in public transportation by retracing the co-evolution of physical waiting environments and social practices at railway stations. Beyond that, he is Project Leader at Technical University Berlin, coordinating a research project on the future of urban airspace as a new dimension of parcel and passenger transportation (*Sky Limits*).

Sven Kesselring holds a PhD and a doctoral degree in Sociology. He is Research Professor in 'Sustainable Mobilities' at Nuertingen-Geislingen University, Germany (HfWU) and director of the international research network Cosmobilities. He runs the PhD programme 'Sustainable Mobilities in Metropolitan Regions' with TU Munich, and he is the director of the HfWU master programme in 'Sustainable Mobilities'. Publications: Mobilities and Complexities (2019; with Mimi Sheller, Ole B. Jensen) and Exploring Networked Urban Mobilities (2018; with Malene Freudendal-Pedersen).

Francisco Klauser is Professor in Political Geography at the University of Neuchâtel, Switzerland. His work explores the socio-spatial implications, power and surveillance issues arising from the digitisation of present-day life. Main research topics include video surveillance, mega-event security, smart cities, civil drones, and big data in agriculture. Aiming at investigating and conceptualising the intersections between power and space in the digital age, his work also contributes to German, Francophone and Anglophone socio-spatial theory.

Sanneke Kloppenburg is Assistant Professor at the Environmental Policy Group at Wageningen University, the Netherlands. Her background is in Sociology and Science and Technology Studies. She holds a PhD in Social Sciences from University of Amsterdam and has held postdoctoral positions at Maastricht University and Wageningen University. Her current research interests revolve around digitalisation, social practices and sustainability transitions in the domains of energy, food and mobility.

Ida Sofie Gøtzsche Lange is Assistant Professor at the Department of Architecture, Design and Media Technology, Aalborg University (Denmark). She holds a BA in Architecture and Urban Design, an MA in Urban Design and a PhD in Urban Design and Planning. She is a board member at the *Center for Mobilities and Urban Studies* (C-MUS). Her main research interests are within port city relationships, urban design, urban mobilities, planning, place theory and mobilities design.

Gunvor Riber Larsen is Assistant Professor at the Department of Architecture and Media Technology, Aalborg University. Current research is focused on aeromobilities and airport studies. Other research interests include the challenges of being mobile in rural areas with little or no public transport service provision, and the role of distance in international tourist travel.

Jonas Larsen is Professor in Mobility Studies at Roskilde University. He has a long-standing interest in tourist photography, tourism and mobility more broadly. More recently, he has written extensively about urban cycling and is now also conducting research on walking, running mobilities, urban marathons and sport tourism.

Claus Lassen is Associate Professor and Director of the Centre of Mobility and Urban Studies (C-MUS) at Aalborg University. His research analyses changing social relations in the light of international air travel, and he has published a number of articles and book chapters on business travel, aeromobilities and airports.

Christian Licoppe is Professor of Sociology at the Department of Social Science in telecom Paristech in Paris. Trained in history and sociology of science and technology, he has worked for a stretch in industrial research, where he managed social science research at Orange R&D, before taking his current academic position. Among other things he has worked in the field of mobility and communication studies for several years. He has used mobile geolocation and communication data to analyse mobility and sociability patterns of mobile phone users. He has studied various phenomena related to the proliferation of mediated communication events and 'connected presence'. He has also studied extensively the uses of location aware games and proximity-aware mobile technologies communities. His recent work in mobile communication has focused on the development of methods to record and analyse the use of mobile communication in 'natural' situations (such as mobility and transport settings) and on the study of mobile dating applications, video-mediated communication (Skype, Periscope) and surveillance (location-based monitoring of offenders).

Jean-François Lucas, PhD, Sociology, explores the socio-technical issues and the imaginaries of the Smart City. He is interested in the evolution of urban practices (living, dwelling, mobilities, etc.) and in the governance of the urban fabric (public/private relations, civic tech) through digital and Big Data. For several years, he has also been working in the field of design and management of innovative research projects.

Joël Meissonnier is Research Fellow at CEREMA since 2009. He is Socio-anthropologist for Transport and Mobility in Lille (France). He supported his Doctorate of Sociology at Paris V in 2000. His studies focus on changes in daily mobility behaviours from a qualitative point of view. Joël Meissonnier is part of the ESPRIM research team focused on disruption and resiliency of mobility systems.

Nicholas Molyneaux, born in Switzerland in 1989 holds a MSc in Computational Science and Engineering from the Ecole Polytechnique Fédérale de Lausanne (EPFL). After his BSc degree in Civil Engineering from EPFL, he worked on pedestrian flow modelling in 2013 and 2014 in Zürich, Switzerland. He is currently undertaking his PhD, again at EPFL, in a similar field: the design of control and management strategies for pedestrian traffic.

Lesley Murray is Associate Professor in Sociology at the University of Brighton, where her research centres around urban mobilities. Lesley has published extensively in the field of mobilities, including on the intersections between mobile and visual methods and on gendered mobilities, children's mobilities. She has co-authored a book on *Children's Mobilities* (Palgrave Macmillan) and co-edited three collections, *Mobile Methodologies* (Palgrave Macmillan, 2010), *Researching Mobilities: Transdisciplinary Encounters* (Palgrave Macmillan, 2014) and *Intergenerational Mobilities* (Routledge, 2016).

Petter Næss is Professor in Planning in Urban Regions, Department of Urban and Regional Planning, Norwegian University of Life Sciences. His research interests include urban sustainability, built environment impacts on travel, assessment methods, driving forces of urban development and philosophy of science.

Marija Nikolić finished her PhD and post-doc at EPFL, in the Transport and Mobility Laboratory. Her thesis entitled 'Data-driven models for pedestrian movements' was supervised by Professor Michel Bierlaire. She has worked as Data Scientist for Data, Analytics & AI at Swisscom, Swiss telecommunication company, where she was developing algorithms, models and software solutions to extract mobility-related insights from large amounts of anonymised data. Her research interests include probabilistic modelling, machine learning and algorithm design.

Denver Nixon is currently an Honorary Research Associate of the Transport Studies Unit at the University of Oxford, having recently completed a three-year postdoctorate there. He also teaches in the Department of Geography at the University of British Columbia. Denver's work at Oxford investigates community-led walking and cycling infrastructural initiatives for marginalized communities in London and São Paulo to critically evaluate their nature, challenges and potential contributions to just transitions in urban mobility. His doctoral project studied how people's modes of mobility shape their understandings of their social and physical environments and in this way reproduce or transform dominant transport systems. More broadly, Denver is interested in how environmentally (un)sustainable and socially (un)just practices are formed and maintained through embodied experiences and particular social and material contexts.

Yuki Oyama is Research and Teaching Associate at the Transport and Mobility Laboratory at EPFL, Switzerland. He completed his PhD in 2017 at the University of Tokyo, Japan, where he has developed methodologies for route choice modelling and pedestrian activity modelling. At EPFL, he is currently working on network traffic assignment and optimal use of urban spaces, in particular for the mixed traffic in city centres.

Lynne Pearce is Professor of Literary and Cultural Theory in the Department of English Literature and Creative Writing at Lancaster University (UK) where she has worked for nearly 30 years. Her recent books include *Drivetime: Literary Excursions in Automotive*

Consciousness (Edinburgh University Press, 2016) *and Mobility, Memory and the Lifecourse in Twentieth-Century Literature and Culture* (Palgrave Macmillan, 2019). She is currently Director for the Humanities at Lancaster's Centre for Mobilities Research [CeMoRe].

Emmanuel Ravalet is an Engineer and Doctor in Economics (University of Lyon, France). He also holds a PhD in Urban Studies (INRS-UCS, Canada). He works at the Institute of Geography and Sustainability of Lausanne University and at Mobil'Homme, where he is a founding partner. His research focuses on work-related mobility, energy consumption, new mobility services and local economic development.

Patrick Rérat is Full Professor in Geography of Mobilities and Head of the Sustainable Urban Planning Masters' Programme at the University of Lausanne, Switzerland. He has done research and published on various forms of mobilities such as residential mobility (with a focus on gentrification), internal migration (as in the case of young graduates), temporary mobility (such as linguistic stays) and everyday mobility (and more specifically the practice of cycling).

Riccardo Scarinci is Research Scientist at the Transport and Mobility Laboratory at École Polytechnique Fédérale de Lausanne (EPFL), Switzerland, working on modelling and simulation of transportation systems. He is currently collaborating with the World Bank to improve accessibility to schools in developing countries. He received his PhD at University College London (UCL), United Kingdom, in the field of Intelligent Transport Systems (ITS) in collaboration with the Technical University of Delft (TUD), the Netherlands.

Tim Schwanen is Associate Professor in Transport Studies and Director of the Transport Studies Unit, which is one of the constituent institutes of the School of Geography and the Environment at the University of Oxford. He is also Fellow in Geography at St Anne's College, Oxford. Tim's research is concerned with the relationships of the everyday mobilities of people, goods and information with climate change, urbanisation, social inequality and demographic, political and cultural transformations across the global North and South. He has published widely in journals in geography, urban studies, transport studies and interdisciplinary science. His most recent co-edited book is *Handbook of Urban Geography* (Edward Elgar, 2019).

Nicholas Scott is Assistant Professor of Sociology at Simon Fraser University. His research explores relations between urban mobilities, the production of nature and the common good. His forthcoming book explores contradictory visions of the good cycling city.

Mimi Sheller, PhD, is Professor of Sociology and founding Director of the Center for Mobilities Research and Policy at Drexel University in Philadelphia. She is founding co-editor of the journal *Mobilities* and past President of the International Association for the History of Transport, Traffic and Mobility. She has helped to establish the interdisciplinary field of mobilities research. She is author or co-editor of ten books, including most recently *Mobility Justice: The Politics of Movement in an Age of Extremes* (Verso, 2018), *Aluminum Dreams: The Making of Light Modernity* (MIT Press, 2014) and forthcoming *Island Futures* (Duke University Press).

Shelley Smith, architect, urbanist, PhD, is an independent researcher, lecturer and writer. Her work focusses on the discovery and expression – through words, photos and film – of the potential for sensorial experience in contemporary urbanity. In addition, she develops workshops and methods for mapping, cataloguing and communicating this.

Justin Spinney is an Urban Cultural Geographer and Economic Sociologist broadly interested in the intersections between mobility, embodiment, environmental sustainability and technology. These interests are underpinned by a political-economic focus on the production and maintenance of power and inequality understood through the application of post-structuralist theories including Actor Network Theory, the economy of qualities, and bio-politics.

Mikkel Thelle is Associate Professor of Urban History. He has worked across cultural and technological history with an interest in the industrial and post-industrial city. A research focus has been the body in public space, for example as passenger experiences, consumption and urban mobility around 1900. Recently, research in the entanglement of city and nature has been central. Mikkel is director of Danish Centre for Urban History at Aarhus University.

Phillip Vannini is Professor in the School of Communication & Culture and Canada Research in Public Ethnography at Royal Roads University in Victoria, BC, Canada. He is the author/editor of over a dozen books and numerous articles focusing on the mobilities of people such as ferry boat-dependent islanders, float plane passengers and backcountry walkers.

Gil Viry is Lecturer in Sociology at the University of Edinburgh. His research interests bridge the intersection of spatial mobilities, social networks, family and intimate life. He is especially interested in studying how physical distance and mobility behaviours, such as travelling, commuting, moving places and using mobile technologies, relate to personal relationships and individuals' social and professional integration over the life course. He mainly uses social survey methods, social network analysis and sequence analysis.

Chiara Vitrano is Postdoctoral Researcher at K2 – The Swedish Knowledge Centre for Public Transport – at Malmö University, Department of Urban Studies. Her current research interests include access inequalities and transport disadvantage, urban rhythms and time sovereignty, and new mobility services.

Zhengchao Wang, born in China in 1992, holds a BEng in Civil Engineering from Harbin Institute of Technology and MSc in Energy Management and Sustainability from the Ecole Polytechnique Federal de Lausanne (EPFL). After his MSc degree, he worked in Transport and Mobility Laboratory, EPFL, as research and teaching assistant from September, 2017 to June, 2019. Currently, he is undertaking a PhD course at Imperial College London in Operations Research.

Jian Zhuo is Professor of City and Regional Planning at the Tongji University (Shanghai, China). He is Deputy Head of the Urban Planning Department, Director of the Research Center of Urban Regeneration, Director of the national researcher network 'Urban Mobility in Historical Cities', editorial board member of *Urban Planning International*, PTSC member of CODATU. He has published three books and a dozen articles on the relationship between transportation infrastructure, land use and urban development.

INTRODUCTION

Malene Freudendal-Pedersen, Ole B. Jensen, Ida Sofie Gøtzsche Lange, Vincent Kaufmann and Claus Lassen

If you closed your eyes and imagined a city, whether small or large, what would very likely pass through your head is movement: people driving, cycling, walking, using the bus, metro or train; people on their cell phones, computers or other digital media; and behind the surface commercials, urban renewal projects, stores with a global supply of goods, and so forth. Today the urban is increasingly being (re)produced by what flows *through* it rather than what is fixed *within* it (Ritzer 2010; Sheller & Urry 2006). This is not to underestimate the dwelling that the urban also entails: what defines movement is just as much the stasis; in other words, mobilities is also about immobilities. The urban is about the rhythms and temporalities that create spaces in constant movement and change.

They function as 'spatial fixes' (Harvey 1990) in a space of global accumulation, constant mobilities and transformation. As Urry reminds us in his book *Global Complexities*:

> It is the dialectics of mobility/moorings that produces social complexity. If all relationality were mobile or 'liquid' then there would be no complexity. Complexity, I suggest, stems from this dialectics of mobility and moorings.
>
> *(Urry 2003: 126)*

The relationship between the fixed and the fluid is, furthermore, a special case when it comes to cities. Cities are full of speed and moving people, data, goods and vehicles; they also represent important frictions. The 'armatures' (or channels) and the 'enclaves' are equally important for cities to perform (Shane 2005).

Beyond these considerations, the dialectic between the fixed and the fluid is fundamentally related to the question of speed differentials of change, and it is thus possible to conceptualise what seems fixed as falling within long temporalities, while what is fluid refers to short times. Thus, the urban metabolism is composed of multiple rhythms and temporalities to more or less swaying and dissonant choreographies. The analysis of urban mobility is thus at least as much a matter of time as a matter of space.

This vision of the city is, of course, not new, and the idea that the urban phenomenon is intrinsically linked to exchange and movement can be found in the work of historians such as Braudel (1985) or Bairoch (1988), as well as among the researchers of the first school of

Chicago, who define the object of urban ecology as 'the man endowed with locomotion' (Park, McKenzie & Burgess 1925).

The Turn towards Mobilities

In this handbook, we will illustrate 'the urban' through the now well-consolidated research field of mobilities research or what has been termed the 'New Mobilities Turn' (Urry, Sheller & Hannam 2006). This field of study and research has over the past decade or more identified the importance of material, embodied and communicative movement of people, vehicles, information and goods to societies around the world. In the book *Sociology Beyond Societies*, John Urry (2000) argues that the magnitude of virtual and physical movement is 'materially reconstructing the "social as society" into the "social as mobility"' (2). Inspired by this understanding, mobilities research has developed into a transdisciplinary research field drawing on a wide range of theoretical orientations, methodologies and empirical fields, such as, to mention only a few, technologies, transportation, communication, planning, globalisation and urbanisation (Adey, Bissell, Hannam, Merriman & Sheller 2014).

What these contemporary research works have in common is that an analysis based on mobility is a heuristic and stimulating angle of attack, which facilitates fully restoring the dynamics of the city and the urban (Viry & Kaufmann 2015). Such an approach also facilitates a renewal of the theoretical and methodological approaches of urban research (Kaufmann 2010), which are very marked by concepts describing what remains and what is fixed rather than what is changing and what is moving.

This handbook perfectly illustrates the way in which mobilities as an overall theoretical inspiration, or call it an 'umbrella' if you like, provides a different understanding of the urban. Cities of today are composed of complex settings of social, technological, geographical, cultural and digital networks of mobilities (Graham & Marvin 2001; Jensen, Lassen & Lange 2019; Sheller & Urry 2006) (see also Figure 0.1). Throughout the more-than-ten-year-long gestation period, the 'urban issue' and the theme of urbanisation has only partly been touched upon and framed, as very few publications have had this as a distinct and focused theme. This singling out of 'urban' mobilities, which this handbook represents, is important for a number of reasons. Firstly, the United Nations (UN) reports that the world's population is becoming increasingly urbanised. Accordingly, more than 50 per cent of the world's population now live in cities, and it is estimated that the urban population is growing by 1.5 million people every week. Furthermore, this increasing trend of urbanisation shows no sign of subsiding (although it has decelerated in some Western countries). The social, environmental, economic and cultural effects of these continuing transformation processes make a distinct and focused urban mobilities agenda more pertinent and relevant than ever. Secondly, the 'urban question' as a distinct phenomenon has been a predominant focus in social theory already in the works of Simmel, Durkheim and Weber and in the early work of the Chicago School, but the neo-Marxist critique of urban studies as a discipline in and on its own in the 1970s and 1980s placed the urban question on the sidelines. The 'spatial turn' in the mid-1980s brought the urban issue back, especially within geographies and planning. More recently, the 'Grand Old Man' of the Los Angeles School of urban geographers, Ed Soja, turned our attention to the fact that cities are what they are due to their density and multiplicity. Soja (2000) reinvigorated the classic Greek word 'Synekism' for this argument:

Figure 0.1 Modern society is increasingly getting more and more mobile, physically, technologically, virtually etc. Comprehensive traffic systems are orchestrating pedestrians and vehicles of many kinds and with different needs, some using phones, GPS or other technological tools while on the move through urban complexities.

Source: Photo by Bradley Schroeder, used under Creative Commons, desaturated from original

> Synekism is directly derived from *synoikismos*, literally the condition arising from dwelling together in one house, or *oikos*, and used by Aristotle in his *Politics* to describe the formation of the Athenian *polis* or city-state ... Synekism thus connotes, in particular, the economic and ecological interdependencies and the creative – as well as occasionally destructive – synergisms that arise from the purposeful clustering and collective cohabitation of people in space, in a "home" habitat.
>
> *(12, italics in original)*

Spoken in plainer terms, the logic of the city is that two plus two equals five! The agglomeration effects of densities and the ways in which city networks through infrastructural landscapes have created the contemporary multi-scalar urban complexity are testament to this important dimension. The key idea to keep in mind when thinking about cities is therefore that we need to pay close attention to the fluid and the fixed as well as the near and the far.

The Turn towards Urban Mobilities

With this handbook, the mobilities turn will re-engage these theoretical discussions to penetrate a future where the urban question seems unavoidable (se Figure 0.2). The need to refocus on the interdependence between cities and mobilities is not least due to the speed of the implementation and development of large-scale technological transitions in transport and communications.

Figure 0.2 Mobilities as a plural concept is broad and comprising many fields of study. With this book we emphasize a significant aspect of mobilites, that is how mobilities are as a phenomenon highly urban. It seems that mobilities are often articulated as a self-perpetuating pattern of urban components, setting the scene for social life to unfold.

Source: Photo by Jesper Pagh

This has historically been significant when mobility has contained the idea and promise of frictionless speed (Jensen & Freudendal-Pedersen 2012; Kaufmann 2011; Urry 2007) as that which would lead to better and happier lives. Moreover, new visions of the city today, for example, smart and automated cities, continue to have mobilities as their centre for future transformations for the good and liveable city (Freudendal-Pedersen, Kesselring & Servou 2019).

As this handbook also shows, mobilities is a highly ambivalent phenomenon. It has throughout history exerted positive economic and social effects, and increased wealth, flexibility and exchange within and in-between urban agglomerations are reliant on the acceleration and improvement of diverse mobilities systems. Simultaneously, however, these urban mobilities have engendered issues of increased inequality, environmental degradation, acceleration and an increased sense of a loss of control (Birtchnell & Caletrío 2013; Cresswell & Merriman 2011; Freudendal-Pedersen 2009; Pooley, Turnbull & Adams 2006). The realisation of urban visions of 'seamless mobility' and a 'zero-friction society' has produced issues such as congestion, noise and environmental problems that are very visible in everyday urban life, planning and politics (see Lassen & Galland 2014 on this issue). The mobilities of everyday urban life play an important role in urban identities (Urry 2000, 2007; Beckmann 2001; Kaufmann 2002; Kesselring 2006; Freudendal-Pedersen 2009). The ontology of mobilities research provides a framework where urban mobilities are not purely understood as individual choices, technological transformations or economic forces, but there is also a focus on how practices and networks are culturally assembled in producing and performing the urban (Jensen, Sheller, & Wind 2015). In the Foreword of *The Cultures of Alternative*

Mobilities (Vannini 2009), Jensen (2009) stresses that 'we constitute places, and make sense of our environment as we move' (xviii).

It is not only the actual mobilities that are important and deserve attention, however. The abilities to move – the motility, the potential mobilities offered by the environment, and resources to travel – are significant to how the urban, and the sense of belonging to it, develops (Jørgensen 2010; Arp Fallov, Jørgensen & Knudsen 2013; Kaufmann 2011).

In this way, mobilities alters sedentary notions of urban dwellings (Adey et al. 2014; Sheller & Urry 2006) by not purely assuming that the 'spatial fixity' to nation states, villages and cities constitutes the urban (Bauman 1998, 2001; Freudendal-Pedersen 2009; Putnam 2000). Over the years, the work by scholars such as John Urry (e.g. *Sociology Beyond Societies*, Routledge 2000; *Mobility*, Polity 2007, and *Climate Change and Society*, Routledge 2011), Tim Cresswell (*On the Move*, Routledge 2006), Peter Adey (*Mobility*, Routledge 2010), Peter Merriman (*Driving Spaces*, Blackwell 2007), Vincent Kaufmann (*Re-Thinking Mobility*, Ashgate 2002; *Re-Thinking the City*, Routledge 2011), Ole B. Jensen (*Staging Mobilities*, Routledge 2013) and Malene Freudendal-Pedersen (*Mobility in Daily Life*, Ashgate, 2009) has laid out the foundation of a new way of thinking about spaces, cultures and societies. Furthermore, the field has matured, with a number of dedicated journals, such as *Mobilities*, *Transfers*, and *Applied Mobilities* creating an intellectual infrastructure and backbone. Likewise, the sister *Routledge Handbook of Mobilities*, edited in 2014 by Adey, Bissell, Hannam, Merriman and Sheller for Routledge, and the four-volume book *Mobilities*, edited by Jensen in 2015, are important state-of-the-art reference works for all scholars with an interest in the 'mobilities turn'. Moreover, the new *Handbook of Applications and Methods for Mobilities Research* by Büscher, Freudendal-Pedersen and Kesselring (forthcoming) delivers an important overview of the methodological work and reflections that have developed within this new field of mobilities over the past two decades.

This handbook aims to further encourage a cross-disciplinary and empirically focused exchange between such diverse fields as geography, political science, sociology, anthropology, urban planning, traffic and transportation planning, architecture and urban design around the issues of urban mobilities. Based on an empirical and theoretical need for a more explicit and focused discussion of the relationship between mobilities and urbanism, this volume provides such a space of reflection. The aim is to give the reader a comprehensive understanding of the contemporary state of the art within the mobilities research that has cities and 'the urban' as its pivotal axis. The book covers theories and concepts enabling scholars and researchers to understand, observe and analyse the world of urban mobilities. It also discusses the multitude of methods utilised in the research of urban mobilities.

Themes and Sections of the Urban Mobilities Handbook

As in any scientific venture, the devising of conceptual systems, nomenclatures and categories will be unavoidable. Sorting and ordering the complexity of any empirical field is part of the task when one undertakes the task of collecting comprehensive areas of knowledge. Hence, a handbook such as this, with the focus on mobilities and urbanism, must also bring some element of 'order' to the material. Needless to say, we shall be the first ones to acknowledge that 'bringing order' to a multiplicity of texts and reports from research is far from an exact science. Choices must be made and selections implemented, at times in accordance with a clear strategic vision for framing the material and at other times due to the pragmatics of the authors' willingness to participate. Furthermore, the knowledge horizon of the editors trying to bring clarity to such encyclopaedic complexity also matters. In other words, the themes of this book and the detailed content of the chapters are results of a rather complex gestation process. Having said this, we obviously feel that we can justify both the themes and the chapters of this handbook.

We have chosen to organise the 41 chapters after the following seven themes: Histories, Concepts and Theories; Methods, Tools and Approaches; Commutes, Modes and Rhythms; Everyday Life, Bodies and Practices; Power, Conflict and Social Exclusion; Urban Planning, Design and Governance; Infrastructure, Technologies and Sustainable Development. We do not perceive of these themes as random, nor do we think of them as given. Rather, they perform as a heuristic tool for organising a rather complex body of material and a large number of texts. In order for the reader to receive an even more detailed overview of what awaits, we will in the following sections briefly introduce the various themes and present under each theme section a list of chapters that shows how the different chapters are distributed across the seven sections.

Histories, concepts and theories

The attention paid to mobilities over the past decade represents a novel research agenda in many ways. The phenomenon of mobilities and its impact on cities, however, is, of course, not new (Cresswell & Merriman 2011). In Section I, the understanding of urban mobilities is nested firmly within a historic conscience of urban societal transformation (Cresswell 2006; see also Thelle, this volume). Moreover, our theoretical and conceptual understanding is particularly important to the way we understand, analyse and examine the various forms of urban mobilities and the interplay between them (Urry 2000). As Urry argues, 'a mobilities "lens" provides a distinctive social science that is productive of different theories, methods, questions and solutions' (2007: 18). Therefore, Section I also explores key concepts and theories within the 'mobilities turn', with a particular urban framing. In total, there are five chapters on histories, concepts and theories in the first section:

Mobility justice in urban studies
Modern urbanization
Networks, flows, and the city of automobilities
Mobility capital and motility
Co-design times and mobilities

Methods, tools and approaches

All research fields are marked and defined by the ways in which the research is conducted. As the new field of mobilities research has slowly emerged in the past two decades, there has been a growing interest in exploring mobile methods (Büscher, Urry & Witchger 2011; Fincham, McQuinness & Murray 2010; Merriman 2013) and the various forms of approaches and tools used by mobilities scholars (see also the work of Büscher et al. 2019 on this issue). Generally, there are important (and often overlooked) relations between the ontological, epistemological and methodological dimensions of any research field, and this is especially true for a field such as mobilities research, which remains young (Urry 2007: 60). In Section II, we therefore turn to 'ways of doing' research on urban mobilities by exploring methods, tools and approaches in play when empirical urban mobilities research is undertaken. This section consists of five chapters:

Mobile ethnographies of the city
Mobile futures through present behaviours and discourses
Digital approaches and mobilities in the big data era

A methodological hybridization to analyse orientation experience in the urban environment
Methodologies for understanding and improving pedestrian mobility

Commutes, modes and rhythms

The mundane and ordinary dimension of urban mobilities is reaching beyond commutes and aggregated rhythms (Lefebvre 1996). People's lives, the actual embodied practices, and multi-sensorial engagement with the city and its movement spaces and systems are related to 'the geography closest in', or the human body. Section III explores urban mobilities practices related to commutes, particular modes of moving and the recurrent rhythms. There are seven chapters in Section III:

The walking commute: gendered and generationed
The future of the car commute
Ups and downs with urban cycling
The train commute
Waiting (for departure)
Moving and pausing
Providing and working in rhythms

Everyday life, bodies and practices

As we start the move into the empirical realm of what may constitute a research field of urban mobilities, we encounter the mundane and ordinary practices of everyday life (Freudendal-Pedersen 2009; Wind 2014). Research into urban mobilities is an empirical endeavour, and it pivots around ordinary spaces and sites of repetitive practice. Examination of mundane and ordinary practices of everyday life, however, can not only teach us something at the 'micro' level, but the 'immediacy and inherent indexicality of all human existence [also] means that the fine, fleeting, yet essentially social moments of everyday life anchor and articulate the modern macro-order' (Boden & Molotch 1994: 277). Millions of people are living their daily lives on the move in and between cities (from the body to the globe); therefore, in Section IV, we focus on the exploration of everyday life and embodied practices as they manifest themselves within urban mobilities. Section IV has seven chapters:

Life course and mobility
Urban pram strolling
The video-ethnography of embodied urban mobilities
Routine and revelation: dis-embodied urban mobilities
Habit as a better way to understand urban mobilities
Residential mobility
Urban mobility and migrations

Power, conflict and social exclusion

The phenomenon of urban mobilities lies at the heart of how contemporary societies are functioning. As most people would realise, this also means that mobilities has to do with social interests, with decisions about access and rights to mobilities. As Bauman (1998) and Cresswell

(2006) both emphasise, the various forms of mobilities are closely linked to power, hierarchies and stratification mechanisms in modern society (see also Kloppenburg, this volume). Urban mobilities is thus also 'restricted' (Olesen & Lassen 2012), associated with different patterns of inequality (Banister 2018) and associated with different forms of inequality 'tensions, conflicts, or compromises' (see Nixon and Schwanen, this volume). In Section V, we therefore, through four chapters, explore such themes of power, conflict and social exclusion:

Mobility and social stratification
The conflicted pedestrian: walking and mobility conflict in the city
Transitions: methodology and the marginalisation of experience in transport practice
Social implications of spatial mobilities

Urban planning, design and governance

Urban regions, cities and areas are today, as mentioned above, closely linked to various forms of flows and movements (Hall & Hesse 2013). The importance of urban mobilities for economies, cultures and life directs attention to how these elements are decided upon. Systems of regulation and cultures of planning and governance condition the frames for urban mobilities, as do architecture and design decisions and interventions (Freudendal-Pedersen, this volume; Grieco & Urry 2011; Jensen & Lanng 2017). Section VI hosts a discussion of urban planning, public control and governance and has six chapters:

Planning for urban mobilities and everyday life
Mobilities design: cities, movement, and materialities
The movement of public space
Urban tourism
The airport city
Surveillance and urban mobility

Infrastructures, technologies and sustainable development

Cities have grown into immense, complex socio-technical systems. As they represent some of the most complex human-made artefacts, the need for understanding the role of infrastructures and technologies becomes key (Röhl 2019; see also Elliot 2019 on the digital revaluation), especially at the current moment in history where human imprints on the world threaten to jeopardise the development for future generations (as well as for the globe itself). Sustainable development must be addressed in any serious attempt to account for urban mobilities (Banister 2008; Høyer 2000; Næss, this volume). In the seventh and final section, the focus is on infrastructure and technology; the key theme of sustainable development is also addressed in this section. Section VII has seven chapters:

3D printing and the changing logistics of cities
Mobility as a service: moving in the de-synchronized city
Understanding multimodality through rhythm of life: empirical evidence from the Swiss case study
Rethinking the large-scale mobility infrastructure projects in sustainable smart city perspective
Smart cities
Sustainable mobility
Terminal towns

Each of the theme sections is provided with a short theme introduction, which contextualises the selected chapters of the section and gives the reader a quick overview. With a task of this magnitude, we realise that selections could have been made differently and other choices could have been made. We are, however, confident in saying that, to our knowledge, the present selection represents the state of the art within the field of urban mobilities, and we hope the reader will reach a similar conclusion upon exploration of the rich selection of texts.

Bibliography

Adey, P., Bissell, D., Hannam, K., Merriman, P., & Sheller, M. (Eds.). (2014). *The Routledge handbook of mobilities*. London, UK: Routledge.

Arp Fallov, M., Jørgensen, A., & Knudsen, L. B. (2013). Mobile forms of belonging. *Mobilities*, *8*(4), 467–86.

Bairoch, P. (1988). *Cities and economic development: From the dawn of history to the present*. Chicago, IL: University of Chicago Press.

Banister, D. (2008). The sustainable mobility paradigm. *Transport Policy*, *15*(2), 73–80.

Banister, D. (2018). *Inequality in transport*. Marcham, UK: Alexandrine Press.

Bauman, Z. (1998). *Globalization: The human consequences*. New York, NY: Columbia University Press.

Bauman, Z. (2001). *Community: Seeking safety in an insecure world*. New York, NY: Polity.

Beckmann, J. (2001). Risky Mobility. *The filtering of automobility's unintended consequences*. Ph.D.-thesis. Copenhagen: Department of Sociology.

Birtchnell, T. & Caletrío, J. (Eds.). (2013). *Elite mobilities*. London, UK: Routledge.

Boden, D. & Molotch, H. L. (1994). The compulsion of proximity. In R. Friedland. & D. Boden. (Eds.), *Now here. Space, time and modernity*. Berkeley: University of California Press. pp. 257–286.

Braudel, F. (1985). *La dynamique du capitalisme*. Paris, France: Arthaud.

Büscher, M., Freudendal-Pedersen, M., & Kesselring, S. (Eds.). (Forthcoming). *Handbook of applications and methods for mobilities research*. Gloucestershire, UK: Edward Elgar Publishing.

Büscher, M., Urry, J., & Witchger, K. (Eds.). (2011). *Mobile methods*. London, UK: Routledge.

Cresswell, T. (2006). *On the move. Mobility in the modern western world*. London, UK: Routledge.

Cresswell, T. & Merriman, P. (Eds.). (2011). *Geographies of mobilities: Practices, spaces, subjects*. Surrey, UK: Ashgate.

Elliot, A. (2019). *The culture of AI. Everyday life and the digital revolution*. London, UK: Routledge.

Fincham, B., McQuinness, M., & Murray, L. (Eds.). (2010). *Mobile methodologies*. Basingstoke, UK: Palgrave Macmillan.

Freudendal-Pedersen, M. (2009). *Mobility in daily life: Between freedom and unfreedom*. Surrey, UK: Ashgate.

Freudendal-Pedersen, M., Kesselring, S., & Servou, E. (2019). What is smart for the future city? Mobilities and automation. *Sustainability*, *11*(1), 1–21.

Graham, S. & Marvin, S. (2001). *Splintering urbanism: Networked infrastructures, technological mobilities and the urban condition*. London, UK: Routledge.

Grieco, M. & Urry, J. (Eds.). (2011). *Mobilities: New perspectives on transport and society*. London, UK: Routledge.

Hall, P. & Hesse, M. (Eds.). (2013). *Cities, regions and flows*. London, UK: Routledge.

Harvey, D. (1990). *The condition of postmodernity: An enquiry into the origins of cultural change*. Hoboken, NJ: Wiley-Blackwell.

Høyer, K. G. (2000). *Sustainable mobility – The concept and its implications* (Unpublished doctoral dissertation). Roskilde, Denmark, Roskilde University.

Jensen, O. B. (2009). Flows of meaning, cultures of movements: Urban mobility as meaningful everyday life practice. *Mobilities*, *4*(1), 139–58.

Jensen, O. B. & Freudendal-Pedersen, M. (2012). Utopias of mobilities. In M. Hviid Jacobsen & K. Tester (Eds.), *Utopia: Social Theory and the Future*. Farnham, UK: Ashgate. pp. 197–217.

Jensen, O. B., Sheller, M., & Wind, S. (2015), Together and apart: Affective ambiences and negotiation in families' everyday life and mobility. *Mobilities*, *10*(3), 363–82.

Jensen, O. B. & Lanng, D. B. (2017). *Mobilities design. Urban designs for mobile situations.* London, UK: Routledge.

Jensen, O. B., Lassen, C., & Lange, I. S. G. (2019). Material mobilities. In O. Jensen., C. Lassen. & I. S. G. Lange. (Eds.), *Material mobilities.* London, UK: Routledge. pp. 1–15.

Jørgensen, A. (2010). The sense of belonging in the new urban zones in transition. *Current Sociology, 58* (1), 3–23.

Kaufmann, V. (2002). *Re-thinking mobility: Contemporary sociology.* Surrey, UK: Ashgate.

Kaufmann, V. (2010). La théorie urbaine en devenir. In J. Stébé. & H. Marchal. (Eds.), *Traité sur la ville* (pp. 640–65). Paris, France: PUR.

Kaufmann, V. (2011). *Rethinking the city: Urban dynamics and motility.* Lausanne, Switzerland: EPFL Press.

Kesselring, S. (2006). Pioneering mobilities: New patterns of movement and motility in a mobile world. *Environment and Planning A, 38*(2), 269–79.

Lassen, C. & Galland, G. (2014). The dark side of aeromobilities: Unplanned airport planning in Mexico City. *International Planning Studies, 19*(2), 132–53.

Lefebvre, H. (1996). Seen from the window. In E. Kofman. & E. Lebas. (Eds.), *Writings on cities: Henri Lefebvre.* Hoboken, NJ: Wiley-Blackwell. pp. 219–227.

Merriman, P. (2013). Rethinking mobile methods. *Mobilities, 9*(2), 167–87.

Olesen, M. & Lassen, C. (2012). Restricted mobilities: Access to, and activities in, public and private spaces. *International Planning Studies, 17*(3), 215–32.

Park, R., McKenzie, R., & Burgess, E. (1925). *The city: Suggestions for the study of human nature in the urban environment.* Chicago, IL: University of Chicago Press.

Pooley, C., Turnbull, J., & Adams, M. (2006). The impact of new transport technologies on intraurban mobility: A view from the past. *Environment and Planning A, 38*(2), 253–67.

Putnam, R. (2000). *Bowling alone.* New York, NY: Simon and Schuster.

Ritzer, G. (2010). *Globalization: A basic text.* Hoboken, NJ: Wiley-Blackwell.

Röhl, T. (2019). From structure to infrastructuring? On transport infrastructures and socio-material ordering. In O. Jensen., C. Lassen. & I. S. G. Lange. (Eds.), *Material mobilities.* London, UK: Routledge. pp. 16–30.

Shane, D. G. (2005). *Recombinant urbanism.* Hoboken, NJ: John Wiley & Sons Ltd.

Sheller, M. & Urry, J. (2006). The new mobilities paradigm. *Environment and Planning A, 38*(2), 207–26.

Soja, E. W. (2000). *Postmetropolis: Critical studies of cities and regions.* Hoboken, NJ: Wiley-Blackwell.

Urry, J. (2000). *Sociology beyond societies. Mobilities for the twenty-first century.* London, UK: Routledge.

Urry, J. (2003). *Global complexity.* Cambridge, UK: Polity.

Urry, J. (2007). *Mobilities.* London, UK: Routledge.

Urry, J., Sheller, M., & Hannam, K. (2006). Editorial: Mobilities, immobilities and moorings. *Mobilities, 1*(1), 1–22.

Vannini, F. (Ed.). (2009). *The cultures of alternative mobilities. Routes less travelled.* London, UK: Routledge.

Viry, G. & Kaufmann, V. (Eds.). (2015). *High mobility in Europe, work and personal life.* Basingstoke, UK: Palgrave Macmillan.

Wind, S. (2014). *Making everyday mobility. A qualitative study of family mobility in Copenhagen.* Ph.D.-thesis. Aalborg University, Aalborg, Denmark.

SECTION I

Histories, concepts and theories

The attention paid to mobilities over the last decade represents a novel research agenda in many ways. However, the phenomenon of mobilities and its impact on cities, is of course, hardly a new one. Therefore, in the first section the focus is on the history of mobilities, as well as on the conceptualisation and the theorisation within the so-called 'mobilities turn'. The section opens with **Chapter 1** and Sheller's discussion of the emerging concept of mobility justice as a crucial addition to urban studies. Urban mobilities encompasses multiple scales, which this chapter illustrates through three examples: (a) embodied mobilities and the differential power of raced/gendered/classed urban accessibility and (dis)abilities; (b) transportation planning and the impact of uneven and splintered urban mobility systems on 'carbon gentrification'; and (c) climate justice on a planetary scale and the circulations of resources, energy and 'carbon colonialism' within the plans and visions of sustainable mobilities. In this chapter, Sheller shows how these new approaches to mobility justice within urban studies are connected with the social movements, design, planning and policies for promoting mobility justice. **Chapter 2** follows Mikkel Thelle and the tracks of history. This chapter examines the different overall mechanisms and tendencies in the history of urban mobility. First, the railways connected larger cities across the nation, also transgressing borders. Inside the cities and towns, this meant a concentration of activity around the main stations, which for a while became hubs for modern mobility-enabled practices. When, from the 1860s, the Underground in London began to spread, followed from the 1880s onwards by the electric tramcar, mobility as well as the built-up environment developed around a more fine-grained network of hubs. This also contributed to the functional division of the city and, as we can see in case studies, the mobility networks to some degree determined the expansion of the urban fabric. A contemporary urban mobility issue is the forced emigration from northern Africa and the Middle East into European cities that has increased due to the conflicts in these regions. Nearly two million people have fled in recent years, many across the Mediterranean, to escape wars or the like. This mobility has been enabled by small boats or simply by foot, yet it has caused the most significant political crisis in Europe in decades. The routes, borders, possessions and family relations of these refugees are the objects of intense attention from the European public. The chapter closes by reflecting on the question of whether modern urbanisation has an end. It appears the urbanising movements we have seen during industrialisation have continued globally in post-industrial times. In this reflection, the idea of planetary urbanisation is proposed. In **Chapter 3**, Sven Kesselring discusses how cities have always been the product of different flows. Even the earliest forms of urbanisation have been located at the intersections and crossroads of different travel routes. Kesselring utilises the notion of 'networked urban mobilities' that characterises the highly fluid and reticular character of the forces that shape modern cities and their networks. A new political dimension even comes into play. The chapter discusses American sociologist George Ritzer's distinction of four types of flows which are simultaneously connected with different networks: interconnected, multi-directional, conflicting and reverse flows. In particular, the chapter discusses these different flows in respect to the relevance of labour mobilities and the question of whether structurations of global labour markets and global exchange relations between cities can be identified. In **Chapter 4**, Kaufmann and Audikana focus on the notion of 'motility' as a potential for mobilities. Motility can be considered a form of capital in itself, one that is constituted as a resource for social integration. Individuals can have strong or weak motility skills but, above all, skills of a varied nature. Motility is seemingly indispensable for playing with spatial and temporal friction. The ingenuity of the solutions envisaged and applied in this area often depends on the actor's quality of life and the possibilities of changing their social status. Motility is a form of capital whose existence and application does not necessarily correspond to or reflect income, training level or social networks. It is a specific resource that can be acquired through primary or secondary socialisation processes and used for urban living. As an analytical tool, the concept of motility allows us to better describe and conceptualise the interrelation between the dynamics of spatial and social mobility. In this respect, motility provides the key to better understanding the ambivalence and limits as regards the necessity of mobilities in contemporary societies. Section I closes with **Chapter 5**, Luc Gwiazdzinski on different temporalities and mobilities. The observation, analysis, and construction of cities and forms of mobility are not always in step with the complexity of urban systems and changes in contemporary societies. It is possible to use time as a key to reading the functioning and dysfunctions of cities and as a springboard for new mobilities and sustainable urban development policies. This chronotopic perspective makes it possible to approach the complexity of urban systems and reflect on ways of 'indwelling' times and mobilities. The researcher, the urban planner, the builder and the citizen are invited to change their outlooks; to think, design and manage the city by simultaneously considering urban physicality, flows and timetables in order to imagine, together, a 'malleable city': a more human, accessible and hospitable one.

1

MOBILITY JUSTICE IN URBAN STUDIES

Mimi Sheller

Introduction

Cities are formed by mobilities. Located at the confluence of rivers, roadways, ports, rail termini, highways, and airports, cities have long been understood as a space of flows of people, goods, information, and ideas. At the same time, everyday mobility practices, infrastructural systems, and associated mobility regimes are in turn formed by processes of urbanization, which lock in various kinds of immobilities such as buildings, walls, roadways, gates, checkpoints, and fixed infrastructure. Cities and mobilities, in other words, are two sides of the same coin, co-constituted with and through each other.

Cities around the world are currently facing the urgent question of how to make the transition to more sustainable mobilities and more generally to "sustainable cities" or "ecological urbanism" (Mostafavi and Doherty 2013). Facing the effects of climate change, congestion, pollution, health inequity, and insufficient accessibility, many city governments, urban planners, designers, and dwellers increasingly realize that old ways of moving around are broken. The Intergovernmental Panel on Climate Change (IPCC) identifies transportation (of both freight and people) as responsible for around one-quarter of energy-related CO_2 emissions globally and notes that, despite increases in efficiency of vehicles, transport-related GHG emissions have continued to grow (Sims et al. 2014: 603). The most recent 2018 report notes that:

> Transport accounted for 28% of global final-energy demand and 23% of global energy-related CO_2 emissions in 2014. Emissions increased by 2.5% annually between 2010 and 2015, and over the past half century the sector has witnessed faster emissions growth than any other. The transport sector is the least diversified energy end-use sector; the sector consumed 65% of global oil final-energy demand, with 92% of transport final-energy demand consisting of oil products … suggesting major challenges for deep decarbonisation.
>
> *(IPCC SR1.5 2018, Ch. 2, p. 66)*

One key arena of active intervention and experimentation in urban planning and policy is the effort to replace the current dominant "system of automobility" (Urry 2004) (based largely on internal-combustion engines that generate greenhouse gases, private vehicle ownership that contributes to suburban sprawl, and publicly subsidized roads that suck investment away from public transit systems). In its place, we find cities around the world implementing transit-oriented development, building better bicycling infrastructure (bike lanes, bike parking, bike sharing), investing in improved public transit (new metros, bus rapid transit, light rail systems), and promoting active transport (walking, complete streets, new zoning regulations). Some European countries have announced plans to phase out internal combustion engines altogether; some Chinese cities have vast new public transit systems and, for example, 100% electric buses in Shenzhen.

These transformations are also closely tied to an emerging discourse of "smart cities" and new technology applications for "smart mobilities," in particular shared, electric, and potentially autonomous systems for "mobility as a service." Such policies have been extremely mobile, traveling to cities around the world, yet there is still much uncertainty about their overall impact (Cidell and Prytherch 2015). If anything, the growth of transportation network services appears to be contributing to higher rates of vehicle miles traveled (adding 5.7 billion miles of traveling annually in the U.S.) in American cities, while mainly drawing customers away from public transit, biking, and walking (Schaller 2018).

Despite some (relatively weak) promotion of low-carbon modes of transport, therefore, the dominant system of automobility is still expanding worldwide and "carbon capital" continues to be invested in global oil drilling, natural gas fracking, and building new pipeline distribution infrastructure. It is increasingly recognized that sustainability transitions are not simply a technological or economic problem, but are a sociological problem requiring better social science (Dunlap and Brulle 2015). In the face of this intractable dilemma of *how* to advance decarbonization and how to reduce, shift, or avoid transport demand, the mobilities turn in the social sciences offers a more multidimensional and integrated approach to sustainable urban transitions. While the rise and possible demise of automobility has been one important arena for thinking about the relation between cities and mobilities (Sheller and Urry 2000; Featherstone et al. 2005; Dennis and Urry 2009), the field of mobility studies offers far more than this alone. The "mobilities turn" is not a single unified approach, but has multi-faceted dimensions that together offer new ways of thinking about cities, space and mobilities.

This chapter will introduce the wider implications of the mobilities turn for urban studies, including its attention to movement, meaning, and practices within a historically informed analysis of complex relational systems of mobilities and moorings (Cresswell 2006; Hannam et al. 2006; Sheller and Urry 2006). Urban mobilities encompass multiple entangled scales, which each bear on the problem of sustainable urban transitions. First, bodily (im)mobilities can be understood as assemblages of differential power that perform raced/gendered/classed differential accessibility and (dis)abilities. Second, uneven infrastructures, as the material moorings of splintered urban mobility systems, are deeply implicated in transitions toward more sustainable and just cities. And, third, wider extended urban mobility regimes associated with planetary urbanization have implications for the kinopolitical power to control logistical circulations of resources, energy, and communication.

These three dimensions of bodily, infrastructural, and planetary mobilities, each discussed in turn below, illustrate the emergence of a new approach to urban mobilities studies that offers critical interventions in the realms of spatial justice, infrastructural justice, and climate justice.

Bodily (im)mobilities as urban form

Uneven mobilities empower "kinetic elites" with speed, access, and ease of movement while leaving behind others as the "mobility poor" (Cresswell 2010; Sheller 2018). These differential and uneven mobilities prevent urban sustainability transitions because the upper 10% of the kinetic elite use vastly more energy than other people, and produce more than 50% of greenhouse gases according to Kevin Anderson at the Tyndall Centre, University of Manchester. While documents like the IPCC reports sketch out various pathways toward limiting global warming to 1.5 degrees Celsius, ultimately it will require kinopolitical struggles for mobility justice to create more sustainable futures. Critical mobilities studies recognizes differential mobile subjects and the role of uneven urban mobilities in producing unsustainable cities.

At the scale of the human body there is a kind of choreography of human movement in various kinds of assemblages and constellations. Our movements are performed in relation to other people, to vehicles and prosthetic technologies that assist our mobility, and to spatial and material affordances of the built environment that enable or prevent various kinds of movement. Social factors such as gender, race, sexuality, class, age, and ability shape our capacities and styles of movement in relation to other people, to normative social orders, and to socially shaped (and shaping) environments, infrastructures, and places. However, it is the differential capacity for movement which imbues these bodily distinctions in the first place through the empowerment of a kinetic elite who benefit from controlling the mobility of others, whether by blocking, channeling, segmenting, slowing, forcing, or preventing it.

Kim Sawchuk argues that "the term 'differential mobility' is germane for thinking of how some movement-repertoires give preference to bodily norms that create hierarchies of corporeal differences that are structured into the built environment" (Sawchuk 2014: 413). Such structures are also a kind of infrastructure in which some people may "find themselves distanced from the 'able-bodied' and excluded from the world that does not allow them to move through with any ease." Critical disability scholars have shown how such infrastructures are not only "splintered" between favored elites and disfavored non-elite spaces, but also are embedded in deeply "differential mobilities." "We look, we listen, we wend our way through landscapes that continually shape and re-shape our movement-abilities. These environments favor some bodies over others. We are differentially mobile" (Sawchuk 2014: 409).

This understanding draws on Doreen Massey's work on "uneven geographies of oppression" to think through how power is "evident in people's differential abilities to move (Massey 2008: 165, as cited in Sawchuk 2014: 411). Massey observed that "some are more in charge of [their mobility] than others; some initiate flows and movement; some are more on the receiving end of it than others, some are effectively imprisoned by it' (ibid.: 161). Hence, we can refer to different degrees of "mobility power." Histories of colonialism, and its gendered and racialized modes of governing territory and managing mobilities also suggest the connections between intimate moves and transnational flows, which together shape contemporary urban forms. The history of uneven capitalist urbanization is part and parcel of the formation of imperial cities, metropolitan hubs, and colonial ports, which have left behind the "imperial debris" of hierarchies of gender/sexuality, race/ethnicity, class, and nationality which take the form of "managed mobilities" (Stoler 2016).

Embodied differences of uneven mobility and mobility power are orchestrated, choreographed, and governed in ways that produce differences of class, gender, race, ethnicity, nationality, sexuality, and physical ability. In general, we could say that in Europe and North America white, able-bodied, male experts and technicians still dominate transport policy and urban transit agencies, hence policy, planning, and design may overlook the perspectives and experiences of women, people of color, persons living with disabilities, queer folk, and other minorities. Thus, there is an absence of gender awareness and gender analysis in much of the policy discourse surrounding transportation planning. Likewise, there is little racial analysis of differential mobilities, and only a slight awareness of impairment and exclusion of the differently-abled. But more than that, there is a conscious empowerment of the "good" mobile subject who is imagined as the classic "unmarked" liberal subject: male, individual, able-bodied, usually white.

The concept of uneven mobility refers, then, to differential terrains for movement and experiences of moving in which there are specific embodiments, divergent affordances, routes, and pathways, varied access and means of moving, and partial assemblages of (dis) connectivity – in short differential mobility capabilities. These differentiations are literally built into urban form, through the shaping of architectures, infrastructures, standards, regulations, and social practices that allow some ease of movement while denying it to others. Uneven mobility also refers to means or modes of movement that have a greater or lesser degree of ease, comfort, flexibility, and safety with more or less friction, noise, speed, or turbulence. And, finally, it refers to the "splintered" spatial patterns, infrastructural spaces, and control architectures that govern and multiply such relations of mobility and immobility, speed and slowness, comfort and discomfort (Graham and Marvin 2001).

In sum, there are managed mobilities, spatial designs, physical infrastructures, and symbolic impediments that create divergent pathways, differential access, and control architectures for partial connectivity and bypassing while blocking those with less mobility power. We see the emergence of "elite mobilities" (Birtchnell and Caletrio 2014) and "mobile lives" for some (Elliott and Urry 2010), dependent on the controlled (im)mobilities of others without access to the same capabilities and bodily comforts.

Infrastructurescapes

The IPCC report admits that current approaches to climate change mitigation lack the basic social science knowledge to incorporate this into GHG mitigation planning and policy (IPCC 2014: 612–13). They call for additional research on

> the implications of norms, biases, and social learning in decision making, and of the relationship between transportation and lifestyle. For example, how and when people will choose to use new types of low-carbon transport and avoid making unnecessary journeys is unknown.
>
> *(Ibid.: 605)*

The problem is that this framing in and of itself suggests a very limited approach to understanding urban mobilities and how they work. In contrast to this individual "choice" frame, critical mobilities theory offers a more complex understanding of urban sustainability transitions.

The emphasis in low-carbon transitions thinking on personal choices framed by travel time budgets, costs and prices as key drivers of transport decisions is locked within older ways of thinking about transport planning and modeling, including rational choice economic thinking (Shove 2010). In contrast to such individual behaviorist approaches, one could instead argue that any mobility transition is a systemic question that transcends rational choice models, technological availability, or even econometrics. Mobility behaviors, and more widely *mobility cultures*, are more fundamentally embedded in and performed with and through spatial patterns of urbanization and uneven urban form, including histories of gendered and racialized inequity built into existing infrastructure and land use (Sheller 2015, 2018).

The field of transportation equity highlights the inequitable race and class distribution of transport access (Bullard and Johnson 1997; Bullard et al. 2000, 2004). Transport inequities splinter cities and unevenly distribute transport benefits, as well as harms such as air pollution, leaving some people without access to basic services that are considered fundamental to human flourishing (e.g., food, healthcare, education, employment opportunities). More recent literature develops the concept of "transport poverty" (Lucas et al. 2016), which includes a combination of inability to meet the cost of transport, lack of (motorized) transport, lack of access to key activities, and exposure to transport externalities. Transport justice has also been taken up within mobilities theory which focuses more on spatial forms and barriers that create forms of social exclusion (Cass et al. 2005; Preston and Rajé 2007).

Historians of technology have shown us how "infrastructuring" (Star 1999) is an active process of making, embedded in everyday social practices (Guy and Shove 2000), while geographers such as Stephen Graham and Nigel Thrift (2007) have drawn our attention to the behind-the-scenes yet labor-intensive work of maintenance and repair that gives us the smooth surfaces of "ready-to-hand" infrastructure such as electricity, communication systems, roadways, and flows of water and waste through pipes. These infrastructures of urbanism are equally embedded in social practices such as regulations, codes, laws, and everyday activities and interactions that have gradually accumulated over time. All of these elements together constitute what Govind Gopakumar (2020) calls the "infrastructurescape" of urban mobilities, and Keller Easterling refers to as "infrastructure space" (Easterling 2016).

Some of these systems engage physical infrastructure, others concern informational systems; some involve moving things like bodies, vehicles, oil, or water; others involve moving things like data, code, and images (Sheller and Urry 2006: 5–6). These physical and informational mobility systems are being tightly coupled into complex new configurations, such that mobility systems are becoming more complicated, more interdependent, and more dependent on computers and software. Media and communication systems are deeply embedded into specific physical and material contexts, as seen in recent work on the invisible infrastructure of mobile networks and computer servers (Farman 2015), on the laying of undersea cables (McCormack 2014), and the geography of satellites and signal transmission (Parks and Schwoch 2012; Parks and Starosielski 2015). Such hybrid infrastructures of "mobile mediality" (De Souza E Silva and Sheller 2015) are not only horizontal but also vertical, extending from deep beneath the earth to high into the sky and low-earth orbit (Graham 2016), as well as in the middle-range aerial zone of the increasingly prevalent drone-scape (Hildebrand 2019).

Mobilities research also reminds us, moreover, that culture, lived experience, and meanings are all crucial elements of technological systems (Cresswell 2006). Any city is made up of technologies, practices, infrastructures, networks, and assemblages – as well as narratives, images, and stories about them – which together inform its mobility culture. Critical mobility thinking in the field of urban studies calls for "re-conceptualising mobility and infrastructures as sites of (potential) meaningful interaction, pleasure, and cultural production" (Jensen 2009), where people engage in "negotiation in motion" and "mobile sense making" (Jensen 2010). We therefore need a more extensive understanding of how such meaningful interaction, cultural production, and mobile sense-making takes place in the contexts of differential and uneven mobilities and infrastructures. The dominant system of automobility, for example, builds on patterns of kinship, sociability, habitation, and work, all of which are relatively stabilized and locked into built environments and social patterns (Urry 2000; Sheller 2004). Driving cars is implicated in a deep context of affective and embodied relations between people, machines, and spaces of mobility and dwelling, in which emotions and the senses play a key part.

Car consumption is never simply about rational economic choices, but is as much about aesthetic, emotional, and sensory responses to driving which lock in a set of practices. Attention to "automotive emotions" also signals the prominence of embodied aspects of mobilities, and the ways in which elements of race, class, and gendered embodiment reproduce mobility injustice and insufficiencies in accessibility. Car cultures have social, material, and affective dimensions that are overlooked in current strategies to influence car-driving decisions. People (and their feelings) are embedded in historically sedimented and geographically etched patterns of "quotidian mobility" (Kaufmann 2002; Vannini 2009). In contrast to traditional methods of urban planning and transport planning, a critical mobilities approach shifts attention away from the counter-factual "rational actor" and instead recognizes the lived experience of dwelling with cars (or dwelling in alternative cultures of mobility) in its full complexity, ambiguity, and contradiction.

Transitions in mobility systems depend not just on individual choices, new energy sources, or economic forces alone, but also on transitions in entire mobility cultures expressed in political struggles over the re-shaping of "infrastructurescapes," including the ways in which practices and networks are culturally assembled in producing and performing the uneven mobility space of the city. Mobility systems persist and combine into local, national, and even transnational cultural assemblages of mobility that remain very durable over long periods of time. The introduction of cleaner vehicles, alternative fuel vehicles and refueling stations, or bicycling lanes and infrastructure might help to promote sustainability, but only in a very limited sense if it is not coupled with wider reconfigurations of the landscape of mobility systems and discourses that shape the built environment and socio-cultural practices (Shove 2010). While individual level attitudes and mobility choices matter, it is also the wider cultural framings of structured interaction that guide practical behavior, decision making, and urban planning.

A paradigm shift is needed, therefore, to understand the social practices that drive transportation shifts, sustainable mobility transitions and more broadly conceived transitions in urban form. Tinkering with travel time budgets, transit-oriented development, and fuel costs will not be sufficient to generate the massive global infrastructural and behavioral changes needed to bring down global temperatures. This is why mobilities research calls for a far more wide-ranging analysis of complex relational systems of infrastructural and social interaction across multiple scales, including questions that encompass cultural histories of metals and mining (Sheller 2014) and the global scale of climate change (Urry 2011), which I turn to next.

Planetary mobilities

We can think of any historical period as involving specific "assemblages" of human mobility and transport of goods (logistics), media infrastructures (such as telegraph, radio, or satellite communication), and energy circulation to support these routings (i.e., the current infrastructure for liquid hydrocarbons, electric generation, and distribution through national grids and pipelines). Access to energy, and the minerals and metals that make up transportation and communication systems, are a crucial dimension of uneven mobilities and materialities. The potentials for mobility are grounded in where energy is sourced, where it is exported, via what means, and who uses it. Critical mobilities theory has elicited growing attention to the material infrastructures and "moorings" of mobility and communication systems (Hannam et al. 2006; Sheller and Urry 2006), including the global political economies and deep time of media archaeologies based on oil, carbon, and the mining of metals (Urry 2011, 2013).

Energy is consumed by vehicles, buildings, and communication networks, but we can also think of these material objects and infrastructures as temporary embodiments of energy: energy turned into processed metals and combined with other materials (such as cement or glass) used in the construction of particular kinds of energy/object assemblages (e.g., coal-fired steam trains and iron railways, oil-based internal combustion engines and roadways, aluminum MacBook Airs and satellite, wifi and 5G cell systems). There is a raw material basis to energy, transport, and communication infrastructures, which shapes and is shaped by geographies of social relations, lived practices, and meanings. Control over energy involves transferring it into particular objects and moving it through various distribution networks. This infrastructure then supports particular materializations of energy and information that become routinized in the ways people use and access matter in all its forms, such that material cultures embed energy and information in forms that become taken for granted or invisible (Sheller 2014).

Technologies of mobility and dwelling embody (and lock in) energy in their production, in their moment of use, and in their relation to infrastructural "moorings" (Hannam et al. 2006) predicated on unequal access to particular forms of energy such as liquid petroleum or the energy embedded in cars, roads, buildings, etc. Kinetic elites are increasingly monopolizing control over energy, water, and mineral rights, using their offshore financial power (Urry 2014) to control global resources that are becoming increasingly scarce. Uneven mobilities are therefore geo-ecological at their root, as well as geo-political. The unacknowledged responsibility of "high emitters" of carbon dioxide for climate change, created by the splintered provision of unequal infrastructures for mobility and deficits in urban accessibility, has serious implications for the uneven distribution of climate risk vulnerability and resilience.

We therefore need to pay more attention to the interaction between urban infrastructure, land use, and connectivity across "extended" urban systems and "operational landscapes" including mining, oil production, water and energy flows, all of which feed into concentrated urban systems via transportation and logistics networks (Brenner 2014; Brenner and Schmid 2014, 2015). Mobilities research has begun to analyze such systems, and even to conceive of them in a holistic way that acknowledges how multidimensional transport-related energy transitions will have to be if they are to have purchase. Mobility transitions cross and entangle micro, meso, and macro levels, ranging from bodily relations and the circulation of food, to street design and urban scale problems, to transnational and planetary mobilities. This scalar fluency will be necessary not only to address climate change and energy transitions, but also to develop the fundamental bases for mobility justice (Sheller 2018).

There are "network architectures" of "critical infrastructure" including energy grids, pipelines, and communications cables, argues Deborah Cowen, that are "bundled in concentrated form constituting urban areas. Contemporary cities can in fact be understood as *nodes* in transnational networks of critical infrastructure. Infrastructure is not simply proximate to urban centers; it is literally constitutive of the city" (Cowen 2017, n.p.). Given this constitutive role of infrastructure, issues of "splintered urbanism" (Graham and Marvin 2001) – including the potential for economic and political turbulence brought on by climate change and associated urban disasters (Graham 2009, 2011) – will increasingly come to the fore as sites of urban and transnational politics. We already see emerging global struggles over critical infrastructure politics, and the intensification of movements for "infrastructural justice" that are concerned with issues such as access to water, stopping fracking or oil pipelines, or the building of more just roadways, energy grids, or communications systems.

To reduce global greenhouse gases and "decarbonize" our economies therefore will depend on shifting this entire material assemblage for the production of modern urban life, which is not only energy intensive, but increasingly splintered and uneven. We need to redistribute energies and materials in a more just and equitable way, which will require a deeper engagement with mobility justice not just as a question of transport justice within cities, but reaching across every scale from the quotidian embodied practices of urban space to the planetary landscapes, seascapes, and airscapes of extended urbanization (Sheller 2018). The cultural changes needed to transform the underlying technoculture of speed and acceleration controlled by kinetic elites will require a deeper engagement with the materialization of energy in large-scale energy cultures: a full-scale technological, political, *and cultural* transition involving a new social organization of time, space, territory, and mobility.

Conclusion

In this chapter I have reviewed the literature in critical mobility studies on the role of differential embodiment, uneven infrastructurescapes, and planetary urbanization in shaping our current unsustainable forms of urban life. Together these three dimensions can be used to illustrate the emergence of a new approach to urban mobilities studies that sustains critical interventions in the realms of spatial justice, infrastructural justice, and climate justice. I believe that a strong theorization of mobility justice is the best way to bridge these various dimensions of urban inequalities, and show how they are interconnected (Sheller 2018; and see, Cook and Butz 2018).

Responses to climate change and to the wider ills of uneven mobilities and uneven urbanization should entail not only a reduction in energy demand or a shift to non-carbon fuels (the main current policy measures in cities across the world), but a critical stance involving planned reduction in speed and energy usage (carbon pricing and carbon caps), limits on kinetic elite mobilities (limited personal carbon budgets), support for universal basic mobility (ensuring access for the mobility poor), and more reflexive and less destructive forms of mobility (including climate reparations paid by historically high polluting industries). Ultimately this may require a rejection of the modern valuation of speed, acceleration, and limitless growth and its associated infrastructurescapes. The pathway to achieving such radical changes will be through kinopolitical struggles, which will determine the future of urban mobilities, the decarbonization of transportation, and the possibility of limiting global temperature increases.

References

Birtchnell, T. and Caletrio, J. (eds). 2014. *Elite Mobilities*. London: Routledge.

Brenner, N. (ed.). 2014. *Implosions/Explosions: Towards a Study of Planetary Urbanization*. Berlin: Jovis.

Brenner, N. and Schmid, C. 2014. Planetary urbanization. In N. Brenner (ed.) *Implosions/Explosions: Towards a Study of Planetary Urbanization*. Berlin: Jovis, 160–163.

Brenner, N. and Schmid, C. 2015. Towards a new epistemology of the urban? *City* 19 (2–3): 151–182.

Bullard, R. and Johnson, G. (eds). 1997. *Just Transportation: Dismantling Race and Class Barriers to Mobility*. Gabriola Island, BC: New Society Publishers.

Bullard, R., Johnson, G. and Torres, A. 2000. Dismantling transportation apartheid: The quest for equity. In R. Bullard, G. Johnson and A. Torres (eds) *Sprawl City*. Washington, DC: Island Press, 39–68.

Bullard, R., Johnson, G. and Torres, A. 2004. *Highway Robbery: Transportation Racism and New Routes to Equity*. Cambridge, MA: South End Press.

Cass, N., Shove, E. and Urry, J. 2005. Social exclusion, mobility and access. *The Sociological Review* 53 (3): 539–555.
Cidell, J. and Prytherch, D. (eds). 2015. *Transport, Mobility & the Production of Urban Space*. London and New York: Routledge.
Cook, N. and Butz, D. (eds). 2018. *Mobilities, Mobility Justice and Social Justice*. London and New York: Routledge.
Cowen, D. 2017. Infrastructures of empire and resistance. Verso Books Blog, 25 January 2017. Accessed 03 February 2017 at www.versobooks.com/blogs/3067-infrastructures-of-empire-and-resistance
Cresswell, T. 2006. *On the Move: Mobility in the Modern Western World*. London: Routledge.
Cresswell, T. 2010. Towards a politics of mobility. *Environment and Planning D: Society and Space* 28 (1): 17–31.
De Souza, E., Silva, A. and Sheller, M. (eds). 2015. *Mobilities and Locative Media: Mobile Communication in Hybrid Spaces*. New York: Routledge.
Dennis, K. and Urry, J. 2009. *After the Car*. London: Routledge.
Dunlap, R. and Brulle, R. (eds). 2015. *Climate Change and Society: Sociological Perspectives*. Oxford: Oxford University Press.
Easterling, K. 2016. *Extrastatecraft: The Power of Infrastructure Space*. London and New York: Verso.
Elliott, A. and Urry, J. 2010. *Mobile Lives*. London: Routledge.
Farman, J. 2015. The materiality of locative media: On the invisible infrastructure of mobile networks. In A. Herman, J. Hadlaw and T. Swiss (eds) *Theories of the Mobile Internet: Materialities and Imaginaries*. New York and London: Routledge, 45–59.
Featherstone, M., Thrift, N. and Urry, J. 2005. *Automobilities*. London: Sage.
Gopakumar, G. 2020. *Installing Automobility: Emerging Politics of Mobility and Streets in Indian Cities*. Cambridge, MA: MIT Press.
Graham, S. (ed.). 2009. *Disrupted Cities: When Infrastructure Fails*. New York and London: Routledge.
Graham, S. 2011. *Cities Under Siege: The New Military Urbanism*. London: Verso.
Graham, S. 2016. *Vertical: The City from Satellites to Bunkers*. London: Verso.
Graham, S. and Marvin, S. 2001. *Splintered Urbanism: Networked infrastructures, Technological Mobilities and the Urban Condition*. London: Routledge.
Graham, S. and Thrift, N. 2007. Out of order: Understanding repair and maintenance. *Theory, Culture & Society* 24 (3): 1–25.
Guy, S. and Shove, E. 2000. *A Sociology of Energy, Buildings and the Environment: Constructing Knowledge, Designing Practice*. London: Routledge.
Hannam, K., Sheller, M. and Urry, J. 2006. Mobilities, immobilities, and moorings. *Mobilities* 1 (1): 1–22.
Hildebrand, J. M. 2019. Consumer drones as mobile media: A technographic study of seeing, moving and being (with) drones. PhD dissertation, Drexel University.
IPCC. 2014. Mitigation of climate change. Fifth Assessment Report of the Intergovernmental Panel on Climate Change (AR5). Cambridge: Cambridge University Press.
IPCC. 2018. Global warming of 1.5 °C. Special Report. Accessed at www.ipcc.ch/sr15/
Jensen, O. B. 2009. Flows of meaning, cultures of movement: Urban mobility as meaningful everyday life practice. *Mobilities* 4 (1) March 2009: 139–158.
Jensen, O. B. 2010. Negotiation in motion: Unpacking a geography of mobility. *Space and Culture* 13 (4): 389–402.
Kaufmann, V. 2002. *Re-thinking Mobility*. *Contemporary Sociology*. Aldershot: Ashgate.
Lucas, K., Mattioli, G., Verlinghieri, E. and Guzman, A. 2016. Transport poverty and its adverse social consequences. *Transport Missing*. Proceedings of the Institution of Civil Engineers – Transport, Volume 169 Issue 6, December, 2016, pp. 353–365. Themed issue on transport and global poverty.
Massey, D. 2008. Power-geometry and a progressive sense of place. In J. Bird, B. Curtis, T. Putnam, G. Robertson, L. Tickner (eds) *Mapping the Futures: Local Cultures, Global Change*. Routledge: London, 60–70.
McCormack, D. P. 2014. Pipes and cables. In P. Adey, D. Bissell, K. Hannam, P. Merriman and M. Sheller (eds) *The Routledge Handbook of Mobilities*. London: Routledge, 225–232.
Mostafavi, M. and Doherty, G. (eds). 2013. *Ecological Urbanism*. Zurich, Switzerland: Lars Müller.
Parks, L. and Schwoch, J. (eds). 2012. *Down to Earth: Satellite Technologies Industries and Cultures*. New Brunswick, NJ: Rutgers University Press.

Parks, L. and Starosielski, N. (eds). 2015. *Signal Traffic: Critical Studies of Media Infrastructures*. Chicago, IL: University of Illinois Press.
Preston, J. and Rajé, F. 2007. Accessibility, mobility and transport-related social exclusion. *Journal of Transport Geography* 15 (3): 151–160.
Sawchuk, K. 2014. Impaired. In P. Adey, D. Bissell, K. Hannam, P. Merriman and M. Sheller (eds) *The Routledge Handbook of Mobilities*. London: Routledge, 570–584.
Schaller, B. 2018. *The New Automobility: Lyft, Uber, and the Future of American Cities*. Brooklyn: Schaller Consulting, July 2018.
Sheller, M. 2004. Automotive emotions: Feeling the car. *Theory, Culture and Society* 21 (4/5): 221–242.
Sheller, M. 2014. *Aluminum Dreams: The Making of Light Modernity*. Cambridge, MA: MIT Press.
Sheller, M. 2015. Racialized mobility transitions in Philadelphia: Urban sustainability and the problem of transport inequality. *City and Society*, SI on Cities and Mobilities, 27 (1): 70–91.
Sheller, M. 2018. *Mobility Justice: The Politics of Movement in an Age of Extremes*. London: Verso.
Sheller, M. and Urry, J. 2000. The city and the car. *International Journal of Urban and Regional Research* 24: 737–757.
Sheller, M. and Urry, J. 2006. The new mobilities paradigm. *Environment and Planning A* 38 (2): 207–226.
Shove, E. 2010. Beyond the ABC: Climate change policy and theories of social change. *Environment and Planning A* 42: 1, 273–285.
Sims, R., Schaeffer, R., Creutzig, F., Cruz-Núñez, X., D'Agosto, M., Dimitriu, D., Figueroa Meza, M. J., Fulton, L., Kobayashi, S., Lah, O., McKinnon, A., Newman, P., Ouyang, M., Schauer, J. J., Sperling, D. and Tiwari, G. 2014. *Transport In: Climate Change 2014: Mitigation of Climate Change. Contribution of Working Group III to the Fifth Assessment Report of the Intergovernmental Panel on Climate Change*. Cambridge and New York: Cambridge University Press.
Star, S. L. 1999. The ethnography of infrastructure. *American Behavioral Scientist* 43 (3): 377–391.
Stoler, A. L. 2016. *Duress: Imperial Durabilities in Our Times*. Durham: Duke University Press.
Urry, J. 2000. *Sociology beyond Societies: Mobilities for the Twenty-first Century*. London: Routledge.
Urry, J. 2004. The 'system' of automobility. *Theory, Culture & Society* (October 2004), 21 (4–5): 25–39.
Urry, J. 2011. *Climate Change and Society*. Cambridge: Polity.
Urry, J. 2013. *Societies beyond Oil*. London: Zed.
Urry, J. 2014. *Offshoring*. London: Polity.
Vannini, P. (ed.) 2009. *The Cultures of Alternative Mobilities: The Routes Less Travelled*. Farnham and Burlington, VT: Ashgate.

2
MODERN URBANIZATION

Mikkel Thelle

Since 2008, more than half of the world's population lives in cities, and by the middle of this century, the percentage is prognosed to 66 (UN_Habitat, 2016, p. iii). This movement is just the latest in a number of – most agree on three major – waves that have urbanized the world since the first domestication of the hunter–gatherer societies approximately 10,000 years ago. Modern urbanization, spanning from industrial to post-industrial times, spans the planet and mobilizes a multitude of dynamics, networks and movements from national to global scales (Bairoch, 1985; De Vries, Havami and Woude, 1990). A central theme in modern urbanization is the tension between increasingly segregated and regulated spaces, and a growing number of mobile bodies, individual as well as collective, that transgress these boundaries and regulations (Dennis, 2008). Choices have been made due to limited space, why mobilities in and around cities will be prioritized here, without omitting global links altogether.

While in the early modern period, urban cultures such as the Italian city states or the Dutch republic, had gained dominance through trade with specialized, exclusive goods, the base for the next step in urban mobilities was far more complex. Beginning with a rise in urban population in England and Belgium, the period from the late 18th to early 19th century is called the Great Divergence, establishing Europe as the dominant region (Pomerantz, 2000). From the British coal mines serving William Blake's "Dark satanic mills" of early industrialization came the first inventions of steam driven movement, the locomotive and the elevator (Williams, 2008; Graham, 2016). Thus, Stephenson's Rocket of 1829 and Otis' security elevator of the 1850s are examples of the potent alliance between rising city populations and mobile technologies driving one important aspect of what we can term modern urbanization. The British cities saw a shift from 17 to 72% urbanites during the 19th century, making the nation a prime mover and the most transformed society in this period.

Spreading to western, and then eastern and northern Europe, urbanization rates, big cities and urban culture proliferated in an explosive acceleration that established the foundation for an integrated urban system in Europe. In the USA the system, a bit different but still alike, moved from the East Coast through the Midwest to the western shore, so that by 1870, a set of urban systems was established in the western world that should be "filled up" towards World War II (Clark, 2013, p. 14). Aside from new markets and growing populations, these systems were built on the movement of bodies and goods, and the technologies of mobility emerging from this.

The early forms – the steam engine, railroad, elevator and telegraph – were all cornerstones in the forming of a shift in mobility paradigms, both in their own right but not least in conjunction with the radical growth in urban populations in the period (Kern, 2003; Cresswell, 2006). The movement of people and goods by railroad became crucial for the formation of urban systems in the global north, and later in parts of the southern hemisphere also. With British and French capital and know-how to begin with, by 1870 all southern and western European states had national railroad companies, tightly interlinked by 1900 (Kellett, 1969; Hård and Misa, 2008). On a grand scale, projects such as the Trans Siberian Railway, spanning eight time zones, connected empires and brought people across continents. This, as well as national networks, brought along a mechanism of urban centralization. In larger cities, there would be a number of larger stations for intercity connections, but many modern cities would have two or three such nodes, and urban development would concentrate around these, especially if a harbour connection was also at hand. The monumental train terminal buildings of the late 19th century – such as Gare du Nord or Grand Central Station – bear witness to this trend. As Henri Lefebvre pointed out, while these areas of the city were intended to be just beginning or end points for the traveller, from the beginning they became hubs of mobile cultures in themselves: travellers, vagrants and hustlers would fill these spaces, moving along the structures of national or imperial symbolic references, meeting under the clock or simply exploiting the crowd (Löfgren, 2008). With cheap fares and precise timetables, the train networks, with inter-urban systems like the S-bahn of Berlin, could bring people to work and back, facilitating the functional partition of the modern city in residential and industrial areas. The frequent travellers became a large group, and when the railroads added a monthly ticket to the daily fare, a new figure was born, the commuter. Not only being a mobile subject, the commuter came to inhabit the transport space, utilizing the train compartment and waiting room for reading, sleeping, making conversation etc. (O'Dell, 2006).

Centralization trends from railways were met by the opposite dynamic in trams and subways. Until the mid-19th century, horse-driven omnibuses and carriages had been the typical urban transport, but with the rails from the coal mine wagons, a new mobile ensemble was born, introduced in US cities and brought to Europe in the 1860s. The first tram lines would often go from the central railway station through the old centre to the industrial workplaces. With electrification, the system caught attention and spread to new groups, even though tram culture could be exclusive regarding, for example, gender (Schmucki, 2002). Cities such as Vienna and Budapest developed networks, and the frequency of use quadrupled in western cities between 1885 and World War I (McKay, 1976). In 1930, the Paris Metro carried 888 million passengers per year. But this was not only a western story. From the 1870s, privately owned tram systems would spread in Middle Eastern cities such as Istanbul, Cairo and Tehran, distributing populations to new suburbs and contributing to the growth of these cities to metropolitan scale. Thus tram networks, and later bus services, brought about a first wave of suburbanization, which is described vividly for Toronto by an observer: "the spinal column of each suburban extension ... its long fingers stretching out into the countryside" (Carver, 1962, p. 8). The faster and more jumpy expansion of North American cities often happened along such transport lines, "fingers" or tentacles, as Urban thinker Patrick Geddes put it, lamenting the uncontrolled infrastructures making cities into "conurbations" around the fin-de-siècle (Geddes, 1915). In Europe, the first suburban wave happened in a more controlled way, and with a significant higher density of population in both private and public housing (Hohenberg and Lees, 1995, p. 303). The tram and city train lines created an often radial framework for urban expansion, and in a way defined the balance between central and liminal urban space, both physically and in the experience of the citizens (Divall and Bond, 2003).

From the late 19th century until WWII we see the so-called Third Great Age of urbanization (Clark, 2013, p. 14). The rise in population and trade was now followed by a complex set of mobile phenomena. On the national scale, migration from rural to urban areas proliferated, but as migration research suggests, this was not new. Seasonal workers had been migrating for a long time back and forth, but when year-round jobs began to appear in the cities, they tended to stay. This was also the case on a larger scale where, for example, long-term migration across the Atlantic began to change. Earlier, people had been going on long trips to North America, but very often returning to their families in, for example, Southern Europe, nesting old household traditions. However, with faster and cheaper travel, by steam boat for example, the intensity and range of this migration space exploded, contributing to the urban labour market in larger cities.

This was related to one of the oldest urbanizing forces, the ship, becoming transformed. Also, by the late 1870s, all immigration to the US went largely by steam instead of sail (Keeling, 2010). Steamship lines from large port cities such as Hamburg, Marseille and Glasgow multiplied rapidly after 1900, and apart from moving workers and immigrants, these new vessels created a cargo transport system that revolutionized trade. Leisure travel became easier and cheaper, and sailing became an exclusive practice for the new urban elites, such as those aboard Titanic when it sank in 1912. The first wave of modern middle class urban tourism came through an alliance between railways and steamships, and with it came the travel guide and a new way of experiencing cities in "linear" ways (Spring, 2006). With new speed, new connections became pressing, and the opening of the Suez Canal in 1869 was one of the leaps that made global maritime mobility a mass phenomenon, and as such heavily contested by global political actors (Huber, 2013). As harbours had been forming cities since ancient cultures, the industrial port reorganized the urban space it was feeding. Ever expanding areas of docks, warehouses and customs facilities led flows through the inland, often attracting other networks of rails, roads and communication, but also creating certain cultures of cosmopolitanism, cheap labour and exotic goods (Rudolph, 1980; Hoyle, 1988). Coast cities were prospering, and changing, as for example the renowned port cities of the global south, Cape Town and Singapore, representing different tracks of maritime metropolitan development. Throughout the 20th century, dockworkers hired as casual labour would be populating the industrial waterfronts together with sailors, often connected to colourful quarters of bars and prostitution. When the 1970s brought the revolution of the container and emptied the traditional cargo port, this harbour culture became an industry in itself (Weinbold, 2008). As well as other mobility standards, the 20-foot-unit container disrupted transport on many scales from its introduction in the Vietnam war in the 1960s. While at harbour level, unemployment and relocation went hand in hand with gentrification of urban waterfronts, on the global scale a new set of shipping routes bypassed earlier hubs like London or Yokohama, to Rotterdam, Houston, Shanghai and Singapore (Fleming, 1997; Levinson, 2006; Parker, 2013; Klose, 2015).

From the 1920s in the USA and in Europe post-WWII an "unbundling" of urban territory emerged with the advent of the car. From the US, a culture emerged where the individual, mobile subject became disembedded from modernist clock-time to be involved in a complex juggling of "coerced flexibility" of fragmented schedules (Giddens, 1994; Urry, 2006; McShane, 2011). The car transformed the material city in multifarious ways. Streets, especially in the US and some Latin American countries, became contested spaces, where pedestrians were seen as "jaywalkers" (Norton, 2007). With the car, a whole new type of city became possible and, planned or unplanned, sprawl was introduced as the hitherto largest area of urban settlement (Flonneau, 2006). The choice between private and public mobility systems is one of the central issues of modern urban planning as it unfolds in the interwar period, and especially after WWII (Gunn and Townsend, 2019). From global north, car practices and networks spread in imperial, colonial and post-colonial frameworks (McDonald, 2014; Wagner, 2017; Edensor and Kothari, 2018). Since the car and the highway, new large areas have been appropriated as mobility-facilitating structures, both sustaining and locking "supermodern" cities (Augé, 1995; Sieverts, 2003). The ensuing pattern of urban development due to formal or informal peri-urbanization processes is characterized by the displacement of population, industries and services from the city centre to the periphery, and the creation of new centres with their own economic and social dynamics (Hohenberg and Lees, 1995; Lees and Lees, 2007; Clark, 2009). With the car, also, a certain culture of travel and leisure became possible, promoted by actors across the globe, between and within cities (Flonneau, 2006; Urry, 2006; Norton, 2007). Auto clubs and state agencies worked together in making car tourism a mass phenomenon, each in their own national brand, and through mediation car culture became a heavily gendered, utopian imagination (Hildebrand and Sheller, 2018; Redshaw, 2018).

With the automobile regime, the car, aided by the tram, facilitated or even created, suburbanization and thus city formation of the old urban core (Hall, 2014). Thus, the suburb, in its very different manifestations, has potential to be the most significant urban form of the 21st century, not least in the USA (Soja, 2000; Knox, 2008). Already from the late 1920s, the suburban community of the car was providing new recipes for urban mobile living, such as the Radburn-system, separating pedestrian and motor mobility around the front and back door of the single family house. After WWII, suburban populations exploded, both mirroring rising living standards and social inequality, as seen with the favelas and shanty towns in developing countries. In larger metropolitan regions such as Mumbai or Rio, some of these districts are inside city perimeters but most of them far off municipal borders, functioning as containers for cheap labour in metropolitan industry (Glaeser and Kahn, 2003; Davis, 2007). Globally, most of the urban population from the 1980s onwards lives in suburbs, and in many western countries the largest part of new buildings are in this category. As opposed to the "Streetcar suburb" of earlier times, the automobile suburb in many modern states has become part of national urban planning, with leisure and consumption facilities located nearby (Moraglio, 2017). Also, the proliferation of the suburb in the global north depends on cheap land and infrastructures of mobility (Mace, 2009).

The urban region is a phenomenon that has existed for centuries, but it is not controversial to claim that it has taken on a new and significant position since the 1970s. The distinction between region and suburb is collapsing over time, and new urban patterns are seen – edge cities, boomburgs, hybrid cities, citistates and so on. Research points to a "new regionalism" both in terms of these phenomena, but also emphasizing that regional mobility has been a structuring element alongside the waves of urbanization that we already know (Soja, 2015). Around the latest turn of the century, intense debate began around the role of the national state as a primary frame for politics, economy and identity. Supranational institutions such as the EU was promoting new European regions in a so-called "regional panic" at the time, while a process dating from WWII was beginning to yield global economic potential for urban regions (Easterling, 2014). An example is the Pearl River Delta close to Hong Kong, now the largest metropolitan area worldwide, and the Johannesburg-Pretoria agglomeration in South Africa, but also in Europe, regional development is gaining, if on a smaller scale. Here, for example, the Randstad-alliance of the largest Dutch cities forms an economic dynamo of north-western Europe. These regional constructions are based on the connectivity of bridges, roads and airlinks, but also on a mentality of mobility of the groups involved, termed the "regionauts" after the Scandinavian case of the Øresund region (O'Dell, 2003). In urban regions and corridors, power is increasingly related to "soft space", that is planning and decision-making in mixed fora of public institutions and private interests; they tend to have faster growth than the states surrounding them, and environmental agency far beyond their physical boundaries (Segbers, Raiser and Volkmann, 2007; Rosenzweig et al., 2011; Almendinger, 2015).

In the interwar period, commercial air transport emerged, and redistributed urban trade centres. With the jet age, from the 1950s onwards, airports and airlines seriously began to influence the negotiation of the centre and periphery of cities globally. If anything, aerial hubs demanded and produced centrality, and with their location on the outskirts of central cities, airports formed their own infrastructural and mobile worlds (Adey, 2010; Clark, 2013) from the onset, flight hubs had their own dedicated land transport (flight trains, internal shuttles etc.); but later also ideas of an *aerotropolis* grew, as an independent urban form of "boomburbs" (Lang, Nelson and Sohmer, 2008), but also in their own right, such as the smart city of Songdo, planned with a dedicated highway to South Korea's largest airport. Flight and other fast mobility networks proliferated as an exclusive experience, until the 1990s where prices became deregulated. As Sheller, Urry and Hannam note, in late modernity, air terminals are becoming like cities and cities are becoming like airports. Many technologies for networking and surveillance are often "first trialled within airports before moving out as mundane characteristics of cities, places of fear" (Sheller, Urry and Hannam, 2006, p. 6).

From this period on, also, urbanization accelerated in Asia and Latin America while losing pace in Europe and the USA. Especially in Asia, the number of megacities is rising, and concentrations of urbanites on a hitherto unseen scale is taking place. Here patterns can be quite different, with urban regions and planned urbanization playing a far more central role (Heitzman, 2008; Wallace, 2014). While urban concentration becomes extreme in Southeast Asia, for example in Manila, Dhaka or Mumbai, the Chinese rural–urban migration is handled by planning new cities across the country. Here, a central parameter of urbanization mobility is the balance between migration and mortality, the so-called Demographic Transition (Johansen, 2002; Dyson, 2011). By 1850, 50% of the European urban population had been born outside of the city, and this stabilized, so in 1890 the number was 43% – the trend was here that the larger the city the further away the migration came from. With health conditions improving, the urbanization from "inside" has been rising, and in recent decades the pattern of migration has been seen again in Asian megacities, for example (Clark, 2013, p. 468). If this is not regulated, the already large metropolises will become centres in so-called "primate" urban systems, where the largest city is without competition from number two. Such systems have a tendency to be less dynamic (Hohenberg and Lees, 1995).

A contemporary urban mobility issue is the forced emigration from northern Africa and the Middle East into European cities that has risen due to the conflicts there. Nearly two million people have fled in recent years, many across the Mediterranean, to escape conflict. This mobility is enabled by small boats or simply by foot, but has caused the most significant political crisis in Europe in decades. The routes, borders, possessions and family relations of these refugees are objects of intense attention from the European public.

In this period, modern urbanism becomes connected to mobility through notions of circulation, metabolism, fashion and transformation. The modern city as a "machine" is imagined as opposed to the sedentary life of the countryside, and a schism between mobile and fixed becomes one of modernity's deep binaries. At the same time, a "bad" mobility is emerging with the mobile but unidentified subject such as the vagrant or migrant (Amin and Thrift, 2002; Urry, 2006; Cresswell, 2011).

Seemingly, the urbanizing movements we have seen during industrialization have continued globally in post-industrial times, transgressing the town/country distinction, blurring boundaries and disintegrating the notion of the urban hinterland. A next urban leap on the shoulders of global mobility entanglements, and on hitherto unseen scales, could be the one already termed decades ago by Henri Lefebvre, a planetary urbanization (Brenner, Madden and Wachsmuth, 2011; Brenner, 2016).

References

Adey, P. (2010) *Aerial life*. Oxford: Blackwell.

Almendinger, P. (2015) *Soft spaces in Europe: re-negotiating governance, boundaries and borders*. London: Routledge.

Amin, A. and Thrift, N. (2002) *Cities reimagining the urban*. Cambridge: Polity.

Augé, M. (1995) *Non-places: introduction to an anthropology of supermodernity*. London: Verso.

Bairoch, P. (1985) *De Jéricho á Mexico: Villes et économie dans l'histoire*. Paris: Gallimard.

Brenner, N. (ed.) (2016) *Critique of urbanization: selected essays*. Berlin: Walter de Gruyter.

Brenner, N., Madden, D. J. and Wachsmuth, D. (2011) 'Assemblage urbanism and the challenges of critical urban theory', *City*, 15(2), pp. 225–240.

Carver, H. (1962) *Cities in the suburbs*. Toronto: Toronto University Press.

Clark, P. (2009) *European cities and towns: 400–2000*. Oxford and New York: Oxford University Press.

Clark, P. (ed.) (2013) *The Oxford handbook of cities in world history*. Oxford: Oxford University Press.

Cresswell, T. (2006) *On the move: mobility in the modern western world*. London: Routledge.

Cresswell, T. (2011) 'The vagrant/vagabond. The curious career of a mobile subject', in Cresswell, T. and Merriman, P. (eds) *Geographies of mobilities: practices, spaces, subjects*. Farnham: Ashgate, pp. 239–254.

Davis, M. (2007) *Planet of slums*. London and New York: Verso.

De Vries, J., Havami, A. and Woude, A. (eds) (1990) *Urbanization in history: a process of dynamic interactions*. Oxford: Oxford University Press.

Dennis, R. (2008) *Cities in modernity representations and productions of metropolitan space, 1840–1930*. New York: Cambridge University Press.

Divall, C. and Bond, W. (2003) *Suburbanizing the masses: public transport and urban development in historical perspective*. London: Ashgate.

Dyson, T. (2011) 'The role of the demographic transition in the process of urbanization', *Population and Development Review*. [Population Council, Wiley], 37, pp. 34–54. Available at: www.jstor.org/stable/41762398.

Easterling, K. (2014) *Extrastatecraft. The power of infrastructure space*. New York: Verso.

Edensor, T. and Kothari, U. (2018) 'Consuming colonial imaginaries and forging postcolonial networks: on the road with Indian travellers in the 1950s', *Mobilities2*, 16(5), pp. 702–716.

Fleming, D. K. (1997) 'World container port rankings', *Maritime Policy & Management*, 24(2), pp. 175–181.

Flonneau, M. (2006) 'City infrastructures and city dwellers. Accommodating the automobile in twentieth-century Paris', *Journal of Transport History*, 27(1), pp. 93–114.

Geddes, P. (1915) *Cities in evolution: an introduction to the town planning movement and to the study of civics*. London: Williams & Norgate.

Giddens, A. (1994) *The consequences of modernity*. Reprint. Cambridge: Polity.

Glaeser, E. and Kahn, M. E. (2003) *Sprawl and urban growth*. Working Paper 9733.

Graham, S. (2016) *Vertical: the city from satellites to bunkers*. New York: Verso.

Gunn, S. and Townsend, S. C. (2019) *Automobility and the city in twentieth-century Britain and Japan*. London: Bloomsbury.

Hall, P. (2014) *Cities of tomorrow: an intellectual history of urban planning and design since 1880.* Malden, MA: Blackwell.
Hård, M. and Misa, T. J. (2008) 'Modernizing European cities: technological uniformity and cultural distinction', in Hård, M. and Misa, T. J. (eds) *Urban machinery. Inside modern European cities.* MA: MIT Press, pp. 1–22.
Heitzman, J. (2008) *The city in South Asia.* London: Routledge.
Hildebrand, J. and Sheller, M. (2018) 'Media Ecologies of autonomous automobility', *Transfers. Interdisciplinary Journal of Mobility Studies*, 8(1), pp. 64–85.
Hohenberg, P. M. and Lees, L. H. (1995) *The making of urban Europe, 1000–1994.* Cambridge, MA: Harvard University Press.
Hoyle, B. S. (1988) 'Development dynamics at the port-city interface', in Hoyle, B. S. et al. (eds) *Revitalizing the waterfront.* London: Chichester, pp. 3–19.
Huber, V. (2013) *Channelling mobilities : migration and globalisation in the Suez Canal Region and beyond, 1869–1914.* Cambridge: Cambridge University Press.
Johansen, H. C. (2002) *Danish population history 1600–1939.* Odense: University Press of Southern Denmark.
Keeling, D. (2010) 'Repeat migration between Europe and the United States, 1870–1914', in Cruz, L. et al. (eds) *The birth of modern Europe culture and economy, 1400–1800. Essays in honor of Jan de Vries*, pp. 157–186.
Kellett, J. R. (1969) *The impact of railways on victorian cities.* London: Routledge.
Kern, S. (2003) *The culture of time & space.* 2nd edn. Cambridge, MA: Harvard University Press.
Klose, A. (2015) *The container principle.* Cambridge, MA: MIT Press.
Knox, P. (2008) *Metroburbia USA.* New Brunswick, NJ: Rutgers University Press.
Lang, R. E., Nelson, A. C. and Sohmer, R. R. (2008) 'Boomburb downtowns: the next generation of urban centers', *Journal of Urbanism: International Research on Placemaking and Urban Sustainability*, 1(1), pp. 77–90.
Lees, A. and Lees, L. H. (2007) *Cities and the making of modern Europe, 1750–1914.* Cambridge, UK and New York: Cambridge University Press.
Levinson, M. (2006) *The box: how the shipping container made the world smaller and the world economy bigger.* Princeton, NJ: Princeton University Press.
Löfgren, O. (2008) 'Motion and emotion: learning to be a Railway traveller', *Mobilities*, 3(3), pp. 331–351.
Mace, A. (2009) 'Suburbanization', in Thrift, N. (ed.) *International encyclopedia of human geography.* Oxford: Elsevier.
McDonald, K. (2014) 'Imperial mobility. circulation as history in East Asia under Empire', *Transfers. Interdisciplinary Journal of Mobility Studies*, 4(3), pp. 68–87.
McKay, J. P. (1976) *Tramways and trolleys: the rise of urban mass transport in Europe.* Princeton, NJ: Princeton University Press.
McShane, C. (2011) 'Distinctive but not exceptional: innovation in urban transport in the United States', in Mom, G. et al. (eds) *Mobility in history: themes in transport. T2M yearbook 2011.* Neuchatêl: Éditions Alphil, pp. 87–103.
Moraglio, M. (2017) *Driving modernity: technology, experts, politics, and fascist motorways, 1922–1943.* New York and Oxford: Berghahn.
Norton, P. D. (2007) 'Street Rivals: jaywalking and the invention of the Motor Age Street', *Technology and Culture*, 48(2), pp. 331–359.
O'Dell, T. (2003) 'Øresund and the regionauts', *European Studies*, 19, pp. 31–53.
O'Dell, T. (2006) 'Commute (ke-myoot')', *ETN. Etnologisk Skriftserie*, 2, pp. 87–96.
Parker, M. (2013) 'Containerisation: moving things and boxing ideas', *Mobilities*, 8(3), pp. 368–387.
Pomerantz, K. (2000) *The great divergence. China, Europe and the making of the world economy.* Princeton, NJ: Princeton University Press.
Redshaw, S. (2018) 'Combustion, hydraulic, and other forms of masculinity', *Transfers. Interdisciplinary Journal of Mobility Studies*, 8(1), pp. 86–103.
Rosenzweig, C. et al. (2011) *Climate change and cities: first assessment report of the urban climate change research network.* Cambridge: Cambridge University Press.
Rudolph, W. (1980) *Harbour and town. A maritime cultural history.* Leipzig: Edition Leipzig.
Schmucki, B. (2002) 'On the trams: women, men and urban public transport in Germany', *The Journal of Transport History*, 23(1), pp. 60–72.

Segbers, K., Raiser, S. and Volkmann, K. (2007) *The making of global city regions*. Baltimore, MD: Johns Hopkins University Press.

Sheller, M., Urry, J. and Hannam, K. (2006) 'Editorial: mobilities, immobilities and moorings', *Mobilities*, 1(1), pp. 1–22.

Sieverts, T. (2003) *Cities without cities: an interpretation of the Zwischenstadt*. London and New York: Routledge.

Soja, E. (2000) *Postmetropolis*. MA: Blackwell.

Soja, E. (2015) 'Accentuate the regional', *International Journal of Urban and Regional Research*, 39(2), pp. 372–381. doi:10.1111/1468-2427.12176.

Spring, U. (2006) 'The linear city: touring vienna in the nineteenth century', in Sheller, M. and Urry, J. (eds) *Mobile technologies of the city*. New York: Routledge, pp. 21–43.

UN_Habitat. (2016) *Urbanization and development: emerging futures*. Nairobi: UN-HABITAT.

Urry, J. (2006) 'Inhabiting the car', *The Sociological Review*. Blackwell Publishing Ltd, 54, pp. 17–31. doi:10.1111/j.1467-954X.2006.00635.x.

Wagner, L. B. (2017) 'Viscous automobilities: diasporic practices and vehicular assemblages of visiting "home"', *Mobilities*, 12(6), pp. 827–846.

Wallace, J. (2014) *Cities and stability: urbanization, redistribution, and regime survival in China*. Oxford: Oxford University Press.

Weinbold, J. (2008) 'Port culture: maritime entertainment and urban revitalization 1950–2000', in Hessler, M. and Zimmermann, C. (eds) *Creative urban millieus. Historical perspectives on culture, economy and the city*. Frankfurt: Campus Verlag, pp. 3–19.

Williams, R. (2008) *Notes from the underground. An essay on technology, society, and the imagination*. Cambridge, MA: MIT Press.

3

NETWORKS, FLOWS AND THE CITY OF AUTOMOBILITIES

Sven Kesselring

Cities have always been the product of many different flows. Therefore, it is quite a commonplace saying that they are more what flows through them than what is within them. In the global age, flows of capital, energy, labor power, vehicles, tourists, freight, information, data, waste etc. have become the shaping elements of cities, their identities, economic power, their position on a global political scale, within global labor markets and their meaning in the global cultural economy. It is more the question of how cities are connected, how accessible they are and who and what flows in and out of the urban space than what has settled within their buildings and scapes. In other words, multiple movements constitute and mark windows of opportunity for people, businesses, art and culture and the social, economic and cultural networks within these cities. The networks of people, organizations, companies, stakeholders, socio-political, socio-economic and socio-cultural players and initiatives generate a city's mobility capital. But at the same time, the massivity and sometimes even violence of flows threaten urban societies and environments. They can overstretch and overload their systems and capacities to manage movements. In European cities, for example, the daily amount of road traffic is pushing them to the edge when the daily convoy heads out of the city for work while the centers are being hit by constant flows of commuters, cars, train travelers and even air travelers. People commuting into Los Angeles i.e. spend more than 100 hours per year in congestion.

Today, every modern city wants to attract travelers, capital and a huge diversity of products, groceries and all kinds of artifacts. But too many people, goods and resources traveling the city's streets, railways, rivers and airspaces can challenge urban systems to their limits, intervene in public spaces and threaten the quality of life for those living there, seeking recreation, secure spaces, fresh air, peaceful places for children, leisure activities etc.

In this sense, urban formations, agglomerations and megacities in particular have a quite ambivalent relationship with movement and mobility. As German sociologist Georg Simmel clearly spotted at the beginning of the 20th century, both movement and mobility are at the core of what we consider to be urban. Urbanity as a social phenomenon is a very specific form of social life in (often bigger) cities where mobility, speed and social indifference play a key role in how people live, interact, work together and have conflicts. It is an urban form of life that is significantly different from the lives of people in the countryside (Siebel

1998). People living in 'metropolises' (Simmel 1903) cultivate and often also suffer from anonymity; they live in socially complex situations of constant change and mobility where they do not know most of the people around them and who populate urban spaces, buildings, parks and the cities' facilities.

Even early forms of urbanization were built on flows. Settlements were located at ports, intersections, crossroads, or later around railway terminals where international trade and travel routes often met (Benevolo 1980; Braudel 1977). In this sense, mobility has been at the core of the urban fabric even though other flows such as capital or digital mobilities have increasingly gained relevance in history (Pflieger and Castells 2008; Rodrigue, Comtois, and Slack 2005: 171 ff.). Research on the connectivities and accessibilities of cities shows that traffic flows are closely connected to capital flows and the movements of labor power across the globe. Urban sociologists, such as Saskia Sassen (1991) and Neil Brenner (2004), argue from a standpoint of political economy and would always prioritize capital flows over all others, and physical mobility in particular. But seen from a sociology that aims for a deeper understanding of what shapes and structures modern worlds, it needs to be taken into account that not only capital but also the materialization of economic structures within physical flows shape and challenge modern societies and economies. Offshoring, for instance, is not only a matter of cross-border capital transactions, it is at the same time strongly linked and driven by travel routes and connections, the mobility of labor and digital infrastructures and networks (see Bryson and Daniels 2007; Urry 2014).

The work of social and economic geographers like Peter Taylor decipher the intermingling of different flows with the impact of forming the urban fabric and the positioning of cities on a global scale. The illustration in Figure 3.1 shows how the positions of cities on a global scale can be reconstructed based on aeromobilities, namely flows of travelers within a global network of airline connections.

The current discussions on 'airport cities', strongly driven by concepts and ideas such as Karsarda's aerotropolis, the 'aviopolis' and the 'Zwischenstadt' (Cities without cities) (Conway 1954; Fuller and Harley 2005; Kasarda and Lindsay 2011; Sieverts 2003), mark another phase in the process of modernizing the city. Global infrastructures connect 'spaces of territoriality' with 'spaces of globalization' (Brenner 2004) and the urban becomes the connecting interface where the city's materialities, technologies and its stakeholders' networks become embedded in a global social, political and cultural metastructure (Kesselring 2009).

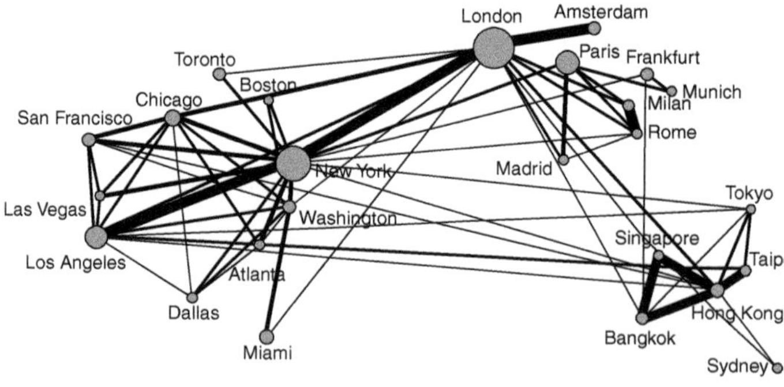

Figure 3.1 Positions of cities on a global scale reconstructed based on aeromobilities.
Source: Derudder and Witlox 2005: 2384

Besides humans, architecture and infrastructure, it is the car, 'the quintessential manufactured object … within 20th century capitalism' (Urry 2004), which has shaped urban spaces by far the most. Endless flows of automobiles enter and leave city spaces worldwide day in day out. Between 2005 and 2015 the number of automobiles worldwide increased from about 1.1 billion to more than 1.3 billion. Cars appear in all conceivable forms, colors and speeds. Packed with more and more technology, increasingly connected through GPS and intelligent transportation systems, and equipped with functional transport-related software and tools as well as those mainly there for entertainment, pleasure and killing time.

Today, Tesla's new electric car models are much more similar to space shuttle cockpits than reminiscent of the origins of the technological artifact itself. It does not have much in common with the famous Benz Patent-Motorwagen no. 3 used by Berta Benz in 1888 for her glorious first 'long distance ride' of 66 miles from Mannheim to Pforzheim in Germany. Even Henry Ford's Model T, colloquially named the Tin Lizzie, shows little resemblance to contemporary cars and even less with cars projected for future use controlled through 5G networks and without any possibility to intervene as a driver.

Nevertheless, since Le Corbusier, the godfather of modern urban planning, proclaimed urban spaces need to be rebuilt for automobiles in 1933 to let them pass through like projectiles (see Hilpert 1978), car-oriented planning has taken over almost all over the world. Paradoxically, German architect Hans Bernhard Reichow's publication 'Die autogerechte Stadt' (The car-oriented city) framed a new era of planning for the car after World War II and in many European cities. Originally, Reichow promoted what he called 'The art of organic building' (Reichow and Baukunst 1949) and even criticized car-oriented architecture in the aftermath of Le Corbusier's work and writing. Instead, he wanted an architecture based on organic and biological principles and in many ways similar to Ebenezer Howard's ideas of the human-scale garden city. The idea of the city as a body reaches from the early beginnings of town and city planning to early modern adaptations of 18th century planning based on William Harvey's medical theories of the blood circulatory and the respiratory system as the foundation of every existence to the contemporary theories of metabolic cycles. Cities are still built on the 'dream of traffic flow' and vehicles sliding through the urban body like blood through arteries and veins. The idea of seamless mobility, of flows of people, goods and vehicles through the city space without interruption, is one of the roots of the far-reaching transformations and problems of modern cities:

> Today, as the desire to move freely has triumphed over the sensory claims of the space through which the body moves, the modern mobile individual has suffered a kind of tactile crisis: motion has helped desensitize the body. This general principle we now see realized in cities given over to the claims of traffic and rapid individual movement, cities filled with neutral spaces, cities which have succumbed to the dominant value of circulation.
>
> *(Sennett 1994: 256)*

Materialities and structurations

Many flows have had significant impacts on the urban form. But the adaptation, rebuilding and redesign of cities towards a high level of efficiency, comfort and usability for the car and its different purposes, the reconstruction of public spaces and the hierarchization of mobility modes and the prioritization of the automobile in built environments has made the most important and durable imprint on urban life and its materialities. In many ways, urban spaces have become manifestations and materializations of the modern concept of automobility today. As sociologist Günter Burkart puts it, modern life today is built on automobilism as a 'normative complex' that relates to a certain lifestyle, which is characterized by the capacity to move and the opportunity to move in an individually motorized way. For Burkart, social mobility and the capacity to move in an individualized-motorized way are interlinked. This is also what Dennis and Urry (2009b) mean when they describe the rise of a 'post-car system' but still based on the capacity of individuals to move freely and independently.

Most modern cities of today have been built to accommodate the car and its needs instead of taking into account what Danish architect Jan Gehl calls the 'human scale' as a principle for sustainable development (Gehl 2010; Max-Neef 1991). Since Le Corbusier and others published the so-called Charta of Athens in the late 1920s, the complexities of human needs and human mobilities disappeared little by little from most planning concepts. Instead, predict-and-provide planning replaced neighborhood-oriented development concepts and those based on the tacit knowledge of planners, architects, local community advocates etc. Expert planning and technocratic knowledge became predominant and increasingly shaped the cities and their social, cultural and ecological environments. Munich, for instance, a city in the south of Germany with almost 1.5 million people, has never been car-centered in the same way as urban agglomerations such as Los Angeles, Tokyo and São Paulo or the rising megacities in Asia like Delhi, Mumbai, Shanghai and the like. But nevertheless, concepts were still discussed in Munich in 1993 on how to transport as many cars as possible into the city center. For this comparably small city with little more than 300 square kilometers, there have been plans to build seven ring roads around the center to guarantee a perfect traffic flow in the streets in and out of the city.

Car producers, urban planners, city politicians and civil societies agree that the limits of growth in car ownership have been reached in many Western countries. Market expansion is taking place somewhere else, i.e. in Asia and mainly in China. But the 'dream of traffic flow' (Schmucki 2001) looks significantly different to the one of the early 20th century engineers and planners. The 'social explosivity' (Beck 1988) of automobility's unintended side effects has left its traces also in the anticipated business models of the future.

Even in China, a country with an aggressive modernization concept and an almost unscrupulous relation to constructing material, technological and infrastructural networks through grown urban and social structures we can see indicators for recognizing the relevance and need to design sustainable urban and mobilities futures. While China's official modernization doctrine aims to build 'a new silk road' to the West, literally overland and oversea, the country's attempts to become the world market leader in electric mobility also hinges on the ecological situation in the cities and the fact that the air in many places is highly toxic with fine dust and other carbon emissions from gas and diesel.

System of automobility

When John Urry published the seminal article 'The system of automobility' in 2004, he pointed out the complexities and cultural embeddedness of the automobile system within modern societies and economies. With his sociological interpretation of the systemic and networked character of automobilism, Urry analyzed automobility as a social phenomenon in the first place, connected and intermingled with almost every sphere of modern human existence. It is not just a material object and entity or an industrial complex where the car itself is one element among many others (including the producing industry and suppliers). Instead, with reference to Niklas Luhmann's social theory, he deciphered the automobile 'as a self-organizing autopoietic, non-linear system that spreads worldwide, and includes cars, car-drivers, roads, petroleum supplies and many novel objects, technologies and signs'. Furthermore, he argued that '[t]he system generates the preconditions for its own self-expansion' (Urry 2004: 27).

In line with Urry, the system of automobility comprises at least six components (Figure 3.2); some of them not directly related to the economic and engineering aspects of the car and other vehicles based on automobile technology such as trucks, auto rickshaws (tuk-tuks) and many more.

The automobile is firstly the iconic manufactured object of the 20th century. Not by coincidence, Antonio Gramsci (1992) coined this phase of industrialization the 'Fordist era' that led to the 'Fordist city' (Harvey 1990). Later on the term Toyotism has been used to describe the manufacturing process but at the same time its social structuration grounded in standardization, efficiency and lean management. Starting from early Taylorist production concepts, Henry Ford's invention of the assembly line had lasting impacts on modes of production but also on societies themselves. The ideas of flexibility, individuality, speed and all-time availability had an enormous structuring effect on modern lives, work and production, the built environments of urban as well as rural spaces and the mobility and time regimes of people, companies and whole nations. In the early 20th century planners, architects and engineers created a 'city machine' (Knie and Marz 1997) as part of the 'industrialization of time and space' (Giedion 1948; Schivelbusch 1979). It changed the character of the 'organized modernity' at its core and of people's professional and social lives. Charlie Chaplin's famous movie 'Modern Times' (1936) with the iconic scene of him getting sucked in from the assembly line to the machine stands for the outreach of a Fordist production system and its comprehensive and peremptory rationalization and functional differentiation of all spheres of life, of 'system and life world' (Habermas).

Secondly, the system of automobility rests on a specific capitalist mode of consumption. The car is still the second largest investment after housing in many people's budgets. It was Ford's declared goal that workers in his company should also be the buyers of the final product. This is one of the reasons why the so-called Tin Lizzie was designed as a multifunctional technology, also enabled to drive a washing machine and many other compatible tools and devices for more convenience in people's households.

Thirdly, Urry considers the system of automobility 'an extraordinarily powerful complex constituted through technical and social interlinkages with other industries' (Urry 2004: 26). It comprises an overwhelming variety of 'car parts and accessories; petrol refining and

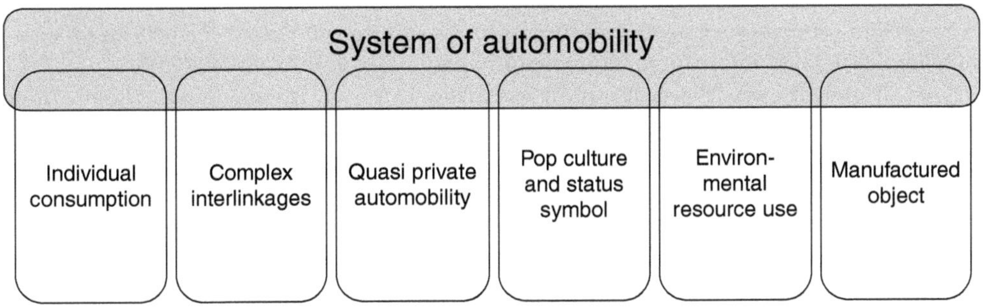

Figure 3.2 System of automobility.
Source: Illustration by the author

distribution; road-building and maintenance; hotels, roadside service areas and motels; car sales and repair workshops; suburban house building; retailing and leisure complexes; advertising and marketing; urban design and planning' (Urry 2004, 26).

In the fourth place, Urry considered automobility a form of a 'quasi-private mobility', based on the social construction of being individually mobile. Automobility has been the epitome of the individualized society and represents the fiction of deciding individually about one's spatial as well as social mobility (Berger 1996). In fact, automobility is increasingly part of a large-scale technological system associated with terms like 'intelligent transport systems', 'car-to-infrastructure communication', 'automated driving' etc. (Freudendal-Pedersen, Kesselring, and Servou 2019) today. In the 'age of digital modernity' (Canzler and Knie 2016) the room for maneuver for individual decisions in car-based traffic is massively decreasing. In fact there has never been much drivers' freedom and individuality but under the rising mobility regime of automatization and fine-grained traffic control systems they are vanishing almost completely. In a multimodal mobility system of quasi-private, private and public modes of transport, the differences and boundaries are becoming fluid, if not 'liquid', to apply Zygmunt Bauman's term. This has serious disruptive effects on the industry's business models since the freedom of choice is one of the basic taken-for-granted, mostly unreflected assumptions of the advantages of the car against other modes of transport.

Finally, the car has been part of the 'script' of popular culture in many ways. As an element of conspicuous consumption, it has been part of constituting people's identities and their communication of social status, economic success and prestige (Daniel 2001). In other words:

> The car is more than just a mode of transport, technical device or artifact, which one can use for the purpose of social actions. The car is an essential part of a modern way of life.
>
> *(Burkart 1994: 220)*[1]

The car has become a 'second skin of social meaning' (Sachs 1981) for the modern mobile human being. Popular culture such as art, literature and music are saturated with automobiles as iconic artifacts of the 20th century. Just to give some examples: Jack Kerouac's classic novel *On the Road* (Kerouac 1957); Roy Lichtenstein's pop art painting *In the Car* from 1963, sold for 16.2 million USD at Christie's; British rock band Queen's song 'I'm in love with my car' (1972) or The Who's gasoline-drunken 'Going Mobile' on the album *Who's Next* from 1971. Road movies from *Easy Rider* in 1969 to *Thelma and Louise* (1991) and the dystopian *Mad Max II* movie with its quest for gasoline and its Heathcote-Williams-like 'Autogeddon' aesthetics are all – be it utopian or dystopian – indicators for how deep automobilities have been rooted in modern popular culture (Flink 1975; Sachs 1992).

Urry names the car 'the single most important cause of environmental resource-use' (2004: 26). Not only do an automobile's emissions have to be counted here but the whole life circle from its conceptual development to the final re-entry into resource use as scrap. Urry considers the whole complex 'scale of material, space and power used in the manufacture of cars, roads and car-only environments, and in coping with the material, air quality, medical, social, ozone, visual, aural, spatial and temporal pollution of global automobility' (2004: 26).

The reason why the system of automobility gets so much attention here is: it has become the essential artifact that structures modern urban lives, communities, sociability and interaction, in general. Mobility, together with communication and the organization of proximity (housing, meeting, community, celebrating etc.), is the fundamental socio-spatial activity constituting 'community and society' (Tönnies 1957).

The automobile and its infrastructures, its cultural and emotional geographies, all the systemic relationships of this globally spread system of material and immaterial structures like roads, gas stations, charging systems, production sites, dealer networks, oil pipelines, global supply chains, labor markets, universities around the globe, knowledge hubs and laboratories and so forth, all this is structuring local, regional, national and global flows. Even airports become part of the system, which is the reason why Cidell (2017) describes the hub-and-spoke structures of 'auto-aeromobilty' where two seemingly different mobility systems appear clearly connected and intermingled with each other – even mobilities.

Transitions

The automobility system with car producers worldwide as its key players has been extremely stable and almost unchallenged at the latest since the 1950s. Even in its early forms of 'Atlantic automobilism' (Mom 2015), the system has been deployed with a historically unique capacity to persist and outreach to many nations and economies beyond the USA. One of the reasons for its strength and resilience was that it coincided with the far-reaching normative social transformations of individualization (Beck 1994; Weber 1978). The car met 'existing cultural values, needs and social practices to which its use could be adapted. ... With the introduction, ... new cultural uses were discovered and invented.' At the end of the day, the car didn't generate anything completely new but rather its existence and design reinforced 'cultural values, especially individuality and mobility, which were folded into a new value pattern' (Burkart 1994: 220).

In the process of industrialization in the late 19th and early 20th century, the car propelled the 'centrifugal tendencies' of modern individuals to transgress traditional boundaries 'spatially, economically and mentally' (Simmel 1890: 47) and helped them gain upward social mobility. This basic internal social structuration of mobility, flows and traffic is often overseen and underestimated. The transformative power of the car and the whole system of automobility comes exactly from this inner social drive: the car opened up new windows of opportunity that could be used and filled by individuals to have success, to climb the social stratification ladder and to realize their own projects and plans. As such, automobility is a principle almost written into the urban fabric. It is the modern promise of individual freedom, development, success and wealth that makes automobility so strong in urban planning and design. It is the mobility potential, the 'motility' stored and represented in this modern iron cage called automobile that gave people the chance to connect a mode of transport with their individual dreams and hopes for a better, a mobile life.

This is completely convincing and highly plausible in the historical and social context of the early 20th century and after WWII where people wanted and needed to escape from precarious social and economic situations. This also explains the massive energy in the process of 'compressed modernity' that is happening in China these days where 600 million people often moved from poverty to the middle class – a process coined as 'elevator effect' (Beck 1992). But how does this look in saturated societies such as Western European, the US and Australia and some Asian countries?

The fact that an increasing number of adolescent people are opting to acquire their driver's licenses later than the age of 17 or 18 does tell a story. This is (probably) less the effect of rising 'post-material values' (Ronald 1977) but rather of the changing mobility systems young people have access to. There are good reasons to assume that the formerly burning love for the car is in a process of becoming an extinguishing love. The driver's license statistics tell a story of shifting priorities rather than making a normative statement for sustainable mobility. But on the other hand this clearly shows the direction for policies and sustainable mobility policies in particular. If young people have alternatives at hand other than the privately owned car they seem to take the opportunity to use their budgets for something else and to realize their projects and plans by using other means. At least this is how we can read recent research on mobility styles and decisions and on the new potential of, for instance, sharing mobilities (Canzler and Knie 2016; Freudendal-Pedersen and Kesselring 2018; Rode and da Cruz 2018).

In fact the system of automobility has reached a certain tipping point. We cannot be sure that there will be a system 'after the car' (Dennis and Urry 2009a); but quite obviously there won't be a system of automobility as we know it but rather a system of mobilities, of multimodal and multi-functional modes of transport. Henning Kagermann, the former president of the German National Academy of Science and Engineering (acatech) wrote the future of mobility will be 'electric, connected and automatized' (Kagermann 2017). This is very much in line with what is coming from car producers worldwide today. All of them consider the mobility of tomorrow as connected, automatized, shared and electric. But this is not a guarantee that the car will be as much in the center of the new system as it was in the past. So far, we do not know yet, and we have serious problems imagining what exactly will be the impacts of these transformations for cities and regions. But so much can be said with certainty: the future of mobilities has not been as open as it is now – for a very long time. And with this openness of mobility systems also comes a window of opportunity to redesign cities and flows. Many stakeholders are still considering replacing the fossil automobile with a car driven by electric energy or by automated vehicles in individual ownership. But considering a post-fossil and sustainable city most likely means planning, designing and constructing for low and renewable energy, shared and post-carbon mobility beyond the contemporary hegemonial position of the car in contemporary cities.

Note

1 Author's translation from German original.

References

Beck, Ulrich. *Gegengifte: die organisierte Unverantwortlichkeit.* Frankfurt/Main: suhrkamp, 1988.

———. *Risk society*. London: Sage, 1992.

———. 'Jenseits von Stand und Klasse?'. In *Riskante Freiheiten*, ed. Ulrich Beck et al., 43–60. Frankfurt/Main: suhrkamp, 1994.

Benevolo, Leonardo. *The history of the city*. London: Scolar Press, 1980.

Berger, Peter A. *Individualisierung, Statusunsicherheit und Erfahrungsvielfalt.* Opladen: Westdeutscher Verlag, 1996.

Braudel, Fernand. *Afterthoughts on material civilization and capitalism. (The Johns Hopkins symposia in comparative history 7).* Baltimore, MD: Johns Hopkins University Press, 1977.

Brenner, Neil. *New state spaces: Urban governance and the rescaling of statehood.* Oxford and New York: Oxford University Press, 2004.

Bryson, John R. and Peter W. Daniels, eds. *The handbook of service industries. Elgar original reference.* Cheltenham, UK: Elgar, 2007.

Burkart, Günter. 'Individuelle Mobilität und soziale Integration. Zur Soziologie des Automobilismus'. *Soziale Welt* 2 (1994): 216–241.

Canzler, Weert and Andreas Knie. 'Mobility in the age of digital modernity: Why the private car is losing its significance, intermodal transport is winning and why digitalisation is the key'. *Applied Mobilities* 1, no. 1 (2016): 56–67. doi:10.1080/23800127.2016.1147781.

Cidell, Julie. 'Aero-automobility: Getting there by ground and by air'. *Mobilities* 12, no. 5 (2017): 692–705.
Conway, Hobart McKinley. *The airport city: development concepts for the 21st century*. Atlanta: Conway Publications, 1954.
Daniel, Miller. *Car cultures*. London: Berg Publishers, 2001.
Dennis, Kingsley and John Urry. *After the car*. Cambridge: Polity Press, 2009a.
Dennis, Kingsley and John Urry. 'Post-car mobilities'. In *Car troubles: Critical studies of automobility*, ed. Jim Conley and Arlene Tigar McLaren, 235–252. Farnham, UK and Burlington, VT: Ashgate Publisher, 2009b.
Derudder, Ben and Frank Witlox. 'An appraisal of the use of airline data in assessing the world city network: A research note on data', *Urban Studies* 42, no. 13 (2005): 2371–2388.
Flink, James J. *The car culture*. Cambridge, MA: MIT Press, 1975.
Freudendal-Pedersen, Malene and Sven Kesselring. 'Sharing mobilities: Some propaedeutic considerations'. *Applied Mobilities* 3, no. 1 (2018): 1–7.
Freudendal-Pedersen, Malene, Sven Kesselring, and Eriketti Servou. 'What is smart for the future city?: Mobilities and automation'. *Sustainability* 11, no. 1 (2019): 1–25.
Fuller, Gillian and Ross Harley *Aviopolis: A book about airports*. London: Black Dog Publishing, 2005.
Gehl, Jan. *Cities for people*. Washington, Covelo and London: Island Press, 2010.
Giedion, Siegfried. *Mechanization takes command: A contribution to anonymous history*. New York: Oxford University Press, 1948.
Gramsci, Antonio. 'Prison notebooks'. In *European perspectives*, ed. Joseph A. Buttigieg. New York: Columbia University Press, 1992.
Harvey, David. *The condition of postmodernity: An enquiry into the origins of cultural change*. Cambridge and Oxford: Blackwell, 1990.
Hilpert, Thilo. *Die funktionelle Stadt.: le Corbusiers Stadtvison. Bedingungen, Motive, Hintergründe*. Braunschweig: vieweg, 1978.
Kagermann, Henning. 'Die Mobilitätswende: die Zukunft der Mobilität ist elektrisch, vernetzt und automatisiert'. In *CSR und Digitalisierung: der digitale Wandel als Chance und Herausforderung für Wirtschaft und Gesellschaft*, ed. Alexandra Hildebrandt and Werner Landhäußer, 357–371. Management-Reihe Corporate Social Responsibility. Berlin and Heidelberg: Springer Berlin Heidelberg, 2017.
Kasarda, John D. and Greg Lindsay. *Aerotropolis: The way we'll live next*. New York: Farrar Straus & Giroux, 2011.
Kerouac, Jack. *On the road*. New York: New American Library, 1957.
Kesselring, Sven. 'Global transfer points: The Making of Airports in the mobile risk society'. In *Aeromobilities*, ed. Saulo Cwerner, Sven Kesselring and John Urry, 39–60. International library of sociology. London and New York: Routledge, 2009.
Knie, Andreas and Lutz Marz. *Die Stadtmaschine: zu einer Raumlast der organisierten Moderne. Veröffentlichungsreihe der Abteilung "Organisation und Technikgenese" des Forschungsschwerpunktes Technik-Arbeit-Umwelt am WZB, Projektgruppe Mobilität FS II 97-108*. Berlin: WZB, 1997.
Max-Neef, Manfred A. *Human scale development: Conception, application and further reflections*. New York: Apex Press, 1991.
Mom, Gijs. *Atlantic automobilism: Emergence and persistence of the car, 1895–1940: Explorations in mobility 1*. New York: Berghahn Books, 2015.
Pflieger, Géraldine and Manuel Castells, eds. *The social fabric of the networked city: Urbanism*. London: Routledge, 2008.
Reichow, Hans B. and Organische Baukunst. *Organische Stadtbaukunst, organische Baukunst, organische Kultur*. Braunschweig: Westermann, 1949.
Rode, Philipp and Nuno F. da Cruz. 'Governing urban accessibility: Moving beyond transport and mobility'. *Applied Mobilities* 3, no. 1 (2018): 8–33.
Rodrigue, Jean-Paul, Claude Comtois, and Brian Slack. *The geography of transport systems*. Abingdon, Oxfordshire: Routledge, 2005.
Ronald, Inglehart. *The silent revolution: Changing values and political styles among western publics*. Princeton, NJ: Princeton University Press, 1977.
Sachs, Wolfgang. *Die Bedeutungshaut des Automobils: annäherung an die Kultur der Hochenergie-Gesellschaft*. Berlin: TU Berlin, 1981.
———. *For love of the automobile: Looking back into the history of our desires*. Berkeley, CA: University of California Press, 1992.

Saskia, Sassen. *The global city*. New York, London and Tokio: Princeton University Press, 1991.
Schivelbusch, Wolfgang. *The railway journey: Trains and travel in the 19th century*. New York: Urizen Books, 1979.
Schmucki, Barbara. *Der Traum vom Verkehrsfluss: städtische Verkehrsplanung seit 1945 im deutsch-deutschen Vergleich. Beiträge zur historischen Verkehrsforschung Bd. 4*. Frankfurt/Main: Campus-Verl., 2001.
Sennett, Richard. *Flesh and stone: The body and the city in Western civilization*. New York: Norton, 1994.
Siebel, Walter. 'Urbanität'. In *Großstadt. Soziologische Stichworte*, ed. Hartmut Häußermann, 262–270. Opladen: Leske + Budrich, 1998.
Sieverts, Thomas. *Cities without cities: An interpretation of the Zwischenstadt*. London and New York: Spon Press, 2003.
Simmel, Georg 'Über sociale Differenzierung.: sociologische und psychologische Untersuchungen. Reprint der Ausgabe von 1890'. In *Staats- und socialwissenschaftliche Forschungen*, ed. Gustav Schmoller, 10. Leipzig: Dunker & Humblot, 1890.
———. *Die Grosstädte und das Geistesleben* [*The Metropolis and Mental Life*]. Erste Auflage. Dresden: Petermann, 1903.
Tönnies, Ferdinand. *Community & society (Gemeinschaft und Gesellschaft)*. East Lansing: Michigan State University Press, 1957.
Urry, John. 'The "system" of automobility'. *Theory, Culture & Society* 21, no. 4–5 (2004): 25–39.
——— *Offshoring*. Hoboken: Wiley, 2014.
Weber, Max. *Economy and society: An outline of interpretive sociology*. New York: University of California Press, 1978.

4

MOBILITY CAPITAL AND MOTILITY

Vincent Kaufmann and Ander Audikana

Introduction

Motility can be considered a form of capital in itself, one that influences social integration. As an analytical tool, the concept of motility allows us to better describe and conceptualize the interrelationship between the dynamics of spatial and social mobility. In this respect, motility provides keys for better understanding the ambivalence and limits as regards the necessity of mobility in contemporary societies.

This chapter contains three sections. The first presents the concept of motility and provides a brief overview of its application in the social sciences. The second discusses why motility can be considered as capital in the sense of Pierre Bourdieu. The third and final section attempts to illustrate the contributions the concept has made by highlighting research findings that support motility as capital relative to urban issues.

Motility as a resource for integration

Motility can be defined as a specific set of characteristics that allows an actor to be mobile. Motility therefore includes (1) social conditions of access (i.e. the conditions under which actors can utilize the offer in the broadest sense), (2) the knowledge and skills required to use the offer and (3) mobility aspirations (i.e. actual use of the offer that enables users to realize these aspirations). Motility is place-based; every territory has a specific range of possibilities and potential for creating opportunities to move. What determines a territory's range of possibilities and receptiveness to projects largely has to do with the material and immaterial conditions the context offers. All action is space specific and requires an environment that provides footholds for realizing it (Gibson 1979).

The concept of motility we have just briefly described has been extensively discussed (Kesselring 2005, 2006, Jensen 2006, Nowicka 2006, Söderström and Crot 2010, Sheller 2011, Merriman 2012). Its use goes far beyond the sociological realm and to raise questions that can be applied to the fields of anthropology (Glick Schiller and Salazar 2012), management (Sergot et al. 2012), human geography (Lévy 2004, Kellerman 2006, 2012a), history (Guigueno 2008, Flonneau and Guigueno 2009) and urban planning (Chalas and Paulhiac 2008, Lord 2011).

Following the publication of *Re-thinking Mobility* in 2002 (Kaufmann 2002), the concept was expanded upon theoretically (Kaufmann et al. 2004, Canzler et al. 2008, Ohnmacht et al. 2009) and methodologically to be applied to individual (Flamm and Kaufmann 2006) and collective actors (Kaufmann 2011). The concept of motility has been used for qualitative research on specific objects such as the use of travel time (Vincent-Geslin and Kaufmann 2012) and time in subway stations more specifically (Tillous 2009), as well as social inequalities (Maksim 2011). Quantitative studies of motility were also done within the framework of the European Job Mobilities and Family Lives in Europe research program (Kaufmann et al. 2010, Viry and Kaufmann 2015) and later to analyze the links between travel practices and the spatialization of social networks (Viry 2011), and finally as part of a survey on the rationale that underpins modal practices in Santiago de Chile (Witter 2012).

Researchers have also adopted the concept in order to conduct empirical research on a wide variety of topics, from the non-use of rights in social policies on accessibility (Féré 2011), social inequalities (Ureta 2008, Oliva 2010), daily routines (Belton 2009, Buhler 2012), social innovation (Brand and Dävila 2011), people's relationship to public space (Jirón 2010), transportation modal choices (Rocci 2007, Vincent 2008, Rivere 2009, Fouillé 2011, Kellerman 2012, De Witte et al. 2013), multi-residence and identity (Halfacree 2011), business travel (Faulconbridge et al. 2009) and even the process of political integration (Kaufmann and Audikana 2017).

What has this work contributed in terms of our understanding of the changes taking place in contemporary societies? One fact stands out: with growing opportunities for travel and mobility, the capacity to be mobile is becoming increasingly important for economic and social integration. This change coupled with the growing demand for flexibility socially reinforces the importance of motility as a resource for social integration.

In this article, we champion the idea that broadening our motility is an inherent part of a major societal transformation. This in turn means that the capacity to be mobile – i.e. motility – is becoming a resource for social integration and a sort of "capital" in its own right. This "capital" is more than a mere combination of income (economic capital), education level (cultural capital) and social networks (social capital). It is this societal transformation that has made motility a form of capital in its own right. Until the 1970s, professional careers were built on trinity (economic, cultural and social). Nowadays, careers also mean being able to play with and master space in the form of motility. More specifically, it seems that motility helps people to activate other capital and thus functions as a "dynamic relay." In many fields, having had one or more professional experiences abroad has become indispensable. For those with professional aspirations, success and motility are intrinsically linked, as are the choice and realization of such trips.

However, mobility in itself is ambivalent: while the ability to wield it in accordance with the dominant values can improve an individual's social status, neglecting it or using it against dominant values can result in a loss of status. For example, Hanja Maksim (Maksim 2011), in her work on the motility of the poor populations, has shown that people with low incomes develop very specific forms of motility to compensate for their economic handicap. However, she demonstrates that these forms do not correspond to the dominant (and hence valued) model of the mobile person who responds to the demand for flexibility that characterizes contemporary Western societies. Hence, certain forms of motility that contribute to social success are valued while others are not.

If spatial mobility is indeed becoming an essential ingredient in the construction of social status, as many works suggest, could we not then consider motility a capital in its own right? Individuals can have strong or weak motility skills but, above all, skills of an extremely varied nature. Motility is seemingly indispensable for playing with spatial and temporal friction. The ingenuity of the solutions envisaged and applied in this area often depends on the actor's quality of life and possibilities for changing their social status.

Why "capital"?

Why consider motility a capital in its own right? By distinguishing between different forms of capital (economic, cultural and social), Bourdieu (1986) demonstrates the many mechanisms that regulate the unequal distribution of resources and social reproduction. While acknowledging the key role of economic capital with regard to the dynamics of social structuring, the consideration of complementary forms of capital makes it possible to cope with "economism," which ignores the "specific efficiency" of the latter. While all capital can ultimately be transformed into economic capital through a process of conversion, Bourdieu's analysis offers a finer, richer understanding of structuring processes based on an "economy of practices," which goes beyond a mere "science of mercantile relations."

The use of different forms of capital responds first and foremost to the researcher's functional need to describe the richness of the social dynamics he or she observes. What justifies this distinction between different forms of capital is the need to be more precise, offer a more detailed description or provide a more complete account of the phenomena analyzed. It is not so much a question of introducing or accumulating forms of capital in a more or less deductive perspective, but of highlighting the variety and detail of practices that contribute to social structuring processes. By describing capital in its various forms, we can consider that Bourdieu lays the theoretical groundwork for carrying out empirical and analytical measurement, the motto being "break down capital in order to better measure and analyze it."

The use of the concept of motility also echoes this desire for measurement and empirical analysis in the area of mobility. The "invention" of mobility capital is simply the end result of a series of empirical studies testing analytic dimensions, statistical indicators, typologies and research hypotheses carried out in recent decades. Such tests are still being carried out today and are helping to renew, nuance and/or reinforce our findings and broaden the initial questions. A far cry from eventual "'capital inventors' that do not really bother proposing specific instruments [...] to measure them" (Neveu 2013: 352, emphasis added), our approach echoes the same empirical effort that inspired Bourdieu, who himself offers a varied and complex definition of capital that goes far beyond the traditional tripartite.

The fundamental issue therefore is understanding under what conditions and through what means mobility contributes to the capitalization process and, in the long term, promotes dynamics of social integration or exclusion.

Bourdieu (1993) introduces the question of mobility as a distinguishing factor in the accumulation process when he asserts that the use of the goods and services distributed in a given space depends both on the capital of the agents and on "the physical distance to these goods" (ibid.: 161) (adding that this distance also depends on their capital). He also speaks of the temporal dimension of mobility when he concludes that, because of physical distances, "the power that capital, in its different forms, gives to space is also ... power over time" (ibid.: 164). However, these analyses, which attempt to clarify the link between the "structures of social space" and "structures of physical space" – versus thinking about places in a "substantialist" way (ibid.: 159) – consider space neither as an object of analysis in itself, nor as an independent variable strictly speaking. The distribution of capital determines the spatial configuration. Economism – the idea that the world can be reduced to money – gives way to sociologism, the idea that the world can be reduced to social relations.

It is not by chance that the analysis of "struggles for the appropriation of space" (ibid.: 164) in general and the influence of spatial mobility issues on the capital accumulation process in particular has systematically attracted the attention of geography researchers well before the mobility turn. Based on the works of Harvey, Swyngedouw (1993) shows that economic and social structuring processes develop "in" and "through" reconfiguration of space and the transportation, communication and mobility patterns associated therewith. He considers that the accumulation of capital correlates with the process of circulation of the latter, and that, far from being a dependent variable, the control of space plays a key role in this dynamic:

> For each individual, increased mobility is essential to maintain positions of social, economic, political and/or cultural power, while, vice versa, we are approaching a condition in which adding to the mobility capacity of one individual may reduce the mobility of all the others.
>
> *(Ibid.: 320)*

Mobility is therefore at the heart of the struggle for the appropriation of space, if space is truly understood as an issue and not simply a framework. Considering mobility as both a commodity and a skill, Swyngedouw (ibid.: 323) concludes that the former is itself one of the arenas in which the struggle for power is carried out: "social power cannot any longer (if it ever could) be disconnected from the power or ability to move quickly over space." Identifying a field of mobility (which is defined as a space of struggle) or a spatial field that corresponds to mobility capital is not possible. However, we are seeing a proliferation of arenas of struggle centered around mobility issues. These arenas are not exclusively mobility arenas and issues; they concern struggles that affect and structure different types of capitalization.

Therefore, considering mobility as capital not only responds to a need for empirical precision (with regard to the dynamics of accumulation); it also highlights this dimension's increasingly dominant, even discriminating role in fueling other capitals (being mobile to earn money, cultivate oneself, or maintain/expand social networks). The fundamental contribution of the mobility turn consists in giving priority attention to mobility due to its structuring, supra-sectoral nature: "There seems [to be] little doubt that mobility is one of the major resources of 21st-century life and that it is the differential distribution of this resource that produces some of the starkest differences today" (Cresswell 2010: 22).

Motility: an analytical tool with a critical vocation

The sociological analysis of mobility we have developed in the past ten years aims to serve social criticism. For the latter to be more effective, theoretical or methodological shortcuts must be avoided by revealing the complexity and ambivalence of mobility-related dynamics. It is through the notion of motility that we are trying to model these types of dynamics. However, motility is also the subject of criticism when linked to the "*range of possibilities*" in terms of mobility: "We must take the social relationships and models of success a particular society proposes and the trials its actors must face in order to succeed into account" (Kaufmann 2008: 30). Motility is becoming central, as the demand obligation to be mobile and flexible themselves become central.

The notion of motility reinforces the links between analysis and social criticism based on three fundamental points.

To begin, it allows us to focus on mobility issues which, today, are fundamental "struggles" for the appropriation of space and thus directly impact social integration. Policies restricting immigration (that is, restricting the motility of certain populations) are a paradigmatic example of this type of rationale.

Secondly, the concept of motility allows us to appreciate the interaction between individuals' adaptation strategies on one hand and collective mobility constraints and opportunities on the other. Mobility analyses highlight the tension between individual actors and the social structure.

Finally, given the excess nature of constructivism (and resulting de-constructivism) that reduce mobility to an ideological issue, an analysis based on motility allows us to identify (more or less latent) concrete situations in which the constraints in terms of mobility play a vital role for social integration.

Sociological analysis is often "re-appropriated" to strengthen social criticism that does not necessarily correspond to that of the researcher. The concept of motility is not safe from such misuse. Yet, it is not by abandoning analytical concerns (and the tools available) that social criticism will become more relevant and, ultimately, more accurate. On the contrary, the study of mobility requires that researchers' rigor and imagination be equal to that of individual and collective choices and the challenges raised by this type of problem.

Applied to issues relating to urban mobility, the concept of motility helps highlight the fact that urban mobility refers to two key skills in particular:

- The ability to uproot and re-root elsewhere: the ability to travel in physical space and to be transformed by these experiences, exploiting the interaction between daily and residential mobility. This skill also encompasses the ability to seize opportunities for migration as well as the ability to build new social networks and relationships while maintaining ties with the social fabric of earlier residential stages.
- The ability to be reversible: the ability to use physical or virtual space to avoid having to move. More specifically, this includes skills for using and mastering various (notably rapid) transportation modes and ICT tools. Reversibility often implies the ability to combine daily life, work, family and leisure roles.

The empirical research on such issues that we have had the opportunity to partake in demonstrates that the two aforementioned urban mobility skills influence social inequalities relative to urban life. The ability to uproot and re-root in another environment is often strongest among immigrants from disadvantaged social groups. This skill affords them strength of adaptation, which allows them to escape certain forms of reproduction in terms of socio-professional or family-related inequalities. This finding emerged more specifically in research on highly mobile people who escape unemployment by moving abroad (Viry and Kaufmann 2015). Reversibility allows people in vulnerable situations to broaden their access to urban resources (e.g. employment opportunities, household provisions, and recreational/social opportunities) geographically. Hanja Maksim (Maksim 2011) made the same observation of poor populations living in the agglomerations of Berne, Clermont-Ferrand, Geneva and Grenoble. The findings demonstrate that when financial resources are lacking, certain individuals use their creativity and resourcefulness in terms of reversibility skills to improve their lives both qualitatively and materially. Mutual aid and free exchange are often at the heart of this process.

These examples clearly illustrate that motility is a form of capital whose existence and application does not necessarily correspond to or reflect income, training level or social networks. It is a specific resource that can be acquired through primary or secondary socialization processes and used for urban living (Viry and Kaufmann 2015).

References

Belton L. (2009) "Mobilités et lien social – Sphères privées et professionnelle à l'épreuve du quotidien." Doctoral thesis. Université de Paris Est.

Bourdieu P. (1986) "The forms of capital", in J. G. Richardson (ed.), *Handbook of Theory and Research for the Sociology of Education*, Greenwood, New York, pp. 241–258.

Bourdieu P. (ed.) (1993) *La misère du monde*, Seuil, Paris.

Brand P. and Dävila J. D. (2011) "Mobility innovation at the urban margins Medellín's Metrocables", City, Ifirstpublished, Vol. 15 (6), pp. 647–661.

Buhler T. (2012) "Eléments pour la prise en compte de l'habitude dans les pratiques de déplacements urbains", Doctoral thesis. ENSA, Lyon.

Canzler W., Kaufmann V. and Kesselring S. (eds.) (2008) *Tracing Mobilities*, Ashgate, Burlington.

Chalas Y. and Paulhiac F. (2008) *La mobilité qui fait la ville*, CERTU Editions, Lyon.

Cresswell T. (2010) "Towards a politics of mobility", *Environment and Planning D: Society and Space*, Vol. 28 (1), pp. 17–31.

De Witte A., Hollevoeta J., Dobruszkes F., Hubert M. and Macharis C. (2013) "Linking modal choice to motility: A comprehensive review", *Transportation Research Part A*, Vol. 49, pp. 329–341.

Faulconbridge J., Beaverstock J. V., Derudder B. and Witlox F. (2009) "Corporate ecologies of business travel in professional service firms: Working towards a research agenda", *Euroean Urban and Regional Studies*, Vol. 16 (3), pp. 295–308.

Féré C. (2011) *Concilier accès à la mobilité pour tous et mobilité durable.* Thèse de doctorat. Université de Lyon 2 – Institut d'Urbanisme de Lyon.

Flamm M. and Kaufmann V. (2006). "Operationalising the concept of motility: A qualitative study", *Mobilities*, Vol. 1 (2), pp. 167–189.

Flonneau M. et Guigueno V. (eds.) (2009). *De l'histoire des transports à l'histoire de la mobilité?* PUR.

Fouillé L. (2010) "L'attachement automobile mis à l'épreuve. Etude des dispositifs de détachement et de recomposition des mobilités", Doctoral thesis. Université de Rennes 2.

Fouillé L. (2011) *L'attachement automobile mis à l'épreuve. Etude des dispositifs de détachement et de recomposition des mobilités.* Thèse de doctorat. Université de Rennes 2.

Gallez C. and Kaufmann V. (2009) "Histoire et sociologie face à la mobilité", in M. Flonneau and V. Guigueno (eds.), *Histoire de la mobilité*, PUR, Rennes, pp. 41–56.

Gibson J. J. (1979) *The Ecological Approach to Visual Perception*, Houghton Mifflin, Boston, MA.

Glick Schiller N. and Salazar N. (2012) "Regimes of mobility accross the globe", *Journal of Ethics and Migration Studies*, Nov 2012 IfirstArticle, Vol. 39 (2), pp. 183–200.

Guigueno V. (2008) "Building a high-speed society: France and the Aérotrain, 1962–1974", *Technology and Culture*, Vol. 49 (1), pp. 21–40.

Halfacree K. (2011) "Heterolocal identities? Counter-urbanisation, second homes, and rural consumption in the era of mobilities", *Population, Space and Place.* Article first published online: 14 March 2011: doi:10.1002/psp.665.

Jensen O. B. (2006) "'Facework', flow and the city: Simmel, Goffman, and mobility in the contemporary city", *Mobilities*, Vol. 2, pp. 143–165.

Jirón P. (2010) "Mobile borders in urban daily mobility practices in Santiago de Chile", *International Political Sociology*, Vol. 4, pp. 66–79.

Kaufmann V., Bergman M. and Joye D. (2004). "Motility: mobility as capital", *International Journal of Urban and Regional Research*, Vol. 28 (4), pp. 745–756.

Kaufmann V., Viry G. and Widmer E. (2010) "Motility", in N. Schneider and B. Collet (ed.), *Mobile Living across Europe II: Causes and Determinants of Job Mobility and Their Individual and Societal Consequences*, Barbara Budrich Publishers, Opladen, pp. 95–112.

Kaufmann V. and Audikana A. (2017) *Mobilité et libre circulation en Europe: un regard suisse*, Economica, Paris.

Kaufmann V. (2002) *Re-thinking Mobility*, Ashgate, Burlington.

Kaufmann V. (2008) *Les paradoxes de la mobilité*. PPUR, Lausanne.
Kaufmann V. (2011) *Rethinking the City*, Routledge and EPFL Press, London and Lausanne.
Kellerman A. (2006) *Personal Mobilities*, Taylor & Francis, London.
Kellerman A. (2012) *Daily Spatial Mobilities: Physical and Virtual*, Ashgate, Aldershot.
Kesselring S. (2005) "New mobilities management: Mobility pioneers between first and second modernity", *Zeitschrift für Familienforschung*, Vol. 17 (2), pp. 129–143.
Kesselring S. (2006) "Pioneering mobilities: New patterns of movement and motility in a mobile world", *Environment and Planning A*, Vol. 38, pp. 269–279.
Lévy J. (2004) "Les essences du mouvement", in S. Allemand, F. Ascher and J. Lévy (eds.), *Les sens du mouvement. Modernité et mobilités dans les sociétés urbaines contemporaines*, Belin, Paris, pp. 298–307.
Lord S. (2011) "Le choix de vieillir à domicile: l'inévitable adaptation des modes de vie – Évolution de la mobilité quotidienne d'un groupe d'aînés de banlieue pavillonnaire", *Retraite et Société*, Vol. 60 (1), pp. 197–213.
Maksim H. (2011) *Potentiels de mobilité et inégalités sociales: La matérialisation des politiques publiques dans quatre agglomérations en Suisse et en France*. Thèse de doctorat EPFL, no 4922.
Merriman P. (2012) *Mobility, Space and Culture*, Routledge, London and New York.
Neveu É. (2013) "Les sciences sociales doivent-elles accumuler les capitaux? À propos de Catherine Hakim, Erotic Capital, et de quelques marcottages intempestifs de la notion de capital", *Revue française de science politique*, Vol. 63 (2), pp. 337–358.
Nowicka M. (2006) *Transnational Professionals and Their Cosmopolitan Universes*, Campus, Frankfurt.
Ohnmacht T. et al. (eds.) (2009) *Mobilities and Inequality*, Ashgate, Farnham.
Oliva J. (2010) "Rural melting-pots, mobilities and fragilities: Reflections on the Spanish case", *Sociologia Ruralis*, Vol. 50, pp. 277–295.
Rivere M. (2009) "Socio-histoire du vélo dans l'espace urbain – d'une écologie politique à une économie médiatique… Toulouse, Genève, Saragosse", Doctoral thesis. Université de Toulouse – Le Mirail.
Rocci A. (2007) *De l'automobilité à la multimodalité? Analyse sociologique des freins et leviers au changement de comportements vers une réduction de l'usage de la voiture. Le cas de la région parisienne et perspective internationale*. Thèse de doctorat. Université de Paris 5.
Sergot B. et al. (2012) "Mobilités spatiales et dynamiques organisationnelles", *Revue française de gestion*, Vol. 2012/7 (226), pp. 77–90.
Sheller M. (2011) *Mobility*, Sociopedia.isa.
Söderström O. and Crot L. (2010) "The mobile constitution of the society: Rethinking the mobility society nexus", Working Paper, Université de Neuchâtel.
Swyngedouw E. (1993) "Communication, mobility and the struggle for power over space", in G. A. Giannopoulos and A. E. Gillespie (eds.), *Transport and Communications Innovation in Europe*, Belhaven Press, London, pp. 305–325.
Tillous M. (2009) Le voyageur au sein des espaces de mobilité. Thèse de doctorat. Université de Paris 1.
Ureta S. (2008) "To move or not to move? Social exclusion, accessibility and daily mobility among the low-income population in Santiago, Chile", *Mobilities*, Vol. 4, pp. 269–289.
Vincent S. (2008) *Les "altermobilités": analyse sociologique d'usages de déplacements alternatifs à la voiture individuelle. Des pratiques en émergence?* Thèse de doctorat. Université de Paris 5.
Vincent-Geslin S. and Kaufmann V. (2012) *Mobilité sans racines*. Descartes, Paris.
Viry G. (2011) *Mobilités spatiales, réseaux sociaux et familiaux: Quels mécanismes d'intégration à l'épreuve de la mobilité?* Thèse de doctorat. Université de Genève.
Viry G. and Kaufmann V. (2015) *High Mobility in Europe. Work and Personal Life*, Palgrave McMillan, London.
Witter R. (2012) "Public urban transport, mobility competences and social exclusion", The case of Santiago de Chile, Thèse EPFL.

5
CO-DESIGN TIMES AND MOBILITIES

Luc Gwiazdzinski

Complexity is a problem word
and not a solution word
Edgar Morin

"Describing, understanding, explaining and regulating mobility requires new cross-cutting approaches" (Kaufmann, 2002). Mobility specialists, researchers, operators and citizens are forced to change their perspectives, methods and tools to address the complexity of a "liquid" society (Baumann, 2000), of "worlds" (Descola, 2014) in a state of constant flux and to design inclusive and intelligent organisation and regulation systems. The "mobility turn" (Sheller and Urry, 2006) is also a "time turn" in social sciences. The integration of the temporal dimension in the observation, analysis and regulation of mobility (Rabin and Gwiazdzinski, 2007; Urry, 2000), the adoption of a "chronotopic" approach (Bonfiglioli, 1990; Drevon et al., 2017) – where the "chronotope" is the "place where the spatial and the temporal dimensions meet" – are interesting ways to approach the figure of the "nimble, malleable, flexible and adaptable city" (Gwiazdzinski, 2007).These are also exciting ways of reflecting on "indwelling" the city and to imagine the contours of an "intelligence of possible forms of mobility.

Opening this chapter on time within a broader reflection on urban mobility requires defining the central terms of the proposal by referring to a number of contributors, for the definition of the city as a "place for the maximization of interactions" (Claval, 1982). Time, "a progressive synthesis of a high level of complexity" (Elias, 1996), can be approached from a social perspective as "the meaning that human communities have given to change" (Tabonni, 2006). "Mobility" – this "new frontier of sociology" (Urry, 2000) – generally refers to changing places (Brunet, Ferras and Thery, 1992) – in geography – and also to a way of experiencing space and time that permits us to overcome these difficulties by articulating them.

A too long forgotten time dimension

In urban research, much work has been devoted to space and very little to time, the space–time relationship and its representation, although in the specific field of mobility, theories such as the "Zahavi conjecture" – to the effect that daily life journeys are made with constant transport time-budgets and that their spatial scope is a function of travel speed – remain stimulating indeed (Crozet and Joly, 2006). For a long time, researchers have favoured the analysis of the modalities presiding over formalizing urban change, the long-time nature of the city's future, *to the detriment of an approach that would have aimed at providing the elements of a typology likely to order the diversities of urban social times and their combination* (Lepetit and Pumain, 1993). Time has long been the orphan child of reflections on the functioning, planning or development of cities and territories, to the advantage of infrastructures. The material aspect has often taken over from the human dimension. *Hardware* has been preferred to *software*. The temporal dimension has been neglected by city councillors and planners alike, although it is an essential aspect of urban dynamics. Up to now, the main reason for this has been to arrange the land so as to "save" more time. In terms of urban planning, the construction of the French *TGV* (high-speed railway) has shrunk maps in France, Europe and elsewhere, allowing those who can afford it to get from point A to point B more quickly. The opposite approach, which consists in arranging time in order to have an effect on the use of space, is less common. Time is nevertheless an essential key to understanding and managing societies and much is at stake collectively for people, organizations and territories. This concept of time, a convention and the *abstract measurement of concrete things* (Sue, 1994) is the product of social activities. The *time system* (Sorokin, 1964), which results from the combination of these activities, is difficult to break down but is essential to explore.

A necessary shift in perspective

To meet the current economic, social, cultural and environmental challenges, it is necessary to change our perspective and "imaginary", *this substrate of mental life and the constitutive dimension of humanity* (Souty, 2006), regarding the city, the territory and forms of mobility. Following the example of architects, we must learn how to consider the city as a three-dimensional entity rather than as a simple plan, if only because our "tridiastatic" agglomerations (Reymond, 1998) are rising, sprawling or burying themselves. Last but not least, a fourth dimension must be added, namely time. Urban materiality evolves over time. The urban offering, populating and occupying spaces also vary on a year-round basis, across seasons, weeks, days and hours. However, cities and territories are not fixed structures. Perpetual changes modify urban materiality, affect the economic and social spaces, as well as the legal or political-administrative spaces. On different scales, social life takes place in multiple times, always divergent and often contradictory, and their relative unification, linked to an often precarious hierarchy, poses a problem to any society (Gurvitch, 1963). The entire city is an ephemeral, fragile and fleeting universe that is difficult to grasp, a labyrinth that evolves through time and space according to daily, weekly, monthly, seasonal rhythms, or together with events. In other words, we are no longer content to work on an "average city", which actually does not exist; instead, we must adapt our observations, analyses and interventions to a city that is not the same by day and by night; on Sundays and during the week; in summer and in winter. Metropolises, "those cities beyond the city", are not entities with strict borders, but rather throbbing pulsations. They attract and reject populations in the distance according to circadian, weekly, seasonal and event rhythms. The city is not only inhabited by permanent dwellers but also welcomes temporary residents who will come to work, stay for a few days or simply visit it without being involved in governance.

This change of perspective on territorial systems also concerns mobility, and it seems useful to think in terms of "chains" and "mobility systems", rather than in terms of isolated modes of transport. It is naturally necessary to adopt a systemic and multi-scalar approach to mobility, from the street to the whole world and conversely. The same is true of users who reason in their daily lives in terms of their journeys. It is possible to conceptualize the linkages rather than the conflicting relations between public transport and the private car – which by the way is changing, consumes less and less energy and will soon be "autonomous" and integrated into a global mobility system. This change of paradigm and imagination can also involve a change in vocabulary: "transport" becomes a "journey", networks turn into the "mobility intelligence" and the user proves to be "mobilian" (Rabin and Gwiazdzinski, 2007), while the urban transport company evolves into a "travel agency". This conceptualization is also integrated into a more general discussion on the permanent changes in society, organizations, territories as well as at individual behaviour level.

Major changes

On almost all roads, the tempo of cities unfolds in the same way, a natural rhythm regulated by the march of men, horses and cars (Reclus, 2005). Such a description has long since ceased to be relevant, and we are not always aware of the upheavals that are affecting our societies and shaping the future of our cities as well as our lifestyles.

The spreading and fragmentation of spaces and times combine with the urgency to recompose new practices, constraints and opportunities: despite the urbanization control discourse, the city sprawls out into space beyond its administrative borders. In France, the equivalent of a *department* (county) disappears every five or six years,[1] as a result of urbanization. Activities also extend over time and, on our action calendars, they eat away at traditional time-offs.

Night work increased from 3.3 million (15.0% of the working population) in 1990 to 4.3 million (16.3%) in 2013 (Cordina-Duverger et al., 2019). The weekend is a moment in time that is consumed by work: 35% of wage earners work on Saturdays, 19% on Sundays.[2] Sleep time has decreased, for the first time ever, falling below the usually recommended minimum of 7 hours a day to recover properly (Léger and Bourdillon, 2019). Lunch time is still shorter: 22 minutes on average today, compared to 1 hour 38 minutes 20 years ago.[3]

Activities are also spreading over time, and the schedules of our lives and cities are getting ever more crowded. The living space is increasingly fragmented, partitioned between places where people sleep, work and get supplies, and the figure generally defined to illustrate it is that of the "archipelago" (Veltz, 1996; Viard, 1994). Fragmentation also means time is breaking down. The great social rhythms that used to mark community life are fading away. In the West, the society that used to live by sunlight, framed with the sound of bells, then of the factory siren, is now in the grips of a new dictator: the smartphone, which we consult on average twenty-three times a day,[4] for a daily 1 hour and 42 minute period. We store time in there, we permanently resynchronize ourselves with individuals who live elsewhere, more or less far away. The third mutation we deem central is urgency or rather, our being egged on into urgency. Sociologists (Aubert, 2003; Rosa, 2005) have highlighted this phenomenon of acceleration and synchronization on a global scale, with a consumer who wants everything, everywhere and at any time. No one can bear queuing and waiting anymore.

These developments have consequences on mobility and vice versa, on individuals, organizations and territories. The average distances and times spent in transport have been increasing for a long time.

Their pattern has also changed. They are spreading out, being ripped apart and less and less regular, more and more peripheral, events – and leisure-driven. Home-to-work commuting mobility, on which many models are built, now accounts for only a quarter of all trips. For managers, mobility becomes less and less manageable when our activities are being decided on at the last minute. The model for the establishment of attractors induces "zigzagging mobility" (Bonfiglioli, 1997) over vast territories, with complicated schedules and timetables. In a strange reversal, it should also be pointed out that mobility has taken root as a new value and as a right. People who never move out of their neighbourhoods or profession are stigmatized. In the schoolyard and later at the company canteen, mobility is often used as a means for self-promotion.

In this system, characterized by spreading, fragmentation and urgency, "hypermodern" individuals (Lipovetsky, 2004) are increasingly mobile, connected, polytopic (Stock, 2006), even "polyactive", fitting in several "trades" at the same time. They are also increasingly unstable, in their families, their jobs, their locations, prove unpredictable and they long for the gift of ubiquity. These mutations, characterized by the temporary, the fragile, the flexible, the mobile and the unstable, are important.

Increased complexity and pressures

Social life flows in multiple, often divergent and contradictory time frames whose relative unification is precarious. There is an increase in complexity, difficulties, increased consumption of space, as well as desynchronizing and difficulties in reconciling work and family life, as well as in controlling time (30% of Europeans say they have no control over their time) and on meeting each other. The acceleration, the emergence of a world time and the fragmentation of social times, as well as desynchronization, urge people, organizations and territories to compete with each other, and thereby generate tensions, conflicts and difficulties. The generalized flexibility of social times combined with the diversification of practices within each social time result in new "time maps" (Asher and Godard, 2003), new temporal regimes that are highly differentiated according to social situations, genders, generations and territories.

While unified by information, people have never experienced such dislocated temporalities. Faced with such de-synchronization, our schedules are breaking down. Everyone juggles with time, torn between their professional, family and social lives, their work and their daily obligations. Information and communication technologies give us the illusion of ubiquity. Faced with increased responsibility and arbitration difficulties, the most vulnerable among us are ambushed by the "fatigue of being oneself" (Ehrenberg, 1998) and the notorious "burnout". At another level, conflicts are multiplying between individuals, groups, territories and neighbourhoods in the "polychronic city", since they no longer live at the same pace, especially at night. More worrying, new inequalities are emerging between populations, organizations and neighbourhoods, as they are unequally equipped to cope with the acceleration and complexity of social times.

Individual and collective adaptations

Faced with these changes and their consequences in terms of tensions, conflicts or inequalities, individuals, groups and territories are getting organized. These are all interesting "signals" for the present and the future of our cities and societies.

Some have decided to take a break from this hustle and bustle and opted for slower leisure activities such as walking, yoga, gardening or a stroll at the flea market. Elsewhere, researchers and essayists praise slowness while networks such as *Slow Food* and *Cittaslow* are flourishing. Besides, in the absence of common meals or work times, items such as the freezer, video recorder, microwave and mobile phone enable each one of us to organize our lives *à la carte* (as they please). The trend is towards the hybridization of practices, times and spaces and new arrangements, alliances and collaborations: co-construction, co-development, co-housing, carpooling or co-design. The boundaries between times and spaces for work and leisure are getting blurred, and "third places" (Oldenburg, 1989) are emerging: library-cafés, café-laundries, entrepreneur-artists incubators, nurseries installed in stations, but also roof gardens or ecomuseums-dwellings. On a different scale, our "urban seasons" schedules are brimming with events, festivals or celebrations. These new rites, which celebrate memory, identity and renewed belonging to the city, make it possible to "do family or territory together", to exist in a context of territorial competition and to maintain an illusion of social cohesion in the face of a diluted daily life. The regime of the "intermittent metropolis", a temporal counterpart of the spatial figure of the archipelago, is essential. The event city, ephemeral and festive, triumphs and unfolds.

In the 1990s, first in Italy, then in France and Germany, public authorities set up structures, platforms for observation, awareness-raising, dialogue, exchange and experimentation, which tried to bring these temporal approaches to the city and territories and to respond to the challenges at hand. Without many resources, they have sought to impose this temporal view of society, proposing new maps, experimenting with new opening hours for services and transport, participating in the debate on issues such as opening nights and Sundays for business, allegedly in order to improve the quality of life, gender equality or the "Right to the City". However, these local initiatives, which concern some thirty local authorities (Mallet, 2011), have not made it possible to implement real public time and mobility policies, though these are essential stakes.

Central temporal issues for mobility

The temporal stakes of mobility in cities – i.e. "what can be gained or lost"[5] – are multifaceted. There is no dismissing the challenge of optimizing flows and traffic in real time, in the logic of *monitoring* mobility and in the context of developing the "Smart city", which usually refers to a city implementing information and communication technologies to improve the quality of urban services or cut costs. But the real spatio-temporal challenges are undoubtedly elsewhere.

The first issue is undoubtedly that of imaginability (Lynch, 1969) and intelligibility of complex spatio-temporal systems for designers, managers and users. This challenge requires observation and representation tools that go beyond the usual static maps and the early contributions of *Time Geography* (Carlstein, Parkes and Thrift, 1978; Hägerstrand, 1975) to read and write about the changing society by thinking and mapping space and time together – and in real time as well. The development of ICT, the trivialization of location technologies, *data mining*, the development of dynamic representation tools have led to significant advances (Drevon, Gwiazdzinski and Klein, 2017). In parallel to this visual approach, it is also important to develop more sensitive exploration approaches in the form of immersions and journeys that make it possible to discover specific temporalities – such as night-time – by summoning body and senses throughout the experience.

Another related issue concerning architectural public space is that of a "distributed intelligence" (Joseph, 1999), combining network connectivity, readability of spaces and services as well as machines interactivity. Naturally, transport operators and citizens need to address the challenge of organizing and synchronizing each other in real-time. When the heart of the city used to beat from 8am to midday and from 2pm to 6pm, it was easy for the transport company to organize regular bus traffic. Less so when everyone works at different paces and makes decisions at the last minute, just in time. Taxis and on-demand transports are then better suited to the demand.

The notion of "long time", dear to the historian Fernand Braudel can be applied to the issue. When we forget the long-time nature of spatial planning, and fail to develop a vision on this scale, mobility policies only approximate policies of adaptation and adjustment on an ad hoc basis, and space consumption can continue around the figure of the "*Cittadiffusa*" (Indovina, 1990).

The issue of participation is essential. The temporal approach is an interesting as well as attested and proven tool for building and exchanging in a cross-cutting and inclusive way. As it is both everyone's and no one's expertise, time is one of the only themes that really makes it possible to engage in a debate with all public and private stakeholders, without tensing or withdrawing behind institutional boundaries. It is in line with people's current expectations in terms of proximity, participation and "concrete" practices when dealing with everyday issues. This requires organizing partnerships from the observation phase through to the experimentation and evaluation phases.

Time is a sensitive dimension, and it naturally places people at the heart of the debate and makes it possible to start thinking in the logic of "innovation through uses" in the sense developed by Von Hippel (2005), a process that values users' place and power in their partnership with research-centres designers and players in the territorial fabric.

The development of a spatio-temporal approach and the implementation of an innovation platform somewhat blurs the boundaries between research and experimentation; between decision-makers and citizen-users, thanks to dialogue interfaces; between public transport and the private car; between different urban operators: transporters, street furniture manufacturers, urban planners, etc. The temporal approach can sometimes act as a *laboratory of illusions* by facilitating work on sensitive questions and suggesting solutions without addressing them head-on. It is part of a design thinking approach, growing analytical thinking and intuitive thinking with feedback towards the end user.

There is also the issue of continuity and enjoying the "Right to the City" 24 hours a day, 7 days a week: spatial continuity with the development of complementary solutions even in low-density locations (transport on demand, car-sharing, etc.); temporal continuity between day and night, week and weekend, school terms and holidays; price continuity with the multi-scalar integration of mobility systems, besides the possibility of paying only one ticket; informational continuity through continuous information for the user – before, during and after mobility; political continuity through overstepping administrative borders; the coherence of policies for the different modes and strategic coherence at different spatial and temporal scales. Another temporal – even rhythmic – challenge of mobility is taking space and stopping times into account. During individuals' journeys, immobility is part and parcel of mobility, just as are tools such as the chair and the bench.

Having the key to times makes it possible to engage in a reflection on the adaptable "malleable city" and on reversibility. The versatility and modularity of public spaces, buildings, networks or means of transport are central to limiting space consumption. A tramway could be used during the day as a means of passenger transport and at night to transport freight, as is done with some aircraft.

The challenge is also qualitative. Mobility time can no longer only be envisioned as time wasted between departure and arrival but more as a quality time, a moment to be valued, and even an experience in its own right. A sign of the times: when a manufacturer used to try to sell a car, he would stress the speed of the automobile. Today, to sell the same vehicle, some automotive groups are emphasizing the quality of the time[6] spent inside the cabin.

Travel time is no longer necessarily a moment of mono-activity but it "hybridizes" (Gwiazdzinski, 2016), thanks to ICTs that make it possible to work, watch a film, play, telephone or listen to music during one's journey in "mobiquity" logic (Greenfield, 2006).

The chronotopic approach involved raises the issue of public space and public time, the very rare one when we find ourselves sharing the city with the others. The street and the tramway embody these places and moments. The time spent in mobility as well as users' needs turn the hospitality, comfort and ergonomics of these spaces and mobility times into major challenges. The temporal approach also makes it possible to uncover and raise many other issues in terms of temporary governance, of reticular territories but also regarding the co-construction of mobility solutions and creative explorations. Finally, one last challenge is in the form of a counterfoil: granting the possibility of doing nothing else than just being on the move, of leaving opportunities for improvisation, distraction and letting go. It is also a matter of seeking rhythm and tempo and of relieving the guilt felt by those who fail to resist such mobilizing injunctions.

More generally, it is about learning to "indwell time and mobility", in Dardel's sense, *a way of knowing the world and a type of emotional relationships far removed from an abstract or technocratic approach to space* (Dardel, 1990). It is the meaning of this *intelligence of mobility* that we long for and that we must mobilize to describe, understand, explain and then regulate mobility.

Towards an understanding of forms of time and mobility

The reflection intersects with "rhythm analysis" at the frontier of science and poetry, which Bachelard (1950) had intuited, and which Lefebvre (Lefebvre, 1993) tried to impose and use for the planning process. In sustainable development logic, the spatio-temporal approach makes it possible to integrate mobility into an "urban planning of times" – "*defined as all the plans, organization of schedules, and coherent actions on space and time that allow the optimal organization of the technical, social and aesthetic functions of the city for a more humane, accessible and hospitable metropolis*" (Gwiazdzinski, 2009).

This soft city building and management, which rests on what is light, dismountable, ephemeral, intelligent and experimental, also raises a series of exciting questions: the more we move towards reflection that integrates space, time and mobility, the more human capital and "collective intelligence" – such as "this intelligence is distributed everywhere, constantly enhanced, coordinated in real time, which results in an effective leveraging of skills" (Levy, 1994, p. 29), prove essential. I propose to use the concepts of "mobility intelligence" to describe this "cognitive capacity"; and "times and mobilities design" to describe the collective process of building a shared reflection, describing, understanding, explaining and regulating mobility, in which spatio-temporal aspects are central. The "co-design" we are talking about here with reference to (space), time and mobility is not just a mere matter of the appearance or optimization of technical transport networks. It is an attitude and a complex task that refers to the essence of products and institutions and to human experience. "Design is thinking in terms of relationships" (Moholy-Nagy, 1993, p. 278) and of "rhythm", in the sense of "how to flow" (Benveniste, 1974).

Beyond the technical aspects, it is based on a higher level of integration by the actors and users of a complex and inter-scalar mobility system in which everyone finds it beneficial to collaborate and where their own performance within the group is better than if they were by themselves. It is a paradigm shift from a single mode approach in terms of managing and resorting to a traditional transport network used by mere user-agents.

The approach adopted here opens up such stimulating concepts as "temporary and mobile living" or "ephemeral and situational citizenship". The design of social times and mobilities, the choice of our temporalities and spaces are great ways of expressing ourselves on the meaning of our cities and lives. The behaviours we adopt at different times and the attention we pay to them are indicative of the changes that are taking place in our societies and of our ability to define a particular "culture of time". It is up to each of us to take part in the definition of the "dance of life" (Hall, 1984) and what we mean by "the city".

Notes

1 Rapport de la FN-SAFER 2016.
2 Dares Analyses, Publication de la direction de l'animation de la recherche, des études et des statistiques, n°030, juin 2018.
3 Étude sur les repas au travail, 2016, Edenred.
4 https://lareclame.fr/omd-infographie-baromobile-193803.
5 www.cnrtl.fr/definition/enjeu.
6 An advert for PSA Peugeot Citroën car-manufacturers "*Quality time*".

References

Asher, F. and Godard, F., 2003. *Modernité: la nouvelle carte du temps*. La Tour d'Aigue: L'Aube.
Aubert, N., 2003. *Le culte de l'urgence*. Paris: Flammarion.
Bachelard, G., 1950. *La dialectique de la durée*. Paris: PUF.
Baumann, Z., 2000. *Liquid Modernity*. Cambridge: Polity Press.
Benveniste, E., 1974. *Problèmes de linguistiquegénérale*. Paris: Gallimard.
Bonfiglioli, S., 1990. *L'architettura del tempo*. Milan: Liguori.
Bonfiglioli, S., 1997. Le politiche dei tempi urbani. *Urbanistica Quaderni* 3, pp. 9–13.
Brunet, R., Ferras, R., and Thery, H. eds., 1992. *Les mots de la géographie*. Paris: la Documentation française.
Carlstein, T., Parkes, D., and Thrift, N., 1978. *Timing Space and Spacing Time*. London: Arnold.
Certeau (De), M., 1980. *L'invention du quotidien*. Paris: Gallimard.
Chesneaux, J., 1996. *Habiter le temps*. Paris: Fayard.
Claval, P., 1982. *La logique des villes*. Essaid'urbanologie. Paris: LITEC.
Cordina-Duverger, E., Houot, M., Tvardik, N., El Yamani, M., Pilorget, C., and Guénel, P., March 12, 2019. Prévalence du travail de nuit en France: caractérisation à partir d'une matrice emplois-expositions. *Bulletin épidémiologique hebdomadaire* (8–9), pp. 168–173.
Crozet, Y. and Joly, I., 2006. La "Loi de Zahavi": quelle pertinence pour comprendre la construction et la dilatation des espaces-temps de la ville? *Plan Urbanisme Construction Architecture* p. 163.
Dardel, E., 1990. *L'homme et la terre: nature de la réalité géographique*. Paris: CTHS.
Descola, P., 2014. *La composition des mondes*. Paris: Flammarion.
Dewey, J., 1934. *Art as Experience*. New York: The Berkeley Publishing Group.
Drevon, G., Gwiazdzinski, L., and Klein, O., 2017. *Chronotopies, Lire et écrire les mondes en mouvement*. Grenoble: Elya.
Ehrenberg, A., 1998. *La Fatigue d'être soi*. Paris: Odile Jacob.
Elias, N., 1996. *Du temps*. Paris: Fayard.
Faste, R., Roth, B., and Wilde, D. J., 1993. Integrating Creativity into the Mechanical Engineering Curriculum. In: C. A. Fisher, ed. *ASME Resource Guide to Innovation in Engineering Design*. New York: American Society of Mechanical Engineers, pp. 93–98.
Greenfield, A., 2006. *The Dawning Age of Ubiquitous Computing*. Berkeley, CA: New Riders.
Gurvitch, G., 1963. La multiplicité des temps sociaux. In: PUF Presses universitaires de France, ed. *La vocation actuelle de la sociologie*. Paris: P.U.F., t.11, pp. 325–340.
Gwiazdzinski, L., 2003. *La nuit dernière frontière de la ville*. La Tour d'Aigues: L'Aube.

Gwiazdzinski, L., 2007. Redistribution des cartes dans la ville malleable. *Espace, Population, Sociétés* 2007 (2–3), pp. 397–410.

Gwiazdzinski, L., 2009. Chronotopies. L'événementiel et l'éphémère dans la ville des 24 heures. *BAGF* 86(3), p. 352.

Gwiazdzinski, L., 2016. *L'hybridation des mondes*. Grenoble: Elya.

Gwiazdzinski, L. and Klein, O., 2014. Du suivi GPS des individus à une "approche" chronotopique, Premiers apports d'expérimentations et de recherches territorialisées. *Netcom, Netcom and Communication Studies* 28(1), pp. 77–106.

Hägerstrand, T., 1975. Space, Time and Human Conditions. In: Karlqvisted, ed. *Dynamic Allocation of Urban Space*. Farnborough: Saxon House, pp. 53–67.

Hall, E. T., 1984. *La danse de la vie. Temps culturels et temps vécus*. Paris: Seuil.

Indovina, F. ed., 1990. *La Cittàdiffusa*. Venise: DAEST-IUAV.

Joseph, J., 1999. *Villes en gares*. La Tour d'Aigues: Éditions de l'Aube.

Kaufmann, V., 2002. *Re-thinking Mobility*. Burlington: Ashgate.

Lefebvre, H., 1993. *Eléments de rythmanalyse*. Paris: Syllepse.

Levy, P., 1994. *L'Intelligence collective. pour une anthropologie du cyberespace*. Paris: La Découverte.

Léger, D. and Bourdillon, F., 2019. Le temps de sommeil, la restriction de sommeil et l'insomnie chronique des 18–45 ans: résultats du baromètre santés. *Bulletin épidémiologique hebdomadaire* 146, pp. 8–9.

Lepetit, B. and Pumain, D., 1993. *Temporalités urbaines*. Paris: Anthropos.

Lipovetsky, G., 2004. *Les temps hypermodernes*. Paris: Grasset.

Lynch, K., 1969. *L'image de la cité*. Paris: Dunod.

Mallet, S., 2011. Que deviennent les politiques temporelles? *Urbanisme* (376), pp. 86–89.

Moholy-Nagy, L., 1993. Le design: une attitude, pas une profession. In: *Peinture Photographie Film*. Nîmes: Éditions Jacqueline Chambon, p. 278.

Oldenburg, R., 1989. *The Great Good Place*. Saint-Paul: Paragon House.

Perec, G., 1974. *Espèces d'espaces*. Paris: Galilée.

Rabin, G. and Gwiazdzinski, L., 2007. *Si la route m'était contée, Un autre regard sur la route et les mobilités durables*. Paris: ÉditionsEyrolles.

Reclus, E., 2005. *L'homme et la terre*. Paris: La Découverte.

Reymond, H., 1998. Approches nouvelles de la coalescence. In: H. Reymond, C. Cauvin, and R. Kleinschmager, eds. *L'espace géographique des villes. Pour une synergie multistrate*. Paris: Anthropos, pp. 21–48.

Rosa, H., 2005. *Die Veränderung der Zeitstrukturen in der Moderne*. Berlin: SuhrkampVerlag.

Saint Augustine, *Confessions*, XI, 14, 17.

Sheller, M. and Urry, J., 2006. The New Mobilities Paradigm. *Environment and Planning* 38(2), pp. 207–226.

Sorokin, P. A., 1964. *Sociocultural Causality Space, Time: A Study of Referential Principles of Sociology and Social Science*. New York: Russel & Russel.

Souty, J., 2006. Gilbert Durand, la réhabilitation de l'imaginaire. *Sciences Humaines* 11(176), p. 11.

Stock, M., 2006. L'hypothèse de l'habiter poly-topique: pratiquer les lieux géographiques dans les sociétés à individus mobiles. *EspacesTemps.net*, Travaux.

Sue, R., 1994. *Temps et ordre social*. Paris: PUF.

Tabonni, S., 2006. *Les temps sociaux*. Paris: Armand Colin.

Urry, J., 2000. *Sociology beyond Societies*. London: Routledge Publishers.

Veltz, P., 1996. *Mondialisation, villes et territoires: une économie d'archipel*. Paris: PUF.

Viard, J., 1994. *La société d'archipel ou les territoires du village global*. La Tour d'Aigues: L'Aube.

Von Hippel, E., 2005. *Democratizing Innovation*. Cambridge: The MIT Press.

SECTION II

Methods, tools and approaches

All research fields are marked and defined by the ways in which the research is conducted. Of course, the theories and concepts are of huge importance in distinguishing a field of research. However, the methods, tools and approaches are at least as important. In fact, there are important (and often overlooked) relations between the ontological, epistemological and methodological dimensions of any research field. Thus, in Section II we turn to 'ways of doing' research on urban mobilities. In **Chapter 6**, Vannini and Scott introduce the principles, procedures and potential of mobile ethnography with a focus on how this research methodology has been applied to the study of urban mobilities. Mobile ethnography shares with other mobile research methodologies an interest in capturing kinaesthetic practices and experiences and in rendering such empirical materials through lively, emplaced and embodied means. As a research strategy, 'on the move' mobile ethnography focuses on the observation of and participation in mobile lifeworlds and social groups, as well as in the ordinary and extraordinary mobile activities of discrete individuals. The chapter reviews the studies of mobile ethnographies in urban environments, reflecting on their methodological challenges, limitations, advantages and opportunities for future research developments. **Chapter 7** by Emmanuel Ravalet focuses on mobile futures through present behaviours and discourses. In mobility studies, the purposes of scientific methods are generally the description and better understanding of the present and the past. There is a gap between such approaches and the need planners have to orientate action towards the future in the short, medium or long term. To fill this gap, researchers have to use and project their knowledge towards the future. This chapter then proposes some learnings on past and present behaviours and discourses on mobility. On the basis of such elements, it becomes possible to consider mobile futures, which is what, in its final part, the chapter proposes to do, via the presentation of three scenarios on mobile futures based on research financed by SNCF in 2015 for the Paris climate conference COP21. The point here is not to forecast mobile futures, but to consider possible futures and policies to accompany or avoid them. **Chapter 8** is Lucas' analysis of digital approaches and mobilities in the big data era. The chapter discusses some of the opportunities offered by Big Data to understand mobility practices. However, beyond its promises, it has to be considered in the light of human and social constructions. Therefore, Lucas underlines the importance of describing all the decisions, operations, constraints, etc. that shape these data and the algorithms they feed in order to identify their limitations. This chapter also highlights the challenges of Big Data for scientific research, in terms of the opportunities it offers to renew the understanding of mobility but also the challenges it poses for the human and social sciences (HSS). In **Chapter 9**, Joël Meissonnier addresses more methodological challenges. Meissonnier discusses how commented walks and obstacle courses are two qualitative methods that belong to the extensive field of mobile methods. The first belongs to the visual sociology tradition; the second to the experimental sciences tradition. The French school of commented walks is close to the walk along methodology. Indeed, unlike the methodological protocol of walk alongs, commented walks are necessarily immortalised by the visual and require a skilled cameraman. The second method, called obstacle courses, comes from anthropology. It is an experimental method because researchers induce a dummy event to focus on unknown routes. The chapter demonstrates how our qualitative methodological hybridisation provides useful answers and several key notes. Thanks to the revealing illustrative images resulting from this method, it can have an important power of persuasion on local officials. The videos resulting from the commented obstacle course methodology are important, firstly to highlight the dysfunctions in the putting in of markers in a given urban area, secondly to inform the authorities and finally to be used as a training tool for mobility managers or bus drivers, for example. Section II ends with **Chapter 10** by Michel Bierlaire, Riccardo Scarini, Marija Nikolic, Yuki Oyama, Nicholas Molyneaux and Zhengchao Wang on pedestrian behaviour and how it can be understood. Pedestrian mobility is at the center of urban transportation systems. Therefore, understanding it is a necessary task to improve the mobility in cities. However, pedestrian mobility is extremely complex, and advanced methods are required to capture its dynamics. This chapter presents the main methods used to describe, model and control pedestrian behaviour. It introduces the fundamental relationships describing pedestrian dynamics, models representing the demand and assignment of pedestrians and, to conclude, innovative management strategies.

6
MOBILE ETHNOGRAPHIES OF THE CITY

Phillip Vannini and Nicholas Scott

Imagine an ethnographer at work, or better yet imagine the stereotypical caricature of an ethnographer at work. With her/his tent pitched among the "natives," she calmly sits at the edge of a field, carefully observing her research subjects' practices, patiently recording notes in her journal. Tomorrow and the day after she will do the same, and likely the next month and the one after. Eventually she will reach the conclusion that she has stayed in one place long enough and she will travel back home to write up her notes from her office desk.

This is a caricature, of course, but like all caricatures it does hold some truth. Traditional (arguably, stereotypical) ethnography has never been overly concerned with the dynamics of travel to and from a field site, or with the intersecting movements of peoples on the go. Indeed much of this ethnography focuses on the "natives" as a tightly encapsulated group, protected by the remoteness of their isolated spaces, and by rigid boundaries of tradition setting them aside from the outer world. How static, sedentary, rooted, and how immobile of a research practice you might be thinking. And in part, you might be right. But only in part.

In actuality, movement has never been entirely excluded from ethnographic practice. Even among the oldest classics there are plenty of instances of researchers following "natives" such as traders, hunters, and nomads along their ordinary and extraordinary journeys. Yet, movement has become a much greater preoccupation for ethnographers over the last three decades. As multiple, complex, and distal mobilities of individuals, groups, images, information, raw materials, and manufactured commodities have increased worldwide, ethnography too had to become more mobile (Büscher & Urry, 2009; Watts & Urry, 2008). A fully mobile ethnography is now deeply invested in addressing translocal and global flows of interconnection and interdependence entangling more-than-just human subjects around the whole planet. Within a fully mobile ethnography the study of movement itself has thus become a central concern, with more and more ethnographers no longer being cavalier about their own ambulations and with a growing number of them actively and innovatively planning for movement as a key component of research design. Ethnographies that are global, multi-sited, and mobile in nature are no longer an exception to the norm, and static ethnographies of secluded "natives" are by now an anachronism.

In this chapter we will survey mobile ethnography as a research methodology, with a concentration on its utilization in urban contexts. In general terms, ethnography can be defined as an inductive, immersive, emic way of researching people within their lifeworlds by way of observing their conduct and participating alongside with them in their daily practices. Mobile ethnography adds two elements to that formula: first, an empirical focus on research participants' mobile practices; second, methodological procedures requiring movement on the part of the researcher. Following Clifford (1997) we might provocatively say that this type of mobile fieldwork is thus less epitomized by roots and more by routes. In this sense mobile ethnography is marked by "travel encounters" and "spatial practices of moving to and from, in and out, passing through" places (1997, p. 67). In short, mobile ethnography can be defined as a research procedure entailing "travelling with people and things, participating in their continual shift through time, place and relations with others" (Watts & Urry, 2008, p. 867).

The aims and types of mobile ethnography

Rather than confining themselves to their own desk to collect data, mobile ethnographers observe mobile subjects first hand by taking part in their journeys. Instead of thin answers to multiple-choice questions, mobile ethnography provides elaborate impressions of how people use, experience, inhabit, and interact with transport infrastructure, thus focusing precisely on what people do, rather than what they say they do. Thanks to such immersive approach, the field of mobile ethnographic research is nowadays rife with rich, in-depth, and thickly-descriptive studies of commuter practices ranging from bus (Clayton et al., 2016), long-distance train (Bissell et al., 2017), subway (Symes, 2013), and cars (Laurier & Lorimer, 2012), to ferry boats (Vannini, 2012), cycling (Spinney, 2010), and walking (Porter et al., 2010) only to name a few. In short, mobile ethnography is kinetic, dynamic, deeply embodied, and closely proximate to the naturalistic contexts where mobile practices unfold.

Mobile ethnography is not just an alternative to the methodological procedures typical of positivist research, but also and more importantly a response to the starting assumptions of that tradition. As Watts and Urry (2008) found, the interpretive and emic orientation of ethnography allowed them to understand commuters' travel as productive and meaningful time, in contrast to the dominant assumption that it was dead, empty, and wasted time. Mobile ethnography's inquisitive focus on the mundane, the habitual, and the otherwise taken-for-granted in fact regularly allows researchers to challenge preconceived notions on the ordinary and the extraordinary, the familiar and foreign (see Orvar Löfgren & Ehn, 2010). By concentrating on what people actually do while on the move, mobile ethnographers uncover the intangible and the ephemeral aspects of movement, shedding light on the sensuous, enskilled, and material dimensions of transport infrastructures.

Several types of mobile ethnography exist, each with their own tools, techniques, and aims. Three of the most common types are the "follow-the-thing" approach, mobile auto-ethnography, and the go-along. Due to limited space we will describe the former two briefly in the paragraphs below, but we will dedicate greater space to the latter because it is the most common approach used by urban mobile ethnographers.

The "follow-the-thing" approach consists of researchers tracing the trajectories of raw materials and manufactured goods, at times following the entire complex global pathways through which raw materials are collected, manufactured, distributed, sold, consumed, and cast away or recycled (see Cook, 2004). One of the most unique and revealing applied examples of this approach comes from a group of filmmakers who follow the technological and social transformation of a school bus in the film *La camioneta* (Kendall, 2012). This captivating ethnographic film tells the story of an old school bus deemed no longer adequate to shuttle kids to and from school in suburban Texas. Auctioned off to a group of enterprising Guatemalans, the yellow school bus eventually reaches Central America, where it undergoes a dramatic transformation. Re-painted, repaired, and re-adapted, it is finally resurrected as a rural transport coach, now facing the new challenges and demands brought about by dramatically different geographic, economic, and political conditions.

While mobile ethnographies writ large see the researcher take a secondary role, mobile autoethnographies cast the researchers themselves as the protagonists. Thus, rather than telling stories of research participants, the researchers themselves turn attention on their own experiences and practices. Autoethnographies are particularly powerful when they shed light on unique or otherwise ephemeral experiences that would be extremely difficult to study through conventional means. For example, through his autoethnography on bike-riding in central London and greater Copenhagen, Larsen (2014) argues that the value of autoethnography lies in its capacity to reflect on the nearly unsayable, such as the habitual experience of riding a bike. Echoing Larsen, Spinney (2011) finds that through its deeply reflexive orientation mobile ethnography and autoethnography can provide us with "an understanding of the less representational – those fleeting, ephemeral and often embodied and sensory aspects of movement," and that such understanding "is vital if we are to fully understand why and how people move around" (Spinney, 2011, p. 162).

The go-along

The go-along is a very popular research methodology that entails a researcher accompanying research subjects on an outing of their choice. Such excursions may vary in length, intensity, locale, purpose, mode of transport, and multiple other variables but they are typically of personal significance to the research subject and taking place in sites relatively unfamiliar to the researcher. During such outings researcher and research participants typically engage in loosely structured conversation about the spaces where the journey takes place, the excursion itself, and anything else related to the experience. This allows the research encounter to focus on the specifics of mobile practices right as they occur in their naturalistic contexts (albeit quite constructed by the research situation). The shared exploration and co-presence of researcher and research subject allows for greater reflection, deeper exchange of knowledge, and therefore a richer understanding of the peculiarities of specific mobilities.

Over the last few years the go-along methodology has rapidly become one of the most popular ways of doing mobile research. Initially designed by Kusenbach (2003) as a walk-along, the research methodology now encompasses a variety of approaches tailored to riding along with drivers, cyclists, boat, train, bus, and even plane passengers. Regardless of the mode of transport, the go-along is believed to open a window into people's everyday lives by letting researchers experience firsthand people's everyday embodied access to places, modes of transport, and mobile infrastructures (see Porter et al., 2010).

One of the great advantages of the go-along pertains to the way it corrects the power imbalance between researchers and research participants. Walking or riding somewhere together, in contexts deeply familiar to the research participant but unfamiliar to the researcher, empowers the former and reduces the authority and social status of the latter. This should come as no surprise to any researcher who has ever conducted an interview on a neutral ground (like a café) or a researcher's "home field" (e.g. a university campus setting). Many people feel uncomfortable on such "terrain" and meeting a stranger for an unfamiliar social occasion such as an interview in an unfamiliar setting is bound to raise feelings of anxiety. In contrast, for example, walking to one's workplace on one's most ordinary surroundings will reduce the anxiety of meeting a stranger (see Christensen et al., 2011; Porter et al., 2010). It is no accident that the go-along method has proven to be very effective in getting access to groups of people like children and youth who would otherwise find it difficult to open up in conventional research settings (see Christensen et al., 2011; Romero Mikkelsen & Christensen, 2009). Uniquely, in the case of go-alongs with children, Christensen and colleagues have also found that parents were much less likely to meddle with the research process and give researchers unmediated access to their children during their shared journeys to and from school (Christensen et al., 2011).

Increasingly, go-along research projects involve the use of digital technologies. Christensen and colleagues (Christensen et al., 2011) for example employed GPS to trace their research participants' journeys to and from school, and simultaneously used SMS to communicate with children. Spinney (2015) has even argued for the advantages of combining video with electroencephalography (EEG) and galvanic skin responses (GSR). Such technologies allow researchers immediate and distant access to research participants' verbal thought processes, and non-verbal information. As Laurier (2010), Spinney (2010) and Brown and Spinney (2010) have argued, such technologies allow researchers to "be there" when co-presence is impossible due to danger, risk, or impracticality.

Video technology, in particular, has profoundly enhanced the way movement can be captured, analyzed, and shared. As opposed to photography, ethnographic video is especially ideal for capturing the kinesthetic properties of skilled and habitual movement (see Vannini, 2017; Vannini & Vannini, 2017). As Spinney (2011, p. 167) writes: "video data embodies the movement which the fixity of photographs and written texts so often fail to evoke." Moreover, when played back later in interview settings video can be used to elicit recollection and reflection (Brown & Spinney, 2010). Lastly, the nature of a walk-along or ride-along is such that for the most part researcher and research participant will move or sit side by side. This reduces the amount of time two people need to make eye contact, lessens the sense of discomfort that making eye contact with a stranger causes, in turn decreasing the amount of time people feel compelled to talk, and thus increasing the possibility of deeper reflection.

Learning from urban mobile ethnographies

The development and doing of mobile ethnography faces unique challenges and vast opportunities in the city. At the time of writing (2018), 55% of humans on earth were living in cities, compared to just 30% in 1950; 68% of humanity is projected to reside in urban areas by 2050 (UN, 2018). This veritable explosion of urbanity on earth animates ever more complex entanglements of relations among differently im/mobile people, things and ideas. Notably, most of the world's walks, and probably its cycle and transit rides, take place in the city (Bates & Rhys-Taylor, 2017)—even as a rapidly urbanizing earth makes way for two billion cars. Density of modal diversity, along with the heavily mediated and noisy nature of cities and commuting (Bissell, 2018), textures an urban terrain for which mobile ethnography is well suited. The multi-sensory demands and nervous stimulations of the city cultivate mobile skills, sensory strategies (Jungnickel & Aldred, 2014) and "affective capacities" (Larsen, 2014)—including stillness and inattention (Bissell, 2010; Jensen, 2006; Jirón, 2010)—whose practice would be impossible to adequately sense and recollect without listening/seeing/feeling there.

Urban mobile ethnographies excel at following pre-reflective flows of environmental feeling, perception, and affect (Spinney, 2015)—in short, the "street phenomenology" (Kusenbach, 2003) that animates diverse urban lifeworlds. As prolific sites of positivistic transport research, the city's naturalistic contexts wherein embodied mobile practices animate lived experience are often poorly understood. By placing themselves in the position of the people and things (see Christophers, 2011; Forman, 2018) in the midst of experiencing (or mediating) movement, mobile ethnographers who go *in situ* often unearth counterintuitive or surprising aspects about the city. For example, in addition to illuminating transit time as productive rather than wasted, urban mobile ethnography has shown how the car, contra its dominant representation as productive and autonomous, can effectively imprison women where they bear the bulk of domestic driving labor (Jirón & Iturra, 2014). Urban mobile ethnographies are particularly effective where they reveal practices of relatively marginalized groups who are often excluded from quantitative models of transport. Teenagers, for instance, as Ocejo and Tonnelat (2013) found by using both group and one-on-one go-alongs on the subway, engage in a wide range of embodied and affective practices for becoming a stranger in the city to maintain social distance from others while dwelling together during public transit.

The phenomenology of contemporary urban mobilities entails varying levels, qualities, and intensities (Bissell et al., 2017) of engagement with the environment. While active aspects of mobile experience are often well documented, recalled, and emphasized by mobile ethnographies, inactive or "quiescent" practices (Bissell, 2010) that escape memory also play a salient role. As Kusenbach (2003, p. 469) observes in her go-along study of life in Los Angeles, sometimes "being in and moving through the world requires a high degree of commitment and concentration, for instance while changing several lanes on a busy freeway. At other times, we are able to (almost) completely withdraw." How we withdraw or acquiesce, of course, depends on the kind of mobility in which we are engaged and our environment. Mobile quiescence, especially outside the car, takes on different qualities and intensities related to, *inter alia*, fatigue, blisters, and changing weather patterns as well as moments of meditation and reverie (Spinney, 2011). For example, Scott (2020), by riding along with people in cities to what they perceive as compelling places of nature, showed how cycling affords quiet "moments of zen" that contest preconceived notions about the city and wilderness.

Reassembling go-alongs

The rise in use of mobile methods, especially go-alongs, can be interpreted as developing mobile ethnography along two distinct spectrums: level of contrivance and level of technological mediation. As originally conceived by Kusenbach (2003, pp. 463–464), the urban go-along did not have much time for contrived or "experimental" versions, which is to say "when researchers take informants into unfamiliar territory or engage them in activities that are not part of their own routines." Kusenbach argues that "natural" go-alongs that closely track only the outings informants would go on anyway are superior for understanding subjects' authentic mobile practices. In this way, go-along ethnographers can remove or at least de-emphasize their "own perceptual presuppositions and biases, which are in the end irrelevant" (Kusenbach, 2003, p. 469). While this approach to go-along methodology resonates with mobile ethnographies zeroing in on "authentic" urban experiences, it too hastily dismisses other possibilities of co-crafting go-alongs that transcend the taken-for-granted routines and environments of participants.

Equally, relatively contrived go-alongs offer a valuable mobile method of interpretive, phenomenologically sensitive ethnography. Putting aside the fact of whether any situation in which a researcher is present (even remotely, e.g. via video camera) counts as a naturally occurring social occasion—as Kusenbach (2003, p. 464) acknowledges—contriving go-alongs whereby researchers (following, to an extent, their own presuppositions and biases) play a role in choosing routes and planning pauses, and ask informants to do "unnatural things" (like leave their phone off), can create valuable ethnographic advantages. For example, Gallagher and Prior (2017, p. 165) illustrated how (phone-less) "listening walks" in the city along sonically significant routes planned largely by researchers offer a multifaceted go-along method whose "open-ended, emergent quality produces unexpected encounters, feelings, thoughts and analyses ... that exceed any prescribed purpose."

Purposeful site visits offer another way of contriving go-alongs. They are particularly useful when they allow researchers and participants to co-produce empathetic understandings of significant places or passages through the "expert eyes" (Horowitz, 2013) and ways of seeing of the participants (see Bates, 2017). For instance, while walking with two informants involved in marketing a future urban redevelopment, Holgersson (2017) found that despite their power over the site, her co-walkers were deeply unfamiliar with its history and inhabitants—leading to insights on the selective and disembodied temporalities of gentrification. As a nice complement to the empowerment of vulnerable people when they get to show the researcher "their space," powerful people, upon leaving their offices and actually visiting up close the products and consequences of their work, lose some of their authority.

A second axis of go-along development relates to technological mediation. As with contrived versus more "natural" go-alongs, technological mediation forms a spectrum rather than an either/or state. Even pre-digital ethnographic procedures of pausing and jotting down notes along the way constitute a form of mediation, one that sets in train a stream of text and "logocentric way of knowing" (Vannini & Vannini, 2017, p. 181). The rapid spread of new mobile technologies such as lightweight video cameras offer deeper layers of mediation that sometimes carry an aura of accuracy and first-hand "up-closeness" with urban lifeworlds. Skepticism such as Merriman's (2014) is therefore warranted, where he cautions against hasty use of film and other mobile technologies under the assumption that they will automatically generate more faithful and authentic understandings of mobile practices.

Mobile video ethnography and cinematic go-alongs, rather than peeling the varnish off a "truthier" urban reality, should instead be seen as opening new avenues for producing knowledge but also creating new reductions, distortions, and abstractions. When deployed in a more-than-representational way (Lorimer, 2005; Simpson, 2015) as an impression of movement, video cameras and other emerging technologies such as over-the-shelf bio-sensors (Spinney, 2015) give ethnographers the power to "bear witness" (Dewsbury, 2003) to intriguing yet fleeting, easily forgotten, and not always conscious phenomena, leading to the creation of what Spinney (2015, p. 236) calls "distortions with a purpose." Whether this purpose is simply to serve as a memory prompt, "seeing there" at a distance without actually "being there" (see Cook et al., 2016) or a more ambitious sensing of intensities, materialities, and affect flowing across intersubjective lifeworlds, mobile video data shows significant potential for feeding the inquisitive imaginations and animations of urban sociology and geography. Nevertheless, doing and sharing cinema in general remains an under-utilized mobile method whose fruits all too often become frozen as stills in articles and books—although this may be starting to change (see Vannini, 2017).

Like a pencil or tape recorder, every audiovisual recording device comes with its own rhythms and affordances to which the mobile ethnographer's own bodily rhythms and attention must become attuned. Enacting walking with a camera (see Vannini & Vannini, 2017), for example, is not "normal walking" so much as a buttocks-driven, lower center of gravity kind of walking to satiate the camera's demand for stability. Vagaries of weather, bodies, and obstacles in the built environment further cajole people walking with video cameras into particular rhythms and arrhythmic (im)mobilities (Edensor, 2010)—the same applies to running with cameras (Cook et al., 2016), driving with cameras (Laurier, 2010), and cycling with cameras (Spinney, 2011).

Filming mobilities in the city, where bodies and obstacles thickly cluster, composes a complex *pas de deux* between filmmaker and camera. For instance, in contriving to ride with people to their idea of wilderness in the city, Scott (2020) found that cycling with a helmet-mounted GoPro camera acted differently than cycling without one. By extending and controlling his head (versus, e.g., his hands and steering on a handle bar-mount), the "camera-head" became freer to follow the sidewise looking of participants and find footage of their wider "cycling habitats" while the rest of his body did the biking. This, in turn, created a reliance on peripheral vision in order to safely navigate city streets and become the shadow (see Jirón, 2011) as slippery precipitation, hilly topography, and car traffic rhythmically (and sensuously) opened and closed space and time between the filmmaker and the filmed like an accordion.

Witnessing group bike rides (McIlvenny, 2015) is even more complicated, where multiple cameras and their configurations can help grasp the unfolding complexity of interactions on the ride-along—or at the least some quick changing of position within the group (see Popan, 2018, p. 93). Interestingly, McIlvenny (2015, p. 63) found that different camera perspectives led sometimes to "competing (and irresolvable) interpretations of what was seeable and hearable," showing the limits of any single angle or configuration of angles. This finding, that any mobile perspective is necessarily partial and may conceal as much as it reveals, also applies to mobile ethnography in general, pointing to the power of mixing mobile ethnography with methodologies other than itself.

Conclusion

In this chapter we examined how a fully mobile ethnography has become deeply invested in understanding the entanglements of people, ideas, nonhuman beings, and infrastructures like never before around a rapidly urbanizing planet. Mobile ethnography not only empirically focuses on research participants' mobile practices, but also literally follows the things, environments, and people under study, offering new ways of participating in city life and understanding urban lifeworlds. In this way, mobile ethnography offers a powerful methodology for sensing the fleeting, quiescent, affective, and other easily forgotten and hard-to-represent dimensions of lived experience, which typically elude more disembodied models of movement found in positivistic transport research and urban planning.

To interpret the methodological elaboration of the most common mobile method, the go-along, we constructed a typology based on relative levels of (1) contrivance and (2) mediation by recording technology. Rather than argue that one particular type is better than the rest, this typology illustrates the different possibilities of the go-along, which early in its development (Kusenbach, 2003) eschewed contrived versions and continues to grapple with how to use new audiovisual technologies. By detailing these possibilities, we mean to emphasize the flexibility and performative potential of go-alongs and, by the same token, highlight the folly of reducing go-alongs to one, standardized kind of "instrumental methodological procedure" (Vannini & Vannini, 2017, p. 187). Ultimately, what kind of go-along or mobile ethnography in general is most appropriate in any given case of research depends on what kind of questions the researcher is asking.

Mobile ethnography as a whole, like any other methodology, is stronger at addressing some questions and topics than others. It is worth noting, the go-along itself came about in part by some innovative technical thinking and practice in the urban field by Kusenbach (2003; and see; Thibaud, 2001) on how to overcome the limitations of both participant observation and interviewing. It is equally worth noting that most studies that use it do not rely on mobile ethnography alone, but rather mix it with other methodologies, which has the effect of diversifying their evidence and enhancing the validity of their conclusions. Its unique strengths, pragmatic genesis, and potential for supplementing other ways of knowing the city should be seen as pillars of mobile ethnography, and should also caution researchers against locking mobile ethnography into only one epistemology or ontology. Perhaps the greatest gain in using mobile ethnography, independently and in tandem with other methods, lies in "shifting stubborn ontologies" (Spinney, 2015, p. 241) and replacing rigid epistemologies, qualitative and quantitative, that continue to overlook and marginalize the fleeting, the felt, and the affective experiences of urban mobilities.

References

Bates, C 2017, 'Desire lines: walking Woolwich', in C Bates & A Rhys-Taylor (eds.), *Walking through social research*, Routledge, London.

Bates, C & Rhys-Taylor, A 2017, 'Finding our feet', in C Bates & A Rhys-Taylor (eds.), *Walking through social research*, Routledge, London.

Bissell, D 2018, *Transit life: how commuting is transforming our cities*, The MIT Press, Cambridge, MA.

Bissell, D, Vannini, P & Jensen, O 2017, 'Intensities of mobility: kinetic energy, commotion and qualities of supercommuting', *Mobilities*, vol. 12, no. 6, pp. 795–812.

Bissell, D 2010, 'Passenger mobilities: affective atmospheres and the sociality of public transport', *Environment and Planning D: Society and Space*, vol. 28, no. 2, pp. 270–289.

Brown, K & Spinney, J 2010, 'Catching a glimpse: the value of video in evoking, understanding and representing the practice of cycling', in B Fincham, M McGuinness & L Murray (eds.), *Mobile methodologies*, Ashgate, Aldershot.

Büscher, M & Urry, J 2009, 'Mobile methods and the empirical', *European Journal of Social Theory*, vol. 12, no. 1, pp. 99–116.

Christensen, P, Romero Mikkelsen, M, Sick Nielsen, TA & Harder, H 2011, 'Children, mobility, and space: using GPS and mobile phone technologies in ethnographic research', *Journal of Mixed Methods Research*, vol. 5, no. 3, pp. 227–246.

Christophers, B 2011, 'Follow the thing: money', *Environment and Planning D: Society and Space*, vol. 29, no. 6, pp. 1068–1084.

Clayton, W, Jain, J & Parkhurst, G 2016, 'An ideal journey: making bus travel desirable', *Mobilities*, vol. 12, no. 5, pp. 706–725.

Clifford, J 1997, *Routes*, Harvard University Press, Boston, MA.

Cook, I 2004, 'Follow the thing: papaya', *Antipode*, vol. 36, no. 4, pp. 642–664.

Cook, S, Shaw, J & Simpson, P 2016, 'Jography: exploring meanings, experiences and spatialities of recreational road-running', *Mobilities*, vol. 11, no. 5, pp. 744–769.

Dewsbury, JD 2003, 'Witnessing space: knowledge without contemplation', *Environment & Planning A*, vol. 35, no. 11, pp. 1907–1932.

Edensor, T 2010, 'Walking in rhythms: place, regulation, style, and the flow of experience', *Visual Studies*, vol. 25, no. 1, pp. 69–79.

Forman, PJ 2018, 'Circulations beyond nodes: (in)securities along the pipeline', *Mobilities*, vol. 13, no. 2, pp. 231–245.

Gallagher, M & Prior, J 2017, 'Listening walks: a method of multiplicity', in C Bates & A Rhys-Taylor (eds.), *Walking through social research*, Routledge, London.

Holgersson, H 2017, 'Keep walking: notes on how to research urban pasts and futures', in C Bates & A Rhys-Taylor (eds.), *Walking through social research*, Routledge, London.

Horowitz, A 2013, *On looking: eleven walks with expert eyes*, Scribner, New York.

Jensen, OB 2006, '"Facework", flow and the city: Simmel, Goffman, and mobility in the contemporary city', *Mobilities*, vol. 1, no. 2, pp. 143–165.

Jirón, P 2010, 'Mobile borders in urban daily mobility practices in Santiago De Chile', *International Political Sociology*, vol. 4, no. 1, pp. 66–79.

Jirón, P 2011, 'On becoming "la sombra/the shadow"', in M Buscher, J Urry & K Witchger (eds.), *Mobile methods*, Routledge, London.

Jirón, P & Iturra, L 2014, 'Travelling the journey: understanding mobility trajectories by recreating research paths', in L Murray & S Upstone (eds.), *Researching and representing mobilities: transdisciplinary encounters*, Palgrave MacMillan, London.

Jungnickel, K & Aldred, A 2014, 'Cycling's sensory strategies: how cyclists mediate their exposure to the urban environment', *Mobilities*, vol. 9, no. 2, pp. 238–255.
Kusenbach, M 2003, 'Street phenomenology: the go-along as ethnographic research tool', *Ethnography*, vol. 4, no. 3, pp. 455–485.
La Camioneta 2012, motion picture, Follow Your Nose Films, Toronto, Canada. Produced by Esther B. Robinson; directed by Mark Kendall.
Larsen, J 2014, '(Auto)ethnography and cycling', *International Journal of Social Research Methodology*, vol. 17, no. 1, pp. 59–71.
Laurier, E 2010, 'Being there/seeing there: recording and analysing life in the car', in B Fincham, M McGuinness & L Murray (eds.), *Mobile methodologies*, Palgrave Macmillan, New York.
Laurier, E & Lorimer, H 2012, 'Other ways: landscapes of commuting', *Landscape Research*, vol. 37, no. 2, pp. 207–223.
Lorimer, H 2005, 'Cultural geography: the busyness of being "more-than-representational"', *Progress in Human Geography*, vol. 29, no. 1, pp. 83–94.
McIlvenny, P 2015, 'The joy of biking together: sharing everyday experiences of vélomobility', *Mobilities*, vol. 10, no. 1, pp. 55–82.
Merriman, P 2014, 'Rethinking mobile methods', *Mobilities*, vol. 9, no. 2, pp. 167–187.
Ocejo, RE & Tonnelat, S 2013, 'Subway diaries: how people experience and practice riding the train', *Ethnography*, vol. 15, no. 4, pp. 493–515.
Orvar Löfgren, O & Ehn, B 2010, *The secret world of doing nothing*, University of California Press, Berkeley, CA.
Popan, IC 2018, 'Utopias of slow cycling: Imagining a bicycle system' PhD thesis, Lancaster University.
Porter, G, Hampshire, K, Abane, E, Munthali, A, Robson, E, Mashiri, M & Maponya, G 2010, 'Where dogs, ghosts, and lions roam. Learning from mobile ethnographies on the journey from school', *Children's Geographies*, vol. 8, no. 2, pp. 91–105.
Romero Mikkelsen, M & Christensen, P 2009, 'Is children's independent mobility really independent? A study of children's mobility combining ethnography and GPS/mobile phone technologies', *Mobilities*, vol. 4, no. 1, pp. 37–58.
Scott, N 2020, 'Ecologizing Lefebvre: urban mobilities & the production of nature', in, ME Leary-Owhin & JP McCarthy (eds.), *The Routledge handbook of Henri Lefebvre, the city and urban society*, Routledge, London.
Simpson, P 2015, 'Atmospheres of arrival/departure and multi-angle video recording: reflections from St Pancras and Gare du Nord', in C Bates (ed.), *Video methods: social science research in motion*, Routledge, London.
Spinney, J 2010, 'Performing resistance? re-reading practices of urban cycling on London's South Bank', *Environment and Planning A*, vol. 42, no. 12, pp. 2914–2937.
Spinney, J 2011, 'A chance to catch a breath: using mobile video ethnography in cycling research', *Mobilities*, vol. 6, no. 2, pp. 161–182.
Spinney, J 2015, 'Close encounters? Mobile methods, (post)phenomenology and affect', *Cultural Geographies*, vol. 22, no. 2, pp. 231–246.
Thibaud, JP 2001, 'La methode des parcours commentes', in M Grosjean & JP Thibaud (eds.), *L'espace urbain en methodes*, Editions Parentheses, Marseille.
Symes, C 2013, 'Entr'acte: mobile choreography and Sydney rail commuters', *Mobilities*, vol. 8, no. 4, pp. 542–559.
United Nations, The Population Division of the Department of Economic and Social Affairs 2018, *World urbanization prospects: the 2018 revision*, viewed 29 July 2018, https://esa.un.org/unpd/wup/
Vannini, P 2012, *Ferry tales*, Routledge, New York.
Vannini, P 2017, 'Low and slow: notes on the production and distribution of a mobile video ethnography', *Mobilities*, vol. 12, no. 1, pp. 155–166.
Vannini, P & Vannini, A 2017, 'Wild-walking: a two-fold critique of the walk-along method', in C Bates & A Rhys-Taylor (eds.), *Walking through social research*, Routledge, London.
Watts, L & Urry, J 2008, 'Moving methods, travelling times', *Environment and Planning D*, vol. 26, no. 5, pp. 860–874.

7
MOBILE FUTURES THROUGH PRESENT BEHAVIOURS AND DISCOURSES

Emmanuel Ravalet

Introduction

What will the future of mobility look like? When it comes to talking or thinking about the future of mobility, in political or technical circles, we often forget that tomorrow, we will walk, we will probably pedal more than today, we will continue to use massive public transport and many of us will still drive their cars. Technological challenges, and autonomous cars, should not blind us to what has made the city and mobility for decades and will continue tomorrow.

Then, what about the balance between all these modes and what place will carpooling, carsharing, services such as Uber, or the autonomous car have? To think about this question, I propose in this chapter to discuss two elements that have great importance in our mobility: cars and communication and information technologies. I will also discuss some results from a survey on people's wishes concerning their future mobility. Finally, I propose to discuss three scenarios that serve to give us inputs on three possible futures, and thus shed light on the political actions they suggest.

Cars and communication technologies in mobile futures

Cars in the past, cars in the future

In the 1960s and 1970s, the transportation system was considered as a condition of the economic development and a symbol of progress. This vision was widely shared by the population, decision makers and experts (Dollinger, 1972; Fichelet, 1979). The car had a special status in this context, in the sense that it freed users from the constraints of specific routes and schedules and allowed for quick and individual travel in very good comfort conditions. Road infrastructures developed considerably during this period, which made the car ever more relevant and ever more necessary to travel in most areas.

In the 1980s, the modal shift from the car to alternative modes of transport was a controversial topic in urban transport policies in Europe (Banister, 2005), mainly supported by environmental movements. But with the growing awareness of pollution problems related to car traffic, this strong and dominant image has begun to change, to worsen. So, modal shift has gradually become the key objective of urban transport policies in most Western countries (Kaufmann, 2003). Such policies addressed several different issues such as road congestion, consumption of public spaces or negative environmental impacts (Lefèvre and Offner, 1990). The initial strategy of the modal shift in the 1980s and 1990s was to develop subways and train networks, S-Bahn and other regional trains that were efficient in terms of speed (Metz, 2008). But these last years, coordination strategies between development and multimodal transport policies approaches have developed (Canzler and Knie, 1998). If car use continued to increase during this period, 2005 marked a turning point as it began to decline in many urban agglomerations. Initially thought to concern only several urban areas, this trend was widespread in all cities from northern and southern Europe (Aguilera et al., 2014). However, such a decrease in car use is measurable by the proportion of trips, but car distances continue to increase.

Today, the individual car continues to play a major role in mobility. Its use is still often necessary for many people to deploy their programs of activities (Dupuy, 2006; Newman and Kenworthy, 1989), because accessibilities are conceived on the time metric scale of the automobile (Sheller and Urry, 2000; Wiel, 1999). This helps to understand why the car is still strongly rooted in the contemporary culture of mobility (Vincent-Geslin, 2012).

The image of the car has changed undoubtedly. Its impact on greenhouse gas emissions and its consumption of space (in circulation and parking) now explain why many politicians seek to limit its use in the densest urban areas. But the development of electromobility on the one hand, and the growth of shared uses, on the other hand, question the still recent weak legitimacy of the car in western city centres. Both justifications – combining shared and electric mobility – are often used to present the autonomous car in its best light.

The advantages the autonomous car can have are great, especially in terms of road safety. It also paves the way for a major reallocation of car parking spaces in dense cities. But behind these perspectives is also the ghost of an individual use of autonomous cars, which would then generate empty traffic. The consequences in terms of congestion of traffic lanes and energy consumption would undoubtedly be negative. More than ever, the question is not "How autonomous cars will transform our cities?" but "How do we want autonomous cars to transform our cities? And how can we achieve our goals?" Technology does not determine the future (Urry, 2016).

Communication technologies and spatial mobilities

Land use and transportation policies are not confined to efforts that aim to make people change the way they travel. The question of the need for travelling is also on the work plan. Information and communication technologies can play a role in this context.

This is an old question. At the arrival of the telephone, even though environmental issues were less prominent in the public space, the idea of a substitution of travel by telephone calls was mentioned. When mobile phones began to spread among the population, the same hopes re-emerged. Yet, travel has continued to increase, both in number and in distance. The Internet network and related equipment such as smartphone, tablet, laptop connected, etc. are opportunities to test the substitution potential of travels in space by virtual exchanges. Possibilities are large: organizing videoconferences rather than grouping several people in one place, shopping on the internet and waiting for delivery rather than moving to the point of sale, taking steps online, rather than in the premises of the institutions concerned, or teleworking at home, in transport, or elsewhere (telecentres and other dedicated or non-dedicated third places) rather than going to the office every day.

Beyond this question of substituting trips by virtual exchanges, the development of information and communication technologies invites thinking about the use and valuation of travel time. A great deal of empirical research directly refers to the use of ICT (Ettema and Verschuren, 2007; Gripsrud and Hjorthol, 2012; Guo et al., 2015; Keseru and Macharis, 2018). Also, the possibilities these digital tools offer to better use transport times cover a broad spectrum that concerns work (e-mails, office automation, etc.), sociability (remote discussions with friends or family members), and recreational activities (e.g. games, movies, music). ICT use is highly dependent on the spatial environment of travel. Thus, the appropriation of travel times is facilitated in public transport as opposed to car journeys during which the traveller must focus on driving (Lyons et al., 2013). In fact, commuters are increasingly using their train travel time to work (Gripsrud and Hjorthol, 2012; Timmermans and Van Der Waerden, 2008).

At the heart of the subject is the question of the productivity such activities, performed during travelling, have (Ettema and Verschuren, 2007; Keseru and Macharis, 2018; Varghese and Jana, 2018). Considering transport time productivity allows an adjustment of models that measure the value of transport time (Ettema and Verschuren, 2007; Pawlak et al., 2017). In fact, using the transport time can reduce the costs related to the actual journey time. This is one of the key advantages of the autonomous car (level of autonomy 5) but the impact on the economic value of time is still difficult to assess (Perret et al., 2017).

Rebound effects: what knowledges? what effects?

Analyses of mobility behaviours have highlighted that people's choices can evolve. Then, a change in the transport offer, or an evolution in the supply of equipment, shops or services, can lead people to adapt their practices and/or their discourses in a way that is sometimes unexpected. This is called a rebound effect. In economics, it has been suggested that "if a good gets cheaper in terms of its price or any effort necessary to obtain it, the demand for this good usually increases" (Hilty et al., 2006). One of the most famous rebound effects concerns technical improvements in fuel efficiency, which lead to decreasing the cost of fuel and cause an increase in vehicle use (Hymel et al., 2010). Here are some examples of other rebound effects in the mobility field.

In the 1970s, Yacov Zahavi, a researcher at the World Bank, demonstrated a constancy of travel time budgets in cities around the world. He formulated what is now called the "Zahavi Conjecture," namely the fact that daily mobility is a function of transport speed for a time budget of approximately one hour (Zahavi, 1979). Zahavi's work highlights a key mechanism of urban development: the time saved thanks to the speed of the transport mode is not used to limit or reduce the time spent travelling, but to travel further (Zahavi and Talvitie, 1980). This is the first rebound effect to keep in mind.

Second point: Since the 1990s, travel time budgets have been increasing throughout Europe. In Switzerland, for instance, 10% of the working population works more than 50 km from their main residence. This trend is developing rapidly due to train travel: the further away from home people work, the more likely they are to take the train. Careful analysis shows that the dispersion of time budgets has become more pronounced. It also shows that train users and walkers have the longest travel time budgets. This is due in particular to the fact that one can now use this time constructively (Jain and Lyons, 2008). With ICTs, travel time is no longer wasted time between activities. Beyond the benefits presented before for ICT in terms of use and valuation of travel times, they also seem to allow people to travel ever longer travel times and distances.

A third example concerns teleworking. At first glance, it is a way to avoid commuting trips. But in the beginning of the 1990s, the hypothesis was made that longer commuting distances would be made acceptable by the possibility of teleworking at home or in a dedicated place next to home (Janelle, 1986; Nilles, 1991). Then, teleworkers appear to live further away from their main place of work than others (Helminen and Ristimaki, 2007; Mokhtarian et al., 2004; Zhu, 2013). The mechanism behind this trend has not yet been fully addressed. It could result from a negotiation with the employer (e.g. in order to take care of children, or for quality of life issues) that took place after the residential choice. However, no research has assessed the possible greater tolerance towards commuting distance induced by the possibility of teleworking. Teleworking could, for example, be seen as an incentive (1) to choose a home that is remote from the work place, or (2) not to relocate, or (3) to accept or keep a job far away from home. In parallel, teleworking also makes more time available. This time can be reinvested in additional work (Bosua et al., 2017), but also in additional trips, which could otherwise not have been possible on a regular commuting day. This is the second possible space–time rebound effect that can be linked with teleworking. According to Walls and Safirova (2004), total kilometres travelled on teleworking days are between 53% and 77% lower than on non-teleworking days in the United States. In South Korea, Kim (2017) shows that non-work-related trips on teleworking days amount to 4 kilometres on average.

These examples of rebound effects are not intended to support the idea that these innovations, these technologies, should not be developed and politically supported. They aim to emphasize that people sometimes appropriate services and technologies in an "unplanned" way. Previous elements also aim to confirm the very strong need for regulation to accompany such innovations. Recent technological developments do not relegate politics to a secondary role, they make this role even more important. Another lesson can be removed from this first section, it concerns the people themselves and how they want to live, consume, work in the future. These elements will depend on the use they want to make of available technologies.

What aspirations of people concerning mobile futures?

I rely here on the notion of aspiration in sociology. Without developing the different meanings this notion has in the sociology literature, it appears however important to define it. Here I mean aspirations as forms of motivations people have to achieve a specific goal. "Aspirations" are considered to be related to preferences or values to which people adhere. Here I deal with people's aspirations for lifestyles in the future, especially at the level of activities and mobility.

Databases usually available to analyse mobility practices in terms of activities or transport do not allow their aspirations to be described and considered. It was therefore necessary to use ad hoc data collection. The quantitative survey I propose to rely on here, funded by the Mobile Lives Forum,[1] was linked to a collection of qualitative data I will not present here.[2]

The survey, organized at the end of 2015, was conducted by Obsoco[3] (Consumption and Society Observatory). It took place in six countries: France, Spain, Germany, the United States, Japan and Turkey. Just over 2,000 people were surveyed in each country. Thus, the database contains more than 12,000 answers. The elements presented here come from the website of the Mobile Lives Forum and from the report of the quantitative synthesis (Descarrega and Moati, 2016).

I highlight here the results that I think are the most important. In terms of the nature of the activities that people would like to achieve in the future, an aspiration related to the desire to work less appears for a small majority of respondents. Rather older and healthier populations are the most likely to opt for a reduction in the time spent working. Spaniards, Japanese and Turks are the most concerned, which is certainly related to the current high working time in their countries.

The second result I wanted to present here concerns the spatial inscription of activities: people want to be based on proximity. This form of spatial withdrawal is expressed by an aspiration to reduce the time spent traveling (thus, 31% of respondents consider it "very important" to spend less time in transport) but also, more generally, by an aspiration to not see activities practised too widely in space. Thus, 44% of those surveyed associate the notion of "ideal mobility" with a greater proximity, breaking with the current model. Lastly, and with regard to the temporal inscription of activities people aspire to, there is a strong aspiration for a form of slowing down of the rhythm of life. This is broken down by three results:

- 74% of respondents believe that the pace of life in today's society is too fast,
- 78% personally wish to slow down,
- 50% say they are running out of time to do what they want or need to do.

These elements show that actual behaviours can be quite different than those of today, even if recent surveys confirm that mobility trips are increasingly longer.

Scenarios: three possible mobile futures

This chapter aims to stimulate reflection on mobile futures. To achieve such an objective, I considered it essential to first present some of our knowledges on the roles the car and ICT play in past and present spatial mobility. Obviously, technical and informational innovations are part of the mobility systems, but I consider here the mobility of persons. And people sometimes surprise transport experts. That is especially what the rebound effects remind us of; and personal aspirations allow us to consider mobility in a more global perspective that concerns ways of life.

All of the elements previously presented are useful to build scenarios, understood as complex systems of the future (Urry, 2016). I now propose to consider three possible mobile futures. They come from a study funded in 2015 by SNCF in preparation for the Paris Climate Conference (COP 21).[4] The study has drawn on the cross-expertise of sociologists, economists, engineers, transport specialists from the Swiss Federal Institute of Technology in Lausanne (including the author of the present chapter), as well as ADEME, Trans-missions and OuiShare. The proposed demand scenarios concerned France. They were built on the basis of a scientific literature review and on an ad hoc survey I decided not to present here. Another element led us: the idea that political action, to be effective and not to be rejected, must resonate with people's practices and aspirations.

The "ultramobility" scenario

This first scenario considers that the car, autonomous or not, will remain a key mode of transport in France in 2050. This does not necessarily imply that its modal share will not decrease. However, the car's resistance relative to other modes of transport implies slow and limited development of alternative modes. This hypothesis suggests more residential locations in sparsely populated areas, which also limits the development of transport alternatives to single occupancy driving.

I gleaned in the previous pages people's strong desire to make more constructive use of travel time, especially those who travel often and/or long distances. Thus, this scenario is slightly more favourable to train/plane travel than car travel for long distance trips, which are considered here on the rise.

Modal shares of other alternative modes have increased very little in this first scenario, which confirms current development trends. However, even in this scenario – which can be described as the current trend – the proportion of cycling trips more than doubled, mainly due to its strong development at the moment, and which is likely to continue in the years to come. The subway's share has changed little, whereas the tram's has developed considerably due to the current urban policy context. The project team then considered a rather marginal increase in carpooling in local mobility, but a more marked one in long-distance mobility.

The "altermobility" scenario

In this second scenario the modal share of alternative modes is greater than in the previous scenario, while the others (cars, two-wheel vehicles and planes) decrease. This scenario is associated with a modal shift from single-occupancy car use to alternative modes, thanks to a high-quality integrated transport offer. The situation is changing, especially for modes that already enjoyed a positive image (cycling and public transport) in 2015. For the underground and Urban Railway Network, however, some city-size thresholds effects have to be taken into account in their development, which naturally limits the increase in vehicle-kilometres. Moreover, little investment is projected for these modes in the coming decades. Trams and public bus transport (single buses, trolley-buses, bus lanes, etc.), on the other hand, can develop in a significant way.

For modal distribution, the project team assumed that a door-to-door offer exists in France that allows alternative mobility solutions to get almost anywhere in the country. To account for this, and given the current lack of capillarity of bus networks, a modal shift to bus modes and shared modes like car sharing and carpooling was simulated for this scenario. We also allowed for a "network effect" of the mobility strategy leading to an increase in use of all public transport modes (a phenomenon observed in several European countries following the introduction of an integrated offer). To establish a modal share for trains, the project team benchmarked countries where the train's modal share is currently the highest (Japan and Switzerland), and thus ascribed it 16% of traveller-kilometres.

For this scenario, significant investments in rail and public transport services overall will be necessary, with the latter becoming a core technical network – like water or electricity networks – and thus irrigating the entire territory, like in Germany and Switzerland.

The "proximobility" scenario

This third scenario is based on the hypothesis of a deceleration of lifestyles, resulting in a decrease in the intensity of travel in traffic volumes associated with a modal shift from single-occupancy car use towards alternative modes. Such a change is made possible thanks to a high-quality integrated public transport.

This scenario takes for its starting point the growing proportion of people under 25 without a driving licence, the quest for quality of life and the development of local mobility around "slow" modes (such as cycling and running) and the revaluing of urban life since the early 2000s. This scenario is also based on the fact that, despite an increase in long-distance mobilities and in average travel speed, proximity continues to be highly valued. That preference is quite developed already in the population, as I showed before when presenting the results from the survey on aspirations.

The re-investment of proximity does not mean coming back to candlelit evenings. Local life can benefit from ICT in a number of ways. The key assumption in this scenario is a change in how mobility is valued. Moving often, far and fast could no longer be a sign of success and progress.

In territorial terms, this scenario is characterized by the development of concentrated decentralization. Until now, lifestyles considered as the most sustainable ones concern dense urban areas. However, cycling also is re-emerging in suburban and rural areas, following its resurgence in big cities. In this third scenario these territories achieve the maturity they lacked previously and, with a stabilizing of populations, develop mutual aid systems and local activities run directly by the people. Suburban areas will be less dispersed and reach critical mass in terms of population through strong policy measures, particularly as regards citizenship rights. So densifying suburban and rural areas is not only possible – it can also contribute to meeting thresholds in many regions that, today, are ill-adapted to public transport, and mesh urban areas to better organize inter-urban travel.

Regarding forms of mobility and their translation in traffic volumes, this scenario postulates a drop in traffic services based on a decrease in travel time budgets, a substituting of reversible mobilities (commuting/high mobility) for irreversible ones (inter-regional migration) and greater appreciation of local life versus fast, far and frequent travel. To implement this scenario, the project team assumed a 20% decrease in traveller-kilometres by 2050.

For the modal attribution, as with the "altermobility" scenario, the project team assumed that a door-to-door offer existed in France, allowing for mobility solutions other than single-occupancy car use to get almost anywhere in the territory. To consider this, and given the current lack of capillarity of bus networks, a modal shift was attributed to bus modes as well as shared modes such as carsharing and carpooling. We also allowed for a "network effect" of the mobility strategy, resulting in increased use for all public transport (an effect seen in several European countries following the introduction of an integrated offer).

In this last scenario, the objective to divide the greenhouse gas emissions by four (from 2013 to 2050) is just barely achieved, even though an overall decline in traveller-kilometres, a break in modal practices and a significant degree of technological progress are considered.

Conclusion

The prospective study presented in this article led to several conclusions, both methodologically and relative to the future of mobility. Methodologically speaking, it demonstrates the advantage of establishing prospective scenarios based on an analysis of scientific researches. This approach has notably helped identify powerful levers that are likely to profoundly change the future of mobility. One important conclusion arising from the literature encourages us to consider that working on mobile futures necessitates a better understanding of people, their practices, their choices, their aspirations to be acquired. As considered by J. Urry (2016), this chapter aims at defending the idea that social sciences have a major role to play just next to engineering sciences. Weak signals, rebound effects, travel demand learnings, etc., that are difficult to highlight through surveys or measures, are often underlying some rifts and revolutions (Kaufmann and Ravalet, 2016).

This chapter concerns mobile futures, but it directly leads to questioning the political actions needed to achieve the goals we want to achieve. More specifically, both a limitation in travel volume and some modal shares less oriented to the car need to be considered in order to achieve some sustainable mobility goals. The European Commission published a white paper on transport development for 2050 (European Commission, 2011) wherein it sets a target: by 2050, more than 50% of medium- and long-distance trips should be made by public transport. But what does such an objective mean in terms of political commitment?

Is it necessary for public authorities to find a way of limiting the distances we used to travel? How is it possible for us to ensure preferable futures become the most probable (Urry, 2016)? Are all kinds of mobilities supposed to be supported? This encourages us to rethink the place of mobility and transport in our lifestyles. What does each of us want? What do public authorities want?

Working on mobile futures is useful, especially when it can help in achieving a political project, a societal project.

Notes

1 http://en.forumviesmobiles.org/
2 http://fr.forumviesmobiles.org/projet/2016/05/23/aspirations-liees-mobilite-et-aux-modes-vie-enquete-internationale-3240
3 http://lobsoco.com/
4 For more information: www.sncf.com/sncv1/fr/presse/fil-information/etude-facteur-4/159951

References

Aguilera, A., Grébert, J. and Nandy Formentin, H., 2014. Passengers transport modes hierarchy and trends in cities: results of a worldwide survey. *Transport Research Arena 2014.*

Banister, D., 2005. *Unsustainable Transport.* London: Spon Press.

Bosua, R., Kurnia, S., Gloet, M. and Moza, A., 2017. Telework Impact on Productivity and Well-Being. In: J. Choudrie, P. Tsatsou and S. Kurnia, Eds. *Social Inclusion and Usability of ICT-Enabled Services.* New York city: Routledge, pp. 201–219.

Canzler, W. and Knie, A., 1998. *Möglichkeitsräume – Grundrisse einer modernen Mobilitäts- und Verkehrspolitik.* Vienne: Böhlau.

Descarrega, B. and Moati, P., 2016. *Modes de vie et mobilité – Une approche par les aspirations. Phase quantitative.* Research report, Mobiles Lives Forum. Paris: Obsoco.

Dollinger, H., 1972. *Die Totale Autogesellschaft.* Munich: Carl Hanser Verlag.

Dupuy, G., 2006. *La dépendance à l'égard de l'automobile.* Paris: La Documentation Française.

Ettema, D. and Verschuren, L., 2007. Multitasking and Value of Travel Time Savings. *Transport Research Record*, 2010, pp. 19–25.

European Commission, 2011. White Paper. Roadmap to a Single European Transport Area: Towards a competitive and resource efficient transport system. https://ec.europa.eu/transport/themes/strategies/2011_white_paper_en

Fichelet, R., 1979. Éléments pour une compréhension des pratiques de déplacement automobile. In: *Transport et société, Actes du colloque de Royaumont.* Paris: Economica.

Gripsrud, M. and Hjorthol, R., 2012. Working on the train: from 'dead time' to productive and vital time. *Transportation*, 39, pp. 941–956.

Guo, Z., Derian, A. and Zhao, J., 2015. Smart Devices and Travel Time Use by Bus Passengers in Vancouver, Canada. *International Journal of Sustainable Transportation*, 9, pp. 335–347.

Helminen, V. and Ristimaki, M., 2007. Relationships between commuting distance, frequency and telework in Finland. *Journal of Transport Geography*, 15, pp. 331–342.
Hilty, L.M., Köhler, A., Von Schéele, F., Zah, R. and Ruddy, T., 2006. Rebound effects of progress in information technology. *Poiesis and Praxis*, 4, 1, pp. 19–38.
Hymel, K.M., Small, K.A. and Van Dender, K., 2010. Induced demand and rebound effects in road transport. *Transportation Research Part B: Methodological*, 44, 10, pp. 1220–1241.
Jain, J. and Lyons, G., 2008. The gift of travel time. *Journal of Transport Geography*, 16, 2, pp. 81–89.
Janelle, D.G., 1986. Metropolitan expansion and the communications: transportation trade-off. In: S. Hanson, Ed. *The Geography of Urban Transportation*. New York city: The Guilford Press, pp. 357–385.
Kaufmann, V., 2003. Pratiques modales des déplacements de personnes en milieu urbain: des rationalités d'usage à la cohérence de l'action publique. *Revue d'Economie Régionale et Urbaine*, 1, pp. 39–58.
Kaufmann, V. and Ravalet, E., 2016. From weak signals to mobility scenarios: A prospective study of France in 2050. *Transportation Research Procedia*, 19, pp. 18–32.
Keseru, I. and Macharis, C., 2018. Travel-based multitasking: review of the empirical evidence. *Transportation Reviews*, 38, pp. 162–183.
Kim, S.N., 2017. Is telecommuting sustainable? An alternative approach to estimating the impact of home-based telecommuting on household travel. *International Journal of Sustainable Transportation*, 11, 2, pp. 72–85.
Lefèvre, C. and Offner, J.-M., 1990. *Les transports urbains en question*. Paris: Celse.
Lyons, G., Jain, J., Susilo, Y. and Atkins, S., 2013. Comparing rail passengers' travel time use in Great Britain between 2004 and 2010. *Mobilities*, 8, pp. 560–579.
Metz, D., 2008. The myth of travel time saving. *Transport Reviews*, 28, 3, pp. 321–336.
Mokhtarian, P.L., Collantes, G.O. and Gertz, C., 2004. Telecommuting, residential locations, and commute distance traveled: evidence from state of California employees. *Environment and Planning A*, 36, 10, pp. 1877–1897.
Newman, P. and Kenworthy, J., 1989. *Cities and Automobile Dependance*. Sydney: Gower Technical.
Nilles, J.M., 1991. Telecommuting and urban sprawl: mitigator or inciter? *Transportation*, 18, pp. 411–432.
Pawlak, J., Polak, J.W. and Sivakumar, A., 2017. A framework for joint modelling of activity choice, duration, and productivity while travelling. *Transportation Research Part B: Methodological*, 106, pp. 153–172.
Perret, F., Bruns, F., Raymann, L., Hofmann, S., Fischer, R., Abegg, C., de Haan, P., Straumann, R., Heuel, S., Deublein, M. and Willi, C., 2017. *Utilisation de véhicules automatisés au quotidien: les applications envisageables et leurs effets en Suisse. Rapport final sur l'analyse fondamentale*, phase A, BaslerFonds, Union des villes Suisses et autres partenaires.
Sheller, M. and Urry, J., 2000. The city and the car. *International Journal of Urban and Regional Research*, 24, 4, pp. 737–757.
Timmermans, H. and Van Der Waerden, P., 2008. Synchronicity of activity engagement and travel in time and space. Descriptors and correlates of field observations. *Transportation Research Record*, 2054, pp. 1–9.
Urry, J., 2016. *What is the Future?* Cambridge: Polity Press.
Varghese, V. and Jana, A., 2018. Impact of ICT on multitasking during travel and the value of travel time savings: empirical evidences from Mumbai, India. *Travel Behaviour Sociology*, 12, pp. 11–22.
Vincent-Geslin, S., 2012. Les altermobilités contre la voiture, tout contre. *Les Annales de la Recherche Urbaine*, 107, pp. 84–93.
Walls, M.A. and Safirova, E.A., 2004. *A Review of the Literature on Telecommuting and Its Implications for Vehicle Travel and Emissions*. Discussion paper 04-44, Ressources for the future.
Wiel, M., 1999. *La Transition urbaine: ou le passage de la ville pédestre à la ville motorisée*. Paris: Mardaga.
Zahavi, Y., 1979. *The UMOT Project*. Washington, DC: USDOT.
Zahavi, Y. and Talvitie, A., 1980. Regularities in travel time and money expenditure. *Transportation Research Record*, 750, pp. 13–19.
Zhu, P., 2013. Telecommuting, household commute and location choice. *Urban Studies*, 50, pp. 2441–2459.

8

DIGITAL APPROACHES AND MOBILITIES IN THE BIG DATA ERA

Jean-François Lucas

The data city

Data for optimization

"In 2008, our global civilization reached three historical thresholds" (Townsend, 2013, 1). First, the world's urban population has become equivalent to the world's rural population. Second, "the number of mobile cellular broadband subscribers surpassed the number of fixe DSL, cable and fiber-optic lines". Finally, during the same year we moved from the Internet of People to the Internet of Things (IoT) (ibid.).

The Internet of Things refers to a network of connected physical objects with their own digital identity and capable of communicating with each other. These objects can be mobile, such as our mobile phones, GPS, connected watches, transport cards with RFID chips. They can also be fixed, such as surveillance cameras or the various sensors that are increasingly hidden in each urban interstice. In 2010, the city of Santander, which has 180,000 inhabitants, deployed nearly 12,000 sensors to measure CO_2, NO_2, noise, light intensity, humidity, the amount of waste in the trash bins,[1] etc. IT is therefore described as "*pervasive*" (Boullier, 2016, 48–49), because connected objects are deployed throughout our entire daily environment, to the point of constituting a "digital skin" (Rabari and Storper, 2014). Paradoxically, their multiplication favors their invisibility, because these technologies are integrated "into the fabric of everyday life until they are indistinguishable from it" (Wieser, 1991, 94).

The data generated by these sensors led to a multiplication of services and new uses related to the different functions of the city: moving around, living, working, lodging, etc. Many mobility applications allow knowing in real time public transport timetables, calculating alternative routes according to incidents taking place, finding a parking space near one's position, identifying points of interest, etc. At the city level, the analysis of mobility flows data allow, for example, the optimization of travel times of city dwellers and reduction of pollution.

Technical and socio-political issues

Digital traces represent an unprecedented potential to understand and manage mobility issues, since they provide access to knowledge of spatial practices at the individual and collective scales.

For users, services that use Big Data are decision support systems based on the promise of a personalized mobility experience that is simpler, more fluid, faster, cheaper and more social.

For companies, these data represent financial opportunities, be it through their monetization as a paid service or through the resale of the collected data to a third party.

For a municipality, Big Data offers the prospect of a better optimization and management of the services offered to users of transport networks, with a potential reduction in the effects of congestion and pollution. However, the plurality of actors and offers at the scale of a territory can sometimes complicate the management and regulation of mobility. This is for example what happens when Waze[2] suggests a user takes a road parallel to a main axis in order to avoid a traffic jam. This individual solution seems relevant, but becomes problematic as soon as users multiply and the application finally redirects an uninterrupted flow of vehicles onto secondary roads not intended to accommodate the flow of a main road (even if it is clogged). In addition to these structural problems, there are also security issues and the nuisances suffered by local residents. The issues and problems related to the use and interpretation of Big Data in the field of mobility are not only technical, they are fundamentally political.

In this chapter, we will discuss some of the possibilities offered by Big Data for capturing and understanding mobility practices. In the next part, we will present the ways in which the properties of these data can profoundly modify the analytical frameworks with which we understand mobility practices. Beyond the promises of Big Data, we will insist on the importance of considering it, like any other technique, as human and therefore social constructions. Therefore, we will emphasize the importance of describing all the decisions, operations, constraints, etc. that shape these data and the algorithms they feed, in order to understand their limitations as well as to accept their part of subjectivity. Finally, we will discuss the challenges of Big Data for scientific research, from the point of view of the possibilities it offers to renew the understanding of mobility, but also the challenges it raises for the humanities and social sciences (HSS).

What Big Data changes

The 3V

A large quantity of various traces of our mobility practices are produced in real time thanks to fixed sensors (cameras, radars, telephone antennas, bicycle terminals, counting sensors …) and mobile sensors (GPS, smartphone …), or through our activities on the Web and social networks. These traces can be generated voluntarily or not, consciously or not. These data are called "Big Data" when they are varied, mass-produced and in real time. Although there are models with 7 or even 12 dimensions (Uprichard, 2013), "volume", "variety" and "velocity" are the three most commonly cited dimensions. This is known as the "3V" of Big Data (De Mauro, Greco and Grimaldi, 2016; Laney, 2001; Wilder-James, 2012). In spite of the criticisms that highlight the weakness of these three dimensions to describe the diversity of what is commonly called "Big Data" (Kitchin and McArdle, 2016), it is interesting to understand how these data, because of the dimensions that characterize them, make it possible to rethink mobility.

Volume and correlation

The generation, collection, storage, processing and analysis of an ever-increasing volume of mobility data allows making spatial and therefore social practices visible, such as routes and traveling times. In addition to being more and more numerous, these data must be precise in terms of geolocation and time stamping to be relevant. These qualities can be specific to the data, or obtained by mixing different datasets, such as satellite data, road network or GSM data.

The analysis of a large number of accidentology or congestion-related data makes it possible, for example, to better understand the reasons (speed, dangerous passage, etc. in the case of accidents; urban planning, vehicle typology, meteorology, etc., for congestion), to predict or prevent occurrences in order to improve user safety and comfort. If certain variables such as driving speed or blood alcohol content seem to be obvious accidentology factors, Big Data makes it possible to identify behavioral models (patterns) based on an ever-increasing number of variables, and therefore weak signals, that were difficult or even impossible to detect before.

The data are no longer used to test hypotheses; they allow the emergence of general laws based on correlations that will not "tell us precisely *why* something is happening, but will alert us to the *fact* that it is happening" (Mayer-Schonberger and Cukier, 2013, 14). In this sense, the inductive logic of the Big Data replaces the traditional deductive logic of the science.

These new possibilities have led some to speak of a new form of "data-driven" society (Pentland, 2012). Hypotheses, models, theories would then be obsolete in the face of data and algorithms that would find "patterns where science cannot" (Anderson, ibid.). All forms of subjectivity are thus evacuated, since, basically, "who knows why people do what they do? The point is they do it, and we can track and measure it with unprecedented fidelity. With enough data, the numbers speak for themselves" (Anderson, 2008).

This proposal elicited many comments, both on the possibilities offered by Big Data to discover new patterns and on the fact that this argument is part of a tradition that has seen this issue emerge each time a new form of "massive data" has appeared. Strasser thus recalls that "perceptions of an 'information overload' (or a 'data deluge') have emerged repeatedly from the Renaissance through the early modern and modern periods and each time specific technologies were invented to deal with the perceived overload" (Strasser, 2012, 85). Some consider that this does not translate a return to determinism but rather a "current mismatch between the characteristics of these massive data and social science's methods and infrastructures" (Plantin and Russo, 2016).

Beyond the fuss that Anderson's "provocation" may have generated and the criticisms that have emerged in the academic sphere (Boyd and Crawford, 2012), the potential of the data and the correlations that can be made massively irrigate the industrial world and the direction that the development of new mobility services takes.

Variety and diversity

Variety illustrates the possibility of combining structured and unstructured data from multiple sources and different formats. It is thus possible to establish cross-references between descriptive data (ticket prices, theoretical timetables, geolocation of infrastructures, etc.), real-time data (geolocation of vehicles, occupancy rates, available fleet, video surveillance, incident alerts, etc.) or statistical data.

Many mobility services aggregate data from public and private actors relating to public transport, bicycle rental services, car sharing or carpooling, for example by compiling data on time (public transport timetables), stock (bicycles available at a terminal), cost (price of booking a car for a specific journey), etc., in order to offer a multimodal mobility service to users.[3]

Users can then compare offers and make choices according to several criteria: mode of transport, travel time, travel cost, type of lanes used, carbon footprint of the journey, or even more subjective criteria such as the type of landscape or odors they want to have on their route[4].

Velocity and real time city

Velocity embodies the high frequency at which data are generated, collected, calculated and then published. While geolocated data are produced at a regular frequency, ranging from a few seconds to several minutes, their aggregation gives the impression that they are produced continuously.

Georeferenced and time-stamped digital traces make it possible to capture and make visible spatial practices over time (Beaude, 2015), i.e. frequencies, habits, rhythms. The visual representation of these flows on topographic maps may reveal the "pulse" of a city, which contributes to exacerbating the event dimension of contemporary urbanity (Picon, 2009).

These flows are grouped together with other data and other types of representation formats (figures, percentages, tables, videos, etc.) within dashboards, such as London's[5] or Dublin's,[6] in order to facilitate the real-time management and regulation of the city and its flows (transport, energy, water, people, goods, etc.).

Rio de Janeiro's emblematic Operations Center undoubtedly embodies the pinnacle of the city "dashboard". Inaugurated in 2010, as a result of a partnership between IBM and the municipality of Rio, this project initially consisted in predicting and responding effectively to natural disasters. However, thanks to the interconnection of information from multiple sources (sensors, cameras, mobile phones, etc.), the intelligent operations center is able to assist the city in its daily operations and in many sectors (water and sanitation management, urban transport, traffic conditions, health system, civil security, law enforcement, waste collection, street lighting, housing, tourism, education system, etc.) as well as in emergency situations: heavy rainfall, road accidents, power cuts or landslides.[7]. The operations room, made up of hundreds of screens, provides many views of the city, its built environment, its flows, its inhabitants, etc. Like the eye of God in the sky overlooking the city, this project symbolizes the modern version of Bentham's Panopticon, the architectural device that "develops spatial units that allow to see continuously and to recognize immediately" (Foucault, 2016 [1975], 233).

Open the algorithm's hood

Data are "constructed data"

Data are a human production. They do not pre-exist the many human operations that make them. Indeed, people are needed to create sensors and determine what they should pick up, how often, etc.

In fact, data are never "raw", they never speak for themselves (Bowker, 2005; Gitelman and Jackson, 2013). They are a production that makes us say things about the observed world, and produce it in turn. Whatever their origins, they are the result of multiple actions, techniques and ideologies of "data workers" (Bastard et al., 2013) that shape them as soon as they are collected, stored, cleaned, processed, analyzed, shared, visualized … (Bowker and Star, 1999; Ribes and Jackson, 2013).

> What data are generated is the product of choices and constraints, shaped by a system of thought, technical know-how, public and political opinion, ethical considerations, the regulatory environment and funding and resourcing […] Data then are situated, contingent, relational, and framed and used contextually to try and achieve certain aims and goals.
>
> *(Kitchin, Lauriault, and McArdle, 2015, 21–23)*

To account for the goals and issues pursued, it is important to inform and know the choices made for each processing in order to make explicit (Boullier, 2016) how the data that are shared, visualized, consulted, etc., are the result of a co-construction process (Carmes and Noyer, 2014; Latour, 1999).

Big is not all

While the prospects offered by Big Data for studying and understanding mobility are particularly stimulating, they should not hide limitations that are too often forgotten.

First of all, it must be considered that the amount of data obtained, however large, does not mean that it is exhaustive. There are always people who are not taken into account, either because they are not equipped with the appropriate technology or because they use a service different from that used, or because they do not want to, or because the data collection or the processing system have their own failures. Thus, with Big Data, "n=all" in which "all" is related to a particular system (Kitchin and McArdle, 2016, 8), itself "shaped by the technology and platform used, the data ontology employed and the regulatory environment, and it is subject to sampling bias" (Kitchin, 2014, 4).

Moreover, just because we access billions of data representing terabytes of digital information does not mean that we have more representative data. As large as these data may be, they are never global (Beaude, 2015). Moreover, linking digital data to identities, when it is not forbidden, is sometimes very complicated, as for social networks such as Facebook or Twitter where everyone is more or less free to create an alternative identity, with a pseudonym and an age that is not their own (Ravalet, Lucas, and Lohou, 2017). Thus, "by insisting too often on the delicate representativeness of a population by a sample, we do not sufficiently understand what the data represent. Before representing a population, do the data accurately represent the individuals who make up the population?" (Beaude, 2015, 148).

The paradox of transparency

Few people probably know that the price of an Uber ride increases all the more as the battery level of the user's phone is low at the time of booking.[8] This example among many others illustrates the asymmetry that exists

> between operators' demand for greater transparency on the part of users to be able to benefit from digital services, without them reciprocating. Uber knows everything about the identity of both drivers and customers on his platform, the routes they take, the rating they give each other. Conversely, the latter do not have all this information, either on the functioning of pricing algorithms or on those on users' rating. Trust offer is unilateral.
>
> *(Basdevant and Mignard, 2018, 75)*

The opacity surrounding the manufacture of algorithms and data processing complicates the contract of trust between a service provider and a user. However, the quality of the service provided is still the main criterion for membership, maybe because users are often unaware that their data are being sold to advertisers or insurance companies for current and future uses, the impact on personal life of which is still difficult to predict, both from a legal and ethical point of view.

At a time when figures are less and less representative of the social reality they are more and more shaping (Desrosières, 2008), it is becoming urgent to be able to analyze the "black boxes" (Pasquale, 2015) that are algorithms, since they construct forms and deploy statistical representations of society that organize the world in a certain way (Cardon, 2015).

In addition to the opacity of the algorithms, it is complicated to access the data that feeds them, when they are shared.

Researchers who have access to telephone operator data have often obtained a "pass" as part of a specific project. In this case, the rest of the scientific community does not have access to these data, which raises issues in terms of the very foundations of scientific practice, if only from the point of view of its validation or replicability. Academics who do not have an exclusive partnership with the big names in social networks must pay huge sums of money to access data relevant to their research purpose. For others, openly accessible data are often too fragmented to provide convincing results in a scientific study.

In addition, technical access to this type of data, whether through an API or a crawler, requires some specific skills. Finally, the issue of access to data also arises in terms of the balance of power that exists between private approaches within data-holding companies and public research.

Big Data and mobility: challenges for research

Beyond travel, mobility

Beyond the perspectives mentioned in the field of mobility for public and territorial actors, companies and users, few academic actors recur to Big Data to question the concept of mobility, be it because social scientists are still not very interested in these data, because computer scientists who handle these data are not very interested in this concept, or because Big Data is mainly "movement" data.

However, mobility can be understood as the three-dimensional articulation of the realm of possibility to move around (available networks, performance, access conditions, territorial configurations, etc.), the potential for mobility that each individual acquires according to his or her life course, and actual movements in space (Kaufmann, 2008).

Big Data blur the lines between two currents of geography that study transport and "mobilities", where "mobilities" is considered as the central fact of modern and postmodern life (Hannam, Sheller and Urry, 2006; Sheller and Urry, 2006; Urry, 2000). Indeed, Big Data makes it possible to track flows and exchanges of people and goods, regardless of the micro, meso or macro scales. That allows us to capture some aspects of our contemporary, mobile and connected society. However, data collected from individuals or groups of individuals often only describe mappable and calculable movement.

So, while the realm of possibility or actual movements in space can easily be informed thanks to digital techniques, it seems more complex to obtain information relating to the life courses of individuals, defined as mobility capital or motility (Kaufmann, 2002), thanks to Big Data, since its paradigm is that of "the erasure of the subject, the subtle shift from the individual to the statistical individual and, more subtly, the dissolution of the subject in the network" (Beaude, 2015, 152).

From this point of view, qualitative methods

> are the only way to access corners of social reality that are not yet digitized, or whose digital traces are privatized by large commercial companies that are reluctant

> to share them. Qualitative methods also remain of great interest as a complement to work using digital data: they provide a window into the feelings of individuals (Kennedy, 2018), into their experiences and their interpretation of practices of which digital data are the trace, and which are often difficult to understand otherwise.
>
> *(Bastin and Tubaro, 2018, 387)*

Let me quickly point out that I have been referring since the beginning of this article to geographical mobility in an urban context, but that this would need to be extended to larger mobility, for example.

Big Data and mixed methods

Data from social networks can contribute to the analysis of mobility practices and representations, using a combination of qualitative and quantitative methods (Ravalet, Lucas, and Lohou, 2017).

Nevertheless, while the combination of at least one qualitative element with at least one quantitative element is sufficient to characterize a mixed method (Bergman, 2008), several authors emphasize the need to go beyond the articulation of different data collection methods to include how qualitative and quantitative data analysis is conducted (Barbour, 1999; O'Cathain, Murphy, and Nicholl, 2010). Two levels of articulation of a mixed method can thus be dissociated: the *primary level*, which corresponds to the types of combinations made between quantitative and qualitative methods to recover data, and the *synthesis level*, which refers to the level of analysis, depending on whether data processing is qualitative, quantitative or mixed; even though the latter option is not very used (Heyvaert, Maes, and Onghena, 2013).

While some individuals have the skills and knowledge in *computer science*, HSS sufficient to conduct their research on their own, the use of Big Data to study mobility in the HSS field often involves collaborating with data analysts or computer scientists.

Mobility and Big Data at the crossroads of disciplinary fields

Collaboration between researchers in the humanities and social sciences and *computer scientists* is potentially fruitful from a scientific point of view because,

> while computer scientists have produced powerful new tools for automated analyses of such "big data", they lack the theoretical direction necessary to extract meaning from them. Meanwhile, cultural sociologists have produced sophisticated theories of the social origins of meaning, but lack the methodological capacity to explore them beyond micro-levels of analysis.
>
> *(Bail, 2014, 465)*

However, this type of collaboration implies negotiating as well as possible with the different "knowledge positions" (Darré, 1999; Morrissette and Desgagne, 2009) of the different actors, but also agreeing on terminology and a convergent vision (Star and Griesemer, 1989), which requires "articulation work" (Strauss, 1988), i.e. coordination operations to bring together different points of view, and time investments that can be costly. However, the purpose of these investments differs according to the actors involved, because researchers in the humanities and social sciences may seek to understand the meaning of the recovered data, while researchers in the computer sciences field may first seek to optimize algorithms (time gain, robustness and reliability of systems, etc.), without questioning the ontologies from which they condition data recovery.

Conclusion

Big Data is an opportunity to discover new patterns of behavior that had not been possible to observe at the individual and collective level until now. Nevertheless, if the discourses surrounding them praise the many possibilities they open up for understanding such phenomena, particular attention must be paid to their properties, because "big" is not synonymous with exhaustiveness since it always has entities (individuals, objects, etc.) that are not taken into account, and to the algorithms that aggregate and calculate these data, because they are often "black boxes" that do not make it possible to understand how phenomena are analyzed, or even how they prescribed behavior, which raises many questions of equity, ethics and problems related to the general interest.

The data generated and collected in the field of mobility are generally related to travel by individuals or groups of individuals. However, if Big Data puts into question the role of quantitative methods, notably because of the rapid development of the machine learning (Bastin and Tubaro, 2018), qualitative methods still have a whole field of social issues that only they (for the time being?) are able to explore.

In the coming years, we can only expect an increase in collaboration between researchers specializing in the processing of Big Data and researchers in the human sciences field specializing in mobility issues, as it is important to produce knowledge that goes beyond the analysis and understanding of movements, and to provide – particularly in the case of sociologists – reflexivity to decision makers in a political context in which "mobility" is a daily, social, economic, or even psychological issue for millions of people.

Notes

1 SmartSantander: www.smartsantander.eu
2 Mobile GPS navigation application acquired by Google in 2013.
3 "MaaS – Mobility as Services – consists of aggregating public and private transport offers into a single package of mobility services accessible to the user on a unified interface. It is certainly the concept that best embodies the service-switch that takes place in the world of mobility", *in* "Mobility as Networks", Le Lab OuiShare x Chronos, 2018: www.mobilityasnetworks.eu/publications-1
4 Goodcitylife.org: http://goodcitylife.org
5 http://citydashboard.org/london
6 www.dublindashboard.ie/pages/index
7 "Smarter Business for a Sustainable Future ", IBM brochure available online.
8 "This Is Your Brain On Uber", National Pulic Radio, published on 17 May 2016: www.npr.org/2016/05/17/478266839/this-is-your-brain-on-uber?t=1549134490084

References

Anderson, C., 2008. The end of theory. *Wired magazine*, [online] 23 June. Available at: www.wired.com/2008/06/pb-theory [Accessed 12 April 2019].

Bail, C.A., 2014. The cultural environment: measuring culture with big data. *Theory and Society*, 43(3), pp. 465–482.

Barbour, R.S., 1999. The case of combining qualitative and quantitative approaches in health services research. *Journal of Health Services Research & Policy*, 4(1), pp. 39–43.

Basdevant, A., and Mignard, J.P., 2018. *L'empire des données. Essai sur la société, les algorithmes et la loi.* Paris: Don Quichotte éditions.

Bastard, I., Cardon, D., Fouetillou, G., Prieur, C., and Raux, S., 2013. Travail et travailleurs de la donnée. *Internetactu,* [online] 13 December. Available at: www.internetactu.net/2013/12/13/travail-et-travailleurs-de-la-donnee [Accessed 12 April 2019].

Bastin, G., and Tubaro, P., 2018. Le moment big data des sciences sociales. *Revue française de sociologie,* 59 (3), pp. 375–394.

Beaude, B., 2015. Spatialités algorithmiques. In: M. Severo and A. Romele, eds. *Territoires et traces numériques.* Paris: Mines ParisTech, pp. 135–162.

Bergman, M.M., ed., 2008. *Advances in Mixed Methods Research.* London: Sage.

Boullier, D., 2016. *Sociologie du numérique.* Paris: Armand Colin.

Bowker, G.C., 2005. *Memory Practices in the Sciences.* Cambridge, MA: The MIT Press.

Bowker, G.C., and Star, S.L., 1999. *Sorting Things Out: Classification and Its Consequences.* Cambridge, MA: The MIT Press.

Boyd, D., and Crawford, K., 2012. Critical questions for big data. *Information, Communication & Society,* 15(5), pp. 662–679.

Cardon, D., 2015. *À quoi rêvent les algorithmes : nos vies à l'heure des big data.* Paris: Seuil.

Carmes, M., and Noyer, J.M., 2014. L'irrésistible montée de l'algorithmique. *Les Cahiers du numérique,* 10(4), pp. 63–102.

Darré, J.P., 1999. *La production de connaissances pour l'action: arguments contre le racisme de l'intelligence.* Paris: Éditions de la Maison des sciences de l'homme et Institut National de la Recherche Agronomique.

De Mauro, A., Greco, M., and Grimaldi, M., 2016. A formal definition of big data based on its essential features. *Library Review,* 65(3), pp. 122–135.

Desrosières, A., 2008. *Gouverner par les nombres.* Paris: Presses de l'Ecole des mines.

Foucault, M., 2016. *Surveiller et punir. Naissance de la prison,* 1st ed. Paris: Gallimard. 1975.

Gitelman, L., and Jackson, V., 2013. Introduction. In: L. Gitelman, ed. *"Raw data" Is an Oxymoron.* Cambridge, MA: The MIT Press, pp. 1–14.

Hannam, K., Sheller, M., and Urry, J., 2006. Mobilities, immobilities and moorings. *Mobilities,* 1(1), pp. 1–22.

Heyvaert, M., Maes, B., and Onghena, P., 2013. Mixed methods research synthesis: definition, framework, and potential. *Quality & Quantity,* 47, pp. 659–676.

Kaufmann, V., 2002. *Re-thinking Mobility: Contemporary Sociology.* Farnham: Ashgate Publishing Company.

Kaufmann, V., 2008. *Les paradoxes de la mobilité, bouger, s'enraciner.* Lausanne: Presses polytechniques et universitaires romandes.

Kennedy, H., 2018. How people feel about what companies do with their data is just as important as what they know about it. *LSE Impact Blog,* [online] 29 March. Available at: https://blogs.lse.ac.uk/impactofsocialsciences/2018/03/29/how-people-feel-about-what-companies-do-with-their-data-is-just-as-important-as-what-they-know-about-it/ [Accessed 12 April 2019].

Kitchin, R., 2014. Big data, new epistemologies and paradigm shifts. *Big Data & Society,* 1(1), April–June 2014, pp. 1–12.

Kitchin, R., Lauriault, T.P., and McArdle, G., 2015. Smart cities and the politics of urban data. In: S. Marvin, A. Luque-Ayala, and C. McFarlane, eds. *Smart Urbanism: Utopian Vision or False Dawn?* London: Routledge, pp. 16–33.

Kitchin, R., and McArdle, G., 2016. The diverse nature of big data. *Big Data & Society,* 3, pp. 1–10.

Laney, D., 2001. 3-D data management: controlling data volume, velocity and variety. *META Group Research Note,* [online] 6 February. Available at: https://blogs.gartner.com/doug-laney/files/2012/01/ad949-3D-Data-Management-Controlling-Data-Volume-Velocity-and-Variety.pdf [Accessed 12 April 2019].

Latour, B., 1999. *Pandora's Hope, Essays on the Reality of Science Studies.* Cambridge, MA: Harvard University Press.

Mayer-Schonberger, V., and Cukier, K., 2013. *Big Data: A Revolution that Will Transform How We Live Work and Think.* Boston, MA: Houghton Mifflin Harcourt.

Morrissette, J., and Desgagne, S., 2009. Le jeu des positions de savoir en recherche collaborative: une analyse. *Recherches Qualitatives,* 28(2), pp. 118–144.

O'Cathain, A., Murphy, E., and Nicholl, J., 2010. Three techniques for integrating data in mixed methods studies. *British Medical Journal,* 341, p. c4587.

Pasquale, F., 2015. *The Black Box Society, the Secret Algorithms that Control Money and Information.* Cambridge, MA: Harvard University Press.

Pentland, A., 2012. Reinventing society in the wake of big data. *Edge*, [online] 30 August. Available at: www.edge.org/conversation/alex_sandy_pentland-reinventing-society-in-the-wake-of-big-data [Accessed 12 April 2019].

Picon, A., 2009. Ville numérique, ville événement. *Flux*, 4(78), pp. 17–23.

Plantin, J.C., and Russo, F., 2016. D'abord les données, ensuite la méthode? *Socio*, 6, pp. 97–115. Available at: https://journals.openedition.org/socio/2328 [Accessed 12 April 2019].

Rabari, C., and Storper, M., 2014. The digital skin of cities: urban theory and research in the age of the sensored and metered city, ubiquitous computing and big data. *Cambridge Journal of Regions Economy and Society*, 8(1), pp. 27–42.

Ravalet, E., Lucas, J.F., and Lohou, A., 2017. Les retours d'une exploration méthodologique croisant données Twitter, recrutement via Facebook et questionnaires web. *Netcom*, 31(3/4), pp. 309–334.

Ribes, D., and Jackson, S.J., 2013. Data bite man: the work of sustaining long-term study. In: L. Gitelman, ed. *"Raw Data" Is an Oxymoron.* Cambridge, MA: The MIT Press, pp. 147–166.

Sheller, M., and Urry, J., 2006. The new mobilities paradigm. *Environment and Planning A*, 38(2), pp. 207–226.

Star, S.L., and Griesemer, J., 1989. Institutionnal ecology, "translations", and boundary objects: amateurs and professionals on Berkeley's museum of vertebrate zoologie. *Social Studies of Science*, 19(3), pp. 387–420.

Strasser, B., 2012. Data-driven sciences: from wonder cabinets to electronic databases. *Studies in History and Philosophy of Biological and Biomedical Sciences*, 43(1), pp. 85–87.

Strauss, A., 1988. The articulation of project work: an organizational process. *The Sociological Quaterly*, 29 (2), pp. 163–178.

Townsend, A., 2013. *Smart Cities: Big Data, Civic Hackers, and the Quest for a New Utopia.* New York and London: W. W. Norton & Company.

Uprichard, E., 2013. Big data, little questions. *Discover Society*, [online] 1 October. Available at: http://discoversociety.org/2013/10/01/focus-big-data-little-questions [Accessed 12 April 2019].

Urry, J., 2000. *Sociology beyond Societies: Mobilities for the Twenty-first Century.* London: Routledge.

Wieser, M., 1991. The computer of the 21st century. *Scientific American*, 265(3), pp. 94–104.

Wilder-James, E., 2012. What is big data? an introduction to the big data landscape. *O'Reilly Radar*, [online] 11 January. Available at: http://radar.oreilly.com/2012/01/what-is-big-data.html [Accessed 12 April 2019].

9

A METHODOLOGICAL HYBRIDIZATION TO ANALYZE ORIENTATION EXPERIENCE IN THE URBAN ENVIRONMENT

Joël Meissonnier

Introduction

This chapter shows the advantages of the hybridization of two methods, i.e. *the commented walk* and *the obstacle course*; both come from the French corpus of sociology and ethnology. Therefore, it is not strictly a mixed method like the one described by Bergman. He uses this term exclusively for the hybridization between qualitative and quantitative methods.[1] Both of our methods are in fact qualitative. One comes from visual sociology, the other from experimental ethnology.

The work is organized as follows. We will place our two methods in the wide framework of so-called *mobile methods*. We will first consider the French school of *commented walk* as part of the visual sociology methodological tradition. Then we will present what we call *obstacle course* whose origins come from experimental sciences. So we will show the relevance of merging the two. We had the opportunity to do this several times as part of our research. Selected examples will illustrate the interest of this hybridization and show orientation strategies conceived by people in order to find their way in the urban environment and the difficulties they face. Finally, this methodological hybridization appears appropriate for highlighting large or small issues which can impact on daily travel. This method could be of particular use to town planners, public transport providers and vehicle designers.

Two methods belonging to the large field of *mobile methods*

According to Monika Büscher (2013), *mobile methods* produce insight by moving physically, virtually or analytically with research subjects. She brings together under this term a wide range of methods, qualitative, quantitative, visual or experimental. Without quoting them all, let's mention: walk alongs (Kusenbach, 2003), ride alongs (Laurier, 2004) or shadowing (Czarniawska, 2007).

The French School of *commented walks*

Commented walk *or* walk alongs?

The French School of *commented walks* is quite similar to the *walk alongs* method. *Walk alongs* have long existed and been used by ethnographers as a way of establishing confidence between the researcher and interviewee. According to Jones and colleagues (2008), during a walk along, the actual geographic location should be much more relevant when analyzing interviews results. Authors noticed that, curiously, for a method which takes an explicitly spatial approach, few projects have attempted to rigorously connect what participants say with where they say it. But for Jones, *walk alongs* are still the means to a research end, they are just a tool that should help the researcher to consider locations issues. Thus, many researchers use a *mobile method* without particular interest in mobility. For instance Richard Carpiano (2009) uses go along methods for studying the health issues of a neighborhood because the location can have implications for health and well-being.

So even if *mobile methods* are a broad family of methods that allow us to understand numerous social behaviors, we use them in a more specific way to understand mobility patterns. I place my work in connection with *mobility studies* (Büscher, Urry, and Witchger, 2011; Fincham, McGuinness, and Murray, 2009). Hence, we call *commented walks* (Thibaud, 2001) a qualitative methodology used by sociologists and urbanists to investigate the quality of the urban environment using either photos or videos. This method was popularized in France, by the Research Center on Sonic space & Urban environment (CRESSON, Grenoble). According to Jones, there is the question of how researchers engage subjects verbally and/or non-verbally while going around with them. Verbal approaches range from using fairly structured interview guides to completely open conversations (Jones et al., 2008). A *commented walk* relies on the *thinking aloud* technique devised by Alan Newell and Herbert Simon (1972). There are no or very few prompts by investigators. "This is a journey with a real-time storytelling" (Miaux, 2008). That is, the researcher follows the interviewee interacting with several other actors (e.g. passers-by) and objects (e.g. street furniture) encountered on the route. A route that takes place on a well known and appropriated territory. We seek to access the expertise of interviewees experience of their daily territory. And we take the opportunity that research participants display a certain degree of environmental engagement which routinely happens when people are on the move (Seamon, 1979).

The objective of a *commented walk* on a routine route is first to accompany the interviewee in their daily life, and also to give them the opportunity to describe the territory as they live it. Every intersection, every road, every vehicle used can reactivate pleasant or painful, distant or recent, precise or vague memories. The *commented walk* methodology thus puts the onus on to the individual to state his experience: "We consider emotions such as speech engender and the local atmosphere as good reasons for verbalization" (Miaux, 2008). The researcher considers this experience as potentially revealing an expertise born out of repetition on the one hand, and of the elaborated knowledge of the traveled territory on the other hand. The *commented walk* methodology makes more understandable the impact of town planning on how people behave. It also leads to a more precise analysis of mobility choices.

While walking, the interviewees verbalize what they are doing and what they are thinking. On the road, the researcher gets a subjective point of view (Petiteau and Pasquier, 2001). The path acts as a revealer of the interaction between man and the urban environment to be crossed. Street furniture, for example, becomes either a resource if it constitutes an aid or an obstacle if it introduces a difficulty on the way. "We can't get away with not doing this live, an approach that integrates the point of view of actors on the move" (Lévy, 1999).

During the day of investigation, the interviewees become guides. The exercise, therefore, consists of undertaking with them a "ritual" during which they conduct "the initiation of the researcher" (Petiteau and Pasquier, 2001). The interviewees gradually enjoy sharing their experience of the city. Obviously, the experience is unique and not reproducible. The day of the itinerary places the team (researcher, photographer or cameraman, interviewees) "in a situation of shooting in real scenery". Indeed, unlike the methodological protocol of *walk alongs*, *commented walks* are necessarily immortalized by the image.

The commented walk *as part of visual social sciences*

The *commented walk* is part of a long tradition of anthropology and visual sociology. Photography, video recording or film making are fertile tools for exploring the visual dimensions of social life. If it is static and discreet, video can be adapted to studying the movements of the people in a particular public space and their habits. If it is hand held, it films at close quarter the unreeling scene.

In the beginning of ethnology, photography was used to immortalize the traditional customs of so-called primitive societies. The aim was to record for posterity the way-of-life of societies considered doomed to extinction. Finding words inadequate by themselves, ethnographers added photographs to their analysis (Bateson and Mead, 1942; Harper, 2003). But George and Louise Spindler, writing in the Foreword to John Collier's classic text on visual anthropology, criticized this point:

> Usually an anthropologist takes a photograph to illustrate a finding that he has already decided is significant ... He waits until whatever it is happens, then points his camera at it. ... He uses the camera not as a research technique, but as a confirmation that certain things are so, or as a very selective sample of "reality".
>
> *(Collier, 1967)*

Today, it is widely accepted that the sincerity of the researcher conditions the quality of any qualitative research.

The French school of *commented walk* is born out of this methodological tradition of visual sociology. A cameraman tries to catch reflective pauses, paces, movement variations or noticeable emotional changes. Sarah Pink (2007) also does *walk alongs* interviews with a camera attached to herself. One of her studies is based on the evolution of a resident's reaction when exposed to changes in his environment (in this case, the creation of a new park in the neighborhood). The French school of *commented walks* demands, by contrast, the presence of a cameraman who is equipped with wireless headphones connected to the microphone carried on the interviewee. He follows the scene from behind and will therefore, focus on specific points in the environment relevant to illustrating what is being said in the interview. In fact, he is totally complicit in the *commented walk* data collecting process.

Limits

We are aware of the fact that the presence of the researcher and cameraman will not be without consequences. For example, the interviewee will act as a guide and behave differently when accompanied than when alone. On the other hand the researcher's presence will also prevent the possibility of interactions between the interviewee and fellow travelers.

However, although it may seem artificial to ask someone to verbalize their action and although it is not natural to be filmed in the street, people gradually forget the camera and the context of the research. Thanks to unpredictable and random urban phenomena, most of them end up behaving as if the camera was not there.

The *obstacle course* methodology

The *obstacle course* also belongs to the family of *mobile methods*. But contrary to the *commented walks* which focus on mobility routines, the *obstacle courses* is an experimental method because researchers induce a dummy event to focus on unknown routes.

The idea of analyzing access to public spaces using scenarios was born in the 1990s from a research program coordinated by Anni Borzeix (2001, p. 203) which had a broad problematic aim, in terms of accessibility to spaces and utilities. Taking an anthropological perspective, the chosen field of investigation focused on access to the *Gare du Nord* station in Paris. "The research [should have] contributed to improving the services currently offered to users, by allowing easier access."

Concretely, a realizable mission is assigned to the interviewee. The method is, therefore, a deliberately experimental trial:

> Complicit and consenting, interviewees agreed to be guinea pigs in going from one point to another in the *Gare du Nord* site. The aim was to collect situational and experienced information, as a result of a real active work of orientation from travelers .... Instead of questioning travelers in a classical way, we followed them to observe how they deal with space and paths in the station.
>
> *(Borzeix, 2001, p. 204)*

The work of Chevrier and Juguet (2003, p. 6) focuses on the problem of waiting for transport and uses the same method: "Like detectives, spying on these travelers, we collected data on their orientation strategies. We especially studied the tactics deployed to deal with unexpected events."

The research conducted by Chevrier and Juguet proposed scenarios, possibly drawn by lot to make the situation more fun. For example: "We sent them a personalized instruction: you must go to the lost property office to recover your wallet. Your car has broken down. What do you do ?" The purpose of the scenarios is to compel the individual to adjust his action plans to each situation.

Obstacle course *as part of experimental sciences*

As we said above, the *obstacle course* method uses scenarios. Among human sciences, psychology mainly has put in place experimental protocols. We can remember the well-known Hawthorne Factory Experiments (1924–32) conducted by Elton Mayo to see if workers would become more productive in response to various changes in their working environment. However the *obstacle course* method is not a controlled experiment thanks to stable laboratory conditions. It doesn't consist of testing a scientific hypothesis isolated from its context. Neither is it a question of hiding from the participants the true object of the study, as several psychology experimenters had to do. The *obstacle course* method is closer to the "understanding" form of experiment that has been developed in the classical tradition of the Chicago School of Sociology since the 1890s. Experimentation in society, according to

Addams, always includes an expected element of uncertainty which cannot completely be eliminated by controlled planning (Addams, 1967). In the thought of Robert Park (1915), the city especially was to be treated as a social laboratory.

There are two variants of the *obstacle course* methodology: the *treasure hunt* and the *annoying scenario*. The following will present them more precisely.

First variant: the treasure hunt

That's what Anni Borzeix (2001) did in the Gare du Nord station, even if she didn't use the term *treasure hunt*. "We interviewed people unfamiliar with the Gare du Nord station (more likely to get lost)", she writes. The *treasure hunt* is an experiment carried out with a person who is known to be a novice or without experience of the studied area (a person foreign to the city, a non-user of a network, a tourist). This person is confronted with an unusual situation. This experiment makes it possible to clearly grasp the hesitations, the emotions, the decisions and the questioning that go through the mind of the moving interviewee facing an environment that is unknown to him and that he does not cope with very well: "the *treasure hunt* forces the traveler to combine several means of transport to achieve his goal. What are their abilities in orientation? What are their requirements in terms of comfort, information, speed …?" (Chevrier and Juguet, 2003, p. 18).

The *treasure hunt* is set up as a plausible scenario of possible situations which could occur using a volunteer guinea pig. For example, in a train station such as Gare du Nord, people who enter for the first time might "look for their way, for a relative, for their train, and for reliable information, the right ticket, the right way". A variety of both positive and negative circumstances can occur. People are sometimes "without money, cluttered with luggage or children, stressed …". This kind of *obstacle course* has shown that all station employees, be they transport agents, cleaners or shop workers, are called upon "to produce, transmit, make available to travelers relevant information". However, aside from official resources, written or oral, provided by officials or institutions, informal, unforeseen and improvised resources remain useful. The research conducted by Anni Borzeix intended to identify these resources.

The researchers followed interviewees who did not know the Gare du Nord station in order to quickly grasp any problems of access arising at the station (information markers) and the tactics used to cope with them. "We chose novices, poorly adapted to the situation and lacking the expertise of regular users". The investigation revealed

> a poorly equipped space, with misleading signs, not enough information and sales offices which are difficult to find, pricing traps, entry gates, inhospitable corridors, wrong ways, dead ends, illegible posters, arrows in all directions, and unclear announcements due to bad acoustics.
>
> *(Borzeix, 2001, p. 208)*

The travelers may be shortsighted, but they are the user, and their wandering, and daily difficulties in these large public spaces, must be seen as a warning of the wide gap between the functionalist idea of the engineer and the actual needs of the users. Using the *treasure hunt* methodology, "the sociologist looks at social reality through the eyes of travelers, with their intrinsic myopia" (Lévy, 1999, p. 243).

Second variant: the annoying scenario

The way people follow a route is absolutely not the same depending on whether the journey is routine or completely unknown. Most of the actions performed daily are on an automatic mode. "Usually, the public transport customer operates on autopilot. Guided by their habits, they flow through procedures without questioning what they are doing. They 'physically' know how to deal with instructions for use (of the vehicle, the place ...)" (Chevrier and Juguet, 2003, p. 19). We also can refer to a kind of "embodied awareness which is a sense of the body's position and orientation in a movement-space". The "negotiation in motion" is a significant part of our culture and identity. (Jensen, Sheller, and Wind, 2015, p. 363).

Therefore, frequent users of the transport system are "experts". They need less information to reach their destination but their behavior doesn't help to improve everyone's access to the destination. This is the reason why Chevrier and Juguet imagined the *obstacle course* based on an *annoying scenario*:

> To measure people's adaptability, we deliberately disrupted the "transport chain" by inventing, for example, an accident, a flooding ... As a result, during the trip, the interviewee is invited to get off prematurely from the bus and find an alternative solution depending on the means available in their immediate environment.
>
> *(Chevrier and Juguet, 2003, p. 18)*

This kind of experimentation helps in understanding how these regular users behave. Some are able to draw on previous experiences to cope with this new situation; whereas others are less able to find their bearings because they are prisoners of their usual routines and behavior. The former, by transferring learned skills, manage to quickly find alternative solutions, whereas the latter are soon overwhelmed by the mass of information and mental effort required to find their way.

The *obstacle course* method used here explores the unfolding behavior of the interviewee when faced with the simulated disruption caused by an unexpected event interrupting their journey. The resulting reactions to the upheaval can be observed in:

- the finding of a new route, a new means of transport and the change in the traveler's pace.
- the evolving interaction with people and things encountered along the route.
- the evolving relationship to the ordinary objects carried on the journey.

Limits

The main limitation of the scenario investigation method is the artificial context in which researchers place the interviewee. The stakes (and therefore the stress conditions) differ from reality. There is also no guarantee that the interviewee's attitude observed in this artificial context is really in line with what he would have had in real life. But even with an imagined scenario, because it is taking place in a real environment, the traveler's decisions are nonetheless real and so are the consequences which are triggered by the traveler and not the researcher. The more the traveler enjoys the challenge of the set up situation, the more he enters into the spirit of it and forgets that it is unreal.

The benefit of hybridizing the two methods

There is a lot of knowledge and many clues that are sorely lacking to the tourist, foreigner or novice. The main interest of our methodological work is to help to point them out. We have hybridized the aforementioned methods in order to more deeply explore people's mobility behavior. On the one hand, the *commented walk* method has the disadvantage of only focusing the real life experiences of routine routes. On the other hand, in the *obstacle course* method the journey was not initially designed to be filmed. It was necessary for us to draw inspiration from each of these methodological sources to overcome their disadvantages and to cumulate their advantages. In the end, the mixed *commented walks* and *obstacle courses* that we have been able to achieve have the particularity of being *ad hoc* adaptations and hence not faithful to the initial recommendations of their authors.

In urban environments with high information density (transport hubs, rail stations, crossroads in urban centers), a lot of messages collide. Our methodological proposition helps to identify this mass of information and its impact. To get one's bearings in this sign-saturated environment, the interviewee, especially if they are discovering the place for the first time, must be able to take in the surrounding environment, to collect and prioritize relevant information. The sensorimotor sensors of the human body are strongly called upon to treat these "affordances" (Gibson, 1977) coming out of the ambient environment. Some are needed to find one's way while others, such as advertising displays, are extremely disturbing as highly disruptive cognitive "attractors" (Denis and Pontille, 2010).

Some concrete results

The orientation strategies in the urban space of cyclists

In 2011 we had the opportunity to test this hybrid method for the first time in Lille (France) in a study of strategies used by cyclists to get their bearings in town (Meissonnier and Palmier, 2016). This work shows firstly the lack of resources available to help cyclists find their bearings in what could even be their home town. Furthermore signs and markers have been designed for car users in mind and often direct all road users onto large main roads. These roads are less adapted to cyclists than small ones, where traffic is slower and traffic noise is less stressful and the air cleaner.

The hybrid method used for this study reveals the main orientation rationales employed by cyclists. The main difference depends on whether daily cyclists use their bikes only to commute or for a variety of purposes. While some are averse to trying a new route and often remain prisoners of "carable" paths because they doubt whether their destination is "bikeable", others wholly exploit to their benefit the advantages of the bike and furthermore consider the bike as a guaranteed tool to successfully arrive at their destination. As a result, a public policy to encourage bike use could focus on ways for cyclists to find their way. In that perspective we advised policy makers in using specific markings helping cyclists to become more confident (to dare to use unknown roads) and how to accumulate various experiences (that are clearly useful). Due to lack of confidence, some cyclists were unwilling to venture into quiet streets even if more "bikeable" than the main roads.

The mobility difficulties experienced by people with cognitive disabilities

In 2012, we were looking for a method to give us a better understanding of mobility behaviors, especially with people with disabilities. We were also looking more precisely for a method suitable for participants with verbal or non-structured verbal difficulties. So, we carried out four *commented walks* in the city of Amiens (France) which supported this study (Meissonnier and Dejoux, 2016).

Marie, one of our interviewees, has trisomy 21. As Marie seemed relatively autonomous on public transport, we asked her to add slight challenges. *Commented walks* were followed by an *annoying scenario obstacle course*. In the morning Marie was going to her workplace with her father by car. She was returning alone, by bus. We suggested she tried an outward journey by public transport for the first time.

Usually when Marie returns from work by bus No. 9 and arrives in the downtown interchange bus station of Amiens, she gets off the bus at a stop located near a pedestrian crossing. As she was trying the outward journey by public transport for the first time she placed herself on the other side of the pedestrian crossing. Indeed this is how she had been taught to catch a bus going in the opposite direction. Therefore, she was trying to apply a rule that told her to wait on the other side of the street in order to catch it. But this rule doesn't work in such an interchange station. On the video, Marie is seen waiting with other passersby at the pedestrian crossing. When bus No. 9 passed it didn't stop. And we had to help her to find the right bus stop.

The No. 9 bus line was split into two directions along the route and in order to get the right bus it was necessary to know the bus's final destination. The problem was that Marie couldn't know the precise final destination because she had arrived from the back of the bus where there was only the bus number on display. When Marie realized her mistake in taking the wrong No. 9 bus, she asked for the driver's help. The driver answered that "As a rule, one bus out of two should go to *Zone Industrielle* (her final destination)". Therefore, the driver should have told her to get off the bus and wait for the next one. And when the bus route split the driver didn't think of advising Marie to get off now. The driver failed to do this possibly due to a lack of training. In both situations, Marie tried to carefully follow the rules but she failed in her task due to breakdowns in the information chain. These examples show that both the continuity of the information chain and the standardization of the displayed information are vital.

The videos resulting from the *commented walk* mixed with *obstacle course* are important, firstly to explore the difficulties experienced, then to inform the necessary authorities and finally to feed into training programs. The video shot of Marie during the *obstacle course* is now used by an association for training related to disabled travel and access. Our method manages to explain concretely what those difficulties are for people with mental, cognitive or psychological disabilities. Our results reveal the effectiveness of our hybrid methodology which could be more formally introduced in order to explore in each city the reality and the details of the problems encountered by people with disabilities when trying to orientate themselves. This in turn will have the knock-on effect of helping able-bodied people who are unfamiliar with their environment (foreigners, tourists, etc.).

Conclusion

Commented walks and *obstacle courses* are two qualitative methods that belong to the extensive field of *mobile methods*. The first one belongs to the visual sociology tradition; the second one to the experimental sciences tradition. The French school of *commented walks* is close to the *walk along* methodology. Indeed, unlike the methodological protocol of *walk alongs*, *commented walks* are necessarily immortalized by the visual and require a skilled cameraman. The second method, called *obstacle courses* comes from anthropology. It is an experimental method because researchers induce a dummy event to focus on unknown routes. However, the *obstacle course* method is not a controlled experiment thanks to stable laboratory conditions. It is close to the "understanding" form of experiment that has been developed in the classical tradition of the Chicago School of Sociology.

The results which have come out of our hybridization of *commented walks* and *obstacle courses* methodologies are not based on representative samples. However, the recommendations to improve transport systems resulting from the dysfunctions highlighted by these methods have a wider impact that goes beyond the few experiences of the interviewees presented here, that is to say the few travelers questioned and consenting to make the urban space a playground over a couple of hours. The main strength of our qualitative hybridization is that it is illustrative. We tried to show how exemplification is a powerful tool rather than anecdotal. Quantitative methods alone don't assess to what extent both public transport and public areas create difficulties or not for users and can't precisely identify these difficulties. Therefore, our method can be most successfully employed, especially when used in conjunction with quantitative methods, in carrying out studies aimed at informing policy makers or measuring their impact.

As shown in the aforementioned examples, our method successfully analyzed the orientation rationale of cyclists in town. It pointed out difficulties encountered by people with cognitive disabilities in finding their way. And it showed that, for some commuters, rush hour mobility can sometimes be enjoyable and playful. In the field of *mobility studies* the understanding of orientation tactics and the social aspects of finding one's bearings is relatively unexplored. Sociology tooled with a hybrid method could meet this challenge.

Note

1 "Mixed method research design is one of the fastest growing areas in research methodology today. Its aims and benefits appear rather simple: take the best of qualitative and quantitative methods and combine them" (Bergman, 2008, p. 11).

References

Addams, J., 1967. *Twenty Years at Hull-House*, 1st ed 1910. New York: Macmillan.

Bateson, G. and Mead, M., 1942. *Balinese Character: A Photographic Analysis*. New York: New York Academy of Sciences.

Bergman, M. M., 2008. The Straw Men of the Qualitative-Quantitative Divide and Their Influence on Mixed Methods Research. In M. M. Bergman ed., *Advances in Mixed Methods Research*. SAGE, pp. 10–22.

Büscher, M., Urry, J. and Witchger, K., 2011. *Mobile Methods*. London: Routledge.

Büscher, M., 2013. Mobile Methods, *Mobile Lifes Forum*, 16-04-2013, http://en.forumviesmobiles.org/marks/mobile-methods-697

Borzeix, A., 2001. L'information-voyageurs en Gare du Nord. In A. Borzeix and B. Fraenkel eds., *Langage et travail* [*Communication, Cognition, Action*]. Paris: CNRS Éditions, pp. 203–230.

Carpiano, R. M., 2009. Come Take Awalk with Me: The "Go-Along" Interview as a Novel Method for Studying the Implications of Place for Health and Well-being, *Health & Place*, 15, pp. 263–272.

Chevrier, S. and Juguet, S., 2003. *Arrêt demandé*. Rennes: LARES-Enigmatek.

Collier, J., 1967. *Visual Anthropology: Photography as a Research Method*. New York: Holt, Rinehart & Winston.

Czarniawska, B., 2007. *Shadowing: And Other Techniques for Doing Fieldwork in Modern Societies*. Copenhagen: Liber Copenhagen Business School Press.

Denis, J. and Pontille, D., 2010. *Petite sociologie de la signalétique. Les coulisses des panneaux du métro*. Paris: Presses des mines.

Fincham, D. B., McGuinness, M. and Murray, L., 2009. *Mobile Methodologies*. Palgrave Macmillan.

Gibson, J. J., 1977. The Theory of Affordances. In R. Shaw and J. Bransford eds., *Perceiving, Acting, and Knowing: Toward an Ecological Psychology*. Hillsdale, NJ: Lawrence Erlbaum, pp. 67–82.

Harper, D., 2003. An Argument for Visual Sociology. In J. Prosser ed., *Imaged-based Research – A Sourcebook for Qualitative Researchers*. London and New York: Routledge Falmer, pp. 20–35.

Jensen, O. B., Sheller, M. and Wind, S., 2015. Together and Apart: Affective Ambiences and Negociation in Families' Everyday Life and Mobility, *Mobilities*, 10(3), pp. 363–382. doi:10.1080/17450101.2013.868158.

Jones, P., Bunce, G., Evans, J., Gibbs, H. and Hein, J. R., 2008. Exploring Space and Place with Walking Interviews, *Journal of Research Practice*, AU Press, 4(2), pp. 1–9. Retrieved [2019-04-26], from http://jrp.icaap.org/index.php/jrp/article/view/150/161

Kusenbach, M., 2003. Street Phenomenology: The Go-Along as Ethnographic Research Tool, *Ethnography*, 4(3), pp. 455–485.

Laurier, E., 2004. Doing Office Work on the Motorway, *Theory, Culture & Society*, 21(4–5), pp. 261–277.

Lévy, E., 1999. Saisir l'accessibilité. Les trajets-voyageurs à la gare du Nord. In I. Joseph ed., *Villes en gares*. La-Tour-d'Aigues: L'Aube, pp. 242–258.

Meissonnier, J. and Dejoux, V., 2016. The Commented Walk Method as a Way of Highlighting Precise Daily Mobility Difficulties: A Case Study Focusing on Cognitive or Mental Diseases, *Transport Research Arena Procedia*, 14, pp. 4403–4409.

Meissonnier, J. and Palmier, P., 2016. Chromorientation: vers un jalonnement plus adapté aux logiques cognitives des piétons et cyclistes en ville ? *Association de Science Régionale de Langue Française – ASRDLF international conference*, Gâtineau, July 7–9.

Miaux, S., 2008. Comment la façon d'envisager la marche conditionne la perception de l'environnement urbain et le choix des itinéraires piétonniers – L'expérience de la marche dans deux quartiers de Montréal. In *RTS*, vol. 101. Paris: Lavoisier, pp. 327–351.

Newell, A. and Simon, H., 1972. *Human Problem Solving*. Englewood Cliffs, NJ: Prentice Hall.

Park, R. E., 1915. The City: Suggestions for the Investigation of Human Behavior in the City Environment, *American Journal of Sociology*, 20(5), pp. 577–612.

Pink, S., 2007. Walking with Video, *Visual Studies*, 22(3), pp. 240–252.

Petiteau, J.-Y. and Pasquier, E., 2001. La méthode des itinéraires : récits et parcours. In M. Grosjean and J.-P. Thibaud eds., *L'espace urbain en méthodes*. Marseille: Parenthèses, pp. 63–77.

Seamon, D., 1979. *A Geography of the Life World*. New York: St Martin Press.

Thibaud, J.-P., 2001. La méthode des parcours commentés. In M. Grosjean and J.-P. Thibaud eds., *L'espace urbain en méthodes*. Marseille: Parenthèses, pp. 79–99.

10

METHODOLOGIES FOR UNDERSTANDING AND IMPROVING PEDESTRIAN MOBILITY

Michel Bierlaire, Riccardo Scarinci, Marija Nikolić, Yuki Oyama, Nicholas Molyneaux and Zhengchao Wang

Introduction

Individuals are at the center of every transportation system, therefore understanding pedestrian mobility is of fundamental importance. Global phenomena such as urbanization and population growth lead to a large increase in the number of pedestrians in city centers and passengers in public transport. This creates congestion in sidewalks and urban crossings, as well as overcrowding in public transport stations and vehicles. A vast range of methods is available to support municipalities and transport operators to better understand, predict and manage pedestrian mobility with the final goal of improving the facilities and their operations.

This chapter covers these methods and is organized in the following sections:

1. the *fundamental indicators and relationships* describing pedestrian mobility considering the heterogeneity of the individuals;
2. the models needed to estimate the *aggregate demand* of pedestrians represented by origin-destination matrices including unique generation phenomena of transportation hubs like train or bus arrivals and departures;
3. the behavioral models describing the *disaggregated demand* focusing on the activities performed by individuals within multifunctional facilities offering shopping and leisure possibilities alongside transport services;
4. the appropriate *assignment models* at different aggregation scales for a trade-off between accuracy and computation time;
5. the algorithms to generate and evaluate *management strategies* of pedestrian mobility for improving the level of service perceived by the passengers.

All these methods are interconnected in a unique framework that is useful to investigate pedestrian mobility in multimodal and multifunctional facilities, but also in locations where pedestrian dynamics are at the center of the mobility systems such as city centers and large sport, cultural and religious events.

This chapter does not intend to provide an exhaustive literature review on pedestrian research. Instead, it focuses on the research performed at the Transport and Mobility Laboratory (TRANSP-OR), EPFL, Switzerland.

Fundamental indicators and relationships of pedestrian mobility

Data on pedestrian dynamics and behavior are necessary to understand pedestrian mobility (Hänseler et al., 2016). The most used systems are mechanical counting, cameras, depth and infrared sensors. Surveys, such as revealed and stated preferences, are another source of data. Laboratory experiments provide trajectories in various scenarios, for example, crossings, counter flows and bottlenecks.

Recently, portable devices like smartphones and smart travel cards offer a new type of data. Intrinsic characteristics of transportation hubs, such as train or flight schedules, can provide useful information on the pedestrian movements.

These data are used to extract the fundamental quantities used to observe and to model the mobility of pedestrians: density, flow and speed. These quantities are not directly observed by the data collection techniques previously introduced, and their derivation is not always straightforward. Furthermore, the highly heterogeneous and complex nature of pedestrian movement behavior leads to additional difficulties for pedestrian flow characterization (Nikolic and Bierlaire, 2017). And the relationships between density, flow and speed are referred to as the fundamental relationships. They play an important role in the field (Weidmann, 1992). They are (i) used to predict pedestrian flow under specific circumstances, (ii) useful for planning and designing of pedestrian facilities, and (iii) required input or calibration criteria for models of pedestrian dynamics. In the context of pedestrian traffic, fundamental relations are usually established by fitting deterministic curves to empirical data. Both linear (Fruin, 1970) and nonlinear (Tregenza, 1976) speed–density models have been proposed. The researchers have suggested several explanations for these deviations: the cultural differences, the differences between pedestrian facilities and the effects of the environment, flow composition, measurement methods, etc. (Steffen and Seyfried, 2010).

Simulation-based fundamental relationships are predominantly obtained via cellular automaton models. For instance, Blue and Adler (1998) specify a unidirectional cellular automaton model that produces a speed–density relationship similar to the one proposed by Weidmann (1992). Importantly, empirical analyses reported in the literature reveal significant scatter in the data (Figure 10.1), which eliminates the use of a unique equilibrium relationship (Nikolic et al., 2016). This indicates that in addition to density, other factors are likely to influence the speed of pedestrians. Weidmann (1992) has empirically shown that the trip purpose of pedestrians represents one of the relevant factors. The speed of pedestrians appears to be affected by the age and the gender as well.

There are several ways to account for the observed heterogeneity in speed. A possible approach to capture this complex phenomenon consists in modeling explicitly the exact underlying walking process and the explicit interactions at the disaggregate level. At a more aggregate level, mesoscopic models describe pedestrian behavior in terms of probabilities, e.g. velocity distributions (Hoogendoorn and Bovy, 2000). In the field of vehicular traffic, Wang et al. (2013) propose a model derived by adding Gaussian noise to the existing deterministic relationships.

This can potentially lead to unrealistic outcomes (e.g. negative speed values). Jabari et al. (2014) propose a probabilistic speed–density relationship based on a microscopic car-following model. Probabilistic features are incorporated by introducing random parameters capturing population heterogeneity. However, limited behavioral basis exists to help in the

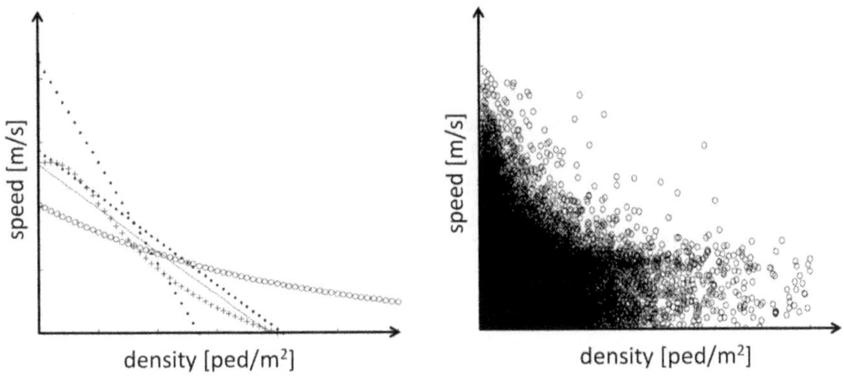

Figure 10.1 Models from the literature (left) versus empirical observations (right) collected in the train station in Lausanne, Switzerland.

specification of the distribution of these parameters. An alternative for dealing with heterogeneity is a two-stage approach, where the data is first segmented based on some observed characteristics (e.g. socio-economic or demographic variables). In the second stage, a separate model is estimated for each predefined segment in the population (Weidmann, 1992). The issue of imprecise parameter estimates may arise due to potentially small sample sizes in some segments. Also, segmentation is usually performed based on a single characteristic and assumed to be error-free. In reality, the heterogeneity may come from multiple factors, which may introduce errors in the second stage.

Nikolic et al. (2016) and Nikolic et al. (2019) present an alternative approach, based on an aggregate representation of the pedestrian traffic. The authors propose methodologies to represent the speed–density relationship of pedestrian traffic in a probabilistic way. The scatter is explicitly represented by relaxing the homogeneity assumption of the equilibrium speed–density relationship.

The probabilistic model presented in Nikolic et al. (2016) implicitly accounts for the heterogeneity of pedestrian flows by bringing together first principles and a data-driven approach. The study assumes that the speed of pedestrians is a random variable, such that for each density level there is a distribution of speed values rather than one deterministic value. The distribution of speed has two properties: (i) it is continuous with positive support, and (ii) it is unimodal. The first property is in accordance with the physical characteristic of the speed, being that the speed is a continuous variable whose values cannot be negative. The second property is introduced with the purpose of limiting the number of model parameters. The model specification ensures the physical correctness of the results, and satisfactory predictive capabilities across case studies. However, it lacks behavior-oriented explanatory power. This issue is tackled in Nikolic et al. (2019), through the latent class methodology.

Aggregate demand

Knowledge of pedestrian demand is necessary for the analysis of pedestrian flows, the design of infrastructure, the optimization of services and the management of flows (Hoogendoorn and Daamen, 2004). The demand can only be observed indirectly, and estimation methods are needed. However, the estimation process is complex, and often, theoretical demand scenarios or simple assumptions are used in practice (Van den Heuvel and Hoogenraad, 2014). Promising studies are based on the estimation of demand using activity-based models, which create a disaggregate demand representation, i.e. for single individuals (Hoogendoorn and Bovy, 2004). In some cases, an aggregate demand representation, i.e. described by flows grouped in origin–destination (OD) pairs, is preferable over a disaggregate demand. Estimation of OD demand does not require disaggregate data, which in practice is rarely available. Moreover, it is less computationally expensive, making it suitable for real-time applications, or large pedestrian infrastructure.

The OD estimation is often based on the methods used in the context of road traffic (Cascetta, 1984). Typically, the OD demand is "reverse engineered" using observed flows and an assignment model. The OD is modified until the assigned flows are similar to the measured ones.

However, pedestrian dynamics are characterized by a higher complexity than vehicular dynamics (Nikolic et al., 2019). For example, individuals have more freedom of movement than vehicles, they may perform unplanned movements, and shopping facilities or public transport stops in a city center induce internal origin–destination sources that should be considered explicitly. For these reasons, it is necessary to provide detailed information on OD demand within pedestrian facilities, and not only at the boundaries.

To address the issues specific to pedestrian dynamics, Hänseler et al. (2017) propose a framework for estimating pedestrian OD. The framework takes into account different data sources such as ridership data and counts collected in Lausanne train station. The walkable space of the infrastructure is represented by a directed graph. The demand is considered in discrete time and at the aggregate level. The passengers are grouped in packages of pedestrians associated with the same origin–destination pair and departure time interval.

A key element of the framework is the use of the train timetable. A significant fraction of the demand comes from passengers alighting from trains (or public transport vehicles in general). Molyneaux et al. (2014) model the train induced pedestrian flows by explicitly taking into account the train timetable and the number of disembarking passengers. This sub-model is referred to as "train induced flows". The embarking/disembarking flow is represented with a function calibrated on the pedestrian trajectory data, infrastructure characteristics, such as the ramp width, and the technical details of the train carriages like the number, position and size of doors.

This approach is appropriate for the estimation of OD demand thanks to its efficient computation times. Moreover, the "train induced flows" model provides a concise yet accurate approach for linking passengers traveling in public transport modes and pedestrians flows, and it has proven accurate for the simulation of the alighting flows at an aggregate level.

Disaggregate demand

Disaggregate demand of pedestrians' mobility is generally described by choices made by each individual (Timmermans, 2009). Examples of these choices are the departure time, the origin and destination location, the activities performed and the duration for each activity. These choices can be captured by activity schedule models, which represent the core of an activity-based demand model of pedestrians.

In literature, these activity schedule models can be divided into utility maximization models and computational models. The utility maximization models adopt a rigorous optimization framework with the assumption that pedestrians choose the choices that allow maximal utility gained from performing each activity. For instances, Hoogendoorn and Bovy (2004) adopt this framework to study the store visit choices of pedestrians in city centers considering random traffic conditions, and Danalet et al. (2013) propose a hierarchy model for the activity sequence, location and duration choices of pedestrians.

However, one criticism of the utility maximization models is that people are not fully rational, and they do not choose the alternative with the highest utility. Instead, people might make their decisions with a bounded rationality, which is at the basis of computational models. This class of model is more flexible in terms of modeling unplanned activities, activity rescheduling behaviors and heterogeneous decision strategies. The perceptual field concept models (Dijkstra et al., 2009) and Monte Carlo simulation (Dijkstra et al., 2014) are examples of computational models which can model unplanned activities of pedestrians.

The next two sections focus on two aspects fundamental for the estimation of the disaggregate demand, namely, (i) the modeling of activities, and (ii) the use of measurement equations and prior information in the models.

Modeling activity choices

The analysis of pedestrian behavior needs to consider a more detailed interaction among activities and the time dimension, in comparison with the traditional travel demand analysis (Bowman and Ben-Akiva, 2001). Danalet and Bierlaire (2014) and Danalet and Bierlaire (2015) model an activity-sequence in a path choice approach, applying methodologies for route choice analysis to an activity network with two dimensions of activity type and time. These studies deal with the large dimensionality of the choice set by using a Metropolis-Hasting algorithm (Flötteröd and Bierlaire, 2013). This activity path choice model allows the analysis of sequence of detailed activities in pedestrian contexts, e.g. buying a ticket, having a coffee and waiting for boarding at a transportation hub.

When considering the utilities of activity choices, the location of an activity plays an important role. Consequently, it is natural to combine the activity choice model with the destination model. Borgers and Timmermans (1986) develop a destination choice model as part of a system of models to predict the total demand of pedestrian shopping behavior in a city center. In some circumstances, it may happen an unplanned activity is performed while walking, stimulated by the attractiveness of streets. This type of behavior is closely linked with the route choice model. To address this, Oyama and Hato (2016) develop an activity path choice model in a time–space network graph and considered the possible myopic decision of pedestrians, based on an implicit approach that avoids the path enumeration. Time–space prism constraints incorporated into the model also help the stability and efficiency of computation.

Measurement equations and priors

Mobile phone data contains rich information of pedestrian behavior. However, such data is often associated with poor quality information on the locations where activities are performed. Therefore, it is useful to include measurement equations into the modeling process. A measurement equation is a model that gives the probability of reproducing a data (e.g. a tracking data) when assuming a state (e.g. a path, an activity location or its sequence) as the actual state. That is to say, it evaluates the likelihood of each possible state being the true state.

WiFi traces is an important data for the analysis of pedestrian activities, especially within facilities. Danalet et al. (2014) introduce a method to detect the sequence of locations where activities are performed specifically using WiFi data. This method is based on a Bayesian approach in which the prior knowledge of the infrastructure, modeled as the attractivity, is taken into account. Introducing the prior is significant when the localization is weak and the pedestrian map is dense. We tested the advantages and limitations of this method using the WiFi data of EPFL university campus.

The use of measurement equations can be extended to other data sources. For example, Bierlaire and Frejinger (2008) propose an estimation procedure for route choice models that does not require a unique path observation and instead considers the measurement equation of possible ones. Probabilistic map-matching models have also been proposed to consider explicitly the measurement errors of GPS data (Ochieng et al., 2003; Bierlaire et al., 2013; Chen and Bierlaire, 2015). They output the path candidates associated with the likelihood of being the actual path. Oyama and Hato (2018) introduce a measurement equation designed for pedestrian path observations in city centers, accounting for the inter-link heterogeneity of the measurement errors.

Assignment models

Pedestrian assignment models determine flows and densities on an infrastructure (Helbing and Molnar, 1995). A multitude of approaches have been proposed ranging from disaggregate to aggregate models. Examples are the social force model (Helbing and Molnar, 1995), behavioral models (Robin et al., 2009) and activity choice models (Hoogendoorn and Bovy, 2004). Aggregate models are computationally efficient, but they often have a low accuracy. On the contrary, disaggregate models offer a good representation of the dynamics at the expense of longer computation time.

In the following, we focus the attention on two models, one aggregate and one disaggregate, that are developed specifically to capture pedestrian dynamics within pedestrian facilities.

Hänseler et al. (2014) develop an aggregate pedestrian network loading model based on an approximation of the continuum theory for pedestrian flow and a discretization scheme similar to the cell transmission model. The model considers discrete-time and discrete-space. The walkable space is partitioned into a set of areas and represented by a directed graph. The flows are associated to stream, which in turn are associated with an area. The interactions of streams within the same area defines the interaction among pedestrians. The result is a fast and accurate model that can be easily applicable to large and complex real-world pedestrian areas. The model is useful as a core component for the estimation of aggregate demand.

Antonini et al. (2006a) investigate the use of discrete choice models to develop a disaggregate pedestrian walking behavior model. The space in front of each individual is discretized in zones. Each zone is an alternative in the individual's choice set for the next step. The probability of choosing a zone is related with the desired speed and direction of the individual, the interactions with other individuals and the surrounding environment. Robin et al. (2009) expand the capability of this model by explicitly capturing leader–follower and collision–avoidance patterns. The use of discrete choice models for pedestrian behavior proved to be useful for detection and tracking of pedestrians in complex scenarios. This has been proven for the tracking data available for Lausanne train station. The application of this model seems nowadays even more relevant due to the need for autonomous vehicles to predict accurately the next step for pedestrians (Antonini et al., 2006b).

Management strategies

The combination of high demand and limited space available for the pedestrians to move leads to congestion, poor level-of-service and even hazardous situations in the worst cases.

One possibility to prevent these issues is the usage of management strategies for controlling pedestrian flows within the infrastructures.

Management strategies for pedestrians can be either static or dynamic. Management strategies which are static generally take place at the planning phase. For example, in Campanella et al. (2013), the authors analyze a highly congested metro station in Copacabana and proposed some modification to the infrastructure (removing sharp corners for example) in order to improve the level-of-service experienced by the pedestrians. On the other hand, dynamic strategies rely on real time measurements of some predefined indicators like density, flow or travel time. A macroscopic pedestrian simulator is combined with a controller which regulates the pedestrian flow through the doorways onto a train platform in Bauer et al. (2007). Some examples of strategies are the control of the pedestrian's walking speeds inside large areas to guarantee a predefined level of service (Zhang et al., 2016) and the optimal timing of the phases of lights at signalized crosswalks to minimize the delay (Zhang et al., 2017).

Molyneaux et al. (2018b) propose two management strategies for train stations inspired from vehicular traffic. The first is gating, based on the ramp metering system (Papageorgiou et al., 1991), which aims at controlling the inflow of pedestrians into the intersections of the pedestrian underpass and the access ways to the platforms. This has proven efficient for preventing congestion without increasing the travel time of pedestrians (Molyneaux et al., 2018a).

The second strategy is the usage of flow separators to prevent counter flow. Unlike vehicular traffic, pedestrians are not constrained by lanes, hence they can interact with pedestrians moving in the opposite direction, causing delay. By allocating a space for each direction of flow, the control strategy aims at preventing these interactions and therefore reducing the travel time of lightly congested flows.

Jeanbart et al. (2018) investigate the use of moving walkways and accelerating moving walkways as an innovative control strategy for pedestrian flows (Scarinci et al., 2017). Moving walkways can support the control of the speed and destinations of pedestrians in transport hubs.

Another advantage of controlling pedestrian flows is the reduction in travel time variability. If the occurrence of congestion can be prevented, or at least decreased, then travel times become more reliable. This aspect is critical to the stability of public transport timetables. As timetables are built such that passengers can catch their connections, a sudden increase in travel time between two platforms makes the passengers miss their connecting train. Such considerations are gaining attention from public transport operators as the passenger experience is now becoming central, especially in multimodal trips.

Conclusions

This chapter presented the key methods needed to understand pedestrian mobility starting from the data and concluding with the management strategies. Each key aspect has strong interactions with the others, and they can be all incorporated into a unified framework (Molyneaux et al., 2017). Figure 10.2 shows a graphical representation of this framework and the interactions among the main components. Demand and assignment are at the center. They both depend on data and the fundamental relationships among density, flow and speed. Pedestrian oriented indicators are the output of the interaction between demand and its assignment on the infrastructure, and they are used to define the management strategies aimed at improving the user experience in pedestrian facilities.

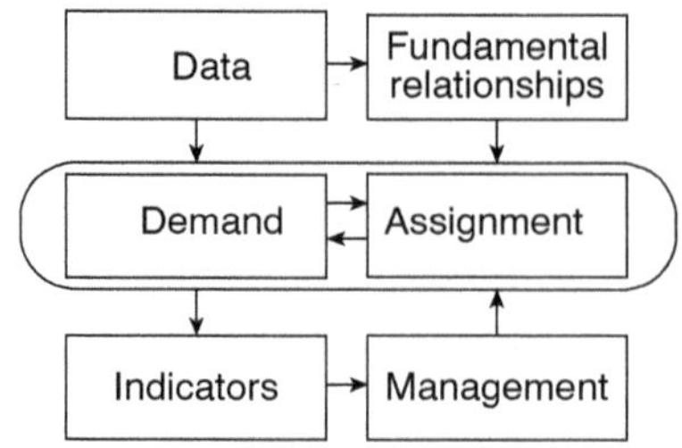

Figure 10.2 Framework linking the main components of pedestrian mobility.

Research in the field of pedestrian mobility is facing several challenges in the coming years. The data provided by modern technologies should be properly analyzed to extract relevant information and undiscovered dynamics while respecting the privacy of the individuals. Active modes like walking and cycling are increasing in the urban environment. This leads to the need for integration with the existing private and public transport system, and the future on-demand and autonomous services.

References

Antonini, G., Bierlaire, M., Weber, M., 2006a. Discrete choice models of pedestrian walking behavior. *Transportation Research Part B: Methodological* 40 (8), 667–687.

Antonini, G., Martinez, S. V., Bierlaire, M., Thiran, J. P., 2006b. Behavioral priors for detection and tracking of pedestrians in video sequences. *International Journal of Computer Vision* 69 (2), 159–180.

Bauer, D., Seer, S., Brändle, N., 2007. Macroscopic pedestrian flow simulation for designing crowd control measures in public transport after special events. In: Proceedings of the 2007 Summer Computer Simulation Conference, Society for Computer Simulation International, pp. 1035–1042.

Bierlaire, M., Chen, J., Newman, J., 2013. A probabilistic map matching method for smartphone gps data. *Transportation Research Part C: Emerging Technologies* 26, 78–98.

Bierlaire, M., Frejinger, E., 2008. Route choice modeling with network-free data. *Transportation Research Part C: Emerging Technologies* 16 (2), 187–198.

Blue, V., Adler, J., 1998. Emergent fundamental pedestrian flows from cellular automata microsimulation. *Transportation Research Record: Journal of the Transportation Research Board* 1644, 29–36.

Borgers, A., Timmermans, H., 1986. City centre entry points, store location patterns and pedestrian route choice behaviour: A microlevel simulation model. *Socio-economic Planning Sciences* 20 (1), 25–31.

Bowman, J. L., Ben-Akiva, M. E., 2001. Activity-based disaggregate travel demand model system with activity schedules. *Transportation Research Part A: Policy and Practice* 35 (1), 1–28.

Campanella, M., Halliday, R., Hoogendoorn, S., Daamen, W., 2013. Managing large flows in metro stations: Lessons learned from the new year celebration in Copacabana. In: 2013 16th International IEEE Conference on Intelligent Transportation Systems-(ITSC), pp. 243–248.

Cascetta, E., 1984. Estimation of trip matrices from traffic counts and survey data: A generalized least squares estimator. *Transportation Research Part B: Methodological* 18 (4–5), 289–299.

Chen, J., Bierlaire, M., 2015. Probabilistic multimodal map matching with rich smartphone data. *Journal of Intelligent Transportation Systems* 19 (2), 134–148.

Danalet, A., Bierlaire, M., 2014. A path choice approach to activity modeling with a pedestrian case study. In: 14th Swiss Transport Research Conference. No. EPFL-CONF-199478.

Danalet, A., Bierlaire, M., 2015. Importance sampling for activity path choice. In: 15th Swiss Transport Research Conference. No. EPFL-CONF-207458.

Danalet, A., Farooq, B., Bierlaire, M., 2013. Towards an activity-based model for pedestrian facilities. In: 13th Swiss Transport Research Conference. No. EPFL-CONF-186042.

Danalet, A., Farooq, B., Bierlaire, M., 2014. A bayesian approach to detect pedestrian destination-sequences from wifi signatures. *Transportation Research Part C: Emerging Technologies* 44, 146–170.

Dijkstra, J., Timmermans, H., Jessurun, J., 2014. Modeling planned and unplanned store visits within a framework for pedestrian movement simulation. *Transportation Research Procedia* 2, 559–566.

Dijkstra, J., Timmermans, H., de Vries, B., 2009. Modeling impulse and non-impulse store choice processes in a multi-agent simulation of pedestrian activity in shopping environments. In *Pedestrian behavior: Models, data collection and applications*. Bingley, Yorkshire: Emerald Group Publishing Limited, pp. 63–85.

Flötteröd, G., Bierlaire, M., 2013. Metropolis–hastings sampling of paths. *Transportation Research Part B: Methodological* 48, 53–66.

Fruin, J., 1970. Designing for pedestrians. A level of service concept. Polytechnical Institute of Brooklyn, Ph.D. thesis, Ph. D.

Hänseler, F., Molyneaux, N., Bierlaire, M., 2017. Estimation of pedestrian origin–destination demand in train stations. *Transportation Science* 51 (3), 981–997.

Hänseler, F. S., Bierlaire, M., Farooq, B., Mühlematter, T., 2014. A macroscopic loading model for time-varying pedestrian flows in public walking areas. *Transportation Research Part B: Methodological* 69, 60–80.

Hänseler, F. S., Bierlaire, M., Scarinci, R., 2016. Assessing the usage and level-of-service of pedestrian facilities in train stations: A Swiss case study. *Transportation Research Part A: Policy and Practice* 89, 106–123.

Helbing, D., Molnar, P., 1995. Social force model for pedestrian dynamics. *Physical Review E* 51 (5), 4282.

Hoogendoorn, S., Bovy, P., 2000. Gas-kinetic modeling and simulation of pedestrian flows. *Transportation Research Record: Journal of the Transportation Research Board* 1710, 28–36.

Hoogendoorn, S., Daamen, W., 2004. Design assessment of Lisbon transfer stations using microscopic pedestrian simulation. *WIT Transactions on the Built Environment* 74, 135–147.

Hoogendoorn, S. P., Bovy, P. H., 2004. Pedestrian route-choice and activity scheduling theory and models. *Transportation Research Part B: Methodological* 38 (2), 169–190.

Jabari, S. E., Zheng, J., Liu, H. X., 2014. A probabilistic stationary speed–density relation based on Newell's simplified car-following model. *Transportation Research Part B: Methodological* 68, 205–223.

Jeanbart, C., Molyneaux, N., Scarinci, R., Bierlaire, M., 2018. Network design of moving walkways in transportation hubs. In: *Proceedings of the 18th Swiss Transport Research Conference, Ascona, Switzerland.*

Molyneaux, N., Hänseler, F., Bierlaire, M., 2014. Modeling of train-induced pedestrian flows in railway stations. *STRC Proceedings.*

Molyneaux, N., Scarinci, R., Bierlaire, M., 2017. Pedestrian management strategies for improving flow dynamics in transportation hubs. In: *17th Swiss Transport Research Conference (STRC).* No. EPFL-TALK-229186.

Molyneaux, N., Scarinci, R., Bierlaire, M., 2018a. Controlling pedestrian flows using a dynamic traffic management system. In: *Proceedings of the 7th International Symposium on Dynamic Traffic Assignment: Smart Transportation, Hong Kong, China.*

Molyneaux, N., Scarinci, R., Bierlaire, M., 2018b. Two management strategies for improving passenger transfer experience in train stations. In: *Proceedings of the 18th Swiss Transport Research Conference, Ascona, Switzerland.*

Nikolic, M., Bierlaire, M., 2017. Data-driven spatio-temporal discretization for pedestrian flow characterization. *Transportation Research Part C: Emerging Technologies* 94, 185–202.

Nikolic, M., Bierlaire, M., de Lapparent, M., Scarinci, R., 2019. Multi-class speed-density relationship for pedestrian traffic. *Transportation Science* (Accepted for publication) 53 (3), 642–664.

Nikolic, M., Bierlaire, M., Farooq, B., de Lapparent, M., 2016. Probabilistic speed–density relationship for pedestrian traffic. *Transportation Research Part B: Methodological* 89, 58–81.

Ochieng, W. Y., Quddus, M., Noland, R. B., 2003. Map-matching in complex urban road networks. *Revista Brasileira De Cartografia* 2, 55.

Oyama, Y., Hato, E., 2016. Pedestrian activity model based on implicit path enumeration. In: *21st International Conference of Hong Kong for Transportation Studies (HKSTS)*, pp. 331–338.

Oyama, Y., Hato, E., 2018. Link-based measurement model to estimate route choice parameters in urban pedestrian networks. *Transportation Research Part C: Emerging Technologies* 93, 62–78.

Papageorgiou, M., Hadj-Salem, H., Blosseville, J.-M., 1991. ALINEA: A local feedback control law for on-ramp metering. *Transportation Research Record* 1320 (1), 58–67.

Robin, T., Antonini, G., Bierlaire, M., Cruz, J., 2009. Specification, estimation and validation of a pedestrian walking behavior model. *Transportation Research Part B: Methodological* 43 (1), 36–56.

Scarinci, R., Markov, I., Bierlaire, M., 2017. Network design of a transport system based on accelerating moving walkways. *Transportation Research Part C: Emerging Technologies* 80, 310–328.
Steffen, B., Seyfried, A., 2010. Methods for measuring pedestrian density, flow, speed and direction with minimal scatter. *Physica A: Statistical Mechanics and Its Applications* 389 (9), 1902–1910.
Timmermans, H., 2009. *Pedestrian behavior: Models, data collection and applications.* London: Emerald Group Publishing.
Tregenza, P., 1976. *The design of interior circulation.* London: Crosby Lockwood Staples.
Van den Heuvel, J., Hoogenraad, J., 2014. Monitoring the performance of the pedestrian transfer function of train stations using automatic fare collection data. *Transportation Research Procedia* 2, 642–650.
Wang, H., Ni, D., Chen, Q.-Y., Li, J., 2013. Stochastic modeling of the equilibrium speed–density relationship. *Journal of Advanced Transportation* 47 (1), 126–150.
Weidmann, U., 1992. Transporttechnik der fussgänger. Schriftenreihe/Institut für Verkehrsplanung, Transporttechnik, Strassen-und Eisenbahnbau 90.
Zhang, Y., Su, R., Zhang, Y., 2017. A macroscopic propagation model for bidirectional pedestrian flows on signalized crosswalks. In: 2017 IEEE 56th Annual Conference on Decision and Control (CDC), pp. 6289–6294.
Zhang, Z., Jia, L., Qin, Y., 2016. Level-of-service based hierarchical feedback control method of network-wide pedestrian flow. *Mathematical Problems in Engineering.*

SECTION III

Commutes, modes and rhythms

As we begin the journey into the empirical realm of what may constitute a research field of urban mobilities, we encounter the mundane and ordinary practices of everyday life. The research into urban mobilities is an empirical endeavour and pivots around the ordinary spaces and sites of repetitive practice. Millions of people are living their lives on the move in and between cities every day, and this mundane realisation is the backdrop for the third section on commutes, modes and rhythms. Section III opens with Murray's **Chapter 11** on the walking commute and how this is significantly both gendered and generational. The walking commute is produced through the intersections of gender and generation and is gendered and generationed as the mobilities of particular social groups, most notably women, children and older people, experience it in particular ways and are consequently disadvantaged. The key argument in this chapter is that the urban walking commute needs to be redefined to move away from the unproblematic promotion of walking as a 'sustainable' form of travel. Viewing the commute in relation to gender and generation reveals the persistent barriers to walking; the urban walking commute is illustrated as an interdependent and intersected urban mobility practice. In **Chapter 12**, Weert Canzler critically discusses the future of the car commute. Commuting is one of the most crucial challenges for the transport systems of urban regions. The future of the car commute is open. One scenario sees a dystopia of full automated personal transport – science fiction-like autonomous vehicles are always on the move, carrying people from A to B. The other scenario could be called integrated intermodal e-mobility: All means of transport in cities are to be electric and operated on a completely renewable energy basis – private availability is generally unnecessary. In the same way that one navigates the Internet via a browser, new services synthesise the various devices and equipment into one continuous journey. The everyday mobility perspective is shifted to that of the bicycle with **Chapter 13**, which features Jonas Larsen on the ups and downs of urban cycling. The chapter explores the dramatic history of mundane urban cycling. The first section documents the overwhelming fall of, and defection from, bike practices worldwide throughout the twentieth century, while the second part provides ethnographic evidence of the distinction of being an urban cyclist and the differences between cycling in a pro-cycling city such as Copenhagen and low-cycling cities such as London and New York. Finally, the third part investigates the quiet retaliation of bike cultures on the streets of Western cities and in urban planning and mobility research. In **Chapter 14**, Hanne Louise Jensen turns to trains and the train commute. The central purpose of this chapter is to demonstrate that by the time the train commuter arrives in the city, ready to work, many meaningful mobile practices have taken place and, in many cases, these practices have become important in the train commuter's everyday life. By focusing on mobile practices inside the train, this chapter moves beyond the functional logic of the train to explore how the particularity of the train and its recurrent rhythms participate in shaping its emotional and social spaces. Robin Kellerman then turns the attention to the immobile practice of waiting and how significant it is to mobility in **Chapter 15**. Pedestrian mobility is at the centre of urban transportation systems. Therefore, understanding it is a necessary task to improve the mobility in cities. However, pedestrian mobility is extremely complex, and advanced methods are needed to capture its dynamics. This chapter presents the main methods used to describe, model and control pedestrian behaviour. The chapter introduces the fundamental relationships describing pedestrian dynamics, models representing the demand and assignment of pedestrians and, to conclude, innovative management strategies. The theme of the immobile is continued in **Chapter 16** when Julie Cidell explores the pause and pausing as an important dimension of mobility. The chapter explores the relationship between the mobile and immobile by looking at pausing. Accordingly, there are three types of pausing in the context of urban areas: the delay, which is an unwanted stoppage; the respite, which is a desired break in movement; and intermission, which is more neutral and scheduled in nature. Pauses all imply a short time span. Mobility will resume soon, and travellers and their belongings will reach their final destinations. This sets the concept of the pause aside from the broader category of waiting, where there might not be an intention to move, or even the ability to do so. The chapter discusses the concepts of delay, respite and intermission in more detail. Following this, it discusses the ways in which places and pauses (of all three types) construct each other, whether spaces designed for waiting or those that become spaces of pausing by default. Furthermore, the chapter explains how the intersection of different speeds and rhythms require or encourage pausing as part of the overall choreography of people, objects and spaces. Here, it addresses in particular the uneven mobilities that arise when some people have to wait for others. Section three ends with Katrine Hartmann-Petersen's **Chapter 17**, which focuses on working rhythms. Understanding urban life through rhythmicity is one way of analysing the pulsating mobile patterns of urbanity. Everyday life and urban living are polyrhythmic – a melting pot of rhythms. Some of these rhythms are caused by mobilities, and some mobilities are consequences of rhythms. The chapter therefore argues that to understand mobilities, it is important to understand how multiple rhythms influence the experience of living and working in the city. This is illustrated by the mobile life of bus drivers in Copenhagen, Denmark. In relation to such theoretical understanding and empirical focus, the chapter discusses what can be learned from such a perspective of rhythms and how these perspectives can improve urban mobilities planning.

11
THE WALKING COMMUTE
Gendered and generationed

Lesley Murray

Introduction

> People are travelling between home and a workplace less often and are more often 'trip-chaining' where people combine two or more trips for differing purposes, such as dropping-off children at school on the way to work.
>
> *(UK National Travel Survey, Department for Transport 2018, 22)*

The walking commute, as journey between home and employment or education, is changing significantly. As well the recognition of 'trip-chaining' the UK's National Travel Survey also ascribes the decline of the 'traditional' commute to the increase in workers without a fixed place of work and the growing levels of home-working. The survey also found an overall decrease in commuting, but in particular in the walking (as well as car) commute, with more people in England commuting by public transport. The walking commute is also unevenly spread across urban areas. It is, understandably given the varying density of the urban form, most common in inner urban areas and least common in outer urban (ibid.). Intersecting with his spatial variation, the walking commute is also socially uneven – particular social groups are more and less likely to commute on foot and the experiences of this mobility practice are highly differentiated. This chapter is concerned with the way in which the walking commute is both gendered and generationed. The commute has most often been referred to as the journey between home and a place of work as discrete social spaces that have particular and distinctive functions. But for the majority of the population, it is not this, but rather a complex and interconnected set of social, temporal and spatial mobile practices.

The walking commute is produced through the intersections of gender and generation. It is also gendered and generationed as the mobilities of particular social groups, most notably women, children and older people experience this form of commuting in particular ways and are consequently disadvantaged. The focus on gender and generation here is not disregarding of race, disability, ethnicity, sexuality and class etc. Indeed, it is recognised that gender and generation intersect with these. But it is nevertheless necessary to focus on particular social groups in order to understand their particular needs, as mobilities scholarship has done so with regard to gender (Priya Uteng and Cresswell 2008), generation (Murray and Robertson 2016a), disability (Parent 2016) and race (Nicolson 2016). The key argument in this chapter is that although walking is encouraged as a 'sustainable' and 'active' form of travel, and despite efforts to revitalise walking in cities, there remain persistent barriers to unproblematic walking and this creates injustice as those with no option are required to traverse the city using pavements that are invariably obstacle-laden, polluted, poorly lit; and are forced to take circuitous routes around cities that prioritise the radial travel of cars and heavy vehicles. This chapter looks at the ways in which through attending to gender and generation, we can challenge established thinking on the walking commute that invisibilises particular experiences. The walking commute is illustrated as an interdependent and intersected urban mobility practice.

The gendering and generationing of the walking commute

As mentioned, like all aspects of urban mobilities, the walking commute is bound up in gender and generation in that it is differentially 'staged' (Jensen 2013) and experienced in ways that exclude and marginalise according to gender and generation. Gender and generation are determined through spatial, social and cultural contexts and practices. The urban walking commute is a mobility practice that is determined by gender and generation (and their intersectionality) and at the same time gender and generation are created through a range of mobilities including the walking commute, albeit that this requires a redefining of this practice, as I will discuss later. Firstly, however, it is useful to consider the concepts of gender and generation in mobility terms, not to review the work being carried out in relation to these concepts (see for example Murray and Robertson 2016a; Priya Uteng and Cresswell 2008), but rather to think about the ways in which they help us understand the changes in the urban walking commute.

There has been some focus in mobilities on the gendering of mobilities (Grieco and McQuaid 2012; Priya Uteng and Cresswell 2008), most recently in relation to mobility justice (Sheller 2018). This work has been mindful of the concept of gender as non-binary so that there is a range of complex relationships that make mobilities. With this in mind, and following on from my previous work, this chapter focuses mainly on one aspect of gender and that is the particular mobilities of women, which are often constrained by the positioning of women in social and mobile space (Law 1999). This is rooted in the historical associations and current practices of childcare and domestic labour, but also the control of women in public spaces (Massey 1994; McDowell 1999). In general, women still tend to do the majority of childcare and domestic labour and so they experience particular 'fixity constraints' (Kwan 2015) that are premised on the relationship between home and work, where work includes domestic labour. Gendered travel is a product of urban form, as low-density developments and urban sprawl means that those with temporal constraints or fixities are less able to access employment opportunities. Numerous studies have found that women experience temporal and spatial constraints due to caring roles, predominantly for children, but also for other people who are dependent on them, including older people. On the other hand, it is argued that the persistence of a gendered division of domestic labour means that compact developments are more likely to promote gender equality in relation to access to employment (Lo and Houston 2018).

Hence, the urban commute remains highly gendered in that in the UK, as in other western countries, men commute farther than women due to their child-caring responsibilities (McQuaid and Chen 2012). The gap between women and men in this regard may be narrowing, but this is happening slowly (Crane 2007). At the same time, in a similar vein to the social world as gendered, classed and raced, it is also generationed. As sociologist Leena Alanen (2010, 9) argues, in the 'system of social ordering' certain generational categories are dominated by others; that there is a 'generational ordering', which is reproduced in mobility practices. There has been some attendance to children's and older people's mobilities – both are invariably left behind in a world of acceleration and speed. The differential mobilities of children and older people also highlights the relationality of age and the wider conceptualisations of generation (Murray and Cortés-Morales 2019). For focusing on generation also allows us to understand changes in the walking commute in a historical perspective. Different generations have experienced this mobility practice in different ways and have pulled these experiences through to different temporal frames – thus the practices and processes of generation create particular mobility practices.

It is the intersection of gender and generation that produces the differing patterns of the walking commute between men and women. Evidence shows that women have different commuting patterns to men in most national contexts (Boarnet and Hsu 2015; Cristaldi 2005; Roberts et al. 2011; Sánchez and Gonzále 2016) and this has been the case over generations. Women's spatial range is smaller, they are less likely engage in work-only related travel and more likely to make multiple trips in one outing: 'trip chaining'. Women's complex mobilities are compounded through intersections with other social categories and identities and with place. For example, McLafferty and Preston (1991) found that black and Hispanic women commute further than white men and women. Women and children are more likely to commute by walking (Department for Transport 2016), to both employment and school. This is a significant aspect of everyday mobilities; indeed, 20 per cent of all walking trips in England were for educational purposes in 2017 (including escorting). Travels to school and workplace are often linked as mothers drop their children off to school on the way to their workplace. The organisation of the commute often involves a complex set of negotiations and adaptations (Murray 2008; Murray and Doughty 2016). These complex mobilities are often the rationale for travelling by car as it is a faster way of moving between multiple destinations and parents, particularly mothers, are more likely to be time-poor. Of course, as discussed later, not everyone has access to a car and so these journeys become particularly difficult when walking is the only option. When children travel to school, without being accompanied by an adult, they are most likely to walk (DfT 2018). Again, this is not always through choice.

The walking commute in urban time and space

Historically, women and children have always had a turbulent association with walking in urban public spaces, both invisibilised and suppressed (Schmucki 2012). From pre-Enlightenment to post-industrialisation, different social groups are marked out in public space according to their mobility practices. Experiences are always intersectional; for example, class and race have throughout history and in different places determined where and when the gendered and generationed body could walk. In recent history, the walking commute is perhaps not entirely as expected. Pooley and Turnbull (1999) found that from the late nineteenth century to the 1930s, women's overall commute was around the same distance as men's. At that time, many men travelled to work by bike so although they had more flexibility than women, their journey was not necessarily longer. Before 1930, walking had been a more important means of travel for both men and women, but was from then on overtaken by car travel. Women at that time were more dependent on walking and therefore travelled at much slower speeds than men. It is really at this stage that the differential commuting mobilities of men and women became marked and travelling more slowly impacted negatively on women's experiences and employment opportunities as they had to juggle childcare and household responsibilities.

There have been changes too in children's commuting patterns with an overall decline in walking, albeit with walking remaining the most practised method of commuting. In their oral history study of mobility trends over time (1940s to 2000s), Pooley et al. (2005) found that the decline in walking to school is partly due to the changes in parental perceptions of harm, the complexities of everyday lives and time poverty. Nevertheless, they found that changes over time were minimal and there is generally continuity in mobility patterns rather than the stark changes that underpin discourse of protectionism. These are in part relatable to the generational rememberings of childhood that underscore many studies of children's mobilities (Murray and Cortés-Morales 2019). There is often an association of walking with past freedoms in public space. Green (2009, 25) captures this as:

> The rather romantic and idealised turn to much of the literature mourning the loss of walking in urban environments, assuming universal health and social benefits of a past golden age in which children walked to school, neighbours greeted each other while walking to local shops and the city was accessible to all.

Although there is also evidence to suggest that the impact of protectionism on children's mobilities is evidence across socio-economic groups, Markovitch and Lucas (2011, 24) argue that middle class parents are more likely to exhibit 'behaviour aversion' as 'part of the "bubble wrap generation" than children living in lower income household'. They argue that 'there is also an important gender and inter-generational component to this type of behaviour aversion, in relation to the adults responsible for chauffeuring children' (Markovitch and Lucas 2011, 26).

Of course, walking has many advantages as a mode of travel, not least because, along with cycling, it is the least variable and therefore most reliable form of transport in terms of journey time (DfT 2016). As an 'active' form of travel it is considered to be free of the precarities of machinery and technologies. Walking is also considered to demand little in terms of infrastructure. These assumptions, however, as discussed later, are based on limited experiences of particular people. Beyond its value as a mode of transport, there is a wealth of literature that advances walking as a healthy form of travel, associated with a reduced incidence of a number of physical problems such as obesity, diabetes, cancer, osteoporosis, cardiovascular diseases, cholesterol level, musculoskeletal problems and psychological wellbeing. In addition, walking for commuting is considered to hold therapeutic and spiritual benefits (Gatrell 2013; Guell et al. 2012). Walking as an 'active' form of commuting is considered to have specific health benefits for older people and children (Bopp et al 2014). The therapeutic benefits of the walking commute for children are especially evident when walking is a social activity, a 'walking bus' and this increases the proportion of children commuting (Gatrell 2013). However, this organised and social form of walking is more likely to be available in affluent areas and so the benefits of walking 'in its very slowness' (Gatrell 2013, 102) are limited by class.

Walking the commute is also considered to be beneficial to cities – 'active' forms of travel are promoted not only for their health benefits but as sustainable modes that neither disgorge air-polluting emissions nor consume the limited physical space of cities. Promoting walking in cities is seen to make them 'liveable', creating cities of bodies rather than cities of machines. Hence, making provisions for walking is central to policies such as 'New Urbanism', 'Healthy-Active City', 'Smart Growth', which have gained much traction in urban policy. There are also global urban policies that are aimed specifically at improving the urban form for older people – 'age-friendly cities' and for children – the 'children-friendly city initiative', but their impact has been limited (Murray 2015). There appears to be little joined-up thinking in determining what makes a city friendly for all generations. Such thinking might attend more carefully to the need to slow down. As Moran et al. (2016, 57) suggest:

> Pedestrians move relatively slow in space, while being open to absorb impressions from the environment. Therefore, they are likely to prefer diverse and complex environments, including multiple buildings of various types and diverse urban design elements (e.g., trees, benches, billboards).

The slowness of walking creates particular sensory experiences and this varies with both generation (Murray and Järviluoma 2019) and gender. In addition, studies have found that children's walking is determined by the level of connectivity and accessibility of streets, the quality of walking infrastructure and the availability of green space for walking. Urban forms that are more compact are unsurprisingly more walkable.

Redefining the walking commute

As discussed, the traditional definition of the walking commute is changing in response to variations in mobility practices. These are themselves a product of changes in employment patterns and the recognition of interdependent mobility practices, which are both connected to gender and generation. However, what is considered to be a divergence from accepted descriptions of the walking commute remains entangled with normative understandings of walking and of commuting, and these understandings give rise to gendered and generationed mobilities. The walking commute, as an interdependent mobility practice, goes beyond the traditional notions of 'walking' and of 'commuting'. Walking in urban areas is more traditionally associated with being outdoors, walking along a pavement or through an urban park. The commute, as a journey from a place of working or of education to a place of 'home' has changed. Looking through a mobilities lens means that both of these concepts have taken on different, more expanded, meanings in contemporary society. In conceptualising the walking commute in a way that is meaningful in understanding its relationship to broader mobility practices, it is useful to look critically at the concepts of walking and of commuting. Attending to gendered and generationed aspects of walking can help highlight some of the ways in which established ways of considering walking can be challenged.

Firstly, walking, as discussed, is predominantly considered to be a means of commuting that is healthy for both people and cities. One of the most poignant studies of gender and generational walking was carried out a number of years ago by Bostock (2001), who set out to challenge prevailing approaches to carlessness and health, which failed to understand the experiences of walking as a mobility practice that has both positive and negative health impacts. Her study of low-income young mothers found that lack of access to private and public forms of automobilisation, which a significant proportion of the population take for granted, gave rise to social and psychological distress. Bostock argues that walking can be emotionally draining if it is the only option available. Often compounded by poor physical environments, the young mothers in her study encountered a number of problems while walking. These included fatigue and stress, negative psycho-social effects of looking after fatigued children, and restrictions to limited geographical areas lacking shops, services and social resources. Access to public transport for all the mothers was limited due to high fares and the prioritisation of other resource demands. As Bostock (ibid., 16) found: 'mothers used their bodies as a means to bridge the gap between responsibilities and resources'. There has been little comparable research since this study, and instead the emphasis has remained on walking as a pathway to healthier lifestyles.

Yet policies to promote walking remain at the centre of strategies aimed at transforming our urban spaces and making them more liveable as walking is unproblematically considered to bring benefits to the health of the walker as well as wider sustainability benefits. The concept of 'walkability' is one of the basic principles of urban policies 'for people' such as New Urbanism. This is important, given the ways in which, for decades, the automobile has shaped the urban form, creating infrastructure that has divided communities and denied access to goods and services for those without a car. So it is not the advent of policies that promote walking per se that is at issue, but rather, as Bostock illuminated, the failure to acknowledge that walking, too, can be excluding. The uncritical advancement of walking (and cycling) as 'active' forms of travel, particularly in times of austerity with piecemeal urban improvement schemes, means that, again, particular interests are privileged over others. For example as Murray and Robertson (2016b) show in their study of 'shared space', a design approach that aims to promote urban streets 'for people', the space is used differentially according to gender and generation. Thus, for those who are not engaged in 'smart' urban development, there are many barriers to walking and walking can create experiences of unliveability. The experience of walking in the city for certain groups of people is far from liveable. Levels of walkability have been found to be associated with 'high-walkable' areas that are in middle class areas. Older people, children and women can be disadvantaged by infrastructure design (Hine 2011). For example, older people find difficulties walking on uneven pavements, hills and ramps, and with traffic and crossing roads. The urban form creates exclusions and injustice in that many opportunities are simply not accessible by walking alone and public transport can sometimes be prohibitively expensive, especially for children in low income families (Mackett and Thoreau 2015). Levels of walkability are marked by social orderings. As Carpenter (2013, 125) argues, we need to think about walking as 'embodied and embedded' socially and culturally and work from a broader frame that incorporates prevailing 'professional, ideological and political agendas'.

However, there is a wealth of evidence to suggest that walking is experienced in very different ways by different social groups. Mackett and Thoreau (2015) found that although walking is beneficial for health, older people are more likely than others to experience embodied limits to walking, both outdoors and indoors, especially in navigating steps and stairs. Children also experience walking in divergent ways. In a study of inequalities in transport in the USA, Sanchez et al. (2003) found evidence that children who could not afford to travel on public transport missed school, especially in winter, due to the length of the walk to school. In cities where the urban form is particularly skewed towards car travel, and where personal safety in public space is a significant issue, particular groups, especially non-whites, were particularly susceptible due to their lack of access to car travel and the need to walk as the only available option. Similarly, in a study of the health benefits of walking in New Zealand, Baig et al. (2009) found that more girls than boys and white than non-white people walked, and that although walking was associated with health benefits for older people it could also be exhausting for this generation. In cities in the global south walking is often the only option, for children in particular, yet Porter et al. (Porter et al. 2011) found that many experienced physical and mental distress on the commute to school. Of course, 'walking' is also often 'wheeling' and studies have demonstrated clearly the limits of the urban form to seamless movement in a wheelchair (Parent 2016) and with a children's pushchair (Cortés-Morales and Christensen 2014).

The second, and under-researched, aspect of the urban 'commute' that is often overlooked relates to the indoor urban spaces that are often overlooked in mobilities studies. The commute goes beyond the outdoor space that is between a place of work or education and the home. The indoor urban spaces are highly relevant to mobilities as not only are they an integral aspect of cities, but they are spaces in which an increasing proportion of the commute is carried out and more workers remain at home to work. They commute between the micro spaces of the home and beyond to virtual spaces in which they connect with colleagues who are located all over the world. The commute no longer begins at the front door. The 'work' aspect of the commute is not always another place, and to classify it as such denies the myriad walking trips of carers, for children, older people, disabled people and people who are ill, in their own homes. The commute is not always carried out twice a day, as also demonstrated by these carers, as they move to and fro, from their 'home' to their 'work', which happen to be in the same building. The indoor places of the urban are often overlooked in urban and mobilities studies. The urban is not only the outdoor spaces of walking but also the indoor micro spaces of the home. Looking at the commute through gendered and generational lens means appreciating the significance of these spaces in people's everyday lives. This means looking more closely at home-working, which is more likely to be practised by women. The commute is also a term applicable to people working outside traditional employment, such as those retired from work who travel to care, volunteer and 'work' outside of formal employment. These micro mobilities are relational to mobilities at wider scales and their neglect means that we are missing the full picture. The walking commute goes beyond 'traditional' conceptualisations and attending to these will produce understandings that are applicable to all aspects of the urban commute.

Conclusion

As I have discussed, the walking commute is undergoing changes and these are gendered and generational. Some of these changes are acknowledged in contemporary studies of transport and mobilities, but there are significant aspects of the walking commute that are obscured by traditional approaches. A mobilities approach allows us to focus on the varying scales of the walking commute, from the micro to the macro and all that is in between. In doing so, we can appreciate the interdependencies between different groups of people and priorities of urban policies that promote walkability without considering how cities both deter walking for a significant proportion of their populations and induce a range of differential experiences so that the opportunities available in cities are not afforded to many. These micro scales of mobilities and their relationalities require further study. In addition, the urban form has not kept pace with changes in social, economic and mobility practices. In order to do so, the walking commute should be understood as not only gendered and generationed but intersectional. This means understanding the ways in which intergenerational knowledge is produced and maintains particular mobility discourses and also understanding how approaching aspects of mobilities like the walking commute can help challenge some of the prevailing approaches.

References

Alanen, L., 2010. Taking children's rights seriously. *Childhood* 17 (1), pp. 5–8.

Baig, F., Hameed, M. A. L., Shorthouse, M., Roalfe, G. and Dale A., A. K., 2009. Association between active commuting to school, weight and physical activity status in ethnically diverse adolescents predominantly living in deprived communities. *Public Health* 123, pp. 39–41.

Boarnet, M. and Hsu, H.-P., 2015. The gender gap in non-work travel: the relative roles of income earning potential and land use. *Journal of Urban Economics* 86, pp. 111–127.

Bopp, M., Der Ananian, C. and Campbell, M., 2014. Differences in active commuting among younger and older adults. *Human Kinetic Journal* 22 (2), pp. 199–211.

Bostock, L., 2001. Pathways of disadvantage? walking as a mode of transport among low-income mothers. *Health and Social Care in the Community* 9, pp. 11–18.

Carpenter, M., 2013. From 'healthful exercise' to 'nature on prescription': the politics of urban green spaces and walking for health. *Landscape and Urban Planning* 118, pp. 120–127.

Cortés-Morales, S. and Christensen, P., 2014. Unfolding the pushchair: children's mobilities and everyday technologies. *Research on Education and Media* 6 (2), pp. 9–18.

Crane, R., 2007. Is there a quiet revolution in women's travel? revisiting the gender gap in commuting. *Journal of the American Planning Association* 73 (3), pp. 298–316.

Cristaldi, F., 2005. Commuting and gender in Italy: a methodological issue. *The Professional Geographer* 57 (2), pp. 268–284.

Department for Transport, 2016. *Commuting trends in England 1988–2015*. London: Department of Transport.

Department for Transport, 2018. *National travel survey: statistical release*. London: Department for Transport.

Gatrell, A. C., 2013. Therapeutic mobilities: walking and 'steps' to wellbeing and health. *Health and Place* 22, pp. 98–106.

Green, J., 2009. 'Walk this way': public health and the social organization of walking. *Social Theory and Health* 7, pp. 20–38.

Grieco, M. and McQuaid, R., 2012. Gender and transport: an editorial introduction. *Research in Transportation Economics* 34 (1), pp. 1–2.

Guell, C., Panter, J., Jones, N. R. and Ogilvie, D., 2012. Towards a differentiated understanding of active travel behaviour: using social theory to explore everyday commuting. *Social Science and Medicine* 75, pp. 233–239.

Hine, J., 2011. Mobility and transport disadvantage. In J. Urry and Grieco, M. (eds.) *Mobilities: new perspectives on transport and society*. London: Routledge, 21–40.

Jensen, O. B., 2013. *Staging mobilities*. London: Routledge.

Kwan, M.-P., 2015. Gender, the home–work link, and space–time patterns of nonemployment activities. *Economic Geography* 75 (4), pp. 370–394.

Law, R., 1999. Beyond 'women and transport': towards new geographies of gender and daily mobility. *Progress in Human Geography* 23, pp. 567–588.

Lo, W.-T. and Houston, D., 2018. How do compact, accessible, and walkable communities promote gender equality in spatial behavior? *Journal of Transport Geography* 68, pp. 42–54.

McDowell, L., 1999. *Gender, identity and place: understanding feminist geographies*. Cambridge: Polity Press.

Mackett, R. L. and Thoreau, R., 2015. Transport, social exclusion and health. *Journal of Transport and Health* 2 (4), pp. 610–617.

McLafferty, S. and Preston, V., 1991. Gender, race and commuting amongst service sector workers. *The Professional Geographer* 43 (1), pp. 1–15.

McQuaid, R. and Chen, T., 2012. Commuting times: the role of gender, children and part-time work. *Research in Transportation Economics* 34 (1), pp. 66–73.

Markovitch, J. and Lucas, K., 2011. *The social and distributional impacts of transport: a literature review*. Oxford: Transport Studies Unit.

Massey, D., 1994. *Space, place and gender*. Cambridge: Polity Press.

Murray, L., 2008. Motherhood, risk and everyday mobilities. In T. Priya Uteng and Cresswell, T. (eds.) *Gendered mobilities*. Aldershot, Hampshire: Ashgate, 47–64.

Murray, L., 2015. Age-friendly mobilities: a transdisciplinary and intergenerational perspective. *Journal of Transport and Health* 2 (2), pp. 302–307.

Murray, L. and Cortés-Morales, S., 2019. *Children's mobilities: interdependent, imagined, relational*. London: Palgrave Macmillan.

Murray, L. and Doughty, K., 2016. Interdependent, imagined and embodied mobilities in mobile social space: disruptions in 'normality', 'habit' and 'routine'. *Journal of Transport Geography* 55, pp. 72–82.

Murray, L. and Järviluoma, H., 2019. Walking as transgenerational methodology. *Qualitative Research*. Published online doi:10.1177/1468794119830533

Murray, L. and Robertson, S. (eds.) 2016a. *Intergenerational mobilities*. London: Routledge.

Murray, L. and Robertson, S., 2016b. Sharing mobile space across generations. In L. Murray and Robertson, S. (eds.) *Intergenerational mobilities*. London: Routledge, 91–104.

Nicolson, J., 2016. 'Don't shoot': black mobilities in American gunscapes. *Mobilities* 11 (4), pp. 553–563.

Parent, L., 2016. The wheeling interview: mobile methods and disability. *Mobilities* 11 (4), pp. 521–532.

Pooley, C. and Turnbull, J., 1999. The journey to work: a century of change. *Area* 31 (3), pp. 281–292.

Pooley, C., Turnbull, J. and Adams, M., 2005. The journey to school in Britain since the 1940s: continuity and change. *Area* 37 (1), pp. 43–53.

Porter, G., Hampshire, K., Abane, A., Robson, E., Munthali, A., Mashiri, M., Tanle, A., Maponya, G. and Dube, S., 2011. Perspectives on young people's daily mobility, transport and service access in sub-saharan Africa. In J. Urry and Grieco, M. (eds.) *Mobilities: new perspectives on transport and society*. London: Routledge, 65–90.

Priya Uteng, T. and Cresswell, T. (eds.) 2008. *Gendered mobilities*. Aldershot, Hampshire: Ashgate.

Roberts, J., Hodgson, R. and Dolan, P., 2011. 'It's driving her mad': gender differences in the effects of commuting on psychological health. *Journal of Health Economics* 30 (5), pp. 1064–1076.

Sánchez, I. O. and Gonzále, E. M., 2016. gender differences in commuting behavior: women's greater sensitivity. *Transportation Research Procedia* 18, pp. 66–72.

Sanchez, T. W., Stolz, R. and Ma, J. S., 2003. *Moving to equity: addressing inequitable effects of transportation policies on minorities*. Los Angeles, CA: UCLA.

Schmucki, B., 2012. "If I walked on my own at night I stuck to well lit areas": gendered spaces and urban transport in 20th century Britain. *Research in Transportation Economics* 34 (1), pp. 74–85.

Sheller, M., 2018. *Mobility justice: the politics of movement in an age of extremes*. London and New Yok: Verso.

12
THE FUTURE OF THE CAR COMMUTE

Weert Canzler

Commuting: the phenomenon and problem

In all countries that experienced early motorisation, commuting to work is a major issue. This form of commuting is defined by the fact that the commuter crosses the border of his/her municipality of residence. The proportion of commuters in the workforce has been increasing for years, as has the average commuting distance. In Germany, for example, 60 per cent of all standard employees (i.e. those subject to social insurance contributions) commuted in 2015, compared to 53 per cent in 2000. The average commuting distance is 17 kilometres, 2 kilometres more than 15 years earlier (BBSR, 2017). At the same time, time series show that the proportion of long-distance commuters has remained fairly constant. In the USA, the proportion of commuters who travelled longer than 60 minutes was 8.1 per cent in 2011, only 0.1 per cent higher than in 2000. The proportion of commuters with a journey time of 90 minutes even fell slightly, from 2.8 to 2.5 per cent, during this period (see McKenzie, 2013). There is also a growing constancy in travel times among commuters with shorter journey times.

The vast majority of commuter journeys are made by car. For decades, the car has dominated everyday commuter traffic, and images of congested streets during rush hour are familiar to everyone. In the USA, more than 85 per cent of all commuter journeys have been car journeys since the beginning of the 1980s; incidentally, only a tenth of these involve carpooling. Three out of four commuters drive alone in their cars (cf. McKenzie, 2015). In Europe, too – with a few exceptions such as Amsterdam, Vienna or Utrecht – the car is also the ubiquitous means of transport for commuters. In 2016, 68 per cent of all commuters in Germany used a car, while 14 per cent used public transport (Statistisches Bundesamt, 2017). The situation is similar in other European countries: Even in the Netherlands, the land of the bicycle, 77 per cent of all commuter journeys are made by car (Netherland Statistics, 2016).

The problematic aspects of commuting by car are a constant topic of public debate. In particular, the environmental impact of pollutant emissions and infrastructure congestion are perceived as serious environmental and transport policy problems. Children, older people and residents on heavily trafficked roads suffer particularly from the poor air quality and noise. But commuters themselves experience stress, which has also been the subject of health research for some time (see Koslowsky, Kluger and Reich, 1995). More recently, the link between daily commuting and an increased risk to physical and mental health has been an important subject for public health research, also with respect to gender and educational background (see, for example, Hansson et al., 2011; Feng and Boyle, 2013). However, the problems of commuting are increasingly being considered in a comprehensive context. The downsides of car traffic and thus also of commuting are more often being linked to fundamental discussions on how the city of tomorrow should be organised and what significance the private car can still have in it. Hence, commuting has become part of a broader discussion on the future of transport and the role of the (private) car in it.

The results of the car society

The issue, then, is the dominance of the car and the consequences of mass motorisation. Motorisation was closely linked to standardised mass production. The concept of Fordism was not accidentally derived from Henry Ford, who not only mass produced the first car, but also set the standards for distribution and sales. This was also aided by the unusually good wages Ford paid by the standards of the period. The idea was that employees should be able to afford the cars they manufactured.

In comparison to the railway, which was the dominant mode of transport in the USA at the time, the Ford model T offered its owners much more scope for individuality. People were no longer restricted to rigid train lines and could depart independently of the timetable and take breaks. In this respect, the car was a strong symbol of a new independence in transport. But there was more to industrial modernity than this, namely the prospect of automotive independence for all. This also implied a reduction in the barrier of geographical distance. Exploring new regions, travelling to distant lands, owning a family home with a garden: these were some of the promises associated with the transition from a railway to a car society. However, first, the enabling conditions for this had to be created. Above all, the infrastructure had to be put in place; roads, bridges and tunnels had to be built. The filling station network, repair shops and an insurance system also developed gradually (cf. Flink, 1975). Cities, in particular, were rebuilt around the car; in everyday urban and increasingly suburban life, the private car was given a privileged position (see various articles in Wachs and Crawford, 1991).

The entire functional framework for mass motorisation corresponded to the pattern of industrial modernity. From the setting up of new car manufacturing companies and the construction and operation of the infrastructure to the establishment of filling station chains and shopping centres, everything was subject to the rules of standardisation and economies of scale. The idealised images of a progressive automotive future in films and the media were also subject to the dictates of standardisation. This pattern remained stable for several decades, and the social consensus about it was almost universal (see Urry, 2007). Only when mass motorisation threatened to take over, when blatant safety deficiencies became apparent and air pollution could no longer be denied did the first fine cracks appear in the seemingly unshakable consensus (cf. Rothschild, 1973). Criticism was almost unheard of for decades. However, there was no renaissance of the railway or other collective means of transport. The pre-modern transport system was not attractive. For its part, after the initial uncertainty, the car industry took advantage of the crises to innovate in order to adapt; it even succeeded in expanding the standardised range of products it had offered over decades by diversifying its models and thus in creating new sales markets (cf. Canzler, 2016, pp. 73f.). Ultimately, however, the now mature car society remained on its well-trodden path, with neither the car's driveline nor its usage forms being subject to radical innovation. Even when the contours of a "second modern age" had long since been delineated, the tried and tested recipe remained the same in the industry shaped by industrial modernity. The car remained the same as it had been for a century, but it was constantly getting bigger, heavier and more expensive. The combustion engine as the technical core also remained the unchallenged standard. All that was individualised was the fixtures and the colour palette.

From a car society to a multimodality society

In an everyday context, transport behaviour is strongly influenced by routines. Routines have the advantage of relieving the pressure to make decisions (cf. Giddens, 1984). Therefore, transport options that can be used routinely have an advantage. This especially applies to the private car as long as finding parking is not too unpredictable or car use is not otherwise restricted. A car can be used "without a second thought", it is a place to store personal belongings, it reflects the owner's standards regarding cleanliness and fittings and it offers its own protection and privacy. Despite these qualities of the private car, which facilitate its routine use, individual transport needs may be better satisfied by other modes of transport. This is the case when owning a car becomes a burden because the places where various activities – even everyday activities – take place are easier, faster or more predictable to reach by public transport or bicycle. This is also the case if an individual sees transport as more than just a means of getting from A to B – i.e. as a way of getting physical activity while walking or cycling.

However, this kind of non-car-fixated transport behaviour is only realistic if there are sufficient alternatives in the first place. Multimodality presupposes the willingness to combine not only different modes of transport, but also sufficiently attractive alternatives. These include classic bus and train connections as well as sharing options of various kinds. In large cities and densely populated regions, this can often be assumed to exist. The device that makes intermodal and multimodal linking easier and routine is the smartphone. Intermodal apps facilitate access to various transport options in real time, even if a real integration of all potential services with the option of booking tickets or of using the app itself as a ticket equivalent has still not occurred. Digitalisation thus offers unprecedented opportunities to routinely use intermodal transport services beyond the private car (cf. Canzler and Knie, 2016).

Is the smartphone the "master key for an intermodal transport service" that the late modern city dweller can use in place of the private car? If the communication and media sociology research is to be believed, the special features of the smartphone allow this. A central, albeit quite general, thesis is: "Since mobile communication changes the social interaction situations of media users, it also transforms people's experiences, individual identity processes and societal socialisation conditions" (Wimmer and Hartmann, 2014, p. 13). The prevalence of smartphones is now almost total and their use in younger age groups is taken for granted.

This high prevalence of smartphones and the permanent online mode for flexibly organising everyday activities that is self-evident for digital natives suggests that simple, reliable digital transport services that span all modes of transport are likely to be used routinely. Individualised profiles even facilitate route chains with different modes of transport. We cannot expect the smartphone to become the "Car 2.0", but it will nevertheless be able to become at least roughly the functional equivalent of a private car. This does not mean that the smartphone will be a substitute for physical transport. However, it expands the space of possibilities enjoyed by each individual, both in terms of changing real physical location and accessing and efficiently combining different transport services. For example, ride sharing and ride pooling options that can be booked via easily accessible platforms and in real-time communication can become attractive. As a result, long-existing standard transport services will be replaced by individualised and yet low-transaction-cost combinations of very different transport options – including ride-sharing options that were previously considered private. In place of the long-dominant model of private car use, a smartphone-supported plural mobility could emerge. This has long been described by John Urry as "mobilities" (cf. Urry, 2007).

The smartphone enables people to dive into their digital sphere and radically reduce their perception of external stimuli, even if they are physically in one specific place together with specific other people. This phenomenon can often be observed in public transport, where passengers sit alone by themselves with their smartphones – with or without earphones – submerged in a virtual, subjective and also partially isolated parallel world. This leads to similar levels of isolation as people may experience in their own cars, even when they are in a vehicle together with others (cf. Canzler and Knie, 2016).

Ambivalences and need for regulation

While the smartphone benefits from and simultaneously reinforces the underlying trend towards individualisation, it also reflects its ambivalence. Isolation in one's own virtual world is one such example. On the one hand, this form of isolation in virtualisation can lend the experience of being and travelling in public and collectively used means of transport a similar quality as car transport. This may be attractive and desirable for many commuters or in certain situations. On the other hand, it threatens to devalue public space and, in the worst case, to lead to an eerie atomisation of users in public transport. This is to say nothing of the fact that spontaneous and potentially interesting contacts may not be able to occur at all.

The rise of the smartphone with the associated forms of interaction has not been uniform, despite the high speed of diffusion to date. There is a simultaneity of non-simultaneity, because the differences between generations and milieus are still great. Even if social exclusion is generally declining as digitalisation progresses, it remains to be seen who and how many people have neither the cognitive abilities nor the communicative skills to cope with the demands of a digitally based transport landscape. The specific ability that ultimately also determines participation in society is captured by the term "motility" (cf. Kaufmann, Viry and Widmer, 2010). It is possible that a "motility gap" will emerge that is associated with new tendencies towards social division. These are closely linked to the ambivalences of individualisation identified some time ago (cf. Beck, 1986).

After all, there is the danger of falling for a simple technological optimism when presented with the much-vaunted opportunities of digitisation and the image of the smartphone as the "key to a post-automobile transport future". What will happen if we fail to guarantee data security and rule out a potentially devastating hacking of digitally based transport services? The unintended side effects of new basic technical and social innovations cannot be assessed, let alone controlled. The opportunities, but also the risks, of digitisation are only beginning to emerge. Future regulation must take into account these imponderables and anticipate possible negative consequences of digitisation in transport. Yet not everything can be anticipated; many things only become visible when you try them out. This is another reason why first gaining experience in "regulatory experimentation areas" and then thinking about suitable regulation may be a reasonable strategy.

In driveline technology, there are also omens of disruption. For reasons of climate protection alone, the burning of fossil fuels must cease; neither coal and gas for power generation nor the use of mineral oil in engines are suitable for the future. Their substitution by biogenic or electricity-based fuels only makes sense where electric propulsion would be almost impossible, as in air transport. This is because the regenerative input energy required for this is several times higher than for direct use (see ICCT, 2018). The electric engine has considerable advantages over the traditional internal combustion engine. Due to its design it is much more efficient than its competitor, it is quiet, emits at least no pollutants locally and is also subject to little wear and tear. Yet it is still unclear what the relationship between the two electric drive variants – either directly powered by a battery or indirectly via a chemical process in a fuel cell – will look like in the future. Similarly, it is currently not possible to predict what role hybrid drives will play. Lastly, it is not possible to conclusively determine whether hybrid drive variants will be used as a transitional technology or whether they will form an independent drive and vehicle segment. But one thing can be regarded as certain: the trend is towards electrification. Additional support comes from the fact that industrial and environmental policy decisions have already been made in China, but also in many other countries and metropolises worldwide. For instance, India and many European countries have announced that they will no longer register vehicles with internal combustion engines from 2030 or a few years later.

However, electrification should not simply be equated with replacing the drive unit. This would be problematic in several ways. Electric cars would need the same space as combustion engine vehicles, and if they were used in the same way as before, they would also be stuck in traffic jams. In addition, battery and fuel cell production would require the use of resources, particularly limited and therefore more expensive materials, on a scale that would be hardly economically and ecologically feasible. Moreover, there would be much greater demand for electricity.

A new non-uniformity

For the transport system of the future, electrification is a necessary but not a sufficient condition. It must also be efficient and, above all, meet the requirement of fairness of space in cities. This means that car traffic in particular will have to make do with less space. The solutions are: networking and sharing. Here, digitisation can open up previously unimagined possibilities (see Canzler and Knie, 2016). It should be understood as a double digitisation, on the one hand, as an automation of processes in the background such as traffic control and on the other hand as a personal digitisation via smartphones and apps. Personal digitisation will likely undermine the current division of labour in the automotive society. It will expedite the next push towards individualisation because it supports individual lifestyles and ways of working in ways never before seen. If you look at the transport behaviour of urban digital natives, you will notice that they are very pragmatic in their use of apps as soon as they appear useful. They appreciate access to cars, but are often indifferent to the model and the brand. Ownership is becoming unimportant and often even annoying. Availability is everything.

Accessibility, real-time information and the transparency of transport services have become taken for granted by many digitally socialised people. Yet the narrative behind it is abstract; it embraces flexible and diverse transport options and avoids standard solutions. It is thus in harmony with the future world of transport. It will be networked several times, namely, first, between different transport modes and options and, second, with the electricity sector. This can be called "networked e-mobility". Or "mobility-as-a-service" (Hietanen, 2014). The future digital transport world will be characterised by the fact that the focus is not on individual transport technologies, but rather on rights and opportunities to use them. Technology will increasingly recede into the background. The car will also lose its special position, becoming a link in a supply chain. The future world of transport will be non-uniform and very diverse.

Transport behaviour will also be non-uniform; for example, it will vary considerably in the various phases of life. This divided picture is already apparent today, and its main features may also define the future world of electric and digital transport.

Carless independence is particularly appreciated by people in education and training. Often, such people do not aim to own a car in the first place, because the possibility of moving away from their current home, the desire for a high degree of locational flexibility and not least the expectations of the peer group make multimodal transport behaviour much more a more obvious choice. It is not just that the proportion of young adults in higher education has increased massively in all countries experiencing early motorisation. At the same time, there has been an increase in the number of people who have experienced multimodal metropolises at some stage, either before, during or after their studies. Someone spending a limited period of time in New York, London or Vilnius would never think of driving there in his or her own car or even buying one there. Also, a student would hardly be able or willing to pay for a private car and possibly a parking space in addition to the costs of their apartment in Barcelona, Maastricht or Vienna. The charm of many university cities lies in the fact that you can get along well in them without a car – by bike or even on foot in campus universities. There are also cheap public transport tickets for many students.

However, after a student or trainee has graduated, things look different. The professional and family formation phase often approximates the traditional "family-home-car model" very closely. The "structural stories" associated with this (Freudendahl-Pedersen, 2005) are still effective. Even though it has become chic in some urban circles to use carrier bikes and bike trailers to transport even small children, for many, travelling the complex route chains between their home, the crèche, their workplace and the supermarket without their own car is nearly inconceivable. This is especially true when it is not possible to organise the various activities in relative proximity to each other.

In retirement, when the children are "out of the house" and daily trips to work are no longer necessary, new freedoms can open up. The pedelec boom is also largely attributable to this age group. However, given that members of this group are financially strong, they are also important buyers of new cars, often with a high seating position and ample safety equipment. SUVs and their derivatives are particularly popular with financially secure older people. At the same time, the gap between the sexes has narrowed compared to earlier generations. Currently, almost as many women over 60–65 have a driving licence as men. Different phases of life are accompanied by different mobility styles, which in turn differ for different milieus and social strata (cf. Cass and Manderscheid, 2018).

The picture is also non-uniform in spatial terms. For instance, dense urban spaces differ greatly from suburban or even rural spaces. In cities, there is great variety; there are different transport options, different transport modes with their strengths and weaknesses. There is actually something to network. Strictly speaking, this applies to the core cities, where space is limited and the distances to be bridged are usually not so long. It is possible to sketch out an inter- and multimodal urban transport system beyond the "car-friendly city" (cf. e.g. Canzler and Knie, 2015). It would be characterised by fast and efficient public transport, bicycles and a wide range of sharing services. The smartphone makes linking different modes of transport and using sharing services simple and reliable. There are already many attractive services, ranging from car-, bike- and scooter-sharing services to various mobility apps and ridesharing. Everyday travel without the private car is easily possible on a routine basis; it is already a matter of course for many city dwellers today. In cities such as Berlin, Vienna or Zurich, this modal split has long since been influenced by the so-called environmental alliance, i.e. the alliance of public transport, walking, cycling and sharing services. The private car is a minority form of transport in many core cities (cf. Eurostat, 2018a).

In the suburbs, by contrast, the settlement and supply structures have, for decades, been determined by and for the private car. The vast majority of journeys are made by private car, and journeys across municipal boundaries – which, besides leisure and shopping trips, are primarily commuting journeys – are for the most part made by car. The proportion of commuters from the suburban settlement belts to the urban centres has hardly changed over many years, but the commuting distances have grown continuously (cf. Eurostat, 2018b). Because the private car dominates everything, and because bus and train services are often only used by students and the poor, the modal share is extremely unbalanced here. An integrated transport service is unrealistic simply because there is hardly anything to integrate. If, for topographical reasons or because there was just no promotion of cycling for a long period of time, the proportion of bicycle journeys is only marginal, the hegemony of the car is unrestricted.

At present, intermodal apps are also of no significance in these suburban areas and there are hardly any sharing services. People there also do not cycle much; they mostly do so on weekends when the weather is fine. At the same time, even in suburban areas and small towns the limits of private car traffic are being reached. Because driving space for private cars cannot be expanded at will, the search for new options and alternative transport services is slowly beginning. Car pooling platforms or even transfer points such as park-and-ride areas are becoming an issue, albeit a marginal one. Carpools for commuters, for example, are nothing new, but so far they have not come out of their niche. E-vehicles are also pretty rare.

At the moment, there is little sign of the transport transition in the suburbs. But that may change. The potential is there and the external pressure – even if it is not evident at present – may develop quickly. Impending driving bans for pollutant-emitting combustion vehicles seeking to enter the city centre and or city tolling would dramatically change the situation. It is not just that e-vehicles would suddenly become attractive to many commuters; sharing services based on profile and evaluation routines that are familiar from social networks would also become more attractive. The speed at which such changes can take place can currently be observed in the case of cycle expressways. They enable commuters to cover moderate or longer distances by bicycle. There are a lot of hopes linked to these expressways because they permit safe and fast cycling, not least with pedelecs, and at the same time accommodate the societal trend towards sporty and health-promoting activities. Experiences from the Netherlands and Denmark have further fuelled this hope.

Against this backdrop, we can expect the pragmatic use of the car to also grow in suburban and small-town areas. As options for moderate distances, ride-sharing platforms, park-and-ride hubs and pedelecs are increasingly becoming alternatives to monomodal car use. The reason for this lies not only in the availability of new services, but also in an attempt to avoid restrictions on private cars in urban areas.

On the precipice of a structural change

Like any structural change, the transport transition is also triggering counterreactions. Many people socialised with cars find it painful to lose traditional privileges; for some it is an actual loss. For example, it is often regarded as a customary right to park a private vehicle in public space for free or for a symbolic price. This is demonstrated by the countless disputes surrounding the elimination of free parking due to the introduction of car parking management zones. There, the emphasis is on the protection of the status quo. A climate of outrage spreads as car owners are "asked to pay" without justification. Frequently, when public parking spaces are scrapped or when charges are introduced for them, for instance, when parking spaces in front of an individual's house or apartment are converted into something else, this is experienced as an insult.

The desired world of networked e-mobility in which owning a private car is obsolete is an empty promise for many die-hard motorists. Many car commuters see no alternative in public transport and in new, networked mobility services, despite congestion and loss of time. However, there is a chicken-and-egg problem here: as long as there are no attractive – i.e. reliable, simple and cost-effective – mobility services, the car will remain dominant. Conversely, as long as the private car dominates public space and enjoys its traditional privileges, there will be neither the space nor the attractive services necessary for sharing and individualised public transport.

The question of who may use public (transport) space and how is a crucial one, a fact that is currently being demonstrated by the debate on the promotion of cycling. If bicycle traffic increases and safe cycle paths are designated, this is generally at the expense of car lanes or parking areas. As a result, the scarce urban space is used differently. The negative reaction to such a redistribution of urban transport space often comes from a noisy minority of car users. It is often interpreted in the municipal and transport policy discussion by the majority parties and public officials as a sign that the broad acceptance that is necessary for the transport transition is lacking. Car owners, so the thinking goes, have long had the red carpet rolled out before them; they should not be expected to change too much. Above all, commuters are cited as potential losers in a transport transition. For transport policy reasons, it is therefore particularly important to actually create good intermodal alternatives to the automobile so that people can thus experience them. Only in this way can we overcome the chicken-and-egg dilemma of traditional car dominance and the lost opportunities of new mobility services.

References

BBSR, 2017. *Immer mehr Menschen pendeln zur Arbeit*. Bonn. Available at: www.bbsr.bund.de/BBSR/DE/Home/Topthemen/2017-pendeln.html [Accessed 24.11.2019].

Beck, U., 1986. *Risk Society. Towards a New Modernity*. London: Sage.

Canzler, W., 2016. *Automobil und moderne Gesellschaft. Beiträge zur sozialwissenschaftlichen Mobilitätsforschung*. Berlin: LIT.

Canzler, W. and Knie, A., 2015. Changes in technologies to meet emerging urban mobility patterns, in: European Parliament, Directorate-General for Internal Policies. *Research TRAN Committee – The World Is Changing. Transport, Too*. Brussels 2016, pp. 53–76.

Canzler, W. and Knie, A., 2016. Mobility in the age of digital modernity: why the private car is losing its significance, intermodal transport is winning and why digitalisation is the key, in: *Applied Mobilities*, Vol. 1. doi:10.1080/23800127.2016.1147781

Cass, N. and Manderscheid, K., 2018. The autonomobility system: mobility justice and freedom under sustainability, in: Cook, N. and Butz, D. eds., *Mobilities, Mobility Justice and Social Justice*. London and New York: Routledge, pp. 101–115.

Eurostat, 2018a. *Urban Europe — Statistics on Cities, Towns and Suburbs — Working in Cities.* Available at: https://ec.europa.eu/eurostat/statistics-explained/index.php?title=Urban_Europe_%E2%80%94_statistics_on_cities,_towns_and_suburbs_%E2%80%94_working_in_cities#Commuter_flows [Acessed 24.11.2019].
Eurostat, 2018b. *Statistics on Commuting Patterns at Regional Level.* Available at: https://ec.europa.eu/eurostat/statistics-explained/index.php/Statistics_on_commuting_patterns_at_regional_level [Accessed 24.11.2019].
Feng, Z. and Boyle, P., 2013. Do long journeys to work have adverse effects on mental health? *Environment and Behavior.* doi:10.1177%2F0013916512472053
Flink, J. J., 1975. *The Car Culture.* Cambridge, MA: MIT Press.
Freudendahl-Pedersen, M., 2005. Structural stories, mobility and (un)freedom, in: Thomsen, T. U., Nielsen, L. D. and Gudmundsson, H. eds., *Social Perspectives on Mobility.* Aldershot: Ashgate, pp. 29–45.
Giddens, A., 1984. *The Constitution of Society.* Cambridge: Polity Press.
Hansson, E., Mattisson, K., Björk, J., Östergren, P.-O. and Jakobsson, K., 2011. Relationship between commuting and health outcomes in a cross-sectional population survey in southern Sweden. *BMC Public Health,* Vol. 11, 834.
Hietanen, S., 2014. Mobility as a service: the new transport model? *Eurotransport,* Vol. 12, 2–4.
ICCT (International Council of Clean Technologies), 2018. *Decarbonization Potential of Electrofuels in the European Union.* Brussels et al. Avalable at: www.theicct.org/sites/default/files/publications/Electrofuels_Decarbonization_EU_20180920.pdf [Accessed 24.11.2019].
Kaufmann, V., Viry, G. and Widmer, E. D., 2010. Motility, in: Schneider, N. F. and Collet, B. eds., *Mobile Living across Europe II.* Opladen and Farmington Hills, MI: Barbara Budrich Publishers, pp. 95–112.
Koslowsky, M., Kluger, A. and Reich, M., 1995. *Commuting Stress: Causes, Effects, and Methods of Coping.* New York: Plenum Publishing.
McKenzie, B., 2013. *Out-of-State and Long Commutes: 2011.* Available at: www2.census.gov/library/publications/2013/acs/acs-20.pdf [Accessed 24.11.2019].
McKenzie, B., 2015. *Who Drives to Work? Commuting by Automobile in the United States: 2013. American Community Survey Reports.* Available at: www.census.gov/content/dam/Census/library/publications/2015/acs/acs-32.pdf [Accessed 24.11.2019].
Netherland Statistics, 2016. *Transport and Mobility 2016.* Available at: www.cbs.nl/en-gb/publication/2016/25/transport-and-mobility-2016 [Accessed 24.11.2019].
Rothschild, E., 1973. *Paradise Lost: The Decline of the Auto-Industrial Age.* New York: Random House.
Statistisches Bundesamt, 2017. *14 Prozent der Beschäftigten nutzen des öffentlichen Verkehr.* Availbale at: www.destatis.de/DE/PresseService/Presse/Pressemitteilungen/2017/08/PD17_288_12211pdf.pdf?__blob=publicationFile [Accessed 24.11.2019].
Urry, J., 2007. *Mobilities.* Cambridge: Polity Press.
Wachs, M. and Crawford, M. eds., 1991. *The Car and the City: The Automobile, the Built Environment, and Daily Urban Life.* Ann Arbor, MI: University of Michigan Press.
Wimmer, J. and Hartmann, M., 2014. Mobilität und Mobilisierung, in: Wimmer, J. and Hartmann, M. eds., *Medienkommunikation in Bewegung. Mobilisierung – Mobile Medien – Kommunikative Mobilität.* Wiesbaden: Springer, pp. 11–30.

13

UPS AND DOWNS WITH URBAN CYCLING

Jonas Larsen

Introduction

There are close ties between cities and bikes. Most cycling takes place in cities, where their compactness and illuminated environments make them ideal for cycling. City-dwellers are often more pro-cycling and less likely to own or drive cars than the rest of the population. This chapter explores the dramatic history of mundane urban cycling. The first section documents the overwhelming fall of, and defection from, bike practices around the world throughout the twentieth century. I explore how forces of modern planning, 'splintering urbanism', and emerging car cultures fatally undermined cycling as a viable and dominating mode of transport in the city. Bikes belonged to the past; motorised movement was the regime of the future. The second part gives ethnographic evidence of the distinction of being an urban cyclist, and the differences between cycling in a pro-cycling city such as Copenhagen and low-cycling cities such as London and New York. The third part investigates the quiet retaliation of bike cultures on the streets of western cities and in urban planning and mobility research. I discuss why urban planners and politicians suddenly venerate, promote, and plan for this old technology and urban practice in very specific ways. I show that this veneration is a tool, serving political goals and having little to do with the lived life of being an urban cyclist. I end by arguing that bike infrastructures (bike lanes and public city bikes) are important for, but do not guarantee, the popularisation of cycling. The making of a high-volume cycling city is much more complex and multifaceted.

Cycling in the pre-automobile city

The history of urban cycling is inextricably linked to the history of competing mobilities. A central tenet of practice theory is that practices compete with other practices for enticing practitioners (Larsen, 2017). Somewhat relatedly, theories of socio-technical change highlight that power relations between mobilities are not static, as once-dominant regimes may be challenged and eventually dethroned by niche innovations if they manage to build up strong internal momentum and external support (Geels, 2012).

This was the case with bicycles, which in the early part of the twentieth century stole an extreme number of practitioners from horse-riding, trams, trains, and walking. Bikes outpaced walking with the same exertion and they were silent, manure-free, and offered easy door-to-door access. In a short span of time, commuters defected from their habitual mobility practices and embraced bicycles' speed, flexibility, and modernity. By 1897, there were some five million urban cyclists in the US. Chicago had as many cyclists as Copenhagen or Amsterdam have today (Henderson, 2017: 224). During the early years of the twentieth century, the bike became the dominant commuting regime in many cities. At the time, many cities were 'bike-friendly' in the sense that they were 'compact' and 'integrated', so commuting distances were relatively short. Cycling outcompeted walking and public transport with regard to speed and flexibility, respectively. Moreover, it was relatively safe to ride a bike. During the 1940s, up to 85 per cent of trips in many European cities were by bike (Shove et al., 2012: 61). I always begin my lecture on urban cycling in Copenhagen by showing a photograph from the 1930s. The photograph shows a horde of Copenhageners commuting home from work on a summer day on the bridge, Dronning Louises, that connects a working-class neighbourhood with the city centre (Nørrebro). Dronning Louises Bro – and the street Nørrebrogade – are now famous for being a car-free (buses are allowed) bridge with very wide bike lanes and benches with 'hipsters' enjoying the spectacle of the many cyclists swirling by. Yet this contemporary rush is nothing compared to the 1930s, when cycling was the strongest regime, as around 90 per cent of the Copenhageners supposedly commuted by bike as was the case in many other cities (Carstensen and Ebert, 2012).

The picture shows that both men and women cycled in big numbers and in their everyday/work clothing. Helmets are nowhere to be seen. Most dramatically, the photograph conveys the feeling that cyclists were 'kings of the roads'; they used the entire width of the paved road as they cycled in layers of up to six people, at ease with each other. There is no car or bus on the horizon. This photograph debunks the prevalent idea that roads were built for cars. This is clearly not the case in many European cities, as most roads were 'laid out' (although not necessarily paved) well before cars began to colonise the streets from the 1950s onwards (earlier in the UK and US). Cyclists were the first to campaign for road improvements like properly paved roads and car drivers were seen as interlopers, not cyclists. As MP Sir Ernest Soares told the British Parliament in 1903: 'Motorists are in the position of statutory trespassers on the road … roads were never made for motor-cars. Those who designed them and laid them out never thought of motor-cars' (cited in Guardian 2014, www.theguardian.com/books/2014/dec/23/roads-were-not-built-for-cars-carlton-reid-review). Yet this street hierarchy soon reversed as car drivers 'flexed their muscles' on the streets and lobbied that cyclists could no longer occupy the middle of the road. They should instead stick close to the 'gutter', or in the bike lane, so that they did not interrupt, or slow down, the normal speed and rhythm of a car. In this perspective, we can understand bike lanes as a defiant response to the new street hierarchy where cars suddenly ruled.

Modern urban planning also supported the dethroning of cyclists. The old compact city that suited cycling was 'spread out' through a ring of residential suburbs, industrial areas, and shopping districts, all linked together by new roads designed for cars. This increased everyday travel distances considerably and biking could no longer compete with the speed and comfort of cars or trains (Dennis and Urry, 2009; Graham and Marvin, 2002; Vivanco, 2013). Moreover, as cars pushed bikes 'into the gutter' cycling became dangerous and intimidating. Biking was no longer safe, desirable, or doable. The fall of the bike regime was indeed brutal. For instance, the number of UK bicycle commuters fell from 40 per cent in the 1940s to a few per cent within a couple of decades (Shove et al., 2012: 71). The bike did not fit into a highly mobile age of great distances and motorised travel, where bodies are moved rather than moving, fuelled by oil rather than human energy.

Ethnographies have demonstrated the pains, risks, and physical intensities associated with cycling in car-dominated cities where 'cyclists are expected to perform largely in the same way as their motorised counterpart despite strikingly different affordances and possessing divergent capabilities' (Spinney, 2010: 114; see also Jones, 2005, 2012; Latham and Wood, 2015). Yet they constantly need to improvise rhythms and ways of cycling that contravene the Highway Code in order to stay safe. Riding in such a hostile and unpredictable sensory environment demands a high level of 'affective capacity'; that is, physical abilities, riding skills, a constant alertness, and nerves of steel (Jones, 2012).

Such 'affective capacity' is found amongst those subcultures that did not defect from cycling and instead developed distinctive meanings and competences. The fearless, playful, and aggressive bicycle messenger on brakeless fixies defined US urban cycling as 'edgy', 'cool', and full of adrenalin rushes (Fincham, 2006). In contrast to the US, inside the UK it is sporty 'Lycra clad warriors' that whizz through the city and traffic that make the urban cycling scene equally 'masculine' and pulse-pounding. This also meant that cyclists were stigmatised as urban outcasts, reckless risk-takers out-of-synch with the rest of traffic. Cycling was for the few and hardened, a serious lifestyle choice with a strong identity attached to it. All this alienated many would-be cyclists.

Or so it was until such cities and urban planning suddenly changed gear and became pro-cycling again (Vivanco, 2013). Much of this inspiration stems from the two famously pro-bike lane cities, Copenhagen and Amsterdam; 'London is going Dutch' and wishes to de-Lycrafy cycling (de Boer and Caprotti, 2017; Goodman et al., 2014) while New York is being 'Copenhagenised' and instructs 'cyclists to save their energy for the spinning classes' (Larsen, 2018).

Cycling in pro-cycling Copenhagen

How does it feel to cycle in a city where cycling has been a 'regime' for a couple of decades and the municipality wishes to be the 'best cycle city in the world'? Almost all Copenhageners own a bike and half of the population cycles more or less daily. The affective capacity required for cycling in Copenhagen is relatively low, as there are separated bike lanes throughout the city on the major roads and they are largely responsible for the fact that most cyclists feel safe. Distances are doable as Copenhagen is compact; the city was only modestly suburbanised and redesigned for cars in the 1960s and 1970s. Moreover, the city is flat and not plagued by extreme temperatures. All this makes Copenhagen suitable for cycling.

Planners and cyclists in Copenhagen state in unison that the bike lanes play a crucial role in 'enlisting' cyclists. Firstly, and most obviously, they make cycling safer as they protect cyclists by increasing visibility; cyclists do not need to rub shoulders with cars, trucks, and buses, or seek a reprieve on the pavement. They normalise commuter cycling by giving cyclists a legitimate space on busy commuter roads. This illustrates that planners see cycling as a serious mode of everyday transport and not merely as leisure touring, as used to be the case in many car-dominated societies.

Cycling is woven into the ecology of many everyday practices, and practitioners ride in their everyday clothing. Cycling makes affective and cultural sense to many Copenhageners; they have the necessary skills to ride safely and conveniently in this cycling-friendly environment where cycling is widely accepted as perfectly normal and respectable for all ages and social classes. There is no stigma attached to cycling and cyclists are not labelled as odd, reckless, or 'left wing'. In contrast to car-dominated societies, biking is primarily a neutral everyday practice rather than a political statement or identity marker. Few are cycling to save the environment, and many bike commuters are car owners. The relationship between cyclists and car drivers in Copenhagen is much less antagonistic than in car-dominated societies. Few Copenhageners are full-time cyclists. Many opt for an alternative if cycling becomes too slow, physically hard, or sensuously unpleasant. This is often the case when people need to travel more than 5 kilometres or to multiple destinations, if their children accompany them, or if the weather is hostile. This also explains why bike commuting is much less prevalent in the suburbs of Copenhagen where many have to commute long distances (Larsen, 2017; see also Heinen et al. (2010) for similar findings in other places).

This further illustrates that the affective and sensuous corporeal experience of cycling from A to B (the tenet of the 'mobilities paradigm'), and not only travel time (the typical argument of transport scholars), is vital for understanding how cycling enlists and loses practitioners in particular situations. The newly erected bike bridges in Copenhagen harbour illustrate this point: they perform 'magic' by concurrently reducing the travel time *and* improving the sensuous experience of biking – and in return the city. They afford a pleasant view of the city and it is appreciated by locals and tourists alike (Larsen, 2017). Yet transport planners and researchers are generally blind to the lived sensuous and aesthetic experiences of movement through a city (Simpson, 2017: 427) as their eye is on utilitarian transport (or commuting) and how it can be made faster and more effective.

Moreover, as I will now discuss, the rekindling of cycling has everything to do with the fact that politicians, planners, and scholars have realised that this old technology cheaply and effectively solves many unexpected problems that car dependency has inflicted on our cities and societies in general. Whether we like cycling or not, we *need* to cycle. Or so we are told by the powerful elite.

'Instrumental love affair'

Within mobilities literature it is now common parlance to highlight how car culture has come to dominate and shape cities according to their specific needs and affordances, and how they have created all sorts of urban problems in the process. Architect Rogers writes:

> It is the car which has played the critical role in undermining the cohesive social structure of the city … they have eroded the quality of public places and have encouraged urban sprawl … The car has made viable the whole concept of dividing everyday activities into compartments, segregating offices, shops and homes.
>
> *(cited in Dennis and Urry, 2009: 42)*

Modern architects planned for cars rather than for people, and they supported the destruction of 'liveable cities' with pedestrian 'life between buildings' (Gehl, 2011). Moreover, congestion, lack of space, air and noise pollution, and obesity personify car-dominated cities. Lastly, such cities are oil-dependent and contribute to mass CO_2 emission.

The political and scientific embrace of cycling must be understood in this post-car context where we are forced to change our mobility practices. Cycling is seen as a cheap way to solve car-inflicted problems and ensure a low-carbon liveable city. Copenhagen municipality writes:

> A bicycle-friendly city is a city with more space, less noise, cleaner air, healthier citizens, and a better economy. It's a city that is a nicer place to be in and where individuals have a higher quality of life … Fortunately it pays off to invest in cycling. Increased cycling levels give society less congestion, fewer sick days, longer life expectancy, less wear and tear on the roads, and less pollution. Cycling initiatives are also inexpensive compared with other transport investments.
>
> *(cited in Larsen, 2017: 880)*

Similarly, bike advocates campaign that bike cultures contribute 'to urban vitality, freedom, invigorated commercial districts, and sociability' (Stehlin, 2014: 22). This illustrates how states and organisations invest in and promote cycling to reduce commuter time, pollution, and obesity and to increase urban productivity, competitiveness, the health of workers, and ensure smooth mobility. In other words, promotion of cycling is a kind of Foucauldian 'dispositif' and neo-liberal mode of governing that is called for and invested in because it is *effective* for modern welfare/capitalist states (Spinney, 2016). The cyclist is the sought-after urban dweller: a person that is active, healthy, and productive on the one hand, and commutes in the desired politically correct fashion on the other hand. While the cyclist is lauded, the inner-city car driver is now considered immoral (Green et al., 2012). This biopolitical production means that we need to investigate the power relations and inequalities involved in urban cycling. Cycling scholars have so far struggled to discuss such issues, perhaps because most are unanimously pro-cycling and active cyclists themselves (Spinney, 2016).

Recent studies discuss how cycling is absorbed by capitalism and gentrifies neighbourhoods. Ironically, given that cycling was long stigmatised as the poor man's vehicle, the bike is now the vehicle of choice for young professionals and 'hipsters', and cycling is now widespread in gentrified and upmarket neighbourhoods (Henderson, 2017; Lam, 2017). As the New York bike author Bike Snob writes:

> One thing that's become an increasingly important part of gentrification, for better or for worse, is the bicycle. The hipster is a particular breed of person, and where there are hipsters, there are bicycles (usually but not always fixed gears). And a hipster on a bicycle can spread gentrification more quickly than a stiff wind can distribute a wind of ragweed pollen. Yes, hipsters on bicycles can cause entire cities to suffer from the itchy eyes and sneezing of trendiness.
>
> *(Snob, 2011: 121)*

Retail developers endorse cycling as bike-friendly neighbourhoods attract young professionals and prevent (often high-income) families from fleeing to the suburbs. In contrast, cycling is often much less prevalent in working-class neighbourhoods where car cultures have a stronger hold. Cycling is increasingly associated with well-educated middle-class people with white skin that live in expensive inner-city areas (Hoffman and Lugo, 2014; Larsen, 2018; Spinney, 2016; Stehlin, 2014).

Contrary to the public myth that all Copenhageners are avid riders, there are very few cyclists with 'dark skin' in this city. Copenhageners of African and Middle Eastern origin prefer walking, taking the bus, and not least driving; they are not buying into the idea that cycling is cool or pleasurable. Indeed, they pity those 'poor cyclists' (Larsen and Funk, 2018). This is also the case in London (Steinbach et al., 2011) and New York where white

Brits and Americans largely dominate the cycling scene – even in low-income and ethnically diverse neighbourhoods such as Bedford-Stuyvesant in Brooklyn (Noyes et al., 2014).

In addition to being predominantly white, New York cyclists tend to be slim despite the fact that more than half of adult New Yorkers are overweight or obese (www1.nyc.gov/site/doh/health/health-topics/obesity.page). While promoted to improve public health and combat obesity, cycling as it is currently construed does not attract many obese practitioners (see explanation below).

Ceteris paribus, a pro-cycling city will include and favour people that are able-bodied, youngish, fit, slim, and live centrally. Conversely, it will potentially have no, or even negative repercussions, for those that are disabled, unfit, obese, and elderly, and for whom cycling is not a realistic option. It may be too much of a physical challenge to ride a bike if you are not able-bodied or habituated to being transported, a form of pointless torture. These people might also be a little embarrassed to cycle in their neighbourhood, as one's body is very exposed when cycling. Nor will it help those that travel with children to different destinations or commute long distances because they cannot afford to live, or find work, in the city. We need to avoid a form of 'cycling fundamentalism' that unreflectively associates cyclists with sustainable living and scapegoats non-cyclists for being immoral (Cupples and Ridley, 2008).

I will now discuss why cycling fails to attract certain people that instead prefer to commute through other means. To understand this, we need to understand cycling as a uniquely corporeal and sensuous mode of movement that is highly moderated by not only the traffic, but also the contours and weather-worlds of a given city. Indeed, my argument will be that 'inexperienced' cyclists and habituated drivers and passengers will often decry the poor level of flexibility, comfort, and convenience that cycling affords. I agree with Dennis and Urry's 'anti-bike' statement: 'we do not think we can turn the clock back and that the individual freedom that the car provides is going to disappear. It is unlikely that everyone will be traveling on foot and on bike' (2009: 1).

The 'nature' of cycling

I will now discuss unique features that identify cycling from other forms of mobility with regards to embodiment as well as perception of the urban landscape and changing weather conditions. Here I follow calls for examining how 'the material, "elementary", molecular and physical aspects' of movement are significant to the affective and experiential qualities of being on the move' (Merriman cited in Simpson, 2018: 4).

For a start, we need to acknowledge that cycling is very different from motorised travel when it comes to the amount of physical work and energy that the practice demands. Whereas cars and trains transport human bodies as sedentary drivers or passengers, cycling reverses that: human bodies are the very motors that ensure the propulsion of this technology. This human-technology hybrid depends upon the energy input of human bodies. How fast, how long, and how often people can ride a bike ultimately depends upon their corporeal fitness and past experiences with cycling (and the quality of their bike). While the bike enables humans to move faster, longer, and easier than on feet, cycling resembles walking and especially running as it brings embodiment to the forefront. As Jones writes:

> Cycling is an embodied experience per se: While all of our activities as humans may be construed as in some ways embodied, there are times when this embodiment is more obvious. Cyclists are engaged in a physically demanding activity,

> exposed to the excesses of climate and always at risk of death and injury beneath the wheels of motorized road users.
>
> *(2005: 814)*

Being physically active means that people may arrive sweating and in need of a shower, and their legs may be sore in the morning. While cycling is potentially physically demanding in all environments, it is particularly so in hilly terrain and when faced by a stiff headwind. Drivers and passengers are largely oblivious to, and unaffected by, the wind direction and small gradients as they infer very little with their rhythmic movement (Nixon, 2012). Cyclists are *never* corporeally oblivious to the environment: 'the nature of cycling means that spaces otherwise rendered homogenous when travelling by car have widely varying characteristics' (Spinney, 2008: 29–30). They experience it as a 'taskscape' and they *feel* energy use (Nixon, 2012). Cyclists are constantly aware of, and sense, the environment and weather as they directly impact their pace and rhythm, or whether they need to push the pedals vigorously or can enjoy a moment of freewheeling (Fajans and Curry, 2001; Nixon, 2012; Spinney, 2010). Cycling in uneven terrain will always be an arrhythmic experience and the habitual commuter ride is always liable to the direction and the intensity of the wind on a particular day. Unlike the heated and air-conditioned comfort of cars and trains that shelter people from the 'excesses' of the weather-world, cyclists are more exposed to boiling hot summer days, freezing cold winter mornings, and thunderous rain throughout the year; they cannot avoid feverish sweat, painful ice-cold fingers, and being soaking wet. It is therefore no surprise that transport studies show that commuter cycling is sensitive to wind and rain and the seasonal rhythms of the year (Böcker et al., 2013).

A red light is annoying to both car drivers and cyclists as they kill momentum. Yet it requires much physical effort to get a bike up to speed again after a full stop, so cyclists are more negatively affected by them than car drivers. This explains 'why cyclists hate stop lights' and they routinely jump them to avoid an arrhythmic halt. Studies in both car-dominated and bike-friendly cities show that all sorts of cyclists often bend the traffic rules to conserve energy and sustain their flow (Fajans and Curry, 2001; Larsen, 2018; van Duppen and Spierings, 2013).

The fact that cycling is a physical activity where one is exposed to the environment and weather-world is both the thrill and pain of cycling. Cycling in headwind on flat terrain with the sun hitting one's back is doable and pleasurable to most, while a stiff headwind on a bitterly cold morning in a hilly city with voluminous clouds of exhaust fumes is off-putting for everybody except hardened practitioners. Here it does not matter much if there are bike lanes or not.

Conclusion

This chapter has explored ups and downs of urban cycling in relation to competing forms of mobility. I end by discussing design implications of the above analysis. Biking is widespread in cities where bike facilities are extensive. Despite these facilities, getting people to cycle more is not only a question of bike lanes as technologies do not work in isolation or guarantee the enticement of cyclists. Singular technologies and designs seldom change social practices, especially not with immediate effect: 'we need to be extremely wary of the dangers of assigning some simple causal or deterministic power to technology or infrastructure networks *per se*' (Graham and Marvin, 2002: 11). Instead, we need to understand cycling as a slowly moving 'socio-technical system' that involves many different people, practices, technologies, designs, urban environments, cultural meanings, and aspirations (Geels, 2012). In a review of the existing literature on bike lanes, Buehler and Dill conclude that 'it remains unclear to what extent bikeways entice individuals to cycle' (Buehler and Dill, 2016: 14). Studies from Johannesburg (Morgan, 2017) and Bogata (Vivanco, 2013: 96) show that newly built bike lanes do not necessarily entice new users if there is no established cycling culture in place, and potential cyclists fear that they are easy targets for criminal activity or stigmatisation. Viewed in this light, it is no coincidence that pro-cycling Copenhagen and Amsterdam are egalitarian cities with little crime and high levels of trust.

I have also argued that we need to pay attention to what cycling affords and requires of the practitioners. Cyclists are vulnerable, and they need some protection and separation from cars. Rather than just settle for bike lanes on car-dominated streets, a more progressive approach may involve reclaiming streets for cycling exclusively. Secondly, planners need to support the unique energy use of cycling by designing environments and laws that facilitate consistent momentum with a minimum of stops and sharp turns, which is seldom the case today. This can be done by building bridges, restricting car access, and reducing the amount of stop lights. More broadly, planners must acknowledge that compact and small/medium-sized cities where work, leisure, and home are close to each other and within cycling distance are well-suited for cycling whereas it is very hard to cycle in hilly and/or 'sprawled' cities.

References

Bike Snob. (2011). *Systematically & Mercilessly Realigning the World of Cycling*. San Francisco: Chronicle Books.

Böcker, L., Dijst, M., and Prillwitz, J. (2013). Impact of everyday weather on individual travel behaviours in perspective: a literature review. *Transport Reviews*, *33*, 71–91.

Buehler, R., and Dill, J. (2016). Bikeway networks: a review of effects on cycling. *Transport Reviews*, *36* (1), 9–27.

Carstensen, T. A., and Ebert, A. K. (2012). Cycling cultures in Northern Europe: from 'golden age' to 'renaissance'. In J. Parkin (ed.) *Cycling and Sustainability*. Emerald Bingley: Insight, pp. 23–58.

Cupples, J., and Ridley, E. (2008). Towards a heterogeneous environmental responsibility: sustainability and cycling fundamentalism. *Area*, *40*(2), 254–264.

de Boer, M. H. M., and Caprotti, F. (2017). Getting Londoners on two wheels: a comparative approach analysing London's potential pathways to a cycling transition. *Sustainable Cities and Society*, *32*, 613–626.

Dennis, K., and Urry, J. (2009). *After the Car*. Cambridge: Polity.

Fajans, J., and Curry, M. (2001). Why bicyclists hate stop signs. *ACCESS Magazine*, *1*(18), 28–31.

Fincham, B. (2006). Bicycle messengers and the road to freedom. *The Sociological Review*, *54*, 208–222.

Geels, F. W. (2012). A socio-technical analysis of low-carbon transitions: introducing the multi-level perspective into transport studies. *Journal of Transport Geography*, *24*, 471–482.

Gehl, J. (2011). *Life between Buildings: Using Public Space*. Washington: Island Press.

Goodman, A., Green, J., and Woodcock, J. (2014). The role of bicycle sharing systems in normalising the image of cycling: an observational study of London cyclists. *Journal of Transport & Health*, *1*(1), 5–8.

Graham, S., and Marvin, S. (2002). *Splintering Urbanism: Networked Infrastructures, Technological Mobilities and the Urban Condition*. London: Routledge.

Green, J., Steinbach, R., and Datta, J. (2012). The travelling citizen: emergent discourses of moral mobility in a study of cycling in London. *Sociology*, *46*(2), 272–289.

Heinen, E., van Wee, B., and Maat, K. (2010). Commuting by bicycle: an overview of the literature. *Transport Reviews*, *30*, 59–96.

Henderson, J. (2017). Book review essay: what is a bicycle city? *Applied Mobilities*, *2*(2), 223–227.

Hoffman, M. and Lugo, A. (2014). Who is "world class"? Transportation justice and bicycle policy. *Urbanities*, *1*, 45–61.

Jones, P. (2005). Performing the city: a body and a bicycle take on Birmingham, UK. *Social & Cultural Geography*, *6*, 813–830.
Jones, P. (2012). Sensory indiscipline and affect: a study of commuter cycling. *Social & Cultural Geography*, *13*, 645–658.
Lam, T. F. (2017). Hackney: a cycling borough for whom? *Applied Mobilities*, 1–18.
Larsen, J. (2017). The making of a pro-cycling city: social practices and bicycle mobilities. *Environment and Planning A*, *49*(4), 876–892.
Larsen, J. (2018). Commuting, exercise and sport: an ethnography of long-distance bike commuting. *Social & Cultural Geography*, *19*(1), 39–58.
Larsen, J., and Funk, O. (2018). Inhabiting infrastructures: the case of cycling in Copenhagen. In M. Freudendahl-Pedersen, K. Hartman-Petersen, and E. Fjalland (eds.) *Experiencing Networked Urban Mobilities*. London: Routledge, pp. 129–134.
Latham, A., and Wood, R. (2015). Inhabiting infrastructure: exploring the interactional spaces of urban cycling. *Environment and Planning A*, *47*(2), 300–319.
Morgan, N. (2017). Cycling infrastructure and the development of a bicycle commuting socio-technical system: the case of Johannesburg. *Applied Mobilities*, 1–18.
Nixon, D. V. (2012). A sense of momentum: mobility practices and dis/embodied landscapes of energy use. *Environment and Planning A*, *44*, 1661–1678.
Noyes, P., Fung, L., Lee, K. K., Grimshaw, V. E., Karpati, A., and DiGrande, L. (2014). Cycling in the city: an in-depth examination of bicycle lane use in a low-income urban neighborhood. *Journal of Physical Activity and Health*, *11*(1), 1–9.
Shove, E., Pantzar, M., and Watson, M. (2012). *The Dynamics of Social Practice: Everyday Life and How It Changes*. London: Sage.
Simpson, P. (2017). A sense of the cycling environment: felt experiences of infrastructure and atmospheres. *Environment and Planning A*, *49*, 426–447.
Simpson, P. (2018). Elemental mobilities: atmospheres, matter and cycling amid the weather-world. *Social & Cultural Geography*, 1–20.
Spinney, J. (2008). Cycling between the traffic: mobility, identity and space. *Urban Design Journal*, *108* (Autumn), 28–30.
Spinney, J. (2010). Improvising rhythms: re-reading urban time and space through everyday practices of cycling. In T. Edensor (ed.) *Geographies of Rhythm: Nature, Place, Mobilities and Bodies*. Aldershot: Ashgate, pp. 113–128.
Spinney, J. (2016). Fixing mobility in the neoliberal city: cycling policy and practice in London as a mode of political–economic and biopolitical governance. *Annals of the American Association of Geographers*, *106*(2), 450–458.
Stehlin, J. (2014). Regulating inclusion: spatial form, social process, and the normalization of cycling practice in the USA. *Mobilities*, *9*(1), 21–41.
Steinbach, R., Green, J., Datta, J., and Edwards, P. (2011). Cycling and the city: a case study of how gendered, ethnic and class identities can shape healthy transport choices. *Social Science & Medicine*, *72* (7), 1123–1130.
van Duppen, J., and Spierings, B. (2013). Retracing trajectories: the embodied experience of cycling, urban sensescapes and the commute between 'neighbourhood' and 'city' in Utrecht, NL. *Journal of Transport Geography*, *30*, 234–243.
Vivanco, L. A. (2013). *Reconsidering the Bicycle: An Anthropological Perspective on a New (Old) Thing*. London: Routledge.

14
THE TRAIN COMMUTE

Hanne Louise Jensen

Introduction

The spatial distribution of home and work ever changes according to the development of infrastructure, technology and residential opportunities and preferences. Railways historically have played an important role in facilitating commuters' travel between home and work, thus creating geographical coherence at regional, national and international scales (Priemus, 2008). At the European scale, the distance travelled between home and work has increased in recent years, while the time spent traveling has remained stable (Viry, Ravalet and Kaufmann, 2015).

Recurring streams of commuters are a well-known phenomenon in larger cities around the world. Early in the morning throughout the year, approach roads are clogged with cars, and bus and train stations are packed full as commuters arrive in cities to work after travelling various distances using different means of transport. In the late afternoon, the stream of commuters reverses as they travel back home. The railway is still important, and although the term *commute* was developed in the United States in the nineteenth century (Bissell, 2018), the practice of train commuting has flourished in most parts of the word, surviving the growth of car ownership.

However, it is not this spatial distribution, the distances travelled every day or the competing modes of transport that are at the centre of this chapter. Instead, we examine how we can understand the practices and experiences of commuting in line with the call to view mobilities as more than travelling between point A and point B (Cresswell, 2006). As Bissell (2018: 30) pointed out, 'the differently designed capsular spaces of trains, buses, and cars can induce very different sensations of being with other people. Understanding the joys and frustrations of commuting requires us to grasp precisely these kinds of transient sociabilities'. By focusing on the experiences and practices of train commuting and the mobile places they produce, it becomes possible to understand not only how long and by which means of transport commuters travel but also what kinds of mobile lives they live on a daily basis before arriving at their destinations.

Becoming a train commuter

Becoming a successful train commuter is by no means an easy task (Löfgren, 2008). Having a commuter in one's family influences the life of the family, and the overall family context influences the decisions regarding mobility the commuter can make (Viry and Lück, 2006). However, even when the family decides that a member will invest time in commuting and provides a supportive family context, much hard work lies ahead. Several strategies, routines and obstacles must be considered and overcome, and even then, there is no guarantee of success (Bissell, 2018). The pleasures and pains of commuting are widely debated in mobility studies. In particular, there are competing views on whether travel time is a waste of time or offers potential for recreation.

Based on data from the Job Mobility and Family Lives in Europe Study, Viry, Ravalet and Kaufmann (2015) showed how emotional perceptions of commuting are related to gender, family situations and age and are highly sensitive to levels of job insecurity and unemployment. Taking a closer look at what happens in train compartments, Lyons, Jain and Holley's (2007) studied rail passengers' time use in England and found that more than half of the passengers read for leisure, half watched other passengers or looked out the windows, a third worked, and a smaller number snoozed, socialised and listened to music. Only 2% were bored, and less than 25% viewed their commuting time as wasted time. Similarly, in a Danish context, Freudendal-Pedersen (2007) reported that commuters experienced their use of time as a valuable space in the middle of busy everyday life. In the following, I present some of my research on how both individual and social train commuters succeed in living appreciated mobile lives on board their daily train journeys between the outskirts of the Danish Region Zealand and the capital Copenhagen.

Individual practices of relaxation, daydreaming and sleeping

Routinisation of travel is at the heart of the successful transformation of a commute into valuable time as it allows the commuter to shift attention from handling travelling to other issues (Edensor, 2011; Ehn and Löfgren, 2010). The commuters in my study use great amounts of their commuting time to daydream, process and plan the day, read, sleep and do crosswords and other recreational activities. As Freudendal-Pedersen (2007) pointed out, it can be difficult to make time for such activities in busy everyday life, or as Ehn and Löfgren (2010) put it, it can be difficult to allow oneself time to do them.

Before the commuters can focus on their individual recreational practices in the train, though, they need to master travelling and make it a routine. Before the commuters enter the train, they are already busy with their mobility routines. They practise an alternative form of wayfinding as they weave efficient routes between home and work. In line with Laurier and Lorimer's (2012) descriptions of co-driving commuters' negotiations of which route to take, the train commuters also find shortcuts—they are only a bit different when the means of transport is the train. Finding shortcuts while commuting on the train is about choosing the departure with the fewest stops, placing oneself at the spot on the platform closest to one's preferred car and, once inside the train, choosing a seat that conveniently facilitates a quick exit. Once the commuters find shortcuts, they practise the shortcuts with attentive care and routinisation. Routinisation of the commuters' 'own route' allows them to have their minds on things other than managing travel. Their specific, routinised routes also create specific places for them to meet the same commuters every day. Routinisation of these wayfinding strategies thus holds great social potential, to which I return in the next section. In the following, we again turn to individual practices inside the train.

To Christina, a mother of two preschool-aged children, the train constitutes one of the few places in her everyday life where she can follow her own desires. She reads magazines, eats and does crossword puzzles. She, in Winther's (2009) words, is 'doing home' on board the train, and to her, the train becomes a place where recreation can occur during the transit between her busy work and family life. This use of the travel time resonates with many commuters who do not engage in interactions on board the train. They perceive the train as a place for processing thoughts related to work and family life and value the separation of work and family life enabled by these reflexive mobile practices. Sleeping while commuting is another important mobile practice and influences home life as it enables having a later bedtime, which allows spending more time at home awake.

Yet another mix of home and train relates to the widespread practice of help from the commuters' parents or their local, working partners—we can name this group of helpers the 'caretakers of mobility'. They take care of the commuters' home-based everyday duties, such as taking children to and from daycare, doing grocery shopping, cooking and so forth. By performing duties difficult for the commuters to manage due to the time spent commuting, the caretakers of mobility enable the commuters to have recreational time on board the train in exchange for home-based duties. Complex family negotiations are part of most commuters' mobility narratives. This illustrates the importance of understanding the significance of the mobile place to social relationships outside the train.

Individual practices of relaxation, daydreaming and sleeping during the transition between home and work are some commuters' preferred ways to live their mobile lives. Others only do these practices early in their commuting careers before they engage in extensive social interactions. In the following, we look at the pattern of how mobile friendships develop on the train and what their importance for commuters is.

Ritual practices and the production of mobile communities

Using the perspectives on mobile relations expressed by Jensen (2010), Laurier et al. (2008) and Laurier, Brown and Lorimer (2007), it is possible to identify four social categories of relationships involving various degrees of community and knowledge sharing on the train. Those in mobile relationships that involve the fewest social interactions are named the Mobile Others and Mobile Acquaintances, while those who engage in more social interactions are named the Mobile Withs and Mobile Friendships. How these relationships can evolve over time is the topic to which we turn here.

Mobile Others perform individual practices on board the train. They either are not invited to participate in social relationships or avoid social contact. The large group of Mobile Others among train commuters shares a lack of social interactions on board the train. A train with groups of predominantly Mobile Others is relatively quiet as they practise individual strategies of interactional avoidance using artefacts such as books, newspapers and headphones. Strategies of interactional avoidance can be practised with small variations as a matter of routine, but they can also be part of a wide range of strategies, including some with higher social content, among which the commuters vary.

The routines and rhythms of the journey play a significant role when Mobile Others become Mobile Acquaintances. As the commuters meet at the same time and place every day over a long period of time, they are able to transcend the group of Mobile Others and become Mobile Acquaintances. Often, they begin to recognise and greet each other. If the train is cancelled or delayed, then the group of commuters is put together in a situation where they share similar experiences and sentiments (e.g. annoyance, anger and frustration), which opens a crack for social possibilities (Jensen, 2018). As Ehn and Löfgren (2010) pointed out, part of this dynamic occurs as a social waiting situation is experienced as shorter than an individual waiting situation. This joint incitement can be the first step to the formation of much tighter social bonds among those dwelling within the same community of commuters every day for long periods of time—for some commuters, until the end of their working life.

Groups of commuters who purposefully locate themselves together in the train I describe as Mobile Withs. They are mobile with each other and often form very engaging social situations on board the train. If we look at groups of Mobile Withs from Collins' perspective on rituals (2004), part of their social success arises from their bodily proximity while sitting on the train. This bodily proximity allows for successful, ritual interactions in which the commuters have a mutual focus of attention and share their moods with each other. When a train car contains one or more groups of Mobile Withs, the noise level increases as laughter and conversations are shared. Due to the public nature of the train, a conversation that begins within a group of Mobile Withs may be experienced by non-members as an invitation to participation and conversation, which becomes a way into an already-constituted group of Mobile Withs.

Some commuters who travel in groups of Mobile Withs start to interact outside the train, and while doing so their social relationships change from Mobile Withs to Mobile Friendships. The formation of these friendships lived both inside and outside the train is a way to give significant time to friendships despite using much time to commute in busy everyday life. Turning Mobile Withs into friends transforms the train into a place where commuters can sustain and develop friendship relationships.

Ritual outcomes of mobile communities

The ritual ingredients of the interactional rituals among Mobile Withs and Mobile Friends, as we saw earlier, involve sitting close to each other and sharing moods, bodily co-presence and a mutual focus of attention. These ritual ingredients create a collective effervescence; accordingly, much of my field notes and transcripts is dominated by 'ha ha ha …'. The ritual outcomes are even richer and, to a large extent, explain why groups of commuters intensify their relationships. From this perspective inspired by Collins (2004), I approach the social interactions in groups of Mobile Withs and Mobile Friends.

One common ritual outcome is that relatively brief, individual emotional energy forms the baseline for new ritual interactions. Another outcome from groups of Mobile Withs and Mobile Friends is that long-term group solidarity works to strengthen the bonds among the participants. Mobile groups thus become tied more closely together the more they interact. Exchanges of laughter within mobile groups increase solidarity, tighten social bonds and reward individuals with increased emotional energy. This combination of ritual outcomes is a likely motive for the commuters to locate themselves with the same mobile group in the same corner of the train every day. The ritual outcomes and emotions created within mobile groups thus contribute to the rhythms and routines in the train.

The sense of belonging to a mobile group is communicated with pride. 'There is no other commuter group like ours', Emil told me on several commutes we travelled together. The 'we' formed within a group contains elements from the negotiation of shared moral standards often related to issues outside the train. 'What do we think of the drowning accident at Præstø?', Trine asked when she returns to the train after three weeks' sick leave. Stories revolving around episodes they have lived together on the train are used in a different manner and, to some extent, become symbols of their sense of belonging. For instance, a group of commuters with a mix of Mobile With and Mobile Friendship relationships had a mobile Christmas lunch as one of their symbolic episodes. On the morning train, they planned what take-away to get, and two of the commuters picked up the food in time to catch their usual afternoon train. The conductor reserved the first-class section for the group, and they decorated it with paper Christmas elves, spruce branches and tablecloths. They drank, sang and ate the whole journey, and to top it off, the conductor had made a song for them—about them. As the individual commuters reached their stations of arrival, they left the party, and at the end station, only two were left. When the members of this commuter group describe themselves as out of the ordinary, they often turn to this episode, telling what happened, how the other passengers reacted to their Christmas party and, especially, the efforts the conductor made to support their party.

When the commuters talk with each other about each other, it becomes clear that they enjoy knowing about each other's peculiarities and the others' knowledge of their own peculiarities. They often predict each other's actions, reactions and sleep patterns. For instance, they know why Henry has that old, crumpled map; they know he is too stingy to buy one, so they are able to deduce that it must have been attached to a weekly magazine or that he often cleans at night. These insights into each other's lives are comparable to knowing shortcuts—they demand a familiarity not everyone has. Realising this shared familiarity is a source of joy for the commuters.

Community and a sense of belonging are important for the commuters who form groups of Mobile Withs and Mobile Friends. Being known by and knowing other commuters is an ongoing invitation for interactions. Not having this familiarity can be a stress factor. 'I feel non-human if I sit between strangers', Jeppe explains to me. Being reduced to an anonymous Mobile Other among other anonymous Mobile Others creates unwanted tension for Jeppe, which is the main reason for his participation in a group of Mobile Withs.

When some Mobile Withs start to also take care of each other outside the train, they become Mobile Friends, as discussed. Manifestations of these Mobile Friendships vary significantly, from practical help to emotional support and birthday celebrations, but they share their roots in local communities outside the train. Friendship relationships developed on the train contribute to local belonging in the commuters' areas of residence as they interact and help each other outside the train. In their local areas of residence, they give each other rides to and from the train, take care of each other's pets, exchange plants, go bowling, lend holiday homes, visit on birthdays and do specialised shopping for each other. This residential belonging is increased by these interactions outside the train and the knowledge sharing inside the train.

The mobile lives lived on board the commuters' train, whether characterised by individual practices or social community, are full of a variety of mobile situations that, using Jensen's (2013) vocabulary can be viewed as staged both from above by the material design of the train and the schedule by which it runs and from below by the embodied performances of both individual commuters and social interactions among the commuters. However, we also need to consider how the mobile lives and situations change during the commute in accordance with both the rhythms of the day and the progress of the journey. Applying the perspective of rhythms, it becomes clear that the train is not a single place but multiple places within the same material setting.

Rhythms and mobile places

Hughes, Mee and Tyndall (2017) showed how quiet is performed in various ways on the train during the morning and evening commutes. The time and the course of the day are important factors to understand the lives on board the train. In commuting by train, as well as other means of transport, the 'rhythms of commuting are exceedingly diverse and shaped by numerous factors, including the mode of transport and its particular affordances, cultural practices and social conventions, modes of regulation, the distance travelled and the specificities of the space passed through' (Edensor, 2011: 193). In addition, Vendemmia (2017) pointed to how commuters' mobile rhythms influence urban rhythms.

Applying the term *everyday rhythms* and the work of Lefebvre and Régulier (2004) using rhythm as a frame of reference, it is possible to sharpen our attention to the link of time, landscapes and the commuters' practices on their daily journey. Most days, related events happen at specific locations on the route, and these rhythms produce a place characterised by a high degree of homeliness and routine. As mentioned, while riding the train, the commuters' ritualised practices weave time and landscape together. At the end of their morning trip, when reaching the second-to-last station before their destination (the Hedehusene station), the commuters from Rødby Færge gather their belongings and put on their jackets. Once the train passes a certain highway between Hedehusene and the end station, Høje Taastrup, the group of commuters enter the exit area of the train car and stand ready to leave. They add the arrival time and calculate possible delays by looking at the clock while passing specific landmarks.

As the train advances, the social practices on board create variations in the rhythms as conversations and sleep patterns increase and decrease over the journey. During the first part of the journey, the commuters usually start by talking when they meet at the train. Within five to ten minutes, the conversation ends and is usually replaced by activities such as eating, sleeping, reading the newspaper, listening to the radio and doing other forms of recreation. During the stops along the way, these rhythms are repeated with various levels of intensity as the train ride progresses, and new groups of commuters meet and depart. At the Næstved station, an hour into the journey, the rhythms of the journey are disrupted. Here, on the morning train ride, a great number of commuters enter the train wide awake, waking up the commuters already on the train. The presence of the Næstved commuters thus changes the last part of the journey. During the afternoon journey home from work, the opposite rhythms occur. The train ride starts out very noisy and busy but ends in almost complete silence at the end station, Rødby Færge. These testimonies of how the activities on the train change due to the progress of the journey underline the dynamic nature of the mobile place.

Homing practices produce a place that loses elements of its public nature and becomes a place one gets to know not so much by contemplation as by participation (Winther, 2009). Homing practices especially evolve around sleep as the commuters use their negotiated right to cut off conversation by referring to the need to sleep—a right most commonly used within families and practised at home. Within other homing practices, some everyday routines from home are brought onto the train as commuters complete their 'work look', putting on the last pieces of jewellery, eating breakfast and drinking coffee. Added to these activities are waking up children by mobile phone, planning shopping and doing other home-related practices. As pointed out by Letherby and Reynolds (2003: 7), 'the train is a place that provides a space for a variety of tasks'.

Commuting time on the train holds social potential for interactions and formation of relationships that are specific to and, to some degree, follow everyday rhythms, as the commuters' rhythms are woven together and dependent on each other. The commuters can perform their ritualised practices only if other commuters also follow theirs (e.g. choose the same set of seats and stay quiet during the same parts of the journey). Some social rhythms, though, go beyond the joint relationships among landscape, time and action and are dominated by the logic of ritual practices. In general, the train is a place where many types of social relationships can occur simultaneously, and the commuters on any one journey can manage a wide range of relations. However, the practised relationships have a great role in determining which places are created at specific departures.

Conclusion

The train becomes a place through the commuters' practices on their daily journey. The mobile place acquires its own characteristics through both what the commuters do and the meanings they ascribe to their activities. As a mobile place with a unique time and space, the train encompasses beginnings, endings, repetitions, routines and renewals within the same materiality (the train car) but with very varied content depending on the location along the route and the presence of other commuters. All the commuters—and the train's other passengers—contribute with their practices to the production of the place at the actual departure. The train can be a place for resting, laughing and storytelling, and it can either be a collage of individual activities or a web of social activities; sometimes, it is both a collage of individual activities and a web of social activities. It can be a place for forming friendships and exchanging favours, and it can be a place for reviewing the day and preparing for the day to come. The mobile place affects the commuters' lives through facilitating the formation of social relationships and other meaningful activities, such as contemplation, reading and regeneration—but only with the predictability of the railway service and the caretakers of mobility.

References

Bissell, D., 2018. *Transit life—how commuting is transforming our cities*. Cambridge, MA: MIT Press.

Collins, R., 2004. *Interaction ritual chains*. Princeton, NJ: Princeton University Press.

Cresswell, T., 2006. *On the move: mobility in the modern western world*. London: Routledge.

Edensor, T., 2011. Commuter: mobility, rhythm and commuting. In: T. Cresswell and P. Merriman, eds. *Geographies of mobilities: practices, spaces, subjects*. Farnham: Ashgate. pp. 189–203.

Ehn, B. and Löfgren, O., 2010. *The secret world of doing nothing*. Berkeley, CA: University of California Press.

Freudendal-Pedersen, M., 2007. *Mellem frihed og ufrihed—strukturelle fortællinger om mobiliteten i hverdagslivet*. Roskilde: Roskilde University.

Hughes, A., Mee, K. and Tyndall, A., 2017. 'Super simple stuff?' Crafting quiet in trains between Newcastle and Sydney. *Mobilities*, 12(5), pp. 740–757.

Jensen, H.L., 2018. On social cracks in train commuting. In: M. Freudendal-Pedersen, K. Hartmann-Petersen and E. Fjalland, eds. *Experiencing networked urban mobilities*. London: Routledge. pp. 38–42.

Jensen, O.B., 2010. Negotiation in motion: unpacking a geography of mobility. *Space and Culture*, 13(4), pp. 389–402.

Jensen, O.B., 2013. *Staging mobilities*. London: Routledge.

Laurier, E., Brown, B. and Lorimer, H., 2007. *Habitable cars: the organisation of collective private transport*. Full research report ESCR end of award report, RES-000-23-0758. Swindon: ESRC.

Laurier, E. and Lorimer, H., 2012. Other ways: landscapes of commuting. *Landscape Research*, 37(2), pp. 207–223.

Laurier, E., Lorimer, H., Brown, B., Jones, O., Juhlin, O., Noble, A., Perry, M., Pica, D., Sormani, P., Strebel, I., Swan, L., Taylor, A.S., Watts, L. and Weilenmann, A., 2008. Driving and passengering: notes on the ordinary organisation of car travel. *Mobilities*, 3(1), pp. 1–23.

Lefebvre, H. and Régulier, C., 2004. The rhythmanalytical project. In: H. Lefebvre, ed. *Rhythmanalysis*. London: Continuum. pp. 71–84.

Letherby, G. and Reynolds, G., 2003. Making connections: the relationship between train travel and the processes of work and leisure. *Sociological Research Online*, 8(3) [online] Available at: www.socreson line.org.uk/8/3/letherby.html. Accessed 29/5 2019.

Löfgren, O., 2008. Motion and emotion: learning to be a railway traveler. *Mobilities*, 3(3), pp. 331–351.
Lyons, G., Jain, J. and Holley, D., 2007. The use of travel time by rail passengers in Great Britain. *Transportation Research Part A*, 41, pp. 107–120.
Priemus, H., 2008. Urban dynamics and transport infrastructure: towards greater synergy. In: F. Bruinsma, E. Pels, H. Priemus, P. Rietveld and B. van Wee, eds. *Railway development—impact on urban dynamics*. Heidelberg: Physica-Verlag. pp. 15–34.
Vendemmia, B., 2017. The length and breadth of Italy: redefining rhythms and territories through mobilities. *Applied Mobilities*, 2(2), pp. 199–214.
Viry, G. and Lück, D., 2006. Conclusion—state-of-the-art of mobility research. *Job Mobility Working Paper*, 2006(1), pp. 205–222.
Viry, G., Ravalet, E. and Kaufmann, V., 2015. High mobility in Europe: an overview. In: G. Viry and V. Kaufmann, eds. *High mobility in Europe*. London: Palgrave Macmillan. pp. 29–58.
Winther, I.W., 2009. 'Homing oneself'—home as a practice. *Home and Space*, 4, p. 2.

15
WAITING (FOR DEPARTURE)

Robin Kellermann

Standing on a platform, waiting for the train to come, sitting in a departure hall, coping with another delay due to strikes or volcanic clouds in the atmosphere: All journeys start with a wait and suddenly, passengers are given time without actually wanting it.

Despite the world's aspiration to move, waiting and interim pausing have at all times formed key constituents of the modern everyday transit experience. Since the implementation of movements, either physically or operationally, require facilities of 'friction', transportation systems involve (and intentionally organize) pre-trip situations of waiting at stops or stations. Owing to a predominant academic attention for 'kinetic' promises of transport and mobility, experiences and practices as well as spatialities of transport-induced waiting have however remained surprisingly trivialized or even unexplored. Comprehending practices as well as material infrastructures of waiting as critical and meaningful indicators for highlighting prevailing notions of time and space, the chapter pays attention to the supposed banality of transport-induced waiting situations, which appear a 'stepchild of mobility'.

Commencing with a cultural historical conceptualization of systemic origins and negative perceptions of waiting within the framework of modernization processes, the second part of the chapter explores transformations of 'handling' waiting passengers through the lens of operational and planning principles in the 19th century railway sector that reflect on waiting as a profound architectural and organizational challenge. That followed, the third part of the chapter examines the subject's rich complexities by presenting multidisciplinary interventional approaches of avoiding, reducing or altering *objective* duration as well as *subjective* perceptions of waiting time.

Inspired by recent reassessments of waiting beyond the primacy of time efficiency, the fourth section challenges conventional notions of travel time and waiting beyond a mere 'waste of time'. The chapter concludes with an outlook on research paths for analysing experiential constellations of passengers in modes of waiting, including methods on how to grasp the 'elusiveness' of transport-related waiting phenomena.

Conceptually, the contribution thereby advocates for a more holistic consideration of the transit experience (including waiting) within the 'mobility turn's' assertive aspirations to acknowledge the irresolvable dialectics of speed and slowness/friction (Urry, 2007).

Tracing the systemic and perceptual origins of waiting

Sociologists define waiting as the experience of 'temporary or provisional interruption' (Gasparini, 1995, p. 31) and a kind of 'stationary abidance' (Paris, 2001, p. 706). Tracing the systemic reasons for such interruptions, waiting phenomena, in the most general sense, originate from the mismatch between a certain demand (goods, services, necessities) and a presently non-available or inaccessible offer (Taylor, 1994). The experience of this mismatch is inherent to human history. Psychology has revealed prominently that, as individuals, we possess a motivation system, which permanently drives us for achieving certain needs (Maslow, 1943). As long as humans have tried to strive for the fulfilment of fundamental or complex needs, humans may have experienced their naturally or socially enforced non-availability or inaccessibility. In this sense, waiting undeniably constitutes an anthropological key aspect of life that has always been experienced as a wait for harvest, religious salvation, access to places and social ranks, or the return of something or someone.

Though waiting may thus not be considered an exclusively modern phenomenon, it was the genesis of modernization processes in the Western World between the 17th and 19th centuries that triggered an unprecedented wave of acceleration (Koselleck, 1985) and prompted an unprecedented wave of occasions and necessities to queue up. Moreover, with the rise of modern societies the mismatch between needs and offers shifted gradually from the natural and religious spheres to the socioeconomic sphere. Modern societies' increased complexity due to longer chains of interdependencies let reasons for waiting shift from an *ecocentristic* (weather and astronomic events) to a *humancentristic* (scarcity of services, goods, devices) system of reference, adding a rather modern type of situational micro-waiting that may be termed 'systemic waiting' (Kellermann, 2019). Compared to traditional waiting situations, systemic waiting was of less existential nature, but no less psychologically demanding because perceptions of and reflections about time inevitably became the focus of attention.

Waiting for transportation

Among the many fields enforcing situations of systemic waiting, the evolution of modern public transport systems can be considered a key constituent of both facilitating and experiencing the temporal regions of situational micro-waiting. Cultural historian Wolfgang Schivelbusch (1986) examined how the evolution of rail travel comprised a radical transformation of time and space. Rail travel, he argued, comprised a psychological need to adopt supplementary perceptual and visual attitudes for the aim of coping with the novel experience of increased *speed*. However, what remained blatantly overlooked in this prevailing focus on speed is the complementary and seemingly mundane novelty of *anticipating* the departure for accelerated transfer in a dedicated transitory environment. Thus, in contrast to pre-modern waiting experiences, the progress of timetable-based railway travel since the 1830s may have induced a novel mode of systemic *in-betweenness*, both *before* departing and *while* travelling.

Reasons for the emergence of the temporal 'niche' *before* departing lay in the operational uniqueness of rail travel that to some point last to this day: Efficient and reliable organization of public transport requires the concentration of huge amounts of passengers before embarking. In other words, because the organization of speed signifies an operational challenge, the temporary stilling of passengers serves as an operational solution. Referring to the very dialectics of mobility and immobility (Adey, 2006; Urry, 2007), accessibility of *flow* was facilitated through spatial infrastructures of *friction*.

Consequently, modern passengers face the imposition of *spatial* (specific locality) and *temporal* (specific point in time) constraints and are obliged to make sense of a transitory stay in transitory environments of waiting halls, platforms, shops or restaurants. Though the stage-coach system involved the similar necessity of grouping passengers before boarding, it was not until the rise of railways that situational micro-waiting became a mass phenomenon and institutionalized the wait as an inevitable key aspect of the transit experience. Accordingly, waiting rooms often represented the largest space group in 19th century railway stations, thus exemplifying the generally increased centrality of systemic waiting in modern everyday life and the transit experience in particular. Hence, forwarding Virilio's (2007) notion of modernity as the unintended discovery of systemic 'crashes', the evolution of accelerated transport gave birth to the (unintended) 'production' and discovery of retardation and delay.

Perceptions of waiting have to be understood in the context of cultural orientations towards time. The genesis of modern Western societies based on an influential cultural change in expectational and (thereby) temporal principles. Religious promises of salvation were gradually replaced by modern aspirations of (accelerated) progress that declared the future to be open and consequently put time (use) centre stage. Habitualization of clock time formed the backbone of synchronizing the increased complexities of modern societies, including a time-sensitive capitalistic economy and a scientific culture that used objective time as a main resource for analysing natural events.

Complementary to cultural paradigms of *efficient* time use (Weber, 1930), situations of standstill were considered increasingly unjustifiable, thus problematical. As speed and time savings manifested key aspirations ('time is money'), waiting was soon considered a *waste of time*. Accordingly, the semantics of waiting started to change by the late 18th century from a formerly neutrally connoted attentive activity (to ward, to protect, to watch) towards a more passively experienced burden. Waiting gradually received adverbial addings such as to wait *passionately*, with *pain* or *agonizingly* (Köhler, 2007), illustrating the semantic transition towards a rather critical connotation. Seeking to quantify this critical connotation, transport planners suggested the value of waiting time to be 2.5 times more 'costly' than time spent moving (Wardman, 2004).

Historical transformations of handling waiting passengers

Today, waiting for a train may seem a trivial necessity but in fact reflected an operational and experiential novelty for railway companies and passengers alike. As architectural historian Carol Meeks argues, 'Neither of the two preceding modes of transportation – the canal and the century-old turnpike system – had developed special buildings for the use of passengers' (Meeks, 1956, p. 27). Consequently, early 19th century architects and engineers faced an unprecedented challenge to erect adequate and functional spatial infrastructures for the provision of smooth, safe and reliable access to the new technical system.

A key problem was handling comparably huge passenger volumes in constricted confines within a short time period. The solution to this problem was a strategic retardation of flows in large-scale assembling rooms. Besides the generalized need to pre-group passengers, such spaces were also considered necessary, because in most countries, except Great Britain, the US or Belgium – for reasons of safety and inexperience in travelling – passengers were not allowed to step on the platforms uncalled (Perdonnet, 1856). Instead, after having purchased tickets and having registered their luggage, millions of passengers, e.g. in Germany, France or Austria, before boarding the train, mandatorily had to gather in the transitory environment of designated and class-specific waiting rooms. Comparable with today's operational principle at airports, it was not before a stroke of a bell (10–15 minutes before departure) that passengers were allowed to enter the platform and to finally board the train (Schivelbusch, 1986).

Seen from an operational standpoint, these waiting rooms served not just as 'control rooms', but as *transmission zones* for *synchronizing* and *attuning* passengers to the organizational and temporal regime of railways. In this sense, waiting rooms reflected fundamental spaces for disciplining, ordering and conditioning where *people* became *passengers*.

Tracing the changes in planning and organizational principles of handling waiting passengers in the first 100 years of railways, Kellermann (2019), based on an investigation of German and English station buildings, proposed two major trajectories of handling passengers: The principle of stilling passengers compulsorily and the principle of self-organization that both may have affected waiting experiences differently.

For the longest part of the 19th century, temporarily stilling of passengers in designated environments was considered an operational precondition for the provision of mass transportation. While the pioneering phase of railway travelling (1830–1850) could witness rather *informal* ways of waiting, the following decades saw an increasing level of *formalizing* the wait in ever-growing spatially bound compartments, thereby establishing a *waiting imperative*. This imperative, functioning as the transition zone for synchronizing passengers to the technical apparatus, was subsequently criticized by its users and was compensated by an increasing provision of service and comfort that allowed for 'equipped waiting' (Gasparini, 1995) and filling time more 'meaningfully'.

However, the period 1880–1914 marked a turning point in both organizing and perceiving the wait, represented by a co-evolutionary atmosphere of problematizing waiting from different angles. First, planners, inspired by industrial organization, obsessively followed the dogma of replacing *controllable stasis* (waiting rooms) by infrastructures of *continuous flow* (concourse). Second, after decades of railway travelling, passengers felt experienced enough to navigate individually and increasingly problematized the compulsory task of waiting, not least against the background of a 'time-pressured' industrializing society. Moreover, the historical retracement of organizational principles indicated that transport-induced waiting in the early 19th century was far more representational or even an exciting experience, thus was connoted more positively. In other words, it was not before the late 19th century that (modern) waiting phenomena became massively problematized as to be avoided by any means.

Planning and management strategies to avoid and reduce *objective* and *perceived* waiting times

Given the modern aspiration for accelerated progress, the history of modernity resembles a history of avoiding or reducing waiting times. Though waiting in the social realm strictly remains an instrument for illustrating individual or group-specific power inequalities or an instrument of protest, commercial sectors such as telecommunication, retail or the transport sector continuously explored strategies to reduce waiting times or improve customers' 'time-sensitive' perception of receiving or consuming services.

Non-transport related waiting

The first explicit academic engagements with problems of congestion and retardation did not evolve in the sphere of transport networks but around questions of efficient telephone network planning. Early telephone networks faced the problem of long waiting times due to capacity problems for telephone exchange. Erlang (1909) developed the first formulas to describe and predict situations of waiting in a network, which solved problems of dimensioning telephone exchanges or optimizing allocation of employees. Having defined principles that are still in use today (e.g. in call centres), the *Queuing Theory* inspired the even more sophisticated *Operations Research*, which aims for the prediction of bottlenecks in all kinds of networks ranging from military planning and transport systems to informational flows in the Internet. From these perspectives, the avoidance of *objective* waiting times remains simply a question of adequate stochastical prediction (Palm, 1953).

Contrary to mathematical descriptions of waiting phenomena, social sciences focused on subjective and societal dimensions. In the 1970s, recognizing the problematic and occasionally pathological condition of waiting, social psychology was spearheading the analysis of people in either enforced or voluntary states of queuing and waiting. Therein, waiting formations were examined as paradigmatic expressions for the social organization of access, as results of social inequalities (Schwartz, 1975) but also for comprising qualities of 'embryonic social systems' (Mann, 1969, p. 340). As a main outcome, time distribution in service reception was found to be socially asymmetric, facilitating the acceleration of specific groups for the price of decelerating others (Schwartz, 1974). Moreover, to wait, from psychological perspectives, was consistently shown to be a source for affective responses such as stress, anger and uncertainty (Osuna, 1985; Taylor, 1994).

Criticizing the unrestricted convertibility of time and money, sociologists and social psychologists also aimed for a more differentiated examination of waiting by separating objective (duration of the wait) from subjective (time estimation) and cognitive (degree of acceptance) factors, that act together and trigger affective responses. Against the background of decreasing time budgets in the Western World since the late 1980s, marketing, management and health studies formed the spearhead of analysing affective responses of customers and patients in the state of being paused temporarily. Consequently, the 1990s marked a peak point of management strategies to avoid or reduce waiting times for customers that were believed to be the main concern of dissatisfaction and disloyalty. Given the assumed unavoidability of waiting for services (Houston et al., 1998; Nie, 2000), most studies tried to elaborate on situational factors that would reduce the negative perception of waiting. Initiated by Hornik's (1984) key study that explored individuals' tendency to drastically overestimate waiting time in service environments, reams of studies researched the influence of music (Hui et al., 1997), light, smells, interiors (Baker and Cameron, 1996; Pruyn and Smidts, 1998), technical facilities, or levels of information (Hui and Tse, 1996). Almost two decades of scholarship tried to operationalize the famous dictum proclaimed by economist Shirley Taylor: 'If you cannot control the actual wait duration, then control the customer's perception of it' (Taylor, 1994, p. 56).

However, the late 1990s left a disenchanted atmosphere since the majority of situational factors had almost no significant influence on waiting time perceptions. Instead of concentrating on situational factors, absolute duration of the wait was found to remain 'the central stimulus' (Durrande-Moreau, 1999, p. 175) for perception. Consequently, mere acceleration of processes has (again) constituted the main aspiration of the service sector. Nevertheless, the established sub-discipline of *perception management* has unveiled a row of relevant key principles. For instance, a central finding of *perception management* was that waiting *before* receiving a service, was most often perceived a tedious waste of time (Dubé et al., 1991). Confirmed by later studies in the transport context, pre-process waiting and delays have been consensually identified as the most critical waiting times

(Wardman, 2014). Another important outcome represents the understanding of people's heterogeneous time perception (due to habitus, different time pressures, mood) that requires differentiated treatments as well as the consideration of cultural differences of waiting time perception (Nie, 2000).

Transport-related waiting

The historical aspiration of modern (public) transport is to achieve 'seamless travel'. However, in practice, travellers regularly face systemic dysfunctionalities, capacity problems or security-imposed bottlenecks, which result in waiting. Therefore, public transport operators, similar to strategies in the commerce sector, have applied 'hard' strategies of reducing *objective* waiting times through systemic interventions as well as 'soft' strategies that aim for smoothing *subjective* waiting time perception by steering passengers' affective responses.

On the network level, reductions of waiting times have been achieved most successfully through the implementation of *Integrated Clockface Timetables* that enable vehicles to arrive at major hubs in almost the same time. Such integrated timetables reduce both the waiting time spent for transfer to connecting means of transport as well as travel time for the entire journey. Having proved successful for rail passengers, e.g. in Switzerland ('Bahn 2000') and the Netherlands ('Spoorslag 70'), other countries (e.g. Germany and Denmark) have recently been developing such integrated strategies, which are assumed to be a better alternative than maximizing travel speeds of selected mainlines (Brezina and Knoflacher, 2014). However, too tight scheduling often increases system vulnerability and may cause even longer waiting times in cases of missed connections, thus illustrating an unsolved conceptual dilemma between *nominal* and *robust* timetabling. Exemplifying the trade-off between modern aspirations (seamless travel) and real-life practice (waiting and delays), today's schedules demand the insertion of 'buffer' and 'supplement times' that facilitate the absorption of complications (Davenport et al., 2001). Nonetheless, from a planning point of view, waiting situations are considered 'stochastic disturbances' (Kroon et al., 2007, p. 41) that can be avoided or reduced by means of mathematics and corresponding infrastructural development.

Beyond that, reduction of objective pre-trip waiting is pursued by installation of ticket selling machines *inside* vehicles, the replacement of paper tickets by electronic tickets and the corresponding potential of automated fare collection, which together shall minimize passengers' objective waiting time by reducing time-consuming tasks before boarding.

In addition to resource-intensive strategies on the network level, transport operators also aim for 'qualifying' the experience and perception of the wait, particularly by means of information and communication technologies (ICT). One of the most prominent measures comprises the installation of telematic-based real-time information systems. Several studies have explored positive effects for both passengers and providers (Brakewood et al., 2015; Dziekan and Kottenhoff, 2007; Watkins et al., 2011), evidencing a reduction of perceived waiting times and increasing ridership.

Focusing on rail services, van Hagen (2011) proposed a balance between different forms of stimulation that correspond inversely to the busy-ness of stations. While commuters in crowded rush hours demand low stimulation levels (to cool them down), off-peak time travellers (e.g. leisure travellers) perceive more pleasant waits through stimulation and distraction. In addition, areas of static waiting should be equipped with more dynamic stimuli than transit areas. Equally important is the preparation of measures in cases of delay, which, in contrast to other generators of waiting, are the least controllable (Taylor, 1994). For airports, Sauter-Servaes and Rammler (2002) therefore proposed the concept of

'Delaytainment', an emergency management catalogue that can provide benefits for both customers and airports (new business segments) alike.

Concerning the impact of material interventions, an urban legend implies that the installation of mirrors in a New York hotel lobby mitigated complaints about slow elevator services because people 'used' the wait for adjustments or observed others while waiting (Sasser et al., 1978). Despite limited transferability to public transport, the provision of basic amenities (seats, benches, shelters) and security improvements was found to reduce perceived waiting times significantly, particularly for women that perceive longer waiting times (Fan et al., 2016). Moreover, studies on passenger behaviour unveiled the tendency to cluster around entrances and seats. Therefore, a more 'waiting-sensitive' design stimulates a more convenient positioning of passengers. Seeking to lower the gap between comfortable equipped in-transit interiors and pre-transit environments, latest informational design interventions comprise the indication of stopping positions of arriving trains and visual information of current capacity levels. Centring on pragmatic questions of which materialities make the mobile situation possible, a recent thread of studies follows the conjunction of mobilities and design (Jensen, 2016).

Challenging conventional notions of travel time and waiting

The majority of approaches implicitly treat travel time as a burden, and, by equalling time with money, as a 'cost'. Countering such conventional notions, studies associated to the 'mobilities turn' reconceptualized passengers' travel time use against the background of opportunities generated by ICT (Lyons and Urry, 2005; Metz, 2008). As ICT allows for both leisure and work activities, rather than a cost, travel time in the information age may be considered a 'gift' (Jain and Lyons, 2008).

ICT transforms our relations to time and space and may involve a re-evaluation of waiting in service environments. From a long-term perspective, the combined use of personal (smartphones) and public (real-time information) devices encompass beneficial implications for passengers (increased certainty, regain of situational control, processual knowledge) that can be considered a seminal moment in the history of passenger experiences.

In this spirit, recent studies emphasized for a stronger differentiation of the temporal region of waiting by revealing the actual social, experiential and bodily complexities (Bissell, 2007; Bissell and Fuller, 2011; Gasparini, 1995; Vannini, 2011), thus challenging the dominant narrative of *mobile* and *active* mobilities as the more desirable relations to the world (Bissell, 2007).

The relevance of these studies emerges from moving the focus from conventional 'productivist' notions of travel time as an economic value to the focus of passengers' perspectives, which include a reassessment of conventional perceptions of waiting as a 'waste of time' and potentially highlighting the phenomenon's positive utilities (e.g. opportunities to relax, reflect, converse, getting surprised or temporal relieve from social duties). Exemplary for these alternative considerations of waiting, Dawes and Rowley (1996) or Friman (2010) explored that in some cases waiting times can even be part of a positive service experience.

Concluding remarks

The chapter explored the temporal phenomenon of waiting through the lens of public transportation. Conventionally treated as a mundane and frustrating practice, the retracement of systemic and perceptual origins of transport-related waiting highlighted public transports' inseparable relationship of speed and friction that has become increasingly tension-filled against the background of modern aspirations of time-efficiency and acceleration. Consequently, the history of modernity can be read as a history of avoiding or reducing waiting times. While mathematics aimed for decreasing objective durations with the help of models and simulations on the network level, social psychology and marketing aimed for reducing perceived waiting times on the subjective level by means of material and psychological interventions.

As a result of both streams, today's transport environments are characterized by an increasing plurality of supplementary technologies that 'moderate', 'direct' and, finally, 'tame' the wait (Kellermann, 2017). ICT plays the crucial role in this development, fostering a reconceptualization of both travel time and waiting that may unveil positive utilities.

Despite unbroken achievements of reducing or 'liquefying' pre-trip waiting, it will remain a central aspect of moving as speed 'needs' friction. Inspired by recent studies highlighting the phenomenon's experiential and corporeal complexities, one might hypothesize that waiting already *is* a way of moving. Acknowledging the dialectics of speed and slowness, future research should aim for a deeper understanding of travel experiences that is sensitive to passengers' temporalities, involving historicizing waiting practices and considering gender, age and cultural factors. Using methods of (auto)ethnography, travel diaries, data sets from electronic ticketing or video surveys, studies should investigate the ways that ICT changes both travel experiences in general and waiting experiences in particular. Profound understandings of waiting phenomena bear the potential of decisive policy implications. As speed may not be any longer the key aspiration, investments in infrastructure might be reassessed for serving reliability or comfort rather than high speeds.

Compared to modes of individual transport, the need to wait for public transport manifests an inherent structural difference. However, as a final note to practitioners, Disney World may provide fruitful inspirations as operators master to absorb long waiting times into the leisure experience. Analogically, transport operators may follow the aim of letting passengers step onto platforms that give them a feeling of having entered the fun ride already.

References

Adey, P., 2006. If mobility is everything then it is nothing: towards a relational politics of (im)mobilities. *Mobilities*, 1(1), pp. 75–94.

Baker, J. and Cameron, M., 1996. The effects of the service environment on affect and consumer perception of waiting time: an integrative review and research propositions. *Journal of the Academy of Marketing Science*, 24(4), pp. 338–349.

Bissell, D., 2007. Animating suspension: waiting for mobilities. *Mobilities*, 2(2), pp. 277–298.

Bissell, D. and Fuller, G. eds., 2011. *Stillness in a Mobile World*. London and New York: Routledge.

Brakewood, C., Macfarlane, G.S., and Watkins, K., 2015. The impact of real-time information on bus ridership in New York City. *Transportation Research Part C: Emerging Technologies*, 53, pp. 59–75.

Brezina, T. and Knoflacher, H., 2014. Railway trip speeds and areal coverage. The emperor's new clothes of effectivity? *Journal of Transport Geography*, 39, pp. 121–130.

Davenport, A., Gefflot, C., and Beck, C., 2001. Slack-based techniques for robust schedules. In: *Proceedings of the Sixth European Conference on Planning*. Toledo (Spain), 12–14 September 2001. Palo Alto: AAAI Press.

Dawes, J. and Rowley, J., 1996. The waiting experience: towards service quality in the leisure industry. *International Journal of Contemporary Hospitality Management*, 8(1), pp. 16–21.

Dubé, L., Schmitt, B.H., and Leclerc, F., 1991. Consumers' affective response to delays at different phases of a service delivery. *Journal of Applied Social Psychology*, 21(10), pp. 810–820.

Durrande-Moreau, A., 1999. Waiting for service: ten years of empirical research. *International Journal of Service Industry Management*, 10(2), pp. 171–194.

Dziekan, K. and Kottenhoff, K., July 2007. Dynamic at-stop real-time information displays for public transport: effects on customers. *Transportation Research Part A: Policy and Practice*, 41(6), pp. 489–501. doi:10.1016/j.tra.2006.11.006.

Erlang, A.K., 1909. The theory of probabilities and telephone conversations. *Nyt Tidsskrift for Matematik B*, 20(6), pp. 87–98.

Fan, Y., Guthrie, A., and Levinson, D., 2016. Waiting time perceptions at transit stops and stations: effects of basic amenities, gender, and security. *Transportation Research Part A: Policy and Practice*, 88, pp. 251–264.

Friman, M., 2010. Affective dimensions of the waiting experience. *Transportation Research Part F: Traffic Psychology and Behaviour*, 13(3), pp. 197–205.

Gasparini, G., 1995. On waiting. *Time & Society*, 4(1), pp. 29–45.

Hornik, J., 1984. Subjective vs. objective time measures: a note on the perception of time in consumer behavior. *Journal of Consumer Research*, 11(1), pp. 615–618.

Houston, M.B., Bettencourt, L.A., and Wenger, S., 1998. The relationship between waiting in a service queue and evaluations of service quality: a field theory perspective. *Psychology & Marketing*, 15(8), pp. 735–753.

Hui, M.K., Dube, L., and Chebat, J.-C., 1997. The impact of music on consumers' reactions to waiting for services. *Journal of Retailing*, 73(1), pp. 87–104.

Hui, M.K. and Tse, D.K., 1996. What to tell consumers in waits of different lengths: an integrative model of service evaluation. *The Journal of Marketing*, 60(2), pp. 81–90.

Jain, J. and Lyons, G., 2008. The gift of travel time. *Journal of Transport Geography*, 16(2), pp. 81–89.

Jensen, O.B., 2016. Of 'other' materialities: why (mobilities) design is central to the future of mobilities research. *Mobilities*, 11(4), pp. 587–597.

Kellermann, R., 2017. The final countdown. Ambiguities of real time information systems 'directing' the waiting experience in public transportation. In: Freudendal-Pedersen, M., Hartmann-Petersen, K., and Fjalland, E.L.P. eds. *Experiencing Networked Urban Mobilites*. London: Routledge. pp. 19–26.

Kellermann, R., 2019. Waiting for railways (1830–1914). In: Singer, C., Berwald, O., and Wirth, R. eds. *Time-Scapes of Waiting: Spaces of Stasis, Delay and Deferral*. Leiden: Brill. pp. 35–57.

Köhler, A., 2007. *Lange Weile: Über das Warten*. Frankfurt am Main: Insel Verlag.

Koselleck, R., 1985. *Futures Past: On the Semantics of Historical Time*. Cambridge: MIT Press.

Kroon, L.G., Dekker, R., and Vromans, M.J.C.M., 2007. Cyclic railway timetabling: a stochastic optimization approach. In: Geraets, F., Kroon, L., Schoebel, A., Wagner, D., and Zaroliagis, C.D. eds. *Algorithmic Methods for Railway Optimization*. Heidelberg: Springer. pp. 41–66.

Lyons, G. and Urry, J., 2005. Travel time use in the information age. *Transportation Research Part A: Policy and Practice*, 39, pp. 257–276.

Mann, L., 1969. Queue culture: the waiting line as a social system. *American Journal of Sociology*, 75(3), pp. 340–354.

Maslow, A.H., 1943. A theory of human motivation. *Psychological Review*, 50(4), pp. 370–396.

Meeks, C.L.V., 1956. *The Railroad Station: An Architectural History*. New Haven, CT: Yale University Press.

Metz, D., 2008. The myth of travel time saving. *Transport Reviews*, 28(3), pp. 321–336.

Nie, W., 2000. Waiting: integrating social and psychological perspectives in operations management. *Omega*, 28(6), pp. 611–629.

Osuna, E.E., 1985. The psychological cost of waiting. *Journal of Mathematical Psychology*, 29(1), pp. 82–105.

Palm, C., 1953. Methods of judging the annoyance caused by congestion. *Tele*, 4, pp. 189–208.

Paris, R., 2001. Warten auf Amtsfluren. *KZfSS Kölner Zeitschrift für Soziologie und Sozialpsychologie*, 53(4), pp. 705–733.

Perdonnet, A., 1856. *Traité Élémentaire des Chemins de Fer*. Paris: Langlois et Leclercq.

Pruyn, A. and Smidts, A., 1998. Effects of waiting on the satisfaction with the service: beyond objective time measures. *International Journal of Research in Marketing*, 15(4), pp. 321–334.

Sasser, W.E., Olsen, R.P., and Wyckoff, D.D., 1978. *Management of Service Operations*. Boston, MA: Allyn & Bacon.

Sauter-Servaes, T. and Rammler, S., 2002. *Delaytainment an Flughäfen: Die Notwendigkeit eines Verspätungsservices und erste Gestaltungsideen*. Berlin: Wissenschaftszentrum Berlin für Sozialforschung (WZB).

Schivelbusch, W., 1986. *The Railway Journey: The Industrialization of Time and Space in the Nineteenth Century*. Berkeley, CA: University of California Press.

Schwartz, B., 1974. Waiting, exchange, and power: the distribution of time in social systems. *American Journal of Sociology*, 79(4), pp. 841–870.

Schwartz, B., 1975. *Queuing and Waiting: Studies in the Social Organization of Access and Delay*. Chicago, IL: University of Chicago Press.

Taylor, S., 1994. Waiting for service: the relationship between delays and evaluations of service. *The Journal of Marketing*, 58(2), pp. 56–69.

Urry, J., 2007. *Mobilities*. London: Polity Press.

van Hagen, M., 2011. *Waiting Experience at Train Stations*. Ph.D. Delft.

Vannini, P., 2011. Mind the gap: the tempo rubato of dwelling in lineups. *Mobilities*, 6(2), pp. 273–299.

Virilio, P., 2007. *The Original Accident*. London: Polity Press.

Vozyanov, A., 2014. Approaches to waiting in mobility studies: utilization, conceptualization, historicizing. *Mobility in History*, 5(1), pp. 64–73.

Wardman, M., 2004. Public transport values of time. *Transport Policy*, 11(4), pp. 363–377.

Wardman, M., 2014. *Valuing Convenience in Public Transport: Roundtable Summary and Conclusions*. Paris: OECD/ITF.

Watkins, K.E., Ferris, B., Borning, A., Rutherford, G.S., and Layton, D., 2011. Where is my bus? impact of mobile real-time information on the perceived and actual wait time of transit riders. *Transportation Research Part A: Policy and Practice*, 45(8), pp. 839–848.

Weber, M., 1930. *The Protestant Ethic and the Spirit of Capitalism*. London: Unwin University Books.

16

MOVING AND PAUSING

Julie Cidell

Introduction

From the start, mobilities and immobilities have been considered together, especially the "moorings" that provide the necessary fixity that enables mobility (Hannam, Sheller, and Urry 2016). Urban infrastructure that is fixed in place to allow the rapid movement of people, goods, and resources has profound effects in terms of connecting distant locations and enabling rapid movement, while blocking or inhibiting movements within other places. Some social groups are able to move freely, with others' movements restricted by the same socio-political processes (Salter 2013). The movement of some requires the immobility of others, and vice versa.

However, there is a phase in between mobility and immobility that deserves attention as well: pausing. Briefly, pausing refers to a temporary cessation of motion with the intention of resuming that motion. This can be wanted or unwanted, intended or disruptive. Pausing is related to the concept of stillness, which Cresswell (2012) has emphasized is not about a return to sedentarism, but a necessary part of motion or mobility (see also Bissell 2007). I consider pausing as orthogonal to stillness in that one can pause in an activity without being physically still; alternatively, one might be immobile but not "still" in a mental or emotional sense. Interestingly, research on pausing is quite often self-reflexive or autobiographical, considering the researcher's journey either alongside of or as the main subject of study. It therefore provides the opportunity to consider the personal aspects of mobility as part of the social.

This chapter considers three types of pausing in the context of urban areas: the delay, which is an unwanted stoppage; the respite, which is a desired break in movement; and intermission, which is more neutral and scheduled in nature. I emphasize that pauses all imply a short time span. Mobility will resume soon, and travelers and their belongings will reach their final destinations. This sets the concept of the pause aside from the broader category of waiting, where there might not be an intention to move, or even the ability to do so. "Chronic waiting" has been termed a condition that particularly unemployed young people in disadvantaged areas are subject to (Jeffrey 2008). With nowhere to go and nothing to do, waiting becomes a "structuring condition of urban life" (Wafer 2017, p. 406). Though it has a more negative connotation, loitering is similar in that there is no expected resumption of movement (and indeed, that is often the "threat" that loitering is seen to pose). Waiting can also be forced upon someone (Joronen 2017) with no expectation of resumed movement, merely forcing others to wait for an unknown period of time. So while a broader literature on waiting exists, I do not consider it here.

The first three sections discuss the concepts of delay, respite, and intermission in more detail. The subsequent section on places and rhythms discusses the ways in which places and pauses (of all three types) construct each other, whether spaces designed for waiting or those that become spaces of pausing by default. The fifth section explains how the intersections of different speeds and rhythms require or encourage pausing as a part of the overall choreography of people, objects, and spaces. In the process, it discusses the uneven mobilities that result when some people have to wait for others.

Delay

Delay is the type of pause that probably most readily comes to mind when thinking about mobility. If we focus on travel, flow, or movement—either our own or that of another person or object—delay is the rupture that temporarily halts the mobile subject against their will. For example, traffic delay is a taken-for-granted aspect of urban life. As with other types of pauses, delay usually lasts for only a short time, but it is often a source of frustration because of its unpredictability, making it difficult to take advantage of travel time to accomplish other things (Jain and Lyons 2008). Transportation planners and engineers focus on reducing delay, with transportation economists calculating how much it costs us in terms of lost productivity. At the same time, delay can change how travelers plan their journeys, avoiding past trouble spots as much as possible (Cidell 2014). Delay is therefore a type of pause to be minimized or reduced if not eliminated altogether.

Social status can be connected to the extent to which one can expect delays. For example, one of the ways in which the "kinetic elite" are set apart from the average air traveler is that they experience less delay. They check in, go through security, and board through shorter, selective lines that reduce the waiting time they experience in moving through the air transportation system (Bissell 2007; Hannam, Sheller, and Urry 2006). Sometimes they even use air travel on the small scale of a city to avoid being delayed on the ground (Cwerner 2009). When their flights are delayed or canceled, the elite are prioritized within airlines' customer service systems to re-route them more quickly to their destinations and minimize their disruption. For other types of travelers, delay is built into the work that they do. Gregson's study of UK truckers highlights how the global logistics network that consists of seamlessly-moving packages and parcels depends on the ability of individual workers to accept delay (Gregson 2017). These truck drivers spend about 75 percent of their time being immobile: in congestion on the freeway, arriving before a delivery site opens in the morning, or waiting for the next container to clear customs and be loaded onto their vehicle. Here, human bodies take up the delay that global freight cannot afford to experience.

Other delays are longer term and more disruptive. One of the prime examples in the literature on the self-reflective nature of research on moving and pausing is the 2011 eruption of Eyjafjallajökull. This Icelandic volcano completely disrupted air travel between Europe and North America for nearly a week, with the cascading effects on air travelers lasting even longer. Coincidentally, the American Association of Geographers' Annual Meeting was taking place in Washington, DC, and many European mobilities scholars were delayed in their return home after the conference. A special issue of the journal *Mobilities* considered the implications of this particular delay on the air transportation network (Adey and Anderson 2011; Budd, Griggs, Howarth, and Ison 2011; O'Regan 2011) and on individual travelers' journeys (Barton 2011; Guiver and Jain 2011; Jensen 2011). Communication networks, vehicles, new friendships, emotions, and, of course, volcanic ash, were all part of the assemblages created as a result of the severe disruption and delay. This extreme example highlights delay as one particular type of pause, namely an unwanted rupture that can produce an emotional response or affect and which is to be avoided if at all possible.

An even longer unwanted pause is that experienced via transit migration (Collyer, Düvell, and de Haas 2012; Düvell 2012; Schapendonk 2012), meaning migrants who are temporarily located in Country B on their way between Countries A and C. For these people, trying to enter a country or region such as the EU is their final goal, but they must wait in countries such as Turkey, Morocco, or Greece in the meantime (Franck 2017). Individuals might be housed in Country B for a few days, weeks, months, or longer, shaping the places of transit in the process (Schapendonk 2013; Zeleke 2017). They must maintain hope that the delay will be short and that they will continue on to their hoped-for destination. However, such delays are often unpredictable, depending on local or national politics (Tudoroiu 2017), and multiple delays or stages might be involved before travel resumes—if it does at all.

Respite

Other kinds of pauses are more positive. A respite refers to a type of pause that is very much wanted, a break in traveling before moving on again. This can include a pause from strenuous physical activity or the surprisingly tiring activity of sitting in a vehicle all day. As some authors note, travel itself can be a respite from other activities, a pause in the daily flow of life that still involves moving. Like delay, a respite also produces affect, but of a more positive nature.

Perhaps because respites are not a problem in the same way as delays are, they have been studied considerably less. For example, while mobilities research has including running and cycling as areas of study, this work is usually focused on the mobile part, not the pauses. For example, Spinney's (2006) description of the physical effort of moving a bicycle–human hybrid up a mountain implies that one must always keep moving. Similarly, work on runners focuses on where running takes place (Hitchings and Latham 2016), with whom (Hitchings and Latham 2017), or in the special cases of events (Cidell 2016; Edensor and Larsen 2018). Guides for beginners emphasize that walk breaks are an expected part of running, and even marathon training plans incorporate walking as a respite from running. Nevertheless, these respites remain understudied in the literature.

It is not only traveling under one's own physical power that requires the occasional pause. For example, one might get out of one's car at a rest stop or gas station not only to refuel, but to stretch and move one's body (Normark 2006). John McPhee's description of UPS's global network includes the tidbit that lobsters need to rest along their journey from farm to table, which is why they are transported by truck and not plane (McPhee 2007). Airlines in the Gulf States and Southeast Asia try to attract travelers between Europe and Asia by offering them a stopover of many hours or a day in order to break up their long journeys (and spend money in Dubai or Bangkok in the process). In all of these cases, even though the individual's body might have been still for hours, taking a break to move may itself be a respite.

In this way, a respite may actually come *through* mobility. Multiple authors have noted the irony in how an individual may be sitting still for hours even as the vehicle carrying them flies along at a high rate of speed (e.g., Bissell 2007). Getting up to walk around the train car or airplane cabin is therefore a pause of sorts, briefly exchanging one state of mobility for another. On a larger time scale, travel itself can function as a respite from other activities (Watts and Urry 2008). Many commuters prefer to have some distance to travel between work and home to mentally set aside their work day and prepare for the interactions they'll have at home (and vice versa). Solo drivers in particular often cherish the solitary trip home as a respite from the demands made on them the rest of the day. Walking may allow time for thinking and reflecting which is unobtainable in other situations (Middleton 2009). Respites may therefore involve pauses in mobility, or mobility *as* a pause from something else, but they are always welcomed and anticipated.

Intermission

The third type of pausing is more neutral than a delay or a respite. *Intermission* is a necessary part of the journey, often due to transfers and scheduling, that temporarily slows forward progress but is neither unexpected nor a break. For example, what Normark (2006) calls "tending to mobility" refers to the need to pause and take care of one's vehicle and/or one's self as part of travel. The "mundane work and interactions" (p. 241) that occur at a petrol station are often taken for granted or hardly noticed, but they are a necessary part of automobility. As an intermission, a stop to refuel demonstrates the cooperation necessary among mobile and fixed people and things to keep cars going.

In the context of travel, intermission can consist of short pieces of time that are not easily used in ways other than waiting. In contrast to a seated train journey that can be used for more productive purposes, intermission is found in waiting for the train or bus to arrive, standing in line for a ticket, or waiting to exit a ferry (Bissell 2007; Jirón, Imilan, and Iturra 2016; Vannini 2011). This kind of waiting is itself a social relation to the world, as Bissell argues, with the expectation that is part of a pause reflected in the actions of individuals and the atmospheres they create around themselves (Jirón, Imilan, and Iturra 2016). Even though inactivity is inherently more difficult to study than a body in motion, Bissell reminds us that waiting is a fundamental part of travel and should not be left out.

Increasingly, the owners and designers of the spaces in which intermission takes place are taking advantage of the short pauses that travelers experience. Airport terminals in particular have been enhanced with retail, restaurants, and "experiences" meant to fill in the intermission time while waiting for a flight (Elliott and Radford 2015; Lloyd 2003). This derives partly from the privatization of airport terminals and the need to generate revenue from more than landing fees, but also because intermission is a necessary part of air travel. Airport amenities that make pauses more like respites and less like intermissions can bring in revenue and perhaps make travelers more likely to choose a specific routing on another occasion. They can also be used to build place identities in what would otherwise be a non-place (Augé 1995), including building a stronger brand for the city that hosts the airport (McNeill 2014).

Places of pausing, intentional or otherwise

Delay, intermission, and respite all take place in a variety of settings. The characteristics of these places of pausing have led some to argue that they are non-places, interchangeable and detached from their surroundings (Augé 1995). However, others have argued that there are definitely important characteristics of these places that both embed them in localities and provide a sense of place for the people passing through them.

For example, a considerable amount of literature looks at terminals and the ways in which they have been constituted as places. This includes the connection between airports, air travel, and national identity, even for non-travelers who might be using the airport for recreational or social purposes (Adey 2006; McNeill 2014). As security lines have grown longer and airports have become more profit-oriented, passengers are encouraged to arrive well in advance of their flights, often giving them a significant amount of intermission with which to deal. Airport authorities have responded with retail, restaurants, and art galleries to fill the time, rebranding themselves as destinations in and of themselves, to the extent that the "airport city" has become a common catchphrase. (Of course, the negative affect that unanticipated flight delays cause might not result in as pleasurable a shopping/gazing experience, another potential topic for further research.) Even more negatively, the design of security facilities within terminals may be intended to induce compliance among travelers, discouraging them from pausing in sensitive locations (Adey 2008).

Work on the production of places of pausing is not limited to air travel. Vannini's work on ferry waiting lines shows that residents who rely on them for transportation play with the spaces of intermission, such as trying to minimize their time by knowing exactly when to line up for the ferry, a type of local knowledge that tourists are not privy to (Vannini 2011). Similarly, social interactions among people waiting for bus or metro transfers shape not only the site of transfer, but the entire journey of a commuter through the pleasure of conversations (Jirón, Imilan, and Iturra 2016). If the transit system is reshaped for purposes of moving people and vehicles more efficiently through the city, that social interaction might be lost (ibid.) Train stations as activity spaces have their own specific patterns of mobility that shape the places and the people who travel through them (Bissell 2007). Their contributions to urban life are not only in terms of monumental architecture or instrumental functionality, but as social places in and of themselves. In all of these cases, authors are discussing places that are expected or planned to be sites of pausing, showing how their affordances enable and construct social interactions and spaces.

Other kinds of spaces are inadvertently created from delays, respites, or intermissions. Merriman's work on urban car parks shows how a space intended for storage of personal vehicles that are on an intermission in their travels can be used by people in all different sorts of ways (Merriman 2016). Informal markets held from the backs of cars both take advantage of the space and produce new kinds of uses and social interactions. Like a car park, a shipping container storage facility is a place of mobility and stillness all at the same time (Cidell 2012). Individual objects move in and out, pausing for varying amounts of time that might be intended or unintended. From an observer's perspective, however, the lot is always occupied with parked objects, no matter if individuals boxes or cars are moving in and out. The space therefore might not be intended as a pausing place, with consequent negative impacts on neighboring residents who do not appreciate the visual clutter of pausing objects, nor the traffic they supply to neighborhood roads.

Rhythms and interactions

Thus far, our consideration of pauses has been largely that of the individual, caught up in their own journeys. However, mobility studies emphasizes the social nature of mobility, including the ways in which our journeys intersect. We move at different paces and with different rhythms, with some authors discussing the choreography that enables us to all fit together, especially within urban settings (Jonasson 2004; Middleton 2009). Sometimes these intersecting rhythms cause pauses as part of the choreography. These pauses can be unwanted delay, anticipated respite, neutral intermission, or somewhere along the continuum. At the same time, mobility studies has been criticized for focusing on agents rather than on overall structures (Salter 2013). This section therefore considers the interactions between not only individuals but larger structures that lead to and arise from pauses of various sorts.

Surveillance and security studies is one way to approach interacting and sometimes conflicting rhythms. For example, Salter argues, "The widening of the mechanisms for trusted travelers and subjects of the extraordinary rendition process are exactly the same: because of what the state already knows about those individuals, regular security procedures can be withheld and special facilitation made" (Salter 2013, p. 13). Larger structures of circulation have been built to allow some travelers to move with minimal pausing, while others are delayed indefinitely, but all under the same structures. Bureaucratic procedures that ostensibly are intended to promote the smooth functioning of household livelihoods can be drawn out in order to induce delay to the point that livelihoods are threatened, all under cover of a functioning state (Joronen 2017). The role of the state is obviously important in setting rhythms for a whole host of social actors, with methods of surveillance pioneered in transportation terminals later being rolled out onto city streets in a variety of ways.

Another structural approach to understanding rhythms and interactions and the pauses they produce is through the production of uneven rhythms (Hubbard and Lilley 2004). One aspect of mobility justice (Sheller 2018) is to consider how resources are unevenly distributed when it comes to mobility systems. For example, the emphasis on high-speed, high-cost transit systems for well-off commuters in Singapore ensures that lower-income people travel more slowly, including more intermissions while waiting for transfers or slow-moving vehicles (Lin 2012). I have already mentioned Gregson's work on truckers and their embodied role in absorbing the pauses of the global logistics system. Drivers in the gig economy, whether for ridesharing or delivery services, attempt to coordinate their own life rhythms with the demand for their services (Christie and Ward 2019). This may result in moving faster than they would like, but also in unwanted delays as they pause for a delivery to be ready or a customer to arrive: "if I'm waiting at a restaurant for food or I'm waiting for the customer to come to the door, I don't get paid for any of that time because it's not an hourly wage" (ibid., p. 118). The urban dwellers who do not want to make the trip to the restaurant or grocery store themselves produce the uneven rhythms of their drivers in the process.

Uneven rhythms are also produced through social understandings of who and who doesn't belong, including perceptions of others' mobility and pauses. For example, those who do not move enough in highly-visible areas such as retail districts may be asked to move along as part of "cleaning up" the neighborhood (Rink and Gamedze 2016). Homeless people are often asked to keep moving regardless of the nature of their surroundings, especially if pauses on park benches or under overpasses turn into something more like waiting or loitering. Alternatively, people might stand out because they move too much or in the wrong manner. For example, Turkish immigrants to Germany continued their practice of promenading as a means of socializing; without the public squares of their homeland, their walking had to take place in front of the Frankfurt train station. Here, walking in a place intended for only short-term visitors, they were considered out of place and unwelcome (Suzuki 1976). These cases highlight that while pauses are sometimes expected as part of common mobility practices, pauses can also mark individuals as out of place, even within cosmopolitan urban areas.

Concluding remarks

Elucidating different kinds of pauses within mobility studies, particularly in the context of urban areas, gains us a few key considerations. First, pauses go beyond the mobility/immobility binary to demonstrate how intertwined stillness and motion are. In fact, movement may itself serve as a respite, especially within a bustling city. Second, pauses can be found along a continuum of desirability, from unwanted delay to neutral intermission to positive respite. Each of these types of pause produces different affects and atmospheres, each worthy of study and comparison, and each contributing to the multiple rhythms of the city. Finally, the autobiographical nature of much of the work on pauses demonstrates the intensely personal nature of mobilities. Researchers have focused on the types of mobilities with which they are the most familiar to highlight the complexities therein. This is where the predominantly Anglo-American nature of mobilities studies raises some concerns. Incorporating the perspectives of a diversity of individuals in a variety of places when it comes to not only mobilities, but pauses, is necessary to strengthen our understanding of the social aspects of mobility and how cities are constructed in the process.

References

Adey, P. 2006. Airports and air-mindedness: spacing, timing and using the Liverpool Airport 1929–1939. *Social & Cultural Geography* 7, 343–363.

Adey, P. 2008. Airports, mobility and the calculative architecture of affective control. *Geoforum* 39, 438–451.

Adey, P. and Anderson, B. 2011. Anticipation, materiality, event: the Icelandic ash cloud disruption and the security of mobility. *Mobilities* 6:1, 11–20.

Augé, M. 1995. *Non-Places: Introduction to an Anthropology of Supermodernity*. London: Verso.

Barton, D. 2011. People and technologies as resources in times of uncertainty. *Mobilities* 6:1, 57–65.

Bissell, D. 2007. Animating suspension: waiting for mobilities. *Mobilities* 2:2, 277–298.

Budd, L., Griggs, S., Howarth, D. and Ison, S. 2011. A fiasco of volcanic proportions? Eyjafjallajökull and the closure of European Airspace. *Mobilities* 6:1, 31–40.

Christie, N. and Ward, H. 2019. The health and safety risks for people who drive for work in the gig economy. *Journal of Transport and Health* 13, 115–127.

Cidell, J. 2012. Flows and pauses in the urban logistics landscape: the municipal regulation of shipping container mobilities. *Mobilities* 7:2, 233–245.

Cidell, J. 2014. Spoke airports, intentional and unintentional ground travel, and the air travel decision-making process. *Transportation Research Part A* 69, 113–123.

Cidell, J. 2016. Time and space to run: The mobilities and immobilities of road races. In Hannam, K., Mostafanezhad, M. and Rickly-Boyd, J.M., eds., *Event Mobilities: The Politics of Place and Performance*. London: Routledge, pp. 82–94.

Collyer, M., Düvell, F. and de Haas, H. 2012. Critical approaches to transit migration. *Population, Space and Place* 18, 407–414.

Cresswell, T. 2012. Mobilities II: still. *Progress in Human Geography* 36:5, 645–653.

Cwerner, S. 2009. Helipads, heliports and urban air space: Governing the contested infrastructure of helicopter travel. In Cwerner, S., Kesselring, S. and Urry, J., eds., *Aeromobilities*. Abingdon, VA: Routledge, pp. 225–246.

Düvell, F. 2012. Transit migration: a blurred and politicised concept. *Population, Space and Place* 18, 415–427.

Edensor, T. and Larsen, J. 2018. Rhythmanalysing marathon running: 'A drama of rhythms'. *Environment and Planning A* 50:3, 730–746.

Elliott, A. and Radford, D. 2015. Terminal experimentation: the transformation of experiences. Events and escapes at global airports. *Environment and Planning D* 33:6, 1063–1079.
Franck, A. 2017. Im/mobility and deportability in transit: Lesvos Island, Greece, June 2015. *Tijdschrift voor Economische en Sociale Geografie* 108:6, 879–884.
Gregson, N. 2017. Logistics at work: trucks. containers and the Friction of circulation in the UK. *Mobilities* 12:3, 343–364.
Guiver, J. and Jain, J. 2011. Grounded: Impacts of and insights from the volcanic ash cloud disruption. *Mobilities* 6:1, 41–55.
Hannam, K., Sheller, M. and Urry, J. 2006. Editorial: mobilities, immobilities, and moorings. *Mobilities* 1:1, 1–22.
Hitchings, R. and Latham, A. 2016. Indoor versus outdoor running: understanding how recreational exercise comes to inhabit environments through practitioner talk. *Transactions of the Institute of British Geographers* 41:4, 503–514.
Hitchings, R. and Latham, A. 2017. How 'social' is recreational running? Findings from a qualitative study in London and implications for public health promotion. *Health and Place* 46, 337–343.
Hubbard, P. and Lilley, K. 2004. Pacemaking the modern city: the urban politics of speed and slowness. *Environment and Planning D* 22, 273–294.
Jain, J. and Lyons, G. 2008. The gift of travel time. *Journal of Transport Geography* 16:2, 81–89.
Jeffrey, C. 2008. Waiting. *Environment and Planning D* 26, 954–958.
Jensen, O. 2011. Emotional eruptions. Volcanic activity and global mobilities – A field account from a European in the US during the eruption of Eyjafjallajökull. *Mobilities* 6:1, 67–75.
Jirón, P., Imilan, W.A. and Iturra, L. 2016. Relearning to travel in Santiago: the importance of mobile place-making and travelling know-how. *Cultural Geographies* 23:4, 599–614.
Jonasson, M. 2004. The performance of improvisation: traffic practice and the production of Space. *ACME* 3:1, 41–62.
Joronen, M. 2017. Spaces of waiting: Politics of precarious recognition in the occupied West Bank. *Environment and Planning D* 35:6, 994–1011.
Lin, W. 2012. Wasting time? The differentiation of travel time in urban transport. *Environment and Planning A* 44, 2477–2492.
Lloyd, J. 2003. Airport technology, travel, and consumption. *Space and Culture* 6, 93–109.
McNeill, D. 2014. Airports and territorial restructuring: the case of Hong Kong. *Urban Studies* 51:14, 2996–3010.
McPhee, J. 2007. *Uncommon Carriers*. New York: Farrar, Straus & Giroux.
Merriman, P. 2016. Mobility infrastructures: modern visions. Affective environments and the problem of car parking. *Mobilities* 11:1, 83–98.
Middleton, J. 2009. 'Stepping in time': walking. Time, and space in the city. *Environment and Planning A* 41:8, 1943–1957.
Normark, D. 2006. Tending to mobility: intensities of staying at the petrol station. *Environment and Planning A* 38, 241–252.
O'Regan, M. 2011. On the edge of chaos: European aviation and disrupted mobilities. *Mobilities* 6:1, 21–30.
Rink, B.M. and Gamedze, A.S. 2016. Mobility and the city improvement district: frictions in the human-capital mobile assemblage. *Mobilities* 11:5, 643–661.
Salter, M. 2013. To Make move and let stop: mobility and the assemblage of circulation. *Mobilities* 8:1, 7–19.
Schapendonk, J. 2012. Migrants' im/mobilities on their way to the EU: lost in transit? *Tijdschrift voor Economische en Sociale Geografie* 103:5, 577–583.
Schapendonk, J. 2013. From transit migrants to trading migrants: development opportunities for Nigerians in the transnational trade sector of Istanbul. *Sustainability* 5, 2856–2873.
Sheller, M. 2018. *Mobility Justice: The Politics of Movement in an Age of Extremes*. London: Verso.
Spinney, J. 2006. A place of sense: a kinaesthetic ethnography of cyclists on Mont Ventoux. *Environment and Planning D* 24, 709–732.
Suzuki, P. 1976. Germans and Turks at Germany's railroad stations: interethnic tensions in the pursuit of walking and loitering. *Urban Life* 4:4, 387–412.
Tudoroiu, T. 2017. Transit migration and "Valve States": the triggering factors of the 2015 migratory wave. *Southeastern Europe* 41, 302–322.

Vannini, P. 2011. Constellations of ferry (im)mobility: islandness as the performance and politics of insulation and isolation. *Cultural Geographies* 18:2, 249–271.

Wafer, A. 2017. Loitering: reassembling time in the city-of-the-global-south. *Social Dynamics* 43:3, 403–420.

Watts, L. and Urry, J. 2008. Moving methods. Traveling times. *Environment and Planning D* 26:5, 860–874.

Zeleke, M. 2017. Too many winds to consider; which way and when to sail!: Ethiopian female transit migrants in Djibouti and the dynamics of their decision-making. *African and Black Diaspora* doi:10.1080/17528631.2017.1412928.

17

PROVIDING AND WORKING IN RHYTHMS

Katrine Hartmann-Petersen

At first sight mobilities of urban life might seem chaotic and unpredictable. With a closer look, they work in a systematically organized and coordinated way. The interconnectedness between these perspectives is explored in numerous transdisciplinary scales in mobilities research. Understanding urban life through rhythmicity is one way of analyzing pulsating mobile patterns of urbanity. This chapter takes its starting point in the complexities of everyday life (Jensen et al., 2019). Everyday life and urban living are polyrhythmic – a melting pot of rhythms. Some of these rhythms are caused by mobilities, and some mobilities are consequences of rhythms. We are all surrounded by rhythms that clash and rhythms that provide a sense of flow to our daily lives. In order to understand mobilities in urban life, it is important to understand how multiple rhythms, like the changing of seasons, weekdays, purposes of errands, and agendas, influence the experience of living and working in the city. As an example of polyrhythmicity in the city, this chapter offers an understanding of the mobile life of bus drivers (Hartmann-Petersen, 2009, 2018). Moreover, the chapter discusses what we can learn from a perspective of rhythms within a polyrhythmic everyday life, whether the empirical focus is urban living in relation to running, driving, walking, coping, etc. Finally, it discusses why these perspectives develop and improve urban mobilities planning.

Rhythms in current mobilities research

Rhythm arises from perceptions of time. Time is linked to events and experiences which have a rhythmic, often repetitive structure. Henri Lefebvre states that "[E]verywhere where there is interaction between place, a time and an expenditure of energy, there is rhythm" (Lefebvre, 2004, p. 15). In mobilities research, the concept of rhythm has been used in an indirect sense to perceptions of time (Eriksen, 2001; Urry, 2007). Most recently, it has been explored how rhythms shape human experience in time–space and pervade everyday life and place (Cresswell, 2010; Edensor, 2010a, 2010b, 2011; Adey, 2017; Cook and Edensor, 2017). This also includes exploring the rhythms of mobility: "Rhythm is an important component of mobility as many different [...] rhythms are composed of repeated moments of movement and rest, or, alternatively, simply repeated movements with a particular measure" (Cresswell, 2010, p. 23). Based on this background, applied examples of the understanding of urban life through rhythmicity have been connected to walking (Wunderlich, 2008; Edensor, 2010b), running (Edensor et al., 2018; Edensor and Larsen, 2018), commuting (Edensor, 2011), cycling (Spinney, 2006; Larsen, 2018), coach tours (Edensor and Holloway, 2008), street performances (Simpson, 2008), taxi driving (Notar, 2012), and bus driving (Hartmann-Petersen, 2009, 2018). These areas are examined in different contexts, at different scales and relate to different geographical places providing descriptions of everyday life in urban areas.

Polyrhythmic everyday life

Late modern everyday life is a complex arena. The individual copes with a messy fusion of duties, habits, rhythms, dreams, hopes, longings and aspirations every day. But it is interesting to know how this puzzle is produced and reproduced. Part of the answer is integrated in the need for mobilities. Mobilities are independent variables that individuals see as fundamental necessities in the maintenance of a modern polyrhythmic everyday life (Hartmann-Petersen, 2009; Edensor, 2010a, 2011). Polyrhythmicity as a notion stems from the area of music, where it, in short, can be defined as "the simultaneous use of two or more conflicting rhythms, that are not readily perceived as deriving from one another" (Wikipedia: Polyrhythm). In relation to everyday life, this chapter suggests that it is used as a starting point for understanding the intermingling – sometimes conflicting – multiple rhythms of urban everyday life.

Most people have parts of everyday life that correspond to a solid rhythmic structure. Sennett stresses (cf. Anthony Giddens, 1986) how a certain amount of routine in everyday life is necessary: "To imagine a life of momentary impulses, of short-term action, devoid of sustainable routines, a life without habits, is to imagine indeed a mindless existence" (Sennett, 1998, p. 44). At the same time, every day a number of non-rhythmic – sometimes unpredictable – elements pressure the rhythmic flow that the individual tries to maintain: Stressful conditions at work, children getting ill, overrunning deadlines, congestion on the way from a to b, or too many leisure time activities make it impossible to maintain viable rhythms. Edensor and Larsen distinguish, referring to Lefebvre, between the notions of eurhythmia and arrhythmia: "Eurhythmia is a desired state in which the individual body works harmoniously and bodies collectively rhythmically align, whereas arrhythmia refers to those situations where such rhythms break apart and cause friction and pain (Lefebvre, 2004, p. 78)" (Edensor and Larsen, 2018, p. 732). This distinction can be illustrative not only in relation to embodied, physical movement but also to the perception of rhythms in everyday life – a daily mixture of being in flow and feeling out of sync.

In everyday life research, the concept of rhythms has been used to analyze work–life balances in the sense that (a lack of) rhythm in working life absorbs leisure time and contributes to an unhealthy everyday life (Hochschild, 1997). As a supplement to the mobilities researchers mentioned above with rhythm analysis at their focus, I suggest that inspiration from other research fields can add to the concepts of rhythms. Based on the systemic optics of working life research – here through the oeuvre of the working life researcher, Hvid (2006), it is possible to develop analytic tools to understand and cope with not only everyday life challenges but also perceptions of mobilities. The following theoretical perspectives might broaden the current discussion on rhythm as a constituent part of the politics of mobility (Cresswell, 2010, p. 17).

Hvid's research analyzes conditions around sustainable working life. As a part of this, understanding rhythms is essential to secure the individual in everyday life. According to

Hvid, much that happens around us is framed by rhythms. If everything happened "atypically", equivalent to the notion of arrhythmia (Lefebvre, 2004), everyday life would decompose into incoherent fragments. It would be impossible to identify time in the moments between events and actions. Rhythm is the ongoing process that encircles the unique to stand out from the atypical, and make it possible to reflect upon and articulate the unpredictable:

> Time is dependent on repetitions, but repetitions are also dependent on time. Without time, events would not be segregated, and thus they would not be repetitive. Rhythm is a combination of two inseparable conditions: repetitive events and time.
>
> *(Hvid, 2006, p. 112, author's translation)*

Everyday life rhythms are not stationary or circular even though they sometimes feel that way. They are rather dynamic and helical: "Events are rarely repeated completely. There are always slight modifications, small changes, small adjustments, and therefore there is also a potential for development in rhythmicity" (Hvid, 2006, p. 113, author's translation). Based on this background, Hvid states, that rhythm – in the sense of repetition over time – is a basic condition in a world of defining and establishing social relations, and thus it is also a basic condition of everyday life. Despite late modern variability, everyday life consists of multiple, routinized, rhythmic actions that are relatively predictable from day to day. We go to work, pick up the kids, shop, cook, and so on. Everyday life activities and experiences are linked by mobilities, movement between locations and activities that make a fragmented life possible. Additionally, mobilities are catalysts for individuals' requirements and desires for a good, meaningful life (Freudendal-Pedersen, 2009). For a lot of people (the desire for) rhythmic elements and the potentials of mobilities are interconnected in the daily puzzle: "Being in rhythm is to live. To be outside the rhythm is a threat to both identity and life" (Hvid, 2006, p. 116, author's translation). If this hypothesis is valid, it underlines the individual's need for mobilities to ensure the rhythmic elements in a busy everyday life. That puts emphasis on the relevance for knowledge of the nexus between rhythms, mobilities, and planning.

Hvid points out that well-functioning rhythms are vigilant rhythms. Vigilant rhythms are those that the individual repeats without being reflexive. Such rhythms also have a potential for improvisation, essential because, as Hvid claims, "Rhythm without freedom decomposes one's identity. They seem like a never-ending treadmill" (Hvid, 2006, p. 122, author's translation). In other words, according to Hvid, rhythms may become too perfunctory, whereby the individual is a slave to – and not a co-player in – the rhythm, and this becomes a burden on their identity. In addition, rhythms become more individualized and independent, failing to synchronize with larger social rhythms. This means that if the rhythms characterizing working life do not correspond to the rhythms in other parts of everyday life, there is a risk that working life expands and invades family life. That affects the capability of realizing other tasks and dreams related to, i.e. leisure, hobbies, and social networks. If one specific rhythm becomes dominant, an anomic relation to the surrounding world may develop as one gets sidelined in the part of everyday life that does not fit into the dominant rhythm. Mobilities might evolve or even conflict in these situations.

The distinction between harmonious, viable rhythms (eurhythmia) and anomic identity-threatening rhythms (arrhythmia) is a late modern dilemma. When everyday life structures (at all levels) become liquid (Bauman, 2005) and have not yet found solid form, the possibility of preserving and shaping rhythms decreases. Rhythmic conditions cannot exist if the surrounding social relations are liquid (Bauman, 2005). There are a lot of issues to take into consideration in a modern family, and it is definitely not difficult to identify the risk of clashing, anomic rhythms, or disruptions (Manderscheid, 2014; Murray and Doughty, 2016). Permanent structures and liquid elements connect and disconnect in everyday life planning.

Late modern everyday life is neither rhythmic nor "out of sync" organic or anomic, or challenging or disruptive; it is polyrhythmic because clashes between different rhythms have to coexist within the same individual frame of everyday life – and thus corresponds with other people's polyrhythmic everyday life (Hartmann-Petersen, 2009). It is very likely that it is possible to cope with these clashes periodically, but in the longer term, it is problematic because it erodes the possibilities of maintaining coherence.

The rhythms of urban mobility providers – the case of bus drivers in Copenhagen, Denmark

Historically, the bus has been a gathering point for various kinds of people. In many cities, the increased levels of car ownership have individualized different mobilities (Dennis and Urry, 2009), and of course, automobility has affected and in some ways changed everyday life, not least the bus systems' status and target audience. But despite that, the bus has served as a common point of reference for many travelers. It is a kind of regular, mobile community platform in which the driver is an enduring, authoritative figure (Hartmann-Petersen, 2018).

The illustrative case in this chapter specifically draws on several in-depth qualitative interviews and focus groups (Kvale, 1996; Halkier, 2010) made with experienced drivers working the most central bus routes in Copenhagen, Denmark. The empirical work was conducted as a part of a PhD funded by the Danish Working Environment Research Fund. The aim was, *inter alia*, to investigate perceptions of mobilities in urban everyday life among bus drivers and the interconnectedness between drivers and passengers. In addition to interviews and focus groups, the project also included an extensive participatory action research perspective (for further elaboration, see Hartmann-Petersen, 2009; Drewes et al., 2010).

Since the early 1990s, the public transport system in Denmark has, as in many other countries, been privatized. In Copenhagen, that means that bus routes are outsourced to private companies that now apply for permission to run different routes for a limited period of time. Permission is given to the company that periodically offers the best product at the best price. Bus routes can change contractors without passengers noticing, and the bus drivers' working life and conditions are automatically reorganized. Potentially, this means a change in everyday life rhythms related to new geographical variations in workplaces, schedules, reconfigured work culture, etc. Depending on the specific employer, some drivers work the same route every day while others change routes during the day. One Sunday they might drive through the calm suburbs of the city, and the next morning they may do the chaotic rush hour, driving through busy central streets. The passengers, traffic, strains, and challenges differ depending on the area, bus route, time of day, and time of year – no day is the same.

By analyzing the stories shared by those who work *within* urban mobilities, new perspectives on the importance of the interconnection of contemporary everyday life evolve (Graham and Marvin, 2002; Hanam et al., 2006; Kesselring, 2006, 2015; Urry, 2007; Freudendal-Pedersen, 2009; Monroe, 2011). At first sight, the drivers' lives can seem highly regulated, predictable, and old-fashioned: they drive according to a schedule and follow a specific, predictable, predestined route throughout the city. Not many working lives are comparable to this. However, the drivers' work intermingles with the urban pulse. The city is a myriad of mobilities, rhythms, and diversities that interact with the regularities and irregularities of the bus schedule. Working amidst traffic, mobility is not always easy to predict or provide, and this unpredictability is experienced while the driver is sat "at home" in his or her seat behind the wheel. Working within such mobilities necessitates coping with unpredictability, albeit from a distinct vantage point. This is illustrated in the example below. The voices of four drivers are representing the general findings among the bigger group of interviewees. Curt, Allan, Frank, and Dora are all experienced middle-aged drivers in Copenhagen. They have driven various lines that pass multiple geographical, cultural, and traffic areas every day and night all year round.

Arrhythmia versus eurhythmia in bus driving

Urban traffic flow is unpredictable. In bigger cities, unexpected things happen every day. Roadworks, accidents, public demonstrations, re-routings, weather conditions, random events – unpredictability is normality. But even though it is a fundamental condition, it challenges the bus drivers. Arrhythmia is a working condition for the drivers. Curt says: "It's just too insane [driving a bus in Copenhagen], it's so stressful. You get stomach ulcers and heart attacks. The pressure is so high" (Hartmann-Petersen, 2009, p. 184, author's translation). According to Hvid, unpredictability and too many breaks in rhythms provide stressful working conditions (2006). The drivers agree that urban traffic is the most common stress factor and most routes pass through some of the most congested streets of Copenhagen. In relation to this, Allan says:

> I have seen colleagues who have been so stressed that they drove the bus into the bike path. It is horrible. This must not happen in a job like ours. Traffic is a death machine because of the mix of cars, cyclists and pedestrians.
>
> *(Hartmann-Petersen, 2009, p. 184, author's translation)*

In general, the cycling culture in Copenhagen is widely spread out and integrated with other modes of transport in city planning (Freudendal-Pedersen, 2015). However, this is a source of concern for the bus drivers. They all worry about the daily encounters with vulnerable road users, especially those who do not necessarily follow formal regulation. The drivers' encounters with passengers also contribute to a myriad of (clashing) rhythms or even the notion of arrhythmia. Here exemplified by Frank, who illustrates how the driver is affected by waiting situations and crowding at the bus stop:

> Then, a mum with three kids comes running towards the bus – and then she is searching for the ticket in her bag. Women's bags! It's crazy. Who invented the bag for ladies? [...] so she empties it, and at the very bottom she finds the ticket – and while all this happens, the driver is looking at his watch. He should have left a long time ago. He wants to be somewhere else. He ignores present time. [...] And you feel the denial while you drive [...] I often find myself getting stressed and annoyed by ladies' bags.
>
> *(Hartmann-Petersen, 2009, p. 184, author's translation)*

According to drivers, the fact that traffic and congestion in urban areas increase is not reflected in the timetable. The breaks get shorter and, if the driver gets behind schedule, the break can disappear as they try to make up time. The possibilities of influencing the work rhythm disappear and it becomes an individual challenge. Frank puts it as follows:

> I usually say that as a bus driver your senses work overtime, from the minute you enter the bus. ALL your senses are on duty – all the time. And that is exhausting [...] we don't choose for our senses to work overtime – and that's the main problem.
>
> *(Hartmann-Petersen, 2009, p. 186, author's translation)*

As indicated above, bus driving is not only affected by arrhythmic, stressful rhythms. As a contrast to experiencing unpredictable, anomic rhythms, drivers sometimes just "go with the flow" in a kind of eurhythmia, providing service and mobilities to the passengers. They also stress that it is of great value to recognize and acknowledge social diversity as an urban bus driver: "no two days are alike ... It's always a party", says Dora (Hartmann-Petersen, 2009, p. 187, author's translation). She loves the differences between the various areas of the city: "I love talking to people. Whether it's the drug addict, the alcoholic or the CEO – I really don't care. Everybody is treated equally on my bus. I just love the thing with people" (Hartmann-Petersen, 2009, p. 187, author's translation). Frank agrees:

> There are different categories of people. Some are bad guys, and some are just people going to buy groceries. Some are on their way to work [...] and we meet all age groups [...] I see old people, I meet children. I drive yuppies, drug addicts, and prostitutes. You become an experienced judge of character as a driver.
>
> *(Hartmann-Petersen, 2009, p. 187, author's translation)*

The look and behavior of the passengers reminds Frank of his location on the route at any particular moment. The drivers' stories show how they are able to recognize in which area of Copenhagen passengers live. The drivers' views are concurrent: It is of great value to them that the bus is a melting pot of cultures, habits, lifestyles, and values. In the case of bus drivers, polyrhythmicity is not only (clashing) multiple rhythms, it is also connected to social values and local signifiers reflecting urban life. Being on the move – co-producing the melting pot of rhythms inside the bus – drivers and passengers continuously learn about urban diversity and city life.

The drivers get to know the city and the routines of its citizens. Changing rhythms also has to do with changes in time. Frank observes changing urban impulses every day: "You sense changes all the time; changes in traffic, the variety of passengers entering the bus, changing lights, changing weather conditions, changing seasons – summer, winter. I see changes constantly. There is no steady rhythm. It changes all the time" (Hartmann-Petersen, 2009, p. 201, author's translation). These quotes contribute to the understanding of bus driving as an integrated part of urban polyrhythmicity.

Concluding remarks

Rhythms are connected to the perception of time, space, place, seasons, flow, cultures, etc. Urban living is polyrhythmic, and we can explore urbanity through the understanding of rhythms related to different mobilities. Everyday life is crammed full of rhythms and mobilities that help the individual fulfill his or her tasks and dreams. The stories of the bus drivers indicate that rhythms are important to perceive on many levels. Bus drivers pass through different neighborhoods every day, moving thousands of people around through multiple urban spaces. Driving a bus in Copenhagen maintains an arena of creating and mediating (competing) rhythms individually and collectively. The bus facilitates mobile encounters that contribute to everyday life cohesion (Koefoed et al., 2017). Empirically this chapter suggests how the interconnectedness between the city and urban mobilities can be investigated through the eyes of local bus drivers. The bus drivers' rhythms merge with city flows and everyday life and teach us how urban life is unpredictable and rhythmic, arrhythmic and eurhythmic, at the same time. They cope with urban multi-level polyrhythmicity every day. Their stories also show how they experience the city through their mobile work in order to provide coherent rhythms to the passengers on the move.

Mobilities research seeks to open the black boxes of modern life and unintended consequences of urbanity. By examining the relations between work, leisure, and modes, we find that everyday life in urban areas is not only routinized and predictable but also dynamic and helical (Hvid, 2006). We also investigate tacit knowledge and learn more about the role of networked urban mobilities. With this knowledge, urban planning potentially becomes better informed. Urban planning has to accommodate multiple – sometimes contradictory – needs in the city (liveabilty, growth, sustainability, accessibility, flow, recreation etc.). This potentially involves thorough, transdisciplinary understandings of the rationalities and irrationalities of, e.g. the citizens' everyday life. Urban areas are growing, and sustainable future planning also improves with an understanding of urban rhythms. Increasing congestion is currently a problem in most urban areas. Understanding the multiple factors that lead to congestion also includes understanding the rhythms in everyday life. Developing a coherent public transport system, including the role of buses, and developing the idea of sharing mobilities can be nuanced by concepts of rhythms. Focusing urban analysis on the bus – that serves as a well-known, maybe sometimes taken for granted, transport mode – is one way of evolving better future planning. Furthermore, discussions on flexible working hours and presence also touch upon the perception of rhythms and their interconnectedness to urban everyday mobilities in multiple ways. The knowledge from the people working within and providing mobilities needs to be integrated into specific urban planning, not as the only perspective but to complement existing knowledge of mobile patterns and behavior.

References

Adey, P., 2017. *Mobility*. 2nd ed. London: Routledge.

Bauman, Z., 2005. *Liquid life*. Cambridge: Polity Press.

Cook, M. and Edensor, T., 2017. Cycling through dark space: apprehending landscape otherwise. *Mobilities*, 12(1), pp. 1–19.

Cresswell, T., 2010. Towards a politics of mobility. *Environment and Planning D: Society and Space*, 28(1), pp. 17–31.

Dennis, K. and Urry, J., 2009. *After the car*. Cambridge: Polity Press.

Drewes, L., Nielsen, K.A., Munk-Madsen, E. and Hartmann-Petersen, K., 2010. *Fleksibilitet, flygtighed og frirum: en kritisk diagnose af det senmoderne arbejdsliv*. Roskilde: Roskilde Universitetsforlag.

Edensor, T., 2010a. *Geographies of rhythm: nature, place, mobilities and bodies*. Aldershot: Ashgate Publishing.

Edensor, T., 2010b. Walking in rhythms: place, regulation, style and the flow of experience. *Visual Studies*, 25(1), pp. 69–79.

Edensor, T., 2011. Commuter: mobility, rhythm and commuting. In Cresswell, T. and Merriman, P. (Eds.). *Geographies of mobilities: practices, spaces, subjects* (pp. 189–204). Aldershot: Ashgate Publishing.

Edensor, T., 2016. Introduction: thinking about rhythm and space. In Edensor, T. (Ed.). *Geographies of rhythm: nature, place, mobilities and bodies* (pp. 13–30). New York: Routledge.

Edensor, T. and Holloway, J., 2008. Rhythmanalysing the coach tour: the Ring of Kerry, Ireland. *Transactions of the Institute of British Geographers*, 33(4), pp. 483–501.

Edensor, T., Kärrholm, M. and Wirdelöv, J., 2018. Rhythmanalysing the urban runner: Pildammsparken, Malmö. *Applied Mobilities*, 3(2), pp. 97–114.

Edensor, T. and Larsen, J., 2018. Rhythmanalysing marathon running: 'A drama of rhythms'. *Environment and Planning A: Economy and Space*, 50(3), pp. 730–746.

Eriksen, T.H., 2001. *Tyranny of the moment*. London: Pluto Press.

Freudendal-Pedersen, M., 2015. Cyclists as part of the city's organism: structural stories on cycling in Copenhagen. *City & Society*, 27(1), pp. 30–50.

Freudendal-Pedersen, M., 2009. *Mobility in daily life: between freedom and unfreedom*. Farnham, Surrey: Ashgate Publishing.

Giddens, A., 1986. *The constitution of society: outline of the theory of structuration* (Vol. 349). Berkeley, CA: University of California Press.

Graham, S. and Marvin, S., 2002. *Splintering urbanism: networked infrastructures, technological mobilities and the urban condition*. London: Taylor & Francis.

Halkier, B., 2010. Focus groups as social enactments: integrating interaction and content in the analysis of focus group data. *Qualitative Research*, 10(1), pp. 71–89.

Hanam, K., Sheller, M. and Urry, J., 2006. Editorial: mobilities, immobilities and moorings, mobilities. *Mobilities*, 1(1), pp. 1–22.

Hartmann-Petersen, K., 2009. *I medgang og modgang: fleksibilitet og flygtighed i buschaufførers mobile liv* (Doctoral dissertation, Roskilde Universitetsforlag).

Hartmann-Petersen, K., 2018. Solid urban mobilities: buses, rhythms, and communities. In *Experiencing networked urban mobilities* (pp. 58–63). New York: Routledge.

Hochschild, A., 1997. *The time bind: when work becomes home and home becomes work*. New York: Metropolitan Books.

Hvid, H., 2006. *Arbejde og bæredygtighed*. Copenhagen: Frydenlund.

Jensen, O.B., Kesselring, S. and Sheller, M. eds., 2019. *Mobilities and complexities*. New York: Routledge.

Kesselring, S., 2006. Pioneering mobilities: new patterns of movement and motility in a mobile world. *Environment and Planning A*, 38(2), pp. 269–279.

Kesselring, S., 2015. Corporate mobilities regimes. Mobility, power and the socio-geographical structurations of mobile work. *Mobilities*, 10(4), pp. 571–591.

Koefoed, L., Christensen, M.D. and Simonsen, K., 2017. Mobile encounters: bus 5A as a cross-cultural meeting place. *Mobilities*, 12(5), pp. 726–739.

Kvale, S., 1996. *Interviews: an introduction to qualitative research interviewing*. Thousand Oaks, CA: Sage Publications, Inc.

Larsen, J., 2018. Commuting, exercise and sport: an ethnography of long-distance bike commuting. *Social & Cultural Geography*, 19(1), pp. 39–58.

Lefebvre, H., 2004. *Rhythmanalysis: space, time and everyday life*. London: Continuum.

Manderscheid, K., 2014. Criticising the solitary mobile subject: researching relational mobilities and reflecting on mobile methods. *Mobilities*, 9(2), pp. 188–219.

Monroe, K.V., 2011. Being mobile in Beirut. *City & Society*, 23(1), pp. 91–111.

Murray, L. and Doughty, K., 2016. Interdependent, imagined, and embodied mobilities in mobile social space: disruptions in 'normality','habit'and 'routine'. *Journal of Transport Geography*, 55, pp. 72–82.

Notar, B.E., 2012. 'Coming out'to 'hit the road': temporal, spatial and affective mobilities of taxi drivers and day trippers in Kunming, China. *City & Society*, 24(3), pp. 281–301.

Sennett, R., 1998. *The corrosion of character: the personal consequences of work in the new capitalism*. New York: WW Norton & Company.

Simpson, P., 2008. Chronic everyday life: rhythmanalysing street performance. *Social & Cultural Geography*, 9(7), pp. 807–829.

Spinney, J., 2006. A place of sense: a kinaesthetic ethnography of cyclists on Mont Ventoux. *Environment and Planning D: Society and Space*, 24(5), pp. 709–732.

Urry, J., 2007. *Mobilities*. Cambridge: Polity Press.

Wunderlich, M., 2008. Walking and rhythmicity: sensing urban space. *Journal of Urban Design*, 13(1), pp. 125–139.

SECTION IV

Everyday life, bodies and practices

The mundane and ordinary dimension of urban mobilities reaches beyond commutes and aggregated rhythms. People's lives, the actual embodied practices and the multi-sensorial engagement with the city and its movement spaces and systems are related to 'the geography closest in' or the human body. In Section IV, the focus is on everyday life, bodies and practices. It opens with **Chapter 18** and Gil Viry's analysis of life courses and mobility. The chapter briefly presents the key principles of the life course approach applied to spatial mobility and provides an overview of the mobility biographies approach as a recent application in mobility and transport research. It discusses the idea that mobility behaviours across the life course have become less predictable, less stable, more flexible and more individualised due to pronounced changes in advanced societies. These changes are operating at a different intensity and pace, and with a different dynamic, across social groups and across socio-spatial contexts. The evidence suggests that gender, class and race, the institutional, cultural and built environment, and their interrelationships, significantly shape life course mobility. These differences can be successfully captured by using the life course approach. The chapter concludes by critically discussing some deficits of this approach and by suggesting some potential avenues for future research. In **Chapter 19**, Martin Trandberg Jensen explores urban mobilities through the perspective of urban pram strolling. Mobilities in the company of children is a particular form of mobility and in this chapter, Jensen discusses this special form of urban mobility. First, it promotes a new way of studying children–family mobilities through mobilities design thinking and multimodal ethnography. This approach breaks with the discursive, textual and symbolic interpretations of pram strolling and relates it instead to questions of design, materialism and embodied practice. Second, the chapter develops the taxonomic term *homo cura* as a heuristic concept to help understand and classify the large group of urban users whose mobility patterns and practices are shaped by a heightened affective and emotional awareness on behalf of their infant's needs. This argument brings the chapter to a discussion on children–family mobilities as seen through the 'mobility of care'. Finally, to inspire future research on children–family based mobilities, the chapter positions this new generic inhabitant, homo cura, within a larger context of urban politics to raise questions related to urban accessibility and the rights to the city, and not least to inspire future research within the field. In **Chapter 20**, Christian Licoppe explores virtual urban interactions. Moving in the city is a practice that is noticeable. In his chapter, Christian Licoppe provides three examples of embodied mobility within traffic encounters in the city. Each is related to a new form of mobility: electric scooters, meeting strangers first encountered on online platforms (here in the course of organising a car sharing ride), and connected urban denizens who try to balance the demands of urban mobility and smartphone involvement. Each example illustrates some distinctive features and concerns associated with such urban behaviour: the multiplication of the types of vehicular units that claim the right to move in certain ways in public places in the case of electric scooters, the proliferation of meetings with strangers related to the use of social networks and collaborative mobility platforms in the case of the car-sharer, and the development of and problems raised by multi-activity in the case of connected mobile users. In each case, video contains recordings of such encounters are focused on 'single' mobile users. The chapter shows how we, as analysts, are still able to work with such video data to uncover significant phenomena and to reveal the witnessable orderliness of such urban mobility encounters and the ways in which their unfolding may be sensitive to, and inform us on, the moment by moment organisation of urban mobilities as accomplishments. That is, it shows us how to develop a video-ethnographic approach to embodied urban mobilities. Lynne Pearce discusses the realm of literary fiction in **Chapter 21**. While recent mobilities scholarship – drawing upon posthuman methodologies and phenomenological textual practices – has found evocative new ways of capturing the multifarious flows, rhythms and routines of urban life, this has necessarily been at the expense of enquiry into what individual subjects are *thinking* about as they travel through the city. This chapter explores the complex ways in which memory intrudes upon our everyday mobilities via thought-chains that are often profoundly *disembodied*, and – with reference to Manchester's literary fiction – demonstrates how a routine commute can turn into a revelatory experience. In **Chapter 22**, Thomas Buhler investigates the role of habit in our understanding of mobility. Much of the conceptualisation of urban mobilities such as traffic models takes for granted that mobile practices are planned and reflective practices. However, a large amount of the mobile practices in everyday life are better understood through the lens of habits. Bühler shows that repetitive and stable contexts in everyday life are the breeding ground for the formation of strong habits that omit the evaluation and deliberation phases, whether they concern mobility or other aspects of lifestyle such as energy consumption. After a necessary clarification of what habit is and is not, the chapter examines the state of the art in terms of methods for measuring the strength of a habit, with an illustration based on a recent French national survey. The theme of residential mobility is at the centre of Patrick Rérat's **Chapter 23**. Where do people live? Why do they move? How do they choose their dwelling and residential location? Residential mobility may be defined as a household's change of residence over short distances. The chapter first presents the traditional approaches of residential mobility and their concepts. It then argues that residential mobility is a choice under constraints and discusses its underlying dimensions (unit of analysis, profile, trajectory, criteria, decision-making process). It finally provides an overview of some research perspectives and of the current urban phenomena where residential mobility is at stake. Section IV ends with Aurore Flipo's **Chapter 24** on migration. In its first accounts of the city, sociology has paid particular attention to the figure of the stranger, defined by Simmel as one of the sociological forms the urban setting was likely to produce. Developed in reference to rural ethnology, urban sociology has initially conceptualised the city as an environment defined by mobility and social distance, as opposed to the anchorage and community-based relationships of the rural life. Urban ecology has investigated ethnic segregation as a process of urban growth and the assimilation of wide cohorts of new inhabitants coming from distant places and countries. This perspective, intrinsically linked with the historical context of 1920s' Chicago, has contributed to the birth of a sociology of immigration. Practices of transnationalism, although changing in shape and intensity with the advent of the 'global world', have always been part of migration systems. Recent research in the history of the development of cities has also demonstrated that the modern European city is not a city of uprooted rural emigrants. In this wide galaxy of mobilities, some are deemed illegitimate and undesirable, whereas others are encouraged and induced. International mobility in particular reveals the paradox between the simultaneous increase of liberalisation and the coercion of mobility practices. Indeed, the difference between migrants, refugees, asylum-seekers, undocumented migrants, international students, expatriates and so on is not found not so much in the motives or content of their mobility than in the way their mobility is managed and, more importantly, the degree of constraint exerted on them. This differential treatment reflects the social and political construction of wanted and unwanted hosts in the city, and has tangible consequences on individuals' experiences of mobility in the city.

18
LIFE COURSE AND MOBILITY

Gil Viry

Introduction

This chapter is concerned with the life course approach to spatial mobility. Rather than mobility being analysed as a 'behaviour' at one point in time, the life course approach emphasises that movement, whether for work, leisure or social activities, should be understood over the lifetime. It examines the stability and changes in mobility behaviours, its drivers and consequences, over people's lives. The forms of mobility that the researcher considers to be relevant, such as relocation, social visits, migration, business or tourist travel, are conceptualised as individual trajectories, in which past experiences and anticipated future are assumed to influence later experiences of mobility. These mobility trajectories are understood as being constructed dynamically through individuals' choices and actions within the opportunity structure and the socio-spatial context (Elder, Johnson and Crosnoe, 2003; Giele and Elder, 1998). The socio-historical context, such as a recession, may induce distinctive mobility responses to the structural forces operating at that *time and place*, which results in *cohort* and *period effects* on spatial mobility. The determinants and impacts of mobility behaviours are also likely to vary according to the *timing* of mobility in individuals' lives, in relation to age or life course transitions, for example. Finally, the life course approach emphasises that mobility behaviours influence and are influenced by key events in other life domains. It is, for example, well known that people who move to city suburbs or become parents are more likely to increase their car use (Scheiner and Holz-Rau, 2013).

A life course (or more broadly a biographical) approach has been successfully used in the field of migration and residential mobility since the 1980s (see Findlay et al., 2015 or Courgeau and Lelièvre, 1989 for a seminar study). But it is more recently that this approach has received the wide attention it deserves in transport and mobility research (see Müggenburg, Busch-Geertsema and Lanzendorf, 2015 for a recent literature review). While transport studies have stressed the importance of habits in travel behaviour and the relative stability of daily activity-travel patterns (Klöckner and Matthies, 2004), a growing literature examines changes in the ways people travel, such as reducing private car use (Redman et al., 2013; Rocci, 2015). A particularly promising research agenda focuses on mobility behaviour changes in response to major life events like childbirth, home purchase or a job change (Lanzendorf, 2010; Prillwitz, Harms and Lanzendorf, 2006, 2007; Scheiner and Holz-Rau,

2013). These recent developments and the growing availability of longitudinal data allowing researchers to track individual mobility histories, either prospectively (panel studies for example) or retrospectively (biographical studies using i.e. life history calendars or life story interviews), offer promising ways to extend the scope of mobility research and strengthen the importance of the life course approach as an indispensable framework for understanding changes in mobility behaviours.

In this chapter, I begin by briefly presenting the key principles of the life course approach applied to spatial mobility and provide an overview of the mobility biographies approach as a recent application in mobility and transport research. Then I discuss the idea that mobility behaviours across the life course have become less predictable, less stable, more flexible and more individualised due to pronounced changes in advanced societies. Following this, I emphasise that these changes are operating at a different intensity and pace, and with a different dynamic, across social groups and across socio-spatial contexts. Evidence suggests that gender, class and race, the institutional, cultural and built environment, and their interrelationships, significantly shape life course mobility. These differences, I argue, can be successfully captured by using the life course approach. I conclude by critically discussing some deficits of this approach and by suggesting some avenues for future research.

The life course approach to spatial mobility

Elder, Johnson and Crosnoe (2003) recognise five principles of the life course approach: the lifespan, timing, human agency, historical time and place, and linked lives. While Elder et al. give little attention to the mobility or spatial dimensions of the life course, I briefly discuss each of these principles with reference to spatial mobility.

The focus on the *lifespan* means that mobility behaviours at any point in the life course are conceived as the product of earlier experiences and future expectations. A life course perspective points to lifelong dynamics and departs from approaches that emphasise the role of early socialisation in mobility. Through their mobility experiences (including immobility), individuals develop specific skills, values and attitudes (e.g. place attachment) that influence later experiences. A well-known result in migration studies is that people who recently migrated are more likely to migrate again in the near future (e.g. Fischer and Malmberg, 2001). This influence of past experiences (or *path dependency*) has been also observed between different forms of spatial mobility, for example when migration in youth or early adulthood fosters overnight business travel later in life (Vincent-Geslin and Ravalet, 2015). Evidence shows the complex and contingent character of this process, since people exposed to spatial mobility early in life can also develop negative attitudes toward spatial mobility (ibid.). Mobility trajectories are viewed as interdependent with trajectories in other areas of life (principle of *multi-dimensionality*). Spatial mobility behaviours, key life events and transitions in other domains, and their respective outcomes (e.g. associated roles, costs and benefits) may complement or conversely compete. It is a well-known finding that having children can trigger moves outside inner-city areas (e.g. Kulu and Steele, 2013). The lower fertility among women commuting long distances over many years is another example (Huinink and Feldhaus, 2012; Rüger and Viry, 2017).

These studies point to the importance of *timing* in mobility decisions and their relations to life course transitions. In particular, the determinants and impacts of mobility behaviours may significantly vary according to the timing of mobility. Experiencing a life event or transition which deviates from normative life-course patterns may be more problematic or may require more resources than conforming to the expected life transitions. Using the English

Longitudinal Study of Ageing, Vanhoutte, Wahrendorf and Nazroo (2017) showed, for example, that frequent moving in childhood had no association with wellbeing later in life, a positive association in young adulthood and a negative association in midlife. In the authors' view, moving in young adulthood was often associated with favourable life transitions (higher education or family development), whereas later moves were more likely to reflect hardships, such as widowhood, unemployment or divorce. Life course researchers have shown that it is not just the timing of mobility in relation to life course transitions that matters but also the *sequencing*. For example, migrating before or after union formation and childbirth, but also single and multiple migrations, have different consequences for women's occupational achievement (Mulder and Van Ham, 2005).

Elder, Johnson and Crosnoe (2003) stress that individuals possess *agency* and construct their own life course through their actions and choices within the opportunity structure. This agency is situational, bound to the perceived circumstances of a place and time (see below), and with respect to the past and anticipated futures. Mobility biographies are therefore the joint product of structural incentives (e.g. affordable and reliable train services, promotion of rail transport) and individual responses to this external reality to pursue certain goals (e.g. deciding to take the train, moving close to a railway station).

The life course is embedded *historically and geographically*. The spatial context, such as the transport infrastructure or the dispersion of activities and services, and historically specific socio-economic structures, such as a recession or economic globalisation, significantly shape the life-course mobility patterns of a given generation and society. For example, a series of socio-cultural changes over the past decades has offered new possibilities for (especially middle-class) young people to experience 'new' forms of living and mobility arrangements (e.g. independent living or studying abroad) between leaving the parental home and establishing a new household (Galland, 1995).

Finally, studying mobility decisions involves analysis of *linked lives*. According to this principle, the mobility trajectory of an individual develops in close relationship with the life course of household or family members. The partner's career or an elderly parent's health for example are likely to impact on people's mobility behaviours. This has focused attention on how people within a household or a broader family configuration negotiate from different age, gender and class positions the relative desirability and benefits of being mobile. Depending on the research focus, not only are mobility decisions negotiated between the linked lives of household and family members, but they are also shaped by other close relationships, such as employers, friends and neighbours. Using panel data in Germany, Knies (2013) showed, for example, that residential mobility and easy access to a car or public transport is negatively associated with visiting neighbours.

Recent developments in transport geography around the concept of *mobility biographies* offer an interesting framework for examining travel behaviour change and their connections with key life events in a life-course perspective (Lanzendorf, 2003; Müggenburg, Busch-Geertsema and Lanzendorf, 2015; Scheiner, 2007). Lanzendorf (2003) suggests analysing how changes in the mobility domain, including long-term mobility decisions, such as purchasing a car or a season ticket for public transport, are related to changes in the lifestyle domain (fertility, household composition, employment and leisure) and the accessibility domain (access to daily activities, such as places of work and recreation). Scheiner (2007) also considers three main domains of individual trajectories, which are somewhat different from those used by Lanzendorf. He distinguishes (i) the trajectories in the employment sphere, comprising training completion and job changes; (ii) trajectories that belong to the household and family, including changes in co-residence, such as leaving the parental home or a divorce; and (iii) residential trajectories including residential relocations and changes in the environment. While interesting hypotheses can be drawn from the mobility biographies approach, a more integrated theoretical explanation of the interdependencies between these life domains and levels of analysis (individual action and societal opportunity structure) is needed – a point I will come back to in the conclusion.

Mobility behaviours over the life course: greater instability?

For authors such as DiPrete et al. (1997) or Mayer (2004), the deregulation of the labour market and other profound changes in work organisation of contemporary societies resulted in an increased discontinuity of life course patterns. By shifting more frequently between jobs and life projects than in the past, individuals would experience more unstable life trajectories. Some scholars claim that individual spatial mobility reflects and contributes to this trend with less predictable and less stable employment-related mobility behaviours in a context of deregulated labour markets, increased labour flexibility and casualisation (Callaghan, 1997; Ludwig-Mayerhofer and Behrend, 2015). More unstable careers would be accompanied by repeated and irregular spatial mobility experiences, either in the form of relocation or travel (e.g. Jirón and Imilan, 2015). Moving or travelling for work would be encouraged by employers and governments, including those of supra-state institutions such as the European Union, with the goal to increase employability, especially that of vulnerable populations living on welfare (Jensen and Richardson, 2004; Orfeuil, 2004).

This increased instability in spatial mobility behaviours would be made possible by the wider access to high-speed technologies and (increasingly digitised) *mobility systems*, which open up a wider range of mobility choices (Urry, 2007). The widespread use of personal cars, planes and telecommunications has profoundly changed the spatial organisation of human activities, whether for work, leisure, family or social life (Larsen, Axhausen and Urry, 2006). Individuals can develop and maintain long-distance relationships in other parts of the country or the world and exchange goods, information, affection or care through physical travel and telecommunications. Other people choose to travel extensively (rather than relocating) to stay close to their loved ones. Because individuals are less bound by physical proximity, the spatial boundaries of human activities may become more blurred with fluid and changing spatial mobility patterns over the life course.

Demographic and family changes since the 1970s could also contribute to a greater instability of mobility behaviours over the life course. Dual-earner households, delayed and partial marriage, low and late fertility, increased union dissolution and pluralised family and cohabitation forms, such as 'solo living' or 'living apart together' could contribute to more complex life course mobility (Beck and Beck-Gernsheim, 2013; Green, 1997). Less influenced by traditional family norms and values, family responsibilities and relocation decisions would be more a matter of negotiation and choice than strict determination by the social structures and norms (Mason, 1999). This would result in more unstable, more individualised and more differentiated (or 'de-standardised') life course mobility in the sense that mobility experiences would occur at more dispersed ages and with more dispersed durations in the new generations.

Life course mobility across socio-spatial contexts

The changes sketched above are however likely to operate at a different intensity and pace, and with a different dynamic, across socio-spatial contexts of people's lives. The instability of life course mobility may particularly apply to some social groups at specific moments in their lives. Although the previous section discusses some important factors contributing to the instability of mobility behaviours, we can likewise identify some other factors contributing to their stability, including lifelong immobility. In a life course approach, all these factors are however likely to operate differently across people's lives and across historically specific socio-economic and spatial structures.

Research within the 'New mobilities paradigm' (Adey, 2009; Cresswell, 2006; Sheller and Urry, 2006) has stressed the unequal distribution of choice around spatial mobility of all kinds and scales (from everyday movement to global travel and communication) and how the mobility of some depends on the immobility, forced or precarious mobility of others (Bissell, 2016; Hannam, Sheller and Urry, 2006). There is considerable evidence that mobility behaviours over the life course are shaped by spatial structures, such as neighbourhood and accessibility, and social structures along age, class, disability, gender, ethnicity and citizenship lines. These structural forces operate at multiple levels, for example within the household through gender roles as discussed in the feminist geography literature (e.g. Hanson and Pratt, 2003; Uteng and Cresswell, 2008), at the local and regional level in the housing or labour market through class and ethnic divisions as stressed in the literature on spatial mismatch between where people live and where jobs are available (e.g. Gobillon, Selod and Zenou, 2007) or at the national and global levels with international mobility flows, including multiple forms of forced migration engendered by political and economic instability in the most vulnerable parts of the world (see e.g. Castles, 2003). It is therefore not only important to examine continuity and change in mobility behaviours but also how these changes (or their absence) are experienced and what consequences they have depending on the populations concerned and the socio-spatial contexts in which mobility occurs. For instance, the experience and consequences of short and repeated periods of extensive business travel are likely to be different for precarious workers and a global, wealthy kinetic elite (Cresswell, Dorow and Roseman, 2016). Below, I summarise some recent studies on two forms of mobility – employment-related mobility and the residential mobility of families – that provide evidence that life course mobility remains strongly organised by the spatial and social structures within which individuals behave.

Travelling extensively or moving to another region or country for a job strongly depends on individuals' earlier mobility experiences and their positions in both the life course (e.g. parenthood) and social structure (e.g. Schneider and Meil, 2008). Highly mobile workers are often highly qualified young people without children and older men who had been 'on the move' for many years. The willingness to migrate or commute long distances is also influenced by changes in the economic context and people's personal financial situation. A panel study in four European countries showed that people in Spain were significantly more willing to be mobile for a job after the 2008 economic crisis than before the crisis (Viry and Kaufmann, 2015). This was particularly true among those who experienced a deterioration of their financial situation (see also Ahn, De La Rica and Ugidos, 1999 for similar results on unemployed people in Spain). Using data from a sparsely populated Swedish region, Cassel et al. (2013) similarly showed that job seekers' willingness to commute long distances significantly varied with work history and sociodemographics. Women, young parents, people with low education and those with long spells of unemployment had lower intentions to accept long commutes. The propensity to commute long distances has also been recognised to change with the characteristics of the spatial contexts within which people reside and work, such as jobs–housing balance, traffic congestion and accessibilities (Holz-Rau, Scheiner and Sicks, 2014; Horner, 2004). For instance, long-distance commuting is more likely for residents in lower-density areas who commute to larger cities (e.g. Öhman and

Lindgren, 2003). There is less evidence about how changes in the spatial characteristics of home and work location impact changes in journey-to-work patterns. An exception is the study by Prillwitz, Harms and Lanzendorf (2007) using panel data from Germany who found that both professional and residential changes appear to increase the average commuting distance.

Residential trajectories of households and families were also found to be strongly determined by the wider spatial and social structures within which individuals act. Research on the geography of families has long identified the various factors that contribute to the residential proximity between generations, such as intergenerational support (e.g. grandchild care) or transmissions (e.g. housing inheritance) (Hallman, 2010; Imbert, Lelièvre and Lessault, 2018). Despite significant differences across countries, studies show that a large majority of Europeans lives in the same region as their parents (e.g. Hank, 2007) and spatial proximity to parents reduces the chances of moving long distances (Ermisch and Mulder, 2018). Evidence shows that the lack of resources may prevent some disadvantaged social groups from moving away from their family and friends (Fol, 2010; Zorlu, 2009). The local presence of parents and siblings acts as a significant barrier to relocation for young people from poor backgrounds and ethnic minorities. Families who are more scattered are often those of immigrants who have strong incentives to migrate for economic or political reasons. Proximity to parents varies also substantially across the life course, reflecting changing needs of both generations over time. For instance, childbirth may trigger a move closer to grandparents for childcare support (e.g. Blaauboer, Mulder and Zorlu, 2011).

Such evidence does not necessarily deny that mobility behaviours over the family and professional life course have become more unstable due to macro-level changes in technology, demography, culture and the economy. But it suggests that spatial mobility often requires important resources by unequally positioned social actors in raced, classed and gendered relations and remains strongly organised by the institutional, cultural and built environment of the specific place, region or country where people live. The growing availability of longitudinal data opens the prospect of examining more fully the changing nature of life course mobility for different actors and generations, and in different places.

Conclusion

The life course approach has recently gained considerable attention in transport and mobility research. There has been a growing interest by mobility researchers in adopting this approach to study mobility changes over the lifetime. Researchers can use various methods for collecting biographical data on mobility behaviours (e.g. life story interviews, life history calendars, panel survey) and for analysing them (e.g. narrative analysis, event-history analysis, sequence analysis, statistical methods for longitudinal data). Following people in time and space offers a unique way to examine the fluidity of present-day mobility biographies, for work, for pleasure, to sustain family and intimate life and so on.

Research from various fields has provided evidence that individual mobility behaviours vary greatly depending on the life circumstances and the historically specific socio-spatial contexts, to which individuals adapt in their mobility choices. The profound transformations in mobile technology, demography and the organisation of work over the past decades is driving changes in spatial mobility. Yet, these changes are likely to operate at a different intensity and pace, and with a different dynamic, across social groups and environments.

The life course approach is well suited for understanding how spatial mobility changes in relation to these structural changes, by analysing the variation of mobility behaviours over the lifetime, and across generations and socio-spatial contexts. Recent developments of the mobility biographies approach (Lanzendorf, 2003; Müggenburg, Busch-Geertsema and Lanzendorf, 2015; Scheiner, 2007) offer an interesting framework in which to analyse drivers of change in travel behaviours, such as key life events and transitions in various life domains (e. g. housing, work, family, leisure). The life course approach also provides enough flexibility to analyse a wide range of mobility behaviours, socio-spatial contexts and life domains identified as relevant by the researcher.

Like any framework, the life course approach is however limited by its ontological and epistemological foundations. This approach additionally lacks the status of a unified theory (Huinink and Kohli, 2014). The mobility biographies approach, which stems from it, yields a series of propositions and research questions. But a theoretical understanding of the ways in which past experiences and anticipated future, key life events, socio-spatial contexts and linked lives influence mobility-related decisions needs further development. In particular, the social processes by which social actors adjust their mobility choices to structural incentives and the interdependencies across the life course between mobility and other life domains and roles are largely left underspecified.

Despite these theoretical limitations, recent life course studies have improved understanding of mobility behaviour change and much empirical work remains to be carried out in several areas. First, future studies could examine possible cohort and period effects on spatial mobility in relation to structural changes at multiple scales (new transport infrastructure or new transport policy, for example) that promote certain mobility options and preclude others. Second, the determinants and implications of more fluid mobility trajectories need further investigation. Third, the linked lives principle has not been applied extensively in mobility studies due to a lack of appropriate data. Measuring how mobile lives are linked over time and space entails important challenges but offers promising avenues for future research. Researchers interested in studying processes of change in the spatial organisation of households, for example, could use household panel survey data to analyse travel mobility biographies of both partners in relation to their careers or the household's residential history. Researching linked lives beyond the household and neighbourhood is also a very exciting possibility. In studies on social visits, the network dimension is recognised (Axhausen, 2008; Axhausen and Kowald, 2015), but evidence on the relationship between life course mobility and personal networks remains limited. Overall, depending on the research questions to be addressed, there are many possible combinations of the forms of mobility, life domains and contextual scales that have yet to be explored.

References

Adey, P., 2009. *Mobility*. London: Routledge.

Ahn, N., De La Rica, S. and Ugidos, A., 1999. Willingness to move for work and unemployment duration in Spain. *Economica*, 66(263), pp. 335–357.

Axhausen, K.W., 2008. Social networks, mobility biographies, and travel: survey challenges. *Environment and Planning B: Planning and Design*, 35(6), pp. 981–996.

Axhausen, K.W. and Kowald, M., 2015. *Social networks and travel behaviour*. Aldershot: Ashgate.

Beck, U. and Beck-Gernsheim, E., 2013. *Distant love*. New York: Wiley.

Bissell, D., 2016. Micropolitics of mobility: public transport commuting and everyday encounters with forces of enablement and constraint. *Annals of the American Association of Geographers*, 106(2), pp. 394–403.

Blaauboer, M., Mulder, C.H. and Zorlu, A., 2011. Distances between couples and the man's and woman's parents. *Population, Space and Place*, 17(5), pp. 597–610.

Callaghan, G., 1997. *Flexibility, mobility and the labour market*. Aldershot: Ashgate.

Cassel, S.H., Macuchova, Z., Rudholm, N. and Rydell, A., 2013. Willingness to commute long distance among job seekers in Dalarna, Sweden. *Journal of Transport Geography*, 28, pp. 49–55.

Castles, S., 2003. Towards a sociology of forced migration and social transformation. *Sociology*, 37(1), pp. 13–34.

Courgeau, D. and Lelièvre, E., 1989. *Analyse démographique des biographies*. Paris: Ined.

Cresswell, T., 2006. *On the move: the politics of mobility in the modern west*. London: Routledge.

Cresswell, T., Dorow, S. and Roseman, S., 2016. Putting mobility theory to work: conceptualizing employment-related geographical mobility. *Environment and Planning A: Economy and Space*, 48(9), pp. 1787–1803.

DiPrete, T.A., De Graaf, P.M., Luijkx, R., Tahlin, M. and Blossfeld, H.-P., 1997. Collectivist versus individualist mobility regimes? structural change and job mobility in four countries. *American Journal of Sociology*, 103(2), pp. 318–358.

Elder, G.H., Johnson, M.K. and Crosnoe, R., 2003. The emergence and development of life course theory. In: J.T. Mortimer and M.J. Shanahan, eds., *Handbook of the life course, handbooks of sociology and social research*. Boston, MA: Springer, pp. 3–19.

Ermisch, J. and Mulder, C.H., 2018. Migration versus immobility, and ties to parents. *European Journal of Population*, 35(3), pp. 587–608.

Findlay, A., McCollum, D., Coulter, R. and Gayle, V., 2015. New mobilities across the life course: a framework for analysing demographically linked drivers of migration. *Population, Space and Place*, 21 (4), pp. 390–402.

Fischer, P.A. and Malmberg, G., 2001. Settled people don't move: on life course and (im-) mobility in Sweden. *International Journal of Population Geography*, 7(5), pp. 357–371.

Fol, S., 2010. Mobilité et ancrage dans les quartiers pauvres: les ressources de la proximité. *Regards Sociologiques*, 40, pp. 27–43.

Galland, O., 1995. Une entrée de plus en plus tardive dans la vie adulte. *Économie et statistique*, 283(1), pp. 33–52.

Giele, J.Z. and Elder, G.H., 1998. *Methods of life course research: qualitative and quantitative approaches*. London: Sage.

Gobillon, L., Selod, H. and Zenou, Y., 2007. The mechanisms of spatial mismatch. *Urban Studies*, 44 (12), pp. 2401–2427.

Green, A.E., 1997. A question of compromise? case study evidence on the location and mobility strategies of dual career households. *Regional Studies*, 31(7), pp. 641–657.

Hallman, B.C. ed., 2010. *Family geographies: the spatiality of families and family life*. Don Mills: Oxford Unversity Press.

Hank, K., 2007. Proximity and contacts between older parents and their children: a European comparison. *Journal of Marriage and Family*, 69(1), pp. 157–173.

Hannam, K., Sheller, M. and Urry, J., 2006. Editorial: mobilities, immobilities and moorings. *Mobilities*, 1(1), pp. 1–22.

Hanson, S. and Pratt, G., 2003. *Gender, work and space*. London: Routledge.

Holz-Rau, C., Scheiner, J. and Sicks, K., 2014. Travel distances in daily travel and long-distance travel: what role is played by urban form? *Environment and Planning A: Economy and Space*, 46(2), pp. 488–507.

Horner, M.W., 2004. Spatial dimensions of urban commuting: a review of major issues and their implications for future geographic research. *The Professional Geographer*, 56(2), pp. 160–173.

Huinink, J. and Feldhaus, M., 2012. Fertility and commuting behaviour in Germany. *Comparative Population Studies*, 37(3–4).

Huinink, J. and Kohli, M., 2014. A life-course approach to fertility. *Demographic Research*, 30, p. 1293.

Imbert, C., Lelièvre, É. and Lessault, D. eds., 2018. *La famille à distance: mobilités, territoires et liens familiaux*. Paris: Ined.

Jensen, O.B. and Richardson, T., 2004. *Making European space: mobility, power and territorial identity*. London: Routledge.

Jirón, P. and Imilan, W.A., 2015. Embodying flexibility: experiencing labour flexibility through urban daily mobility in Santiago de Chile. *Mobilities*, 10(1), pp. 119–135.

Klöckner, C.A. and Matthies, E., 2004. How habits interfere with norm-directed behaviour: a normative decision-making model for travel mode choice. *Journal of Environmental Psychology*, 24(3), pp. 319–327.

Knies, G., 2013. Neighbourhood social ties: how much do residential, physical and virtual mobility matter? *The British Journal of Sociology*, 64(3), pp. 425–452.

Kulu, H. and Steele, F., 2013. Interrelationships between childbearing and housing transitions in the family life course. *Demography*, 50(5), pp. 1687–1714.

Lanzendorf, M., 2003. Mobility biographies. A new perspective for understanding travel behaviour. 10th international conference on travel behaviour research (IATBR). Lucern: IATBR.

Lanzendorf, M., 2010. Key events and their effect on mobility biographies: the case of childbirth. *International Journal of Sustainable Transportation*, 4(5), pp. 272–292.

Larsen, J., Axhausen, K.W. and Urry, J., 2006. *Mobilities, networks, geographies*. London: Ashgate.

Ludwig-Mayerhofer, W. and Behrend, O., 2015. Enforcing mobility: spatial mobility under the Regime of activation. *Mobilities*, 10(2), pp. 326–343.

Mason, J., 1999. Living away from relatives: kinship and geographical reasoning. In: S. McRae, ed., *Changing Britain: families and households in the 1990s*. Oxford: Oxford University Press, pp. 156–175.

Mayer, K.U., 2004. Whose lives? how history, societies, and institutions define and shape life courses. *Research in Human Development*, 1(3), pp. 161–187.

Müggenburg, H., Busch-Geertsema, A. and Lanzendorf, M., 2015. Mobility biographies: a review of achievements and challenges of the mobility biographies approach and a framework for further research. *Journal of Transport Geography*, 46, pp. 151–163.

Mulder, C.H. and Van Ham, M., 2005. Migration histories and occupational achievement. *Population, Space and Place*, 11(3), pp. 173–186.

Öhman, M. and Lindgren, U., 2003. Who are the long-distance commuters? patterns and driving forces in Sweden. *Cybergeo: European Journal of Geography*, 243. http://cybergeo.revues.org/index4118.html.

Orfeuil, J.-P., 2004. *Transports, pauvreté, exclusions: pouvoir bouger pour s'en sortir*. Paris: Editions de l'Aube.

Prillwitz, J., Harms, S. and Lanzendorf, M., 2006. Impact of life-course events on car ownership. *Transportation Research Record*, 1985(1), pp. 71–77.

Prillwitz, J., Harms, S. and Lanzendorf, M., 2007. Interactions between residential relocations, life course events, and daily commute distances. *Transportation Research Record*, 2021(1), pp. 64–69.

Redman, L., Friman, M., Gärling, T. and Hartig, T., 2013. Quality attributes of public transport that attract car users: a research review. *Transport Policy*, 25, pp. 119–127.

Rocci, A., 2015. Comment rompre avec l'habitude? Les programmes d'accompagnement au changement de comportements de mobilité. *Espace populations sociétés*, (1–2).

Rüger, H. and Viry, G., 2017. Work-related travel over the life course and its link to fertility: a comparison between four European countries. *European Sociological Review*, 33(5), pp. 645–660.

Scheiner, J., 2007. Mobility biographies: elements of a biographical theory of travel demand. *Erdkunde*, pp. 161–173.

Scheiner, J. and Holz-Rau, C., 2013. Changes in travel mode use after residential relocation: a contribution to mobility biographies. *Transportation*, 40(2), pp. 431–458.

Schneider, N.F. and Meil, G., 2008. *Mobile living across Europe I: relevance and diversity of job-related spatial mobility in six European countries*. Opladen: Barbara Budrich.

Sheller, M. and Urry, J., 2006. The new mobilities paradigm. *Environment and Planning A*, 38(2), pp. 207–226.

Urry, J., 2007. *Mobilities*. Cambridge: Polity.

Uteng, T.P. and Cresswell, T., 2008. *Gendered mobilities: towards an holistic understanding*. Aldershot: Ashgate.

Vanhoutte, B., Wahrendorf, M. and Nazroo, J., 2017. Duration, timing and order: how housing histories relate to later life wellbeing. *Longitudinal and Life Course Studies*, 8(3), pp. 227–244.

Vincent-Geslin, S. and Ravalet, E., 2015. Socialisation to high mobility? In: G. Viry and V. Kaufmann, eds., *High mobility in Europe: work and personal life*. London: Palgrave Macmillan, pp. 59–82.

Viry, G. and Kaufmann, V. eds., 2015. *High mobility in Europe: work and personal life*. London: Palgrave McMillan.

Zorlu, A., 2009. Ethnic differences in spatial mobility: the impact of family ties. *Population, Space and Place*, 15(4), pp. 323–342.

19
URBAN PRAM STROLLING

Martin Trandberg Jensen

Introduction

> I notice a tense facial expression on my son ... then the wet eyes ... then a moment of pending silence before the screaming shoots through the air in a relentless force: '*Waaa* ... *waaaa* ... *WAAAA AAA AA*!!!' I nervously shake the pram from side-to-side; look apologetically at the urban commuters next to me on the train. Their looks seem judgmental to me as I embody an affective state of both distress, irritation and guilt ... I lift the small one up; caress him softly and, slowly, the screaming turns into a light coughing followed by relaxed silence. The train wagon is yet again subdued into commuter silence.
>
> *(Impressionistic tale – author)*

Mobility with children is radically different from individual mobility. In many ways, the planning, embodied experience and everyday practice of urban mobility is completely rearranged when one is on the move with children. As conveyed in the opening impressionistic tale, this chapter sets out to discuss the everyday challenges and embodied experiences of urban pram strolling. I use the following pages to discuss how this practice is more than an instrumental act of 'walking while pushing a pram', but a socio-material and intuitive negotiation. Importantly, given the scope and limits of this short chapter the following pages reduce 'children–family mobilities' to urban pram strolling with infants, whereas mobility with older children (i.e. toddlers, preschoolers, young teens), or through different modes of mobility (strollers, bicycles, scooters etc.), arguably offer many different types of mobility challenges and coping strategies.

Urban mobilities research has mainly studied the dominant modes of movements within cities, including for example pedestrian studies (Horton et al., 2014; Middleton, 2011) and automobility (Buchanan, 2015; Bull, 2004), whereas research on the various forms of children-shaped mobility remains relatively neglected (Clement & Waitt, 2018; Jensen, 2017; Sander & Coutard, 2001). That said, throughout the last years researchers have begun to explore issues related to pram strolling and the construction of motherhood (Boyer & Spinney, 2016) as well as pram strolling in relation to mobilities of care (Grant-Smith et al., 2017). Yet the understanding of the embodiment and everyday experience of pram strolling is still limited (Clement & Waitt, 2018). This chapter extends and contributes to the existing literature by exploring the links between mobilities design thinking (Jensen & Lanng, 2017) and urban pram strolling. As such the work responds to calls to engage with mobilities design thinking to the study of families and urban mobility (Jensen, 2017). By so doing, the chapter unpacks how pram strollers develop specific tactics and coping strategies to respond to the conditions of designed urban environments. This account thus opens up a range of new design-oriented questions and material sensitivities otherwise largely neglected in pram strolling studies. In so doing, the chapter makes two contributions.

First, it promotes a new way of studying children–family mobilities through mobilities design thinking and multimodal ethnography. This approach breaks with discursive, textual and symbolic interpretations of pram strolling and rather relates it to questions of design, materialism and embodied practice. Second, I develop the taxonomic term, *homo cura*, as a heuristic concept to help understand and classify the large group of urban users whose mobility patterns and practices are shaped by a heightened affective and emotional awareness on behalf of their infant's needs. This argument brings the chapter to a discussion on children–family mobilities as seen through the 'mobility of care' (de Madariaga, 2016). Finally, to inspire future research on children–family based mobilities, I position this new generic inhabitant, homo cura, within a larger context of urban politics to raise questions related to urban accessibility and the rights to the city, and not least inspire future research within the field.

The following shortly outlines the existing research on children–family mobilities research and relates it to the notion of care.

Children-family mobilities and pram strolling studies

The study of children–family mobilities is multifaceted and emerges from different disciplinary standpoints. McLaren and Parusel (2015) focus on gendered parenting practices in automobilised urban spaces whereas Collins and Freeman (2005) address ways of achieving the child-friendly city. Carver et al. (2013) compare the independent mobility of children across Australia and England, while Barker (2009) explores the mobility deprivation experienced by children and young people in the UK. This growing research on children–family mobilities has led Clement and Waitt (2018) to suggest that the rise of accompanied child mobility is partly attributed to changing cultures of parenting, increased automobility cultures as well as the effects of unwalkable cities.

On this background, research on pram mobility has grown steadily in recent years. Early feminist geography studied pram strolling in relation to motherhood practices and explored it as a gendered mobility 'burden' marginalising mothers in the city (McDowell, 1993). More recently, transport scholars have related pram strolling to discussions on accessibility and urban planning (Grant-Smith et al., 2012; Kenyon et al., 2003). As an inspiration to this work, a strand of recent research has studied the affective, embodied and material dimensions of pram strolling. Boyer and Spinney (2016), for example, explore the material assemblages – including the pram – in the formation of motherhood identities. With a similar material focus, Sander and Coutard (2001) provide a phenomenological account of the situational challenges of moving through the Parisian metro system with a bulky stroller. Finally, Cortés-Morales and Christensen (2015) take an actor network theory approach to discuss how pushchairs and children influence everyday family journeys to playgrounds. Through their accounts they provide telling examples of how pushchair mobility is relationally achieved, through technological, material and social networks, thus illustrating that the notion of 'independent mobility' really never exists (Nansen et al., 2015).

One of the main insights that emerge from these corporeal understandings of pram strolling is that it is a mobile practice formed by intentions of *care*. These practices emerge through a complex, social and material amalgamation between the adult–infant–pram. This human–material constellation can also be seen as an emotional and affective state of 'dwelling-in-motion' (Urry, 2007, p. 11) which influences the ways urban environments are interpreted and negotiated. From a Heideggerian understanding, dwelling involves cherishing, protecting and caring for the surroundings we are part of. Whether you travel with infants as a mother, father, youngster, grandparent or friend, this dwelling-in-motion can be expressed as 'babysitting on the go' (Grant-Smith et al., 2017, p. 202) and entails a long range of coping strategies (for example, nap changing, milk giving, caressing, clothes changing, shade ensuring, noise reducing), and not least mobility tactics for negotiating urban transport systems not always designed for family needs (Sander & Coutard, 2001). Henceforth the concept of 'mobilities of care' (Grant-Smith et al., 2017) is a fruitful notion when discussing the infrastructural prerequisites for children–family mobilities and the construction of the child-friendly cities of tomorrow.

The central argument of this chapter is that the act of 'caring for mobility' cannot be reduced to an emotional or purely corporeal modus of pram strolling. Caring is unavoidably conditioned by everyday materials and designed urban environments. Consequently, to inform existing pram strolling literature this chapter links the relatively new transdisciplinary school of mobilities design thinking (Jensen & Lanng, 2017) to pram strolling studies. It does so because mobilities design thinking offers a pragmatic and situational way of addressing the problems and potentials of mobility as emerging in close ties between the user and the built environment. Architect Juhani Pallasmaa speaks of 'haptic architecture' as a way of understanding how important the senses are in understanding the way humans inhabit spaces and sites (Pallasmaa, 2005). He argues that gravity is 'measured by the bottom of the foot' and the ground itself 'through our soles' (ibid., p. 58). And so it is only through a situational and embodied understanding of the 'body-in-place' can we comprehend the material complexity that conditions everyday mobilities (Jensen, 2016, p. 589). For a more thorough discussion on mobilities design thinking see Jensen (2017) or Jensen et al. (2016); for now, suffice to say that mobilities design thinking is fruitful for its ability to zoom in on the affordances of designed materials (such as the pram) and for acknowledging the importance of designed urban environments and the role of human practices in shaping urban pram strolling. The following sets out to describe how I methodically engaged with the aims of mobilities design thinking in the context of the chapter.

Studying children–family mobilities

One of the challenges that holds back creative research on children–family mobilities is a methodical one: how to come to grips with the rich, multisensorial and mobile experience of pram strolling? How to avoid reducing the intuitive, non-representational and 'felt' dimensions of pram strolling without reducing it to discursive, symbolic or purely textual renditions?

My replies to these questions are based on fieldwork undertaken during the summer months of 2016, primarily in the city centre of Copenhagen. This urban ethnography included more than 50 video recordings, numerous standalone photos as well as a number of informal interviews with local pram users. To explore the role of urban infrastructures in forming pram strolling, I developed a materially sensitive way of engaging with mobile video ethnography that I termed 'surface ethnography' (Jensen, 2017). Surface ethnography reveals the qualitative properties of pavements and textures in ways that traditional camera angles (most often eye-level views) are incapable of. The recorded videos make visible the interplay between different modalities (the visual, the auditory, the kinesthetic), and animate the rhythmicity of different material designs, particularly the textures, shapes and qualitative properties of surfaces as they are moved across and negotiated with a pram (e.g. gravel, asphalt, cobble stones and kerbs). Consequently, it works particularly well for evoking the embodied 'feel' of surfaces while pram strolling.

This visual footage is complemented by written impressions by the author that enrich the videos through accounts of the sounds, movements, feelings and affective atmospheres that characterised particular pram strolls. Much like ethnography then, surface ethnography experiments with new ways through which video can aid the creation of thick descriptions (Spinney, 2009). For the researcher, doing surface ethnography can also be seen as an embodied way of learning and knowing (Pink & Mackley, 2012) and a reflexive participation in the research area. This deep practical involvement is aligned to the pragmatic ethos of mobilities design thinking.

To substantiate the findings from the surface ethnography, the analysis also draws on netnographic findings from online family blogs discussing pram strolling practices and challenges, and contains extracts from interviews with stroller parents from the Greater Copenhagen area. Weaved together, these insights form part of a 'bricolage' approach (Löfgren, 2014) where different multimodal materials shape the analysis and lead to a richer understanding of pram strolling in relation to mobilities design issues. Now, having outlined the methods, the following provides an exploration of pram strolling as an everyday and embodied practice informed by mobilities designs.

Mobilities design: care as materially conditioned

The caring for mobility while pram strolling is conditioned by a range of materialities and their affordances. Most conspicuously, the pram or stroller is a designed 'mobility vehicle' that comes with a range of design functions (see Figure 19.1). Most prams or strollers can be folded together in various ways; others can be entirely packed away to fit a carry-along bag; some have large sunshades; some have front wheels that turn 360 degrees whilst others do not, and some have rubber wheels whereas others are made from hardened polymeric material (plastic).

From this mobilities design perspective, a pram cannot be reduced to a 'dead' thing (Ingold, 2007), but is rather an *engineered and designed artefact* with embedded intentionalities that influence the range and type of dwellings-in-motion (Urry, 2007) during pram strolling. The affordances of prams differ greatly and represent the material conditions through which distinct intentions of mobility care can be cultivated. Over time, pram users become accustomed with the pram functions and slowly transform into knowledgeable and 'able-bodied' pram strollers (Jensen, 2017) with new embodied and material sensitivities. Through this mode of dwelling, pram strollers develop tactics to ensure that their infant is calm and safe during travel. For example, some pram strollers use clothes pegs to hang cloths across the pram to shield the infant from visual impressions; others 'refurnish' the interior of the pram with blankets, pillows and cuddly toys to create a safe and soft space during pram strolling; others even install small sound devices (or use their smartphones) to create soundscapes of relaxing music or soothing white noise recordings to relax their infants during travel.

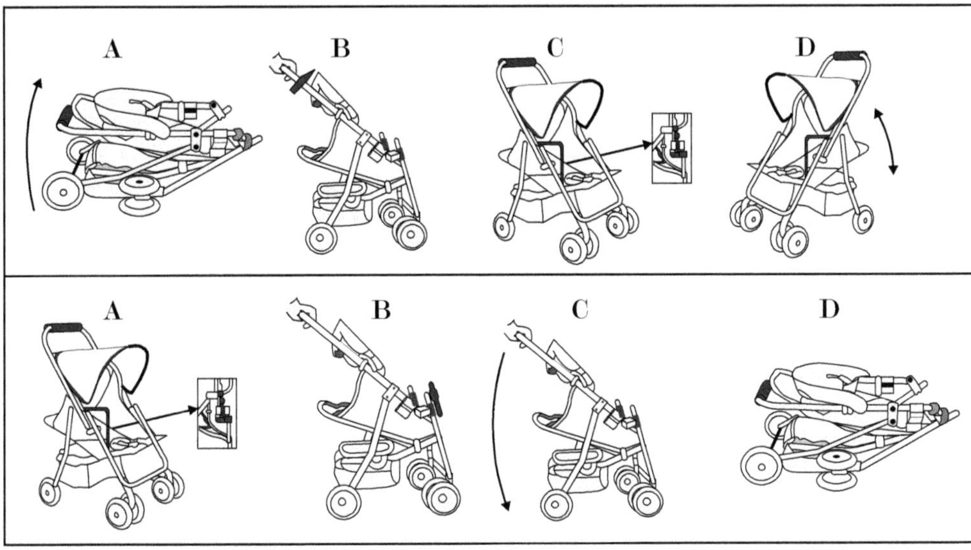

Figure 19.1

A father whom I talked to even changed the wheels of his pram, from spoke air wheels to proper air wheels, as he felt it provided a softer 'mobility feel' during pram strolling.

Accordingly, we cannot talk of pram strolling or mobilities of care without attaining to the *materialism* that holds it together. From this perspective, being a pram stroller cannot be reduced to an instrumental act of 'moving with a pram' but must be understood as a material and embodied practice. Through a mobilities design oriented understanding of the pram's form, shape and design functions I have sought to illustrate how specific material conditions influence intentions of care, but importantly, also how these material conditions are not static, but constantly negotiated, circumscribed and re-appropriated by user practices. A large pram may be practical for a sleeping infant, but troublesome when moving in tight urban spaces; a pram with retro wheels may look beautiful, but does not afford gentle movements across rugged terrain; a fully foldable pram may be practical for mobile transitions, but often not very robust for extended usage (Jensen, 2017). And consequently, as Latour (1992) argues, when we start to see how every material component that holds together pram strolling is folded into contradictory usages, we start to understand the complex socio-material relations that conditions mobilities of care and pram strolling. These material conditions are not reducible to the pram and its designs, but include the greater urban environment as well, which is what the following seeks to shortly address.

Caring for rhythms

Pram strolling is a dwelling-in-motion informed by the embodied understanding of places. One example of this can be seen in the striving for rhythms, different kinds of mobile undulations that function as soothing mobility experiences for the infant. The urge for 'moving-never-stopping' is driven by affective impulses (based on the imagery of a crying infant and all that entails) and requires pram strollers to move in perpetuum: 'When he's finally asleep I shouldn't stop the pram. I must stay on the move.

I have to ensure that the pram is not immobilised! If it happens, he wakes up!' (Pram stroller, Copenhagen). To provide a soothing *rhythm*, pram strollers, over time, develop embodied ways of understanding the potentials of (designed) surfaces:

> I normally enjoy cobblestones. They shape the atmosphere of urban places. Today, however, I am struggling to move across these rectangular interventions with my pram. The standard polymeric wheels of the pram provide little support as I squeeze my way through tightly parked bikes and across the rugged cobblestone pavements.[1] It must be a matter of time before Alexander wakes up, I think, as I suddenly notice two parallel lines of border stones that I immediately aim for with the hope of smooth pram mobility … As I follow the lines, new contemporary obstacles emerge as both sleeping locals and a group of feeding pigeons take up space on these border stones that now function as a safe mobile 'corridor' for my pram … a few minutes I pause despairingly as a large group of tourists suddenly block my route (See Figure 19.2).

When studying the impressionistic tale and the hyperlink, notice the distinct up-and-down movements of the camera, the sound of the creaking plastic and aluminium chassis of the pram, and not least, the sound of the spoke air wheels crossing border stones. While the border stones are laid with the design intention to ease urban access for the mobility impaired, in this example they are re-appropriated into new significances as they function as a relaxing and long-stretched 'riverbed' providing a soothing rhythmscape (Jensen et al., 2015) for the infant. An important insight from this is that we cannot reduce urban stroller parents to docile mobile subjects, rather, they are co-creators and illustrative of how every single mobilities design is open for situated 'interpretative flexibility' (Larsen, 2017) often informed by intentions of care. From this multimodal and everyday example, it becomes clear that pram strolling is informed by the embodiment of places: *Do places 'feel' safe? Do various surfaces 'feel' right for the sort of rhythmicity that you are searching for as a pram stroller?* Yet caring for rhythms as a pram stroller is not a generic tactic, and while some may be searching for relatively smooth surfaces, others look for completely different rhythmscapes. As one pram stroller exclaimed on an online parent's blog:

> If you have a good stroller, with decent air-filled tires, then a bit of bumping isn't going to kill the baby, in fact my kids slept well when things were a bit bumpy, sort of like when you put a car seat on top of a washer or dryer … Some kids find it hypnotic.
>
> *(Victoria, pram stroller)*

Other pram strollers even joke about how their many recurrent pram strolls on cobblestones may have developed, in their child, an embodied response to future cobblestone experiences:

> Living in a city like Istanbul with poor sidewalks and some cobblestone streets makes for a rough ride. We used to joke that Vera was so calmed by rolling over the bumpy streets; she might later have a Pavlovian response to cobblestones and fall asleep when she walks on them.
>
> *(Knocked Up Abroad Travels)*[2]

As seen in these examples, there are various intensities, coping strategies and implications of caring for the rhythms of pram strolling. These decisions are diverse, and depend on the

Figure 19.2

needs of each individual infant, the affordances of the pram design and not least the material environment. This is a dwelling-in-motion (Urry, 2007) that is constantly refined as the pram stroller gets to know and understand the signals and needs of the infant during travel. This short empirical illustration underlines how pram strolling is shaped through a complex coordination between embodied and materially mediated rhythms, repetitions and frequencies. Importantly, this reminds us that the art of pram strolling is not mastered by the eyes alone (Haldrup & Larsen, 2009), it does not 'impoverish the other senses' (ibid., p. 72), but rather emerges through the various ways the perceptive body registers its environments (Pallasmaa, 2005).

Having provided two short examples of how materialism and embodiment inform one another and condition intentions of care, the following sets this discussion within a greater political and urban context.

Homo cura: furthering mobilities of care

As shown in the empirical examples in this chapter, the notion of *care* is central to understand the embodied and practical dimensions of pram strolling. To help further the thinking about mobilities of care in relation to children–family mobilities research I propose a new generic term to the vocabulary of urban planners and civil engineers: *homo cura*. As a combination of the Latin words 'homo' (people, generation) and 'cura' (care), homo cura represents urban inhabitants who are on the move with infants and embody a significant sensuous and affective attentiveness on behalf of their infants. Arguably, homo cura is a taxonomic term (with all the reductionisms that such term entails) that works as an approximation of how people-on-the-move-with-infants obtain the highest possible well-being on behalf of the infant, given specific material and infrastructural opportunities and constraints. It is meant to contribute to the vocabulary of policy-makers and planners when they discuss 'generic users' of the city, and in so doing, nuancing the perspectives through which issues such as access, justice, design and smartness are discussed within urban planning and development contexts.

Homo cura crystallises how a large group of urban inhabitants curate mobility through slow and relatively careful physical gestures, soft and gentle movements, noise and light protection tactics, and so, generally, through substantial practical, affective and emotional efforts. Importantly, the term does not imply that all mobility decisions are rational (in opposition to the idea of the homo economicus in economics), in fact, since no infant gives clear instructions, various modes of caring are relatively intuitive and driven by 'gut feelings'. Accordingly, the act of urban pram strolling is not only a practical amalgamation, or assemblage between the adult–infant–pram (Clement & Waitt, 2018), but an emotional and non-representational dwelling-in-motion (Urry, 2007) which influences the ways urban environments are interpreted and coped with.

Introducing this new mobile archetype to the lexicon of urban studies and mobilities research opens a range of political and ethical questions regarding urban mobility, justice, access, values and power. What material conditions do specific mobile modes of caring, such as the ones embodied by pram strollers, stand within contemporary urban planning visions (often framed by automobility and velomobility doctrines)? What rights to the city should relatively neglected urban users, such as homo cura, have within infrastructural designs of future cities? What role may technologies play in advancing urban pram strolling experiences? How may we design in better ways to cater for the needs of homo cura? In Copenhagen, a city often hailed as a great example of the 'bikeable city' and the 'walkable city', policy-makers will need to think about children–family mobilities as more than instrumental mobility, but as an emotional, affective and practical modus of negotiating the city, and an everyday mobility condition that influences the everyday life of many inhabitants.

Children–family mobilities: future research avenues

While this chapter has focused specifically on urban pram strolling, there is still a range of opportunities for further research on children–family mobilities. For example, questions arising from this work may be connected to larger discussions on how to imagine and built the child-friendly city of tomorrow (Clement & Waitt, 2018). Accordingly, with this new inhabitant, homo cura, we can start to reimagine what large urban visions such as the 'democratic city', safe city, smart city or good city (Amin, 2006) entail in terms of infrastructural design solutions. To inform this process there are a number of areas within children–family mobilities that can be explored. Until now, the majority of research on family mobilities has given voice to the mothers, fathers and grown-up curators of family mobility. There is, still, a lack of research taking into account the thoughts, impulses and wants of children in forming mobility practices (Boyer & Spinney, 2016). Saying this provides openings for methodical innovations within a field of research still largely based on textual representations, symbolism and discursive accounts (for exceptions dealing with the multisensorial dimensions of pram strolling see for example Clement & Waitt, 2018; Jensen, 2017; Sander & Coutard, 2001). Finally, questions related to the creation of parenthood, identity construction, gender and accessibility have informed an abundance of work within this field. This chapter has exposed the promising links between mobilities design thinking and pram strolling, and in so doing promoted a critical material and design-driven understanding of pram mobilities. In this vein, interventionistic and experimental and user-driven research, design workshops or collaborative and cross-sectorial 'pram-athons' (a play on words inspired by hackathons) serve as inspiring ways of engaging anew with children–family mobilities research and change-driven research agenda in the future.

Conclusion

Throughout this chapter I have recoiled from reducing pram strolling to typical representational categories such as gender, identity or class. Rather, I have called into attention the socio-material and embodied practices of pram strolling set within everyday encounters in the city. From this exploration, I contribute twofold to pram strolling studies and children–family mobilities more generally.

Firstly, I have promoted an innovative way of engaging with and understanding children–family mobilities. Inspired by mobilities design thinking (Jensen & Lanng, 2017) I have opened up an avenue for exciting new ways of exploring pram strolling and children–family mobilities. With this chapter I display how mobilities design thinking is relevant in the study of children–family mobilities by addressing exploratory questions such as: How do certain materialities feel? How are they experienced? And since the 'practice of design' is never unintentional, what are the politics of mobilities designs? Who/what are distinct design codes and material intentions catering for? Who are marginalised? These are the type of pertinent design-oriented questions that mobilities design thinking opens up and I believe the time is ripe to ask them in the context of studies on urban children–family mobilities.

Secondly, the chapter has developed the term homo cura as an attempt to give form, visibility, and thus political attention, to a large group of inhabitants whose everyday urban lives are highly shaped by dwellings-in-motion (Urry, 2007) shaped by intentions of care. Further research is needed to understand the organisation of mobilities of care (for example related to everyday materialism, designed infrastructures and interpretive performances). Related to this notion of care, I have finally put forth a range of avenues that future critical research on children–family mobilities may explore. In light of the increasing urbanisation of contemporary societies, the increasing congestion of urban populations and diverse intensities of human flows, these research directions will be topical to explore in order to shape and inform the design directions of the child-friendly city of tomorrow.

Notes

1 https://youtu.be/0UVAzJ0od9Y
2 https://knockedupabroadtravels.com/2013/05/01/gear-what-makes-a-good-travel-stroller/

References

Amin, A., 2006. The good city. *Urban Studies*, 43(5–6), pp. 1009–1023.

Barker, J., 2009. "Driven to distraction?": Children's experiences of car travel. *Mobilities*, 4(1), pp. 59–76.

Boyer, K., & Spinney, J., 2016. Motherhood, mobility and materiality: Material entanglements, journey-making and the process of 'becoming mother'. *Environment and Planning D: Society and Space*, 34(6), pp. 1113–1131.

Buchanan, C., 2015. *Traffic in towns: A study of the long term problems of traffic in urban areas*. New York: Routledge.

Bull, M., 2004. Automobility and the power of sound. *Theory, Culture & Society*, 21(4–5), pp. 243–259.

Carver, A., Watson, B., Shaw, B., & Hillman, M., 2013. A comparison study of children's independent mobility in England and Australia. *Children's Geographies*, 11(4), pp. 461–475.

Clement, S., & Waitt, G., 2018. Pram mobilities: Affordances and atmospheres that assemble childhood and motherhood on-the-move. *Children's Geographies*, 16(3), pp. 252–265.

Collins, D. A. C. F., 2005. 'Creating child friendly cities symposium: Outcomes and directions statement', *Creating Child Friendly Cities (CFCC) Conference, Sydney, 30–31* October *2006.*

Cortés-Morales, S., & Christensen, P., 2015. Unfolding the pushchair. Children's mobilities and everyday technologies. *Rem–research on Education and Media*, 6(2), pp. 9–18.

de Madariaga, I. S., 2016. Mobility of care: Introducing new concepts in urban transport. In: I. S. de Madariaga and M. Roberts, eds. *Fair shared cities: The impact of gender planning in Europe*. New York: Routledge. pp. 51–66.

Grant-Smith, D., Edwards, P., & Johnson, L., 2012. 'Mobility in the child (and carer) friendly city: SEQ vs. Stockholm', *Association of European Schools of Planning (AESOP) 26th Annual Congress, Ankara, 11–15 July 2012.*

Grant-Smith, D., Osborne, N., & Johnson, L., 2017. Managing the challenges of combining mobilities of care and commuting: An Australian perspective. *Community, Work & Family*, 20(2), pp. 201–210.

Haldrup, M., & Larsen, J., 2009. *Tourism, performance and the everyday: Consuming the orient*. New York: Routledge.

Horton, J., Christensen, P., Kraftl, P., & Hadfield-Hill, S., 2014. 'Walking … just walking': How children and young people's everyday pedestrian practices matter. *Social & Cultural Geography*, 15(1), pp. 94–115.

Ingold, T., 2007. Materials against materiality. *Archaeological Dialogues*, 14(01), pp. 1–16.

Jensen, M. T., 2017. Urban pram strolling: A mobilities design perspective. *Mobilities*, 13(4), pp. 584–600.

Jensen, M. T., Scarles, C., & Cohen, S. A., 2015. A multisensory phenomenology of interrail mobilities. *Annals of Tourism Research*, 53, pp. 61–76.

Jensen, O. B., 2016. Of 'other' materialities: Why (mobilities) design is central to the future of mobilities research. *Mobilities*, 11(4), pp. 587–597.

Jensen, O. B., & Lanng, D. B., 2017. *Mobilities design: Urban designs for mobile situations*. New York: Routledge.

Jensen, O. B., Lanng, D. B., & Wind, S., 2016. Mobilities design: Towards a research agenda for applied mobilities research. *Applied Mobilities*, 1(1), pp. 26–42.

Kenyon, S., Rafferty, J., & Lyons, G., 2003. Social exclusion and transport in the UK: A role for virtual accessibility in the alleviation of mobility-related social exclusion? *Journal of Social Policy*, 32(03), pp. 317–338.

Larsen, J., 2017. Bicycle parking and locking: Ethnography of designs and practices. *Mobilities*, 12(1), pp. 53–75.

Latour, B., 1992. Where are the missing masses? The sociology of a few mundane artifacts. In: W. E. Bijker and J. Law, eds. *Shaping technology/building society: Studies in sociotechnical change*. Cambridge: MIT Press. pp. 225–258.

Löfgren, O., 2014. The black box of everyday life entanglements of stuff, affects, and activities. *Cultural Analysis*, 13, pp. 77–98.

McDowell, L., 1993. Space, place and gender relations: Part I. Feminist empiricism and the geography of social relations. *Progress in Human Geography*, 17(2), pp. 157–179.

McLaren, A. T., & Parusel, S., 2015. 'Watching like a hawk': Gendered parenting in automobilized urban spaces. *Gender, Place & Culture*, 22(10), pp. 1426–1444.

Middleton, J., 2011. 'I'm on Autopilot, I just follow the route': Exploring the habits, routines, and decision-making practices of everyday urban mobilities. *Environment and Planning A*, 43(12), pp. 2857–2877.
Nansen, B., Gibbs, L., MacDougall, C., Vetere, F., Ross, N. J., & McKendrick, J., 2015. Children's interdependent mobility: Compositions, collaborations and compromises. *Children's Geographies*, 13 (4), pp. 467–481.
Pallasmaa, J., 2005. *The eyes of the skin: Architecture and the senses*. Chichester: Wiley.
Pink, S., & Mackley, K. L., 2012. Video and a sense of the invisible: Approaching domestic energy consumption through the sensory home. *Sociological Research Online*, 17(1), pp. 1–19.
Sander, A., & Coutard, O., 2001. Zoé dans le métro. *Flux*, 2001(2), pp. 96–98.
Spinney, J., 2009. Cycling the city: Movement, meaning and method. *Geography Compass*, 3(2), pp. 817–835.
Urry, J., 2007. *Mobilities*. Cambridge: Polity Press.

20
THE VIDEO-ETHNOGRAPHY OF EMBODIED URBAN MOBILITIES

Christian Licoppe

Urban mobilities as embodied practices are public and observable. Their 'witnessable' character is a crucial resource for the intelligibility and self-organization of everyday traffic encounters. The contingent and fleeting encounters between embodied/material vehicular units that make up everyday traffic encounters display an emergent order based on the perceptible resources mobile units give off and make available whenever they happen to encounter one another at a given moment. The visually accomplished and visually available orderliness of embodied mobilities and traffic encounters make video recordings a very important empirical resource to analyse them. This chapter will focus on the video-ethnography of embodied mobilities in traffic encounters associated with new forms of mobility.

Two important streams of analytical research have focused on the understanding of coordinated urban mobilities as a lived and embodied experience. There is the interactionist work of Erving Goffman and his successors, who treat urban encounters as an important case of emergent organization in co-present interaction. Such an organization, and particularly in the case of traffic encounters between strangers, rests on mutual displays of perceptible conduct: participants continuously and mutually adjust their behaviour so that it may be observed by others. Embodied mobile behaviour provides information which may be understood on a moment by moment basis by others who 'read' it as potentially indexing intentionality (Goffman, 1963), as forming the basis for potential 'claims' (Goffman, 1971), such as rights of occupation or passage, and as displaying an orientation towards the prohibition of excessive proximity or touch as a constitutive feature of their intelligibility. Such given off information may be used to anticipate and jointly manage coordinated conduct in traffic encounters in a way that is therefore also sensitive to moral expectations regarding propriety in public places. In Goffman's framework, urban mobilities appear as a succession of self-organizing, fleeting encounters enacting a morally laden 'interaction order' (Goffman, 1983).

Ethnomethodology and conversation analysis (EM/CA) has also approached traffic encounters as an orderly phenomenon. It has been used to study phenomena such as crossing the road at complex junctions (Liberman, 2013), going for a run downtown (Smith, 2019), overtaking (Deppermann et al., 2018), looking for a parking space (Laurier, 2005) or pedestrian flow and queues (Watson, 1997). Criticizing Goffman for what it considers a cognitive bias, ethnomethodology has emphasized the importance of reflexivity and categorization analysis. Reflexivity points to the observable way in which actions are made accountable with respect to a relevant context which they in turn ceaselessly transform (Garfinkel, 1967). For instance, 'space' or 'urban spaces' do not 'contain' the traffic encounters that take place 'there'; space is continuously produced by the myriad of practices around which a traffic encounter is organized and which are oriented to it. Membership categorization analysis is also an important EM/CA analytic resource for understanding interactions in public places. It refers to the observable way in which participants may enact categories as relevant to a given encounter. Membership categorization 'does' stuff, because categories fit together in collections, and because categories are tied to activities and claims (Sacks, 1992). For instance, in a queue in a public place, the first in line or the last in line, or the relationship of 'nextness' are turn-generated, witnessable categories with particular implications regarding the ongoing activity (Watson, 1997). Some notions introduced by Goffman may be reinterpreted in categorical terms: his concept of 'with', which consists of mobile units that display that they are moving together in some sense and may thereby claim and be assigned special rights, can be reinterpreted as an interaction-generated category enacted in and as embodied conduct.

Because of the importance of embodied displays in the emergent organization of traffic encounters, video recordings constitute an important resource for understanding coordinated mobility flows in urban public places, and embodied conduct in interaction in general (Heath et al., 2010). I will describe here the combination of video-recordings of mobile conduct and an interaction-oriented analytic frame focused on unpacking the emergent organization of urban mobilities as a lived experience, as the 'video-ethnography' of urban mobilities.

Video-ethnographic analysis may run into various difficulties, however. On the methodological level, because the phenomena of interest involve mobile units, the recording apparatus must also be mobile and may involve specific 'mobile methods' (Buscher et al., 2010), such as cameras affixed to mobile participants or mobile observers 'shadowing' the action. On the analytical level, particularly when participants are moving alone or as 'singles',[1] the video-ethnographic data is mostly visual and may prove difficult to analyse for the ethnographer. EM/CA researchers usually circumvent this difficulty by studying mobile parties moving as pairs or multiples and susceptible to talk about their experiences together, for instance approaching the driving experience from the situated conversation between driver and passenger (Laurier et al., 2008). Then the analyst is able to deploy the resources of multimodal conversation analysis to understand the way in which their talk-in-interaction is articulated to their embodied mobility behaviour (Mondada, 2016). Another approach, outside EM/CA, has been for the ethnographer to 'shadow' the mobile participants and to get them to talk about their mobility experience in the course of its accomplishment (see for instance Pink, 2013), but it is not always possible (it would not work in the electric scooter case below).

What I want to do in this chapter is provide three examples of embodied mobility within traffic encounters in the city. Each is related to a new form of mobility: electric scooters, meeting strangers first encountered on social media, and connected urban denizens who try to balance the demands of urban mobility and smartphone involvement. Each illustrates some distinctive features and concerns associated with such urban behaviour: the multiplication of the types of vehicular units that claim rights of way in public places in the case of electric scooters, the proliferation of meetings with strangers related to the use of social media in the car-sharing example, and the development of problems raised by multi-activity in the case of connected mobile users. In each case, I will introduce video recordings of such encounters focused on 'single' mobile users. I will try to show how it is still possible to work with such video data so as to uncover significant phenomena and to reveal the witnessable orderliness of urban mobility encounters as a moment by moment accomplishment. That is, how to develop a video-ethnographic approach to embodied urban mobilities.

From electric scooter user to pedestrian: 'body gloss' and the reflexive enactment of urban spaces

My first example is based on an ongoing study of the emergent uses of electric scooters in Paris.[2] At the time of the study (autumn 2018), freefloating electric scooters to rent had recently been introduced, their use was spreading, and it was unclear whether they could be ridden on the pavement or whether they should be restricted to cycle lanes. This ambiguity was related to uncertainty regarding the classification[3] of electric scooter users as urban vehicular units (i.e. were they a kind of pedestrian or a kind of cyclist?). As we will see, such an uncertainty in legal and administrative classifications provides resources for flexible embodied behaviours on the part of electric scooter users. In this study, we made users of electric scooters wear camera glasses, and one of us shadowed them with an additional camera to get a perspective on the embodied behaviour of mobile scooter users. In the case below, the scooter user came to a red light (Figure 20.1).

As there was no incoming traffic from the right, she put a foot down onto the ground and then the other (Figure 20.2). She then started walking forwards (Figure 20.3). Even though we are dealing here with a single user, it is possible to make sense of her conduct from an interactional perspective, using the notion of 'body gloss'. Body gloss is 'a means by which the individual can try to free himself from what otherwise would be the undesirable characterological implications of what it is he finds himself doing' (Goffman, 1971, 129). Body gloss can be reinterpreted with membership categorization analysis. Through one's embodied and perceptible conduct, one may make oneself accountable in body gloss as a recognizable incumbent of some category of passer-by, thus making relevant the rights, obligations and activities tied to this category, and allowing for the inferences that can be made on the spot, on the basis of such an incumbency. Coming back to our scooter user, by disembarking and pushing her device on foot, she has 'body glossed' herself into a kind of pedestrian, which makes relevant and accountable in a particular way her walking through the intersection with a scooter at a red light.

However, walking is not the 'official' activity relevant to that urban spatial configuration. She is on the road, where only motorized vehicles should be (there is no cycle lane there), and the light for 'normal' pedestrians wishing to cross the road at this juncture is red. However, it is customary for pedestrians in Paris to try to cross when the light is green for cars

Figure 20.1

Figure 20.2

Figure 20.3

Figure 20.4

Figure 20.5

Figure 20.6

when no cars appear to be coming. What is interesting here is not only the fact that the participant walks through the intersection with her electric scooter (Figure 20.4), but that she swerves a bit to the right, so as to be closer to the pedestrian zebra crossing (Figure 20.5). It is only when she has almost reached the other side that she gets back on her scooter, thus morphing back again into a motorized vehicular unit of sorts (Figure 20.6).

Getting closer to the pedestrian crossing is a way for this woman to reinforce her body gloss and accountability as an incumbent of the category 'pedestrian'. She thus also displays an orientation towards the fact that crossing at a red light on the road without getting off her scooter would appear transgressive to potential onlookers, more so than doing this 'as a pedestrian'. She thus exploits the moral–spatial organization of urban spaces (places in urban spaces are saturated with potential rights, obligations and claims for particular categories of mobile units) as a resource to make herself more readable as a pedestrian, and thus as more strongly entitled to cross the street. However, the moral–spatial organization of space is not only a resource here, but also an outcome of her embodied behaviour. Through the latter, she enacts the zone near the pedestrian crossing as one entitling crossing at that moment (for mobile units recognizable as pedestrian units). She therefore actively constitutes the moral–spatial organization of urban public places through her embodied behaviour, which shows the reflexive relationship between embodied mobilities and urban spaces.

Such a moral–categorical flexibility is related in part to the uncertain status of electric scooters, and in part to the affordances of the device (it is easy to step down from it and push it along). Moreover, such a body gloss highlights the reflexive relationship between the accomplishment of scooter-based mobility as an intelligible activity and the experience of urban spaces as morally organized. Finally, the body gloss and its implications are available in the same way to a passer-by and to the analyst provided they are socialized to everyday urban life. Even though there is no talk of any sort, and no focused interaction with other participants, it is thus still possible for the analyst to unpack the 'witnessable order' of this embodied mobility scene. However, many types of mobile behaviours in the city may be more difficult to interpret and may require additional methodological efforts, as in the next example.

Car sharing: 'approaching' a potential meeting point with a driver and passengers one does not know

This is taken from a study of car sharing practices.[4] Car sharing experiences were recorded by recruiting passengers who were willing to wear camera glasses. One interesting stage for

Figure 20.7

us is the mobility involved in finding the vehicle one is supposed to ride in, within the framework of a pre-arranged encounter (usually online) which will involve a driver and passengers one does not know. In the data shown here, one passenger recorded his approach of such a meeting point.

The unknown car was parked alongside other cars, facing the kerb, with the driver and some passengers standing and talking in front of the car on the side of the road. Our passenger, who we will call Leo, identifies them from the other side of the road as he approaches the group (he glances at them to his right on the video data (Figure 20.7), which he later confirmed when the video was shown to him). However, instead of crossing the road and going straight to them, he approaches them in a quite distinctive way.

He keeps to the left and walks towards the curb six or seven cars away on their left (Figure 20.8), walking in a way and along a trajectory which enacts him as an anonymous pedestrian, unrelated to their group. Once on the kerb, he turns right, and goes past them on the opposite side of their car without stopping (Figure 20.9).

He overtakes it by one car, and only then does he turn right to walk towards the road between two cars (Figure 20.10).

As soon as he is on the road, he turns right to head towards the standing group, looking at the driver's back who is talking to the other passengers (Figure 20.11). As he gets near, the driver turns around (Figure 20.12). He greets her with a '*bonjour*' ('good morning'), and she greets him back.

When I show this in data sessions, participants usually take notice of this embodied approach as remarkable. What makes it a noticeable phenomenon is a common sense expectation regarding mobility that is based on a kind of 'economy maxim': if one must get to a particular location that is visible and there is no obvious obstacle, then it is relevant and expected that one will go straight there so as to minimize one's path. This could have been the case here. In view of such a normative orientation, the participant's actual path appears as a kind of circumnavigation around the parked cars for which there appears

Figure 20.8

Figure 20.9

Figure 20.10

Figure 20.11

Figure 20.12

to be no obvious rationale. This problem relates to the 'unique adequacy requirement of methods' in ethnomethodology (Garfinkel, 2002). An observer relying on basic common sense may notice the phenomenon, but does not have the particular concerns and reflexive adjustment of the participant to this setting. As external observers, such a 'member's perspective' eludes us.

However, we may still provide an ethnographic understanding of what might be going on. We showed the recording to the participant himself, the day after the event. Such a procedure, called '*auto-confrontation*' in French, was developed and refined by French researchers in the Human Factors domain (for instance Theureau, 2010). The idea is that such a showing, accompanied by non-leading questions asked by the analyst, may allow the elicitation of descriptions of his or her own activity by the participant, with the help of the video, which could be deemed 'pre-reflexive' (Vermeersch, 2012), or at least closer to the activity as it was experienced than in traditional interviews. While EM/CA-inspired analyses of 'video-as-data' and 'auto-confrontations' are distinctive empirical methods operating in different conceptual frameworks, they can to some extent be articulated so that they provide complementary accounts of activities as lived experiences (Cahour et al., 2018). In this case, when we showed the video clip to the participant, he said this:

Interviewer: Had you seen them?

Léo: Yeah, so well … it may sound silly but actually, I did not want to attack frontally and I saw this car and er … I have, I had two options, I could either do it like this (points so as to indicate a direct, straight path to the car and the standing group), but I did not want to because it is difficult to bear other people's gaze like that for ten metres, me I can't stand it, like, we are strangers but we will travel together, it's a bit troublesome, I'd rather switch fast from the known to the unknown, rather than sustain their assessing gaze. Maybe they're looking at me from afar and

> they're wondering 'Is it him? Is it not him?' No it's difficult, when one carries a lot, and worse, I had the glasses, and so I did not want that, and so I walked like that 'hop' (draws a roundabout path with his hand) in front of the car, so they did not see me and I preferred to attack them on the side … with hindsight I say to myself I played it well, I came upon them rather naturally, but actually my approach was carefully planned.

This narrative shows how the participant was sensitive to the ways in which various forms of approach projected different forms of interaction with other parties in the car sharing encounter. Going straight towards them would have provided the kind of gradual approach characterizing the initiation of open space encounters: gaze locking and recognition, continuous monitoring, distant and close salutations, as with mutual acquaintances meeting at a party (Kendon, 1990). However, in this setting they do not know one another, and our participant treats this as unsuitable because it would involve categorical recognition issues (is he the new passenger or just a pedestrian?). The potential problem is compounded for him by concerns with his appearance (and its potential assessment), for he is wearing the camera glasses and is preoccupied with how it makes him look. Circumnavigating around the car allows him to maintain the visual appearance of an anonymous passer-by until the very last minute and to escape such issues. It allows him to appear to them as 'the last passenger' only once he is very close to them, at the very last moment, and to design his visual appearance on their interactional scene as sudden enough to make recognition quasi-immediate and to entail a greeting as the immediate next action. Not only dos it minimize the duration of potential scrutiny, but it also allows him to introduce to them the question of the camera glasses and to account for them as a first topic after the exchange of greetings.

His account of the lived work of getting into the encounter makes it into a kind of 'gestalt contexture' (Gurwitsch, 1964; Watson, 1997). Embodied mobility (different approaches), recognizable categories enacted through embodied mobile behaviour (passer-by vs passenger) and the relevance of meaningful response moves (mutual ignorance or the combination of monitoring gaze, greetings and recognition tokens) are reflexively and continuously elaborating one another in the course of his path. Unlike in our first example, such an understanding of his mobile conduct was not directly available to the analyst as an external observer relying on generic and common sense understandings. Unpacking the fine details of the emergent organization of the participant's embodied approach required specific methods of elicitation involving the participant himself watching the video, which proves to be a powerful empirical device to make sense of many forms of urban mobilities.

Figure 20.13

Connected mobiles and multi-activity: driving and texting

New forms of mobility involve connected users and 'messy' urban ecologies (Bell & Dourish, 2011) crisscrossed by heterogeneous infrastructures and offering opportunities for multiple activities on the move. One particularly interesting case is the association of urban mobility with smartphone use, and the forms of multi-activity it may enable. While urban mobilities are embodied and public, smartphone activity is usually not available to co-participants in traffic encounters, or only in part. So a relevant issue becomes to gain an understanding of how participants manage and articulate both streams of activity. As usually in EM/CA, researchers have studied pairs of connected users by video-recording both their embodied mobility and the smartphone screen activity which they may share or discuss as a topic: pairs of pedestrians navigating the city with mobile apps (Laurier et al., 2015), searching the objects on display in museums on their phones (Brown et al., 2015), or moving around taking selfies (Weilenmann et al., 2019).

The following example[5] uses similar video recordings, but to study the behaviour of a *single* car driver who drives and texts (Licoppe & Figeac, 2018). Since she is alone, the video-ethnographic analysis cannot rely on talk-in-interaction to understand her connected mobility. However, the video data allows an analysis of her multi-activity by providing access to the distinct temporal organization and demands of the different streams of activities she is engaged in, and by allowing their articulation to be observed for different activities will generally project different temporal expectations. These may or may not match smoothly, and the consequences of such competing temporal demands may have consequences on the organization of traffic encounters, some of which can be observed in the video data. For instance, it may happen that a relevant transition point (at which some next action in traffic or on Facebook becomes relevant and expected) occurs roughly at the same time that another stream of activity requires an action. Then one action might appear to be delayed to allow for the accomplishment of the other, leading to some kind of observable mismatch.

In our study, a female driver gave us access to a recording (combining video glasses and screen video capture) of her (illegally) surfing on Facebook while driving. She has just come to a traffic light and used this as an opportunity to launch Facebook on her smartphone (Figure 20.13). When the traffic resumes at the green light, her gaze remains on her smartphone, even though some cars that are visible in the camera glasses (and even more so in her peripheral vision) have started to move (Figure 20.14).

That her response is noticeably 'delayed' is not just the analyst's opinion, since independent evidence of this can be found in the video: a driver in the next lane takes advantage of the gap in front of her car (created by her 'late' response) to change lanes and take her 'slot' (Figures 20.14 and 20.15).

Figure 20.14

Figure 20.15

This overtaking makes her 'lateness' concrete and visible. It shows how using a smartphone while driving may have consequences on driving performance, even if in a minor fashion here. Are there more precise reasons for her delay? The screen capture data becomes useful here. When the light turned green, she was in the process of scrolling down her list of Facebook posts with her finger, from the most recent (at the top) to the older ones further down. It is impossible to say whether or not she was looking for a specific post (goal-oriented activity) or just waiting for a post on her list to catch her attention (environment-driven activity). What is significant, however, is the fact that the visual and pragmatic structure of the list does not offer obviously recognizable relevant transition points, the occurrence of which could be taken as an opportunity to gaze away and reorganize one's involvements. One may just scan on and on down the list until an attention-catching post is reached, without the list in itself affording any generic and eventful 'asperity'. In the case of an outside event (such as a light turning green), there is a fair chance that the user may remain absorbed in the scanning activity for an extra moment through the sheer inertia of her focused attention and scrolling, even though these 'outside' events require some kind of response – hence the kind of delay we could observe when traffic actually resumed. The whole thing was treated as unremarkable. Treating such multi-activity events as routine occurrences in this way provides evidence for the claim that using a smartphone while driving is a form of multi-activity that is potentially consequential in terms of social and legal concerns.

Conclusion

Considered from the perspective of lived experience, urban traffic encounters of all sorts are self-organized. Participants rely on their embodied behavioural displays to achieve joint coordination within a normative framework of claims and expectations which are both a resource and an outcome of this emergent process (Goffman's 'interaction order' in urban public places). In this paper we studied three types of traffic encounters that can be argued to be related to forms of mobility that have spread in the last 15 years (encounters with strangers mediated by online platforms, connected mobilities), or even more recently (electric scooters).

First, we showed how these new forms of urban mobility could also be analysed within an interactionist and ethnomethodological framework to account for their witnessable orderliness as a joint accomplishment.

Second, we showed how the combination of video recordings of embodied displays and this analytic framework allows us to make sense of the mobility of 'single' persons in traffic encounters within different types of urban flows (electric scooters in car traffic, pedestrian meetings, a driver active within both car traffic and social networks). The situations we studied differed in terms of the common sense resources the analyst (who is socialized to everyday urban mobility) could bring into the video-ethnographic analysis. While such common sense resources were adequate in the first example (the electric scooter's body gloss), it was not the case in the latter two, which called either for methodological refinements, that is showing the video recordings of their embodied mobility to the participants in the case of the arranged pedestrian meeting with strangers, or respecifying the phenomenon of interest (the organization of multi-activity) in the case of 'connected' mobilities. However, one can still argue that, because of their embodied character, and their visual/normative 'witnessable' orderliness, urban mobilities, whether 'old' or 'new', are situated and public social phenomena, for the study of which qualitative, video-ethnographic methods are especially relevant.

Notes

1 In the sense that they are not moving together with others they can talk to and discuss their mobility with.
2 The study is made in collaboration with Sylvaine Tuncer and Barry Brown from the University of Stockholm, and the data have been recorded by Sylvaine Tuncer.
3 We make a distinction here between classifications, which are typification that stand independently of a given activity and are 'external' to it, and membership categorizations, which are made relevant, in the course of an ongoing activity. They are endogeneous or 'internal' to the mobility as a process.
4 This study was made in collaboration with Beatrice Cahour and Lisa Creno, and the data were recorded by Lisa Creno.
5 This study was made in collaboration with Julien Figeac who recorded the data.

References

Bell, G., & Dourish, P. (2011). *Divining a Digital Future. Mess and Mythology in Ubiquitous Computing.* Cambridge: MIT Press.

Brown, B., McGregor, M., & McMillan, D. (2015). Searchable objects: search in everyday conversations. In *Proceedings of the CSCW 2015 Conference on Computer-Supported Cooperative Work and Social Computing New York: ACM.*

Buscher, M., Urry, J., & Witchger, K. (2010). *Mobile Methods.* London: Routledge.

Cahour, B., Licoppe, C., & Creno, L. (2018). Articulation fine des données vidéo et des entretiens d'auto-confrontation explicitante: Etude de cas d'interactions en co-voiturage. *Le Travail Humain,* 2018/4 (Vol. 81).

Deppermann, A., Laurier, E., Mondada, L., Broth, M., Cromdal, J., De Stefani, E., Haddington, P., Levin, L., Nevile, M., & Rauniomaa, M. (2018). Overtaking as an interactional achievement: video analyses of participants' practices in traffic. *Gesprächsforschung,* 19, pp. 1–131.

Garfinkel, H. (1967). *Studies in Ethnomethodology.* Cambridge: Polity Press.

Garfinkel, H. (2002). *Ethnomethodology's Program. Working Out Durkheim's Aphorism.* Lanham, MD: Rowman & Littlefield.

Goffman, E. (1963). *Behavior in Public Places.* New York: Free Press.

Goffman, E. (1971). *Relations in Public: Microstructure of the Public Order.* New York: Harper & Row.

Goffman, E. (1983). The Interaction Order. *American Sociological Review,* 48(1), pp. 1–17.

Gurwitsch, A. (1964). *The Field of Consciousness.* Pittsburgh, PA: Duquesne University Press.

Heath, C., Hindmarsh, J., & Luff, P. (2010). *Video in Qualitative Research.* London: Sage.

Kendon, A. (1990). *Conducting Interaction: Patterns of Behavior in Focused Encounters.* Cambridge: Cambridge University Press.

Laurier, E. (2005). Searching for a parking place. *Intellectica,* 2–3(41–42), pp. 101–115.

Laurier, E., Brown, B., & Lorimer, H. (2008). Driving and 'passengering': notes on the ordinary organization of car travel. *Mobilities,* 3(1), pp. 1–23.

Laurier, E., Brown, B., & McGregor, M. (2015). Mediated pedestrian mobility: walking and the map app. *Mobilities,* 11(1), pp. 117–134.

Liberman, K. (2013). *More Studies in Ethnomethodology.* Albany, NY: SUNY University Press.

Licoppe, C., & Figeac, J. (2018). Gaze patterns and the temporal organization of multiple activities in mobile smartphone uses. *Human-Computer Interaction,* 33(5–6), pp. 311–334.

Mondada, L. (2016). Challenges of multimodality: language and the body in social interaction. *Journal of Sociolinguistics,* 20(3), pp. 336–366.

Pink, S. (2013). *Doing Visual Ethnography.* London: Sage.

Sacks, H. (1992). Lectures on conversation 1964–1965. *Collected Lectures.* Cambridge, Cambridge University Press.

Smith, R. J. (2019). Visually available order, categorisation practices, and perception-in-action: a running commentary. *Visual Studies,* 34(1), pp. 28–40.

Theureau, J. (2010). Les entretiens d'autoconfrontation et de remise en situation par les traces matérielles et le programme de recherche 'cours d'action'. *Revue d'anthropologie des connaissances 2010/2 (Vol 4, no 2),* pp. 287–322.

Vermeersch, J. (2012). *Explicitation Et Phénoménologie.* Paris: Presses Universitaires de France.

Watson, R. (1997). Some general reflections on 'categorization' and 'sequence' in the analysis of conversation. *Culture in Action. Studies in Membership Categorization Analysis.* S. Hester, & Eglin, P. International Institute for Ethnomethodology and Conversation Analysis & University Press of America, pp. 49–75.

Weilenmann, A., & Hillman, T. (2019). Selfies in the wild: studying selfie photography as a local practice. *Mobile Media & Communication*, 2050157918822131.

21

ROUTINE AND REVELATION

Dis-embodied urban mobilities

Lynne Pearce

By the middle of the nineteenth century, Manchester (England) was the largest and most advanced industrial city in Europe, and although the cotton mills that drove its ascendancy have long since disappeared, its cultural exports – most notably football and popular music – have ensured that it is still a city whose name resonates around the world. From 2006 to 2010, I was the PI on a large funded research project, "Moving Manchester", which explored the ways in which migration has informed the work of the creative writers based in the city from the 1960s to the present.[1] The project worked closely, but not exclusively, with the city's black and Asian communities – in particular those associated with the literature development organisation and publisher, Commonword.[2] This collaboration provided our team with important new insights into the history of grassroots cultural activity in Manchester during the period and the challenges that BME – and non-London based – writers have faced in gaining recognition for their work. Nevertheless, the literature we analysed proved a fascinating historical portrait of the city during a period of major upheaval and redevelopment (see Pearce, Fowler and Crawshaw 2013). For the mobilities scholar, these texts provide equally important insights into the everyday ebb and flow of the city and its people, including those whose voices are not found in mainstream literature.

This chapter will draw upon the "Moving Manchester" archive to make its case for recognising the importance of what I have styled *dis-embodied* mobilities in the everyday experience of urban life. I open with two extracts from Cath Staincliffe's Manchester crime fiction novel, *Missing* (2007), which are illustrative of the twin poles of embodied urban mobilities that I would like to reflect upon in this chapter: namely, the routine and everyday *vs* the unique and revelatory. The first extract sees private investigator, Sal Kilkenny, stopped at a set of traffic lights on her way home:

> It was another fine day, the central reservation was thick with dandelions and daisies, the trees along the edge of the playing field were lush and leafy, spilling shade across the road. An elderly woman, her back bent over a Zimmer frame in her hands, made her way bit by bit along the pavement. I wondered where she was going and whether she was relishing the summer day as I was? Was this a daily

> journey she was making or something out of the ordinary? A chance to take the air and smell the flowers or a tedious chore?
>
> The traffic lights changed and I moved forward towards the junction where another huge apartment block was being thrown up. BUY NOW! screamed the hoardings. Luxury Apartments, only a few remaining …
>
> One other thing struck me about Janet's reported mood. In general, people who leave home tend to do so from a position of unhappiness, despair, a sense of isolation … The picture of Janet, excited and bubbly, suggested she might have been going to something rather than just running away.
>
> *(Staincliffe 2007, p. 68)*

The second extract, meanwhile, sees Kilkenny on a long-distance car journey across the Pennines, reflecting on her first night with Ray – a friend she has known for many years but only now got romantically involved with:

> The Pennine mountain range forms the backbone of England and divides the counties of Lancashire in the west … from Yorkshire in the east. The M62 motorway crosses the hills and to get onto that I had to drive to Stockport and take the M60 near the landmark glass pyramid building …
>
> Traffic doubled once I reached the M62. Lorry land, with articulated trucks travelling the breadth of the country from ports on the west like Holyhead and Liverpool over to Hull on the eastern seaboard … The road runs through the high moorland climbing into Yorkshire and, at the summit, a sign boasts the highest motorway in England, 1221 feet. It's a stunning landscape but I had to concentrate on the traffic, only occasionally noticing the purple shadows of clouds racing across the gold and green moorland. The sweep of sky and land and fall of the valleys gave me that top of the world sensation …
>
> The day was bright and matched my mood. Memories of the night before came back to me sporadically, little moments to savour. Any consideration about whether I'd done the right thing in letting things develop with Ray, was shoved aside by the sheer pleasure of the experience. I felt energised.
>
> *(Staincliffe 2007, pp. 160–1)*

What should be immediately obvious about both extracts is that they combine detailed observation of the urban landscape/roadscape as seen from the driver's seat with thought-chains ostensibly unconnected with the material present but focused, instead, on Kilkenny's reflections on the case she is trying to solve (extract one) and the surprise development in her relationship with Ray (extract two). This perceptual/cognitive relay between the driver's engagement with the world immediately beyond the windscreen, and the interior world of reflection, analysis, memory, fantasy and the imagination is also a central preoccupation in my recent monograph, *Drivetime* (2016), which draws upon literary and autobiographical texts in order to explore the question "what do we think when we drive?" (see also, Pearce 2014, 2017). This focus on the dynamic interchange between outer and inner worlds is, I would suggest, also of importance in gaining a better understanding of embodied urban mobilities more generally, and while some excellent work has been done in recent years by scholars intent on capturing the visceral materiality of urban living though the analysis of its rhythms, patterns and sensations to which I will return (e.g., Bissell 2013; Edensor 2010, 2014; Highmore 2003), most of this has been focused on the subject's experience of the external world, often in the context of posthumanist accounts of everyday living and its multiple, intersecting flows. With one or two exceptions (e.g., Edensor 2003), the connection between the urban subject's apprehension of the material present and his or her personal past – including, most notably, their *memories* – is unspoken, and yet these dis-embodied thoughts and feelings are as surely a feature of everyday urban living as our more immediate sensory saturation in the world of nature, flashing lights, billboards, people, animals, vehicles and the rise and fall of buildings. As we push our way though the crowds on the way to work, or glide peacefully along a newly constructed urban cycleway, our attention is only ever partly focused on the transport or the landscape through which we pass: the immediate future and recent past typically vie for attention, as do more distant memories. Sometimes these are prompted by the sights and experiences we encounter on our journeys, but at other times the connection is less obvious and it is simply the mental space that the journeying has afforded which enables us to day-dream or problem-solve (Pearce 2016).

Needless to say, my own focus on the interiority of subjective experience also raises methodological issues that, at times, create an interesting and productive tension with the widespread adoption of posthuman approaches to both the geographical and the social within the field of mobilities studies. There is, however, a strong case for the humanities and the social sciences coming together in the exploration of the urban, not least because texts – such as the ones I have cited here – often capture, in a paragraph, the very complexity that sociological enquiry has long struggled to express. For the purposes of this short chapter I will focus specifically on *dis*-embodiment in relation to those *auto*-mobilities – i.e., driving, cycling, walking – which require the active participation of the human subject since (as developed in *Drivetime*) one of my particular interests has been the way in which the complex, everyday skills needed to navigate or to operate machinery can, on occasion, prove liberating for other parts of the mind and facilitate both day-dreaming and analytical thought (Pearce 2016, pp. 159–72).

Sensations

One of the key features of work on all types of auto-mobility – post the "mobilities turn" (Sheller and Urry 2006) – has been a new attention to the embodied experience of mobility via the sensorium – i.e., the sights, sounds, smells and bodily sensations of travelling on foot or by vehicle, both leisurely and at speed. With specific reference to driving, John Urry was one of the first to historicise the changing embodied experience of automotive transport as, throughout the course of the twentieth century, and across many different continents, the dirt road gave way to the paved road, and the "open" car was replaced by a "covered" one (Dennis and Urry 2009; Urry 2007). This history is now used to account for the way in which driving has become an experience and practice increasingly removed from the outside world, with particular emphasis being placed on the windscreen as a "glass" comparable to various other mediatising technologies (Urry 2007, p. 130). By contrast, the early years of motoring are now associated with an altogether more visceral, embodied experience that called upon all the senses and – as is still the case with cycling – facilitated a dynamic interaction between drivers and passengers and the landscapes through which they travelled (see Pearce 2016, pp. 59–90). As Peter Merriman's work on motoring in the early twentieth century attests, texts from this period abound with descriptions of the multi-sensory experience of driving and passengering in an open car (Merriman 2012, pp. 73–95).

In *Drivetime*, I nevertheless question the assumption implicit in Urry's (2007) narrative that the increasing automation of the car and "cocooning" of the driver has completely destroyed the sensual experience of driving. The case I make in *Drivetime* – that, even today, drivers frequently roll down their windows and stop to "take the air" (Pearce 2016, p. 130) – is, moreover, especially germane in the case of *urban* driving where the stop–start nature of the travel embeds the car and its drivers/passengers in their environment in a very visceral and immediate way. And while this is typically presented as a negative encounter, drivers can – as evidenced in the first Staincliffe extract – extend themselves into the landscapes in which they are temporarily paused in an evocative and meaningful way also. City driving, with its continual interruptions, diversions and detours is often associated with the most excruciating stress precisely because one's spatial–temporal expectations have been thwarted. However, it is wrong to conflate this frustration with the *absence* of bodily sensation per se: wind down the window and the world is still there in all its heat, chill, darkness, brightness, noise, silence and odour – even if the stimulants are not pleasant ones. At the same time – and with reference back to the second Staincliffe extract – it is clear that the sort of motoring I have described elsewhere as "cruising" (Pearce 2016, pp. 123–55) can also facilitate a profoundly dis-embodied state of being that enables us to muse upon ideas, events and circumstances that have *no immediate association* with the landscape through which we drive other than the euphoria (or other emotional state) that the driving induces. For Kilkenny, the memory of the previous night with Ray conspires with her sensation of speeding along the M62 motorway to create that "top of the world feeling" (Staincliffe 2007, p. 161).

The visceral and haptic sensations of city walking and city cycling have, meanwhile, become the focus of much academic interest in recent times, often in the context of discourses aimed at promoting transport options other than the car. While the obvious physicality of these auto-mobilities foregrounds their embodied experience, it is, however, important to recognise that the sensations associated with both walking and cycling can themselves facilitate complementary cognitive states such as day-dreaming, nostalgia and problem-solving (see for example Nick Dunn's recent autoethnography on "night walking" (Dunn 2016)). In the midst of congested city traffic, cycling also lends commuters a sensation of almost aerial transcendence as they silently "glide" between the stationary cars. This empowerment is captured evocatively in Karline Smith's novel of "gangland" Manchester, *Moss Side Massive* (2000), in which the Rasta teenager, Zuki, is seen to evade not only the traffic but also the drug culture in which the Moss Side estates are mired by keeping moving on his bike: "Zuki glided on the old bicycle, aimlessly cruising out of the Alexandra Park Estate, leaving the blues and the shebeens behind him, leaving the early-morning drug pushers in doorways, like scavenger birds waiting for their prey" (Smith 2000, p. 39).

Rhythms and routines

As mobilities scholars – working in the wake of Lefebvre (2004) – have demonstrated, the rhythms of everyday urban life are themselves an effect of intersecting *repetitions* that may not always be obvious to the person (or thing) performing them (see Edensor 2014). This awareness of the way in which a city's human and non-human bodies and artefacts are engaged in a perpetual, but largely unconscious, dance, or "place-ballet" (Seamon 2016), lies at the heart of David Bissell's article (2013) on the often invisible *mobility loops* that bring commuters into regular proximity with one another as a result of their (repeated) individual daily journeys. Such insights have been invaluable in helping us rethink not only space but also what we understand by "community" and, indeed, "relationships", in contemporary societies perpetually on the move. On the rare occasion I visit a large city, I am always acutely aware of my physical proximity to those I travel with in what are often cramped, intimate spaces, as well as the strategies we adopt to maintain a symbolic distance from those with whom we inadvertently "rub shoulders". For as well as the digital technologies which enable twenty-first century travellers to insulate themselves from the unfamiliar bodies that press against them by withdrawing into virtual space so, too, will the simple act of cognition – pursuing/creating thought-chains which are unique to that particular day – separate us from the crowd and render unique what, for the community at large, is repetitive and routine (see Pearce 2017). Therefore, while I find Bissell's explication of how, through repetition, singular "pointillist" (Bissell 2013, pp. 349ff.) journeys become loops that bind us to others and create "proximate" relations (and communities) thoroughly persuasive in and of itself, I am also interested in the cognitive mechanisms that *shield* us from those proximities and are at odds with the rhythms and routines practised by our bodies on a daily basis.

My (alternative) line of enquiry also highlights, of course, two very different methodological objectives: the one (broadly posthumanist), concerned with the patterning of multiple and contingent bodies in space; the other (notionally humanist), focused on the travels of the individual subject. This said, I do not perceive the two standpoints to be necessarily exclusive of one another; the fact that my travelling subject is pursuing thoughts, or retrieving memories, unique to that particular day does not mean that he or she is not also the product of the multiple material and ideological "flows"; however, it does preserve the notion of the *individual* human consciousness and, with it, the notion that what is mundane and routine for a mobile community at large (e.g., the daily commute) may be anything but for a traveller who has just learnt that (for example) their child has been involved in an accident.

The extent to which our understanding of the rhythms which constitute urban living depend upon our methodological standpoint may be demonstrated by contrasting Edensor's demonstration of a posthuman "rhythmanalysis" of a typical urban scene (following) with my reading of the first Cath Staincliffe extract at the head of this chapter:

> And these walking rhythms co-exist and intersect with a host of other mobile rhythms: the regular timetabled bus, train and tram travel, the pulse of cyclists, cars and motorcycles, and the non-human pulse of electricity, water, gas and telephony. These multiple mobile rhythms of place further supplement seasonal, climatic and tidal rhythms … and the rhythms of plant growth, bird nesting and river flow. Since places are always becoming, walking humans are one rhythmic constituent in a seething space pulsing with intersecting trajectories and temporalities.
>
> *(Edensor 2014, pp. 163–4)*

As should be evident, Edensor's compelling portrait of the multiple rhythms that inform a city rush-hour depends upon him adopting a standpoint in which the human subject becomes but one of many "actors" in a contingent flow of human and non-human bodies and artefacts. Here, as with aerial photographs of city traffic, distance is what makes *pattern* visible. A similar standpoint is arguably adopted by Staincliffe's narrator in extract one (see above) where the traffic, building works, the elderly woman with the zimmer frame, and Kilkenny herself, are all seen to be performing their daily routines at a particular moment in the day and, together, constitute what is sometimes poetically referred to as "the beating heart" of the city; even nature – represented here by the daisies, dandelions and trees – is fulfilling its annual cycle of growth and regeneration. At the same time, as the final paragraph of the extract reveals, the embodied routine of Kilkenny's drive home also facilitates a thought-chain which is unique to that particular day and the crime she is currently attempting to solve. Familiar with the route, her mind drifts away from the perceptual and sensory reality of the road and resumes its tussle with the question of why Janet should have disappeared. Edensor might describe this as an instance of the way in which rhythm is liable to "arrythmic potentialities": "There is no identical absolute repetition indefinitely … there is always something new and unforeseen that introduces itself into the repetitive (Lefebvre 2004, p. 6)." However, while Edensor considers the multiple forms these "eurhythmic" interruptions may take across a range of applications – "somatic, mechanical and spatial" (Edensor 2014, p. 169) – his posthuman standpoint means that the detours, meandering and sudden "about-turns" of the human mind are not included in the list. For me – as an erstwhile "subject-centred" literary scholar – this is a notable silence, especially when – as I now go on to discuss – we shift temporal register and consider the role of the past, both personal and collective, in the (dis)embodied urban everyday.

Memories and revelations

My previous research on automotive consciousness has revealed to me that the role mobility plays in memory-making is rich and various and, in practice (evidenced, in my case, by the study of textual materials), sometimes at odds with the theories most commonly used to explain the connection. The role of the body in the facilitation of memory is, for example, commonly associated with the work of the phenomenologists Edmund Husserl and Maurice Merleau-Ponty, as well as Henri Bergson; however, the main focus of all these thinkers is on "primary memory" in relation to perception (see Husserl 2002) and the way in which memories are re-activated in the present as and when needed (see Bergson 2000). Merleau-Ponty's phenomenology of the body-subject, meanwhile, draws upon Bergson's concept of "habit-memory" to help support his theory of the primacy of the human subject's relationship to the world: our seemingly intuitive orientation in space and/or ability to perform basic tasks depend upon a "haptic" skill-set which circumvents intellectual consciousness. Following Bergson, this is explained by the way in which perceptions are simultaneously laid down as memories and recalled by, and through, the body as and when needed. The fact that such theorising is focused on memories associated with the recent past (which Husserl refers to as *retentions* (Husserl 1992, p. 116) or basic "motor skills" (Merleau-Ponty 2002), has nevertheless made it harder to account for the role of the body in our memories and recollections of the more distant past (i.e., "secondary memory"). A connection can be found through a careful reading of Bergson's later work (see Pearce 2019), but psychologists and philosophers have long associated "secondary memory" with the intellect rather than the body, and with visual – or, indeed, filmic – images rather than bodily sensations (see Casey 1987, p. 213).

However, these recollections – especially to the extent that they involve the intervention of fantasy, delusion, narrative and all that figures under the umbrella of the imagination – have long been the privileged object of literary criticism, especially in the context of modernism and postmodernism where the authors' intense fascination with memory's ability *to deceive* has, in turn, given rise to many volumes of commentary and analysis. In my own research, therefore, I have found it necessary to disregard this privileging of "primary memory" in order to explore the ways in which mobility informs, and facilitates, all manner of cognitive recall including that which is involved in future-oriented day-dreaming and reveries (Pearce 2016, pp. 156–200) focused on the distant past. Further, I have sought to draw connections between the operation and function of "primary memory" in an everyday context – e.g., the way in which the body subject exercises "habit-memory" – and the way in which related "body-space routines" (Seamon 2016) become part of a (typically unconscious) *memorialisation* process (see Pearce 2019).

With specific reference to urban mobilities, I would suggest that there are multiple ways in which memory is an integral part of everyday experience:

1. The role of "habit-memory" in the performance of complex everyday skills such as driving and cycling;
2. The role of "retention", or "primary memory", in the phenomenological experience of the present;
3. The intrusion of memories of the recent past and/or anticipations of the near future while otherwise focused on the present through the practice of mobility (e.g., driving, cycling, walking);
4. The recollection of distant memories facilitated by mobile practices (i.e., the "freeing" of mental space as the result of cruising in a car, gliding along on a bike or riding on a train or other public transport);
5. The arousal of memories – both recent and distant, embodied and dis-embodied – as the direct result of encounters with things and/or events in the present (including, in particular, *discontinuities* between how things are now and how they were in the past).

In the first of the Staincliffe extracts cited above we see examples of my third category here – i.e., the intrusion of memories of the recent past. Briefly stopped at the traffic lights, Kilkenny is, at first, absorbed by the place in which she temporarily finds herself, but then slips to thinking about the case she is currently working on. While it is possible that something she has seen, heard or smelt while parked at the junction stimulates this memory, it could equally have been inspired by the simple process of stopping and starting; the interruption causes her to briefly pause her mentalist and dis-embodied thoughts and focus on the external world while, in the process of letting off the handbrake and pulling away, they are once again resumed – but now from a fresh perspective. In other words, the fluid, eventful cityscape has stimulated her thought processes, but indirectly. In the second extract, a different type of driving – cruising (see Pearce 2016, pp. 123–55) – facilitates Kilkenny's memories of her night with Ray. Viewed from one perspective, this erotic meditation may be seen to take the form of an (already) "distant" memory disconnected from the immediate present and is, in that sense, dis-embodied; however, the comfort and relaxation afforded by this sort of driving is experienced by the body with a sensuality that is arguably conducive of reveries (Pearce 2016; Sheller 2004) as well as providing the mind-space necessary for analytic thought.

The dis-embodiment induced by motorway driving may also be seen to have a role in the resurrection of more distant memories/recollections, as evidenced in the autoethnographies of myself (2000) and Tim Edensor (2003). While my reflections in "Driving North, Driving South" focus on my long-haul journeys to Cornwall, Edensor shows how even an everyday commute from Junction 19 to Junction 16 of the M6 can inspire reveries of past times and past relationships, especially when the drive is accompanied by nostalgic music. As with Kilkenny's more recent memories of her night with Ray, the roadscape may not be directly associated with the day-dream in question but the bodily sensation of "flying" (Pearce 2016, pp. 156–62) undoubtedly contributes to the euphoria associated with this sort of recall – and also accounts for why it is much more difficult to access such fantasies while sat in a traffic jam (Pearce 2016, pp. 171–7).

Where the urban landscape is most likely to figure in our (dis)embodied memories of place, however, is via a chance re-encounter with a person or thing from the past that has been *transformed or modified* in some way. Edensor has argued that such "sensitivity to change" is a crucially "under appreciated" aspect of urban living and an intrinsic part of its rhythms: "The sudden road accident, newly painted house, unusual bird or animal, peculiar vehicle or passer-by stand out in relief to the usual" (Edensor 2014, p. 165). Literary texts abound with passages in which authors and characters reflect upon the often "uncanny" experience of encountering an altered landscape in this way, and they are a distinctive and recurrent feature of Manchester writing as seen in the following extract from Joe Pemberton's *Forever and Ever Amen*:

> Only last week it had been a row of houses. Fairlawn Street next to St. Bees Street next to another road James couldn't remember the name of. Each house had three stories and a million stairs to the top. Only last week the street had been full of kids playing catch and cars driving past the women gossiping on the corner. But not any more, all that was gone and all that was left was a pool the size of a school yard and a reflection of a church spire.
>
> *(Pemberton 2000, p. 52)*

In these encounters – which may also be thought of as memorials to the city – mobility is centred not in the moving subject (walking, cycling, driving etc.) but rather in the city itself whose architecture and communities appear to find no peace. In such episodes, the author/narrator is rather the stationary presence who bears witness to the change he or she perceives on account of the disjunction between perception and memory. It is a change, moreover, that is felt very viscerally through the body, but rather as an absence: everything is displaced, all the old landmarks gone.

Conclusion

In this short chapter, I have sought not to question, but rather to supplement, the recent work of mobilities scholars who – drawing on posthuman methodologies and phenomenological textual practices – have discovered evocative new ways of capturing the multifarious flows, rhythms and routines of urban life. By adopting a more subject-centred standpoint, it can nevertheless be seen that our everyday transit through city space – whether by car, bicycle or on foot – typically *combines* intense sensory experiences with dis-embodied thought processes. Indeed, as theorists across several different disciplines have argued, it is the mundane and repetitive nature of movement that ultimately frees the mind to think of other things; however, this should not be characterised simply as cognitive transcendence given that the body is centrally involved in the process (hence my decision to hyphenate *dis-embodiment*). A major consequence of putting the subject back into the mobile experience in this way is to acknowledge how quickly the routine becomes the exception; how an everyday commute to work can deliver revelation (both analytic and emotional) and – as for Sal Kilkenny (see extract two) – change one's life forever.

Notes

1 "Moving Manchester: How Migration informed Writing in Manchester 1960–the present". AHRC-funded research project, Lancaster University, 2006–10. For the archived website see: wp.lancs.ac.uk/transculturalwriting/.

2 The Commonword website address is: www.cultureword.org.uk/.

References

Bergson, H. 2000 [1908]. Memory of the Present and False Recognition. In: R. Durie, ed. *Time and the Instant: Essays in the Physics and Philosophy of Time*. Manchester: Clinamen Press, pp. 36–63.
Bissell, D. 2013. Pointless Mobilities: Rethinking Proximity through the Loops of Neighbourhood. *Mobilities*, 8 (3), 349–367.
Casey, E. 1987. *Remembering: A Phenomenological Study*. Bloomington, IN: University of Indiana Press.
Dennis, K. and J. Urry. 2009. *After the Car*. Cambridge: Polity.
Dunn, N. 2016. *Dark Matters: A Manifesto for the Nocturnal City*. London: Zero Books.
Edensor, T. 2003. M6: Junction 19–16: Defamiliarising the Mundane Roadscape. *Space and Culture*, 6 (2), 151–168.
Edensor, T. 2010. Walking in Rhythms: Place, Regulation, Style and the Flow of Experience. *Visual Studies*, 25 (1), 69–79.
Edensor, T. 2014. Rhythm and Arrhythmia. In: P. Adey, D. Bissell. K. Hannam, P. Merriman and M. Sheller, eds. *The Routledge Handbook of Mobilities*. London and New York: Routledge, pp. 163–171.
Highmore, B. 2003. *Cityscapes: Cultural Readings in the Material and Symbolic City*. Basingstoke: Palgrave Macmillan.
Husserl, E. 2002 [1964]. The Phenomenology of Internal Time Consciousness. In: D. Moran and T. Mooney, eds. *The Phenomenology Reader*. London and New York: Routledge, pp. 109–123.
Lefebvre, H. 2004. *Rhythmanalysis: Space, Time and Everyday Life*. London: Continuum.
Merleau-Ponty, M. 2002 [1945]. *The Phenomenology of Perception*. London and New York: Routledge.
Merriman, P. 2012. *Mobility, Space and Culture*. London and New York: Routledge.
Pearce, L. 2000. Driving North/Driving South: Reflections on the Spatial/Temporal Co-ordinates of Home. In: L. Pearce, ed. *Devolving Identities: Feminist Readings in Home and Belonging*. Aldershot and Burlington, VT: Ashgate, pp. 162–178.
Pearce, L. 2013. In Search of What We're Thinking When We're Driving. In: J. Stacey and J. Wolff, eds. *Writing Otherwise: Experiments in Cultural Criticism*. Manchester: Manchester University Press, pp. 95–105.
Pearce, L. 2014. A Motor-Flight through Early Twentieth-Century Consciousness. In: L. Murray and S. Upstone, eds. *Researching and Representing Mobilities*. Basingstoke: Palgrave Macmillan, pp. 78–98.
Pearce, L. 2016. *Drivetime: Literary Excursions in Automotive Consciousness*. Edinburgh: Edinburgh University Press.
Pearce, L. 2017. Driving-as-Event: Re-thinking the Car Journey. *Mobilities*, 12 (4), 585–597.
Pearce, L. 2019. *Mobility, Memory and the Lifecourse in Twentieth-Century Literature and Culture*. New York: Palgrave Macmillan USA.
Pearce, L., C. Fowler and R. Crawshaw. 2013. *Postcolonial Manchester: Diaspora Space and the Devolution of Literary Culture*. Manchester: Manchester University Press.
Pemberton, J. 2000. *Forever and Ever Amen*. London: Hodder Headline.
Seamon, D. 2016 [1979]. *A Geography of the Lifeworld: Movement, Rest, Encounter*. London and New York: Routledge Revivals.
Sheller, M. 2004. Automotive Emotions: Feeling the Car. *Theory, Culture and Society*, 21 (4–5), 221–242.
Sheller, M. and J. Urry. 2006. The New Mobilities Paradigm. *Environment and Planning A*, 38, 207–226.
Smith, Z. 2000. *Moss Side Massive*. London: X-Press.
Staincliffe, C. 2007. *Missing*. London: Allison & Busby.
Urry, J. 2007. *Mobilities*. Cambridge: Polity.

22

HABIT AS A BETTER WAY TO UNDERSTAND URBAN MOBILITIES

Thomas Buhler

Introduction

As a result of the long domination of economics and engineering over urban transport issues, mobility practices have come to be thought of as the product of prior reflection and choice by individuals (Kaufmann, 2000). Many aspects of mobility behaviour have long been ignored, including its habitual dimension, which is discussed in this chapter (Bissell, 2012; Buhler, 2015; Goodwin et al., 1987; Schwanen et al., 2012). As Verplanken and Orbell (2003, p. 4) suggest by their rhetorical question, mobility studies cannot ignore the rather repetitive nature of mobility behaviour: "When was the last time you performed a new behaviour? In everyday life, it does not occur so often that we do something for the first time: repetition is the rule, rather than the exception." A substantial body of literature shows that repetitive and stable contexts in everyday life are the breeding ground for the formation of strong habits that omit the evaluation and deliberation phases (Hodgson, 2004) whether they concern mobility (Gärling and Axhausen, 2003) or other aspects of lifestyle such as energy consumption (Maréchal, 2009, 2010).

Although the earliest definition of *habit* dates back to Aristotle (e2005), the concept has long been marginalized in the social sciences primarily because of: (1) the multiple meanings of the term in everyday language which maintain a degree of fuzziness about it (Héran, 1987; Kaufmann, 2001; Lahire, 1998); and (2) certain questionable uses in the history of science, particularly by behaviourists, and the subsequent embarrassment caused by it for researchers in the social sciences. In spite of this, over the past 20 years or so, there has been a renewed interest in the concept of *habit* when it comes to daily mobility practices. Indeed, increasing empirical evidence in social psychology, first, and then in geography, sociology, and economics supports the idea that habit plays a major role in mobility behaviour (e.g. Bissell, 2012; Brette et al., 2014; Schwanen et al., 2012).

After a necessary clarification of what habit is and is not, I will look at the state of the art in terms of methods for measuring the strength of a habit, with an illustration based on a recent French national survey (Buhler and Signoret, 2018).

Habit: what it is and is not

Departing from a common but misleading definition

To fully understand the value of the concept of habit for urban mobility studies, it is necessary to depart from the different definitions that ordinary language associates with this term in English as in other languages. A dictionary of synonyms is a relatively simple and exhaustive way of capturing the multiple meanings associated with the term *habit* in ordinary usage. Under the entry "habit", apart from words associated with "habit" as clothing, the Collins Thesaurus (2019) proposes four groups of words that may be considered relatively close synonyms:

- "a tendency to act in a particular way": mannerism, custom, way, practice, manner ...
- "an established custom or use": custom, rule, practice, tradition, routine, second nature ...
- "an addiction (to drug)": addiction, weakness, obsession, dependence, compulsion ...
- "an irresistible urge to perform some action": urge, need, obsession, necessity, preoccupation ...

These four groups are made up of words that refer respectively to styles or manners, social and collective habits, pejorative considerations associated with dependence, and the inability to control oneself and one's urges. These groups of synonyms underline the multiplicity of notions similar to the term "habit" in everyday usage and testify to its very strong polysemy and the diversity of its uses in research, specialist areas, or everyday life. Kaufmann (2001) identifies three main points that are problematic with the common definition for turning the term "habit" into a scientific concept in the social sciences. As it is commonly defined, *habit* supposedly concerns minor actions only, characterized by their regularity and repetitiveness, and it generally has negative connotations (see Table 22.1 below).

To answer these three points that characterize habit as ordinarily defined, it should be pointed out first that habit does not refer to the category of minor actions only. Habit is associated with the entire spectrum of patterns of acting and thinking. In other words,

Table 22.1 Three problematic points with the common definition of habit (Buhler, 2015)

Habit as commonly defined	*Habit as a concept in the social sciences*
[**1**] Habit supposedly concerns "minor actions" only (*e.g. nail biting, table manners, etc.*)	[**A**] Habit is associated with the entire spectrum of patterns of thinking and doing. All such habits are socially constructed by training or learning.
[**2**] Habit is supposedly characterized by its regularity (in time) and its repetitive nature (in its content)	[**B**] Regularity and repetitiveness characterize only the formation phase of the habit. Once formed, habit is a propensity for behaviour, the regularity of which may vary. Habit is not just the repetition of similar actions; it is also the imprinting of new patterns.
[**3**] Habit is supposedly associated with minor actions of no interest or small acts worthy of disdain	[**C**] Habit is a major anthropological fact that is not subject to ethical evaluation. The content of the habit may be, but not the habit itself.

Source: T. Buhler

"habit" cannot be contrasted with reflectiveness. Habitual behaviour may concern body habits, gestures, sensorimotor habits, as well as reflective, deliberative, rational, or even calculating habits (Lahire, 1998). Indeed, the second type of habit is just as much "socially constructed, in formal or informal repetition and training" (Lahire, 1998, p. 89).

Then, with regard to the "repetitive" and "regular" character attributed to habit, we can argue that the repetition of a behaviour is characteristic of the phase of habit formation, but not of the habit itself. Following this phase of formation, the habit may be reactivated simply from time to time, when the context and conditions are appropriate. For example, a "habitual journey" may be an annual solo hiking trip at the beginning of summer, in a particular place every year. This example shows that habit is a propensity to behave in a certain way, but habit does not determine the frequency of that behaviour. There is a whole range of frequencies between (multi-) annual habits and daily habits. As regards the "repetitiveness" (in content) of behaviour, habits are not confined to reproducing the "old", but they may also record "new" content (Kaufmann, 2001).

Finally, as ordinarily defined, habit may be associated with minor actions of no importance or no interest, or that are worthy of disdain. A rigorous definition of habit would underline its nature as a "major anthropological fact" (Héran, 1987): being only the operative schema, not its content, habit in general cannot be the subject of any ethical evaluation (idem).

Distinguishing habit from near synonyms: repeated behaviour, routine, and instinct

First of all, habit is not the equivalent of repeated behaviour. It must be understood as a propensity to a certain type of behaviour, based on scripts (Gärling et al., 2001; Verplanken and Aarts, 1999), that is modifiable over time and learnt either formally or informally. This propensity (or potential) can be passed on to others socially through education, writing, oral transmission, etc. Habit is therefore something different from repeated behaviour. It is a propensity to behave (act or think) in a particular way and in a particular class of situations (Hodgson, 2010).

Secondly, it is also possible to distinguish between *habits* and *routines*. Although they are sometimes taken to be synonymous, *habits* and *routines* are of different natures. The latter are a series of actions repeated and triggered under the control of habits (Buhler, 2015). They remain "simple" combinations of repeated actions. Habits are very different in nature, as they are a series of internalized or embodied schemas that compose specific (mobility) skills and know-how.

Lastly, *habits* and *instincts* are things of two really different natures, even though they are both strongly associated with automatic behaviour and sometimes considered synonymous. Instincts are inherited behavioural dispositions (Hodgson, 2004) that are culturally and/or biologically hard-wired and make it possible to respond to stable and relatively simple problems (escaping from a predator, etc.). Socially, instincts allow the transmission of certain behaviours that are necessary for the formation of habits (Hodgson, 2004). However, habits are of a different nature in that they are exclusively cultural (Leroi-Gourhan, e1993). They can be thought of as propensities and dispositions that are established by repetition and socially supervised, whether institutionally or not. Collectively, habits are much more adaptable than instincts and enable us to cope with greater cognitive complexity as well as greater instability due to changing problems and environments. Habits, then, can be thought of as entities that can serve as a mode of reproducing social behaviour. Habits can be viewed as the counterpart to biology in regulating behaviour – they are "second nature" as Aristotle put it (Kaufmann, 2001).

Habit in the social sciences: renewed interest in an old concept

Habit: where does it come from?

In order to provide a brief history of the concept of habit in the social sciences and philosophy, it is necessary to return to Aristotle's definition. He introduced it in the "Nicomachean Ethics" (Aristotle, e2005). By deductive reasoning, Aristotle shows that virtue is the result of habit. He then extends the observation to all acquired characters and skills:

> for the things we have to learn before we can do them, we learn by doing them, e.g. men become builders by building and lyreplayers by playing the lyre; so too we become just by doing just acts, temperate by doing temperate acts, brave by doing brave acts.
>
> *(Aristotle, e2005, p. 23)*

He extends this reasoning to the government of the city state:

> This is confirmed by what happens in states; for legislators make the citizens good by forming habits in them, and this is the wish of every legislator, and those who do not effect it miss their mark, and it is in this that a good constitution differs from a bad one.
>
> *(Aristotle, e2005, p. 24)*

As a conclusion to Book II, Chapter 1, Aristotle remarks on the weight of acquired habits in future behaviour: "It makes no small difference, then, whether we form habits of one kind or of another from our very youth; it makes a very great difference, or rather all the difference" (Aristotle, e2005, p. 24).

In these three excerpts we find the fundamental characteristics of *habit* as a behavioural, intellectual, or moral disposition. They set out the dual character of habit for Aristotle. The "active" pole of habit (*ethos*) makes it possible to internalize new patterns through legislation, education, mimicry, or habituation. These patterns will tend, through repetition, to become stable, to form dispositions, which are "possession and power" (*hexis*). The stores of past experience that are laid down are converted into provisions for the future (Héran, 1987).

In the final excerpt from Aristotle, the idea of habit as "second nature" emerges. The reference to legislators making their citizens contract certain habits also shows the transferability of habit, as well as its possible instrumentalization by various institutions for the exercise of government. Building good habits from an early age is of paramount importance for legislators. Aristotle's definition of habit is undoubtedly one of the best ever formulated, since it enables us to understand habit as a mediator between categories that we frequently think of as antagonistic. Habit, as a concept, makes it possible to overcome the cleavages between *mechanism* and *spontaneity*, *freedom* and *determinism*, *object* and *subject*, *agency* and *structure*, *individual* and *society* (Héran, 1987).

Numerous contributions in sociology and in the history of science have underlined the very limited use of the concept of *habit* in the social sciences after the behaviourist works of the early twentieth century (Hodgson, 2004; Kaufmann, 2001; Lahire, 1998). The concept of *habit* has long suffered from its use in the writings of Watson (Watson, e1970) and of Ivan Petrovič Pavlov. The "S–R" (stimulus–response) concept put forward by Watson views "habit" as an automatic response or a biological reflex. "Habit" was relegated to the state of simple reflex, stripped of any social dimension, so that the difference between humans and animals disappeared. For example, Pavlov saw no problem in generalizing results obtained with animals to humans (Kaufmann, 2001).

Other earlier uses of the term "habit" have also caused unease in the social science community. In addition to the behaviourists, the spiritualist philosopher Felix Ravaisson (1813–1900), drawing on metaphysical writings by St. Thomas Aquinas (1224–1274), associated habit and especially *ethos* with "obscure thought", with the "primordial law of being", in other words with the "divine" (Ravaisson, e2008; Schwanen et al., 2012). This was the beginning of a long phase of marked discomfort for sociologists over questions relating to *habit*, which became "data of no scientific interest" (Kaufmann, 2001, p. 114).

A linguistic barrier arose in the work of sociologists in the first part of the twentieth century. This uneasiness is reflected, as in the example of Mauss (e1973), by the use of Latin ("*habitus*") or of quotation marks as devices to circumvent an embarrassing obstacle, as the term was then connoted very negatively.

Habit: why is it relevant to an understanding of contemporary urban mobilities?

The concept of habit was revived in the early 2000s. This revival was connected with renewed interest in social psychology when facing cognitive dissonance and the "value–action gap". In other words, in the late 1990s and early 2000s, in many countries and in many sectors of daily life (mobility, food, energy, waste disposal), individuals were increasingly torn between values and attitudes favourable to "environmental" or "sustainable" solutions on the one hand and practices that remained closely linked to energy-consuming solutions – the car in the case of urban mobility – on the other hand.

Habit has been identified as a concept with strong potential to explain this value–action gap. More specifically, habit has been recently identified as a source of intense interference between two couples: between intentions and behaviours (Gardner, 2009; Verplanken et al., 2008), and between social norms and behaviours (Klöckner and Matthies, 2004).

In other words, a person may have firm intentions to change (e.g. mode of transport, from car to bike), but might only be able to alter their behaviour to the extent that their current (car) habit is weak or moderate. Likewise, a person using the car on a daily basis, surrounded by a social environment that values walking or cycling, will tend to change their mode of travel if and only if their car habit is weak or moderate. Habit is a relevant point of entry for explaining resistance to change, and more specifically for the difficulties many public policies experience in reducing car use in cities.

Beyond the idea of a force of resistance to change, it is also worth looking to habit to explain the skills deployed in everyday travel. One of the reasons why habit is a "major anthropological fact" (Héran, 1987) is that it frees up cognitive resources that can be used for other things (Hodgson, 2004; Leroi-Gourhan, e1993). In a relatively stable context, the habits acquired make it possible to detach oneself from the activity of travelling and engage in other more reflective practices, sometimes requiring a certain dexterity. In the case of daily urban mobility, a well-formed habit allows a more intensive use of time, "enhanced" by the use of mobile terminals, reading, listening to music or programmes, conversation, etc. (Brette et al., 2014).

How can we measure the strength of mobility habits?

The renewed interest in the concept of habit very quickly came up against a major difficulty: How could the strength of a habit be measured? This is the subject of much ongoing debate in social psychology. The first attempt to measure habit was proposed by Verplanken et al. (1994). The "Responses Frequency Measure" (RFM) protocol makes use of the advanced automatic character of a habit. The investigator presents a series of 12 fictitious situations (e.g. going downtown for a coffee, going to work, …). The respondent is expected to state as quickly as possible the first transport mode that springs to mind. If the respondent takes too long to answer, the answer is ignored. The RFM protocol has been tested on people with different habits, in terms of transport modes (cycling or driving, in this case) and the results show significant differences. That said, this protocol has major pitfalls. First, several difficulties arise in operationalizing this protocol (12 situations, time management with a stopwatch while conducting the survey) and secondly, biases occur due to virtual situations that are not necessarily widespread within the population (e.g. having a coffee downtown). Lastly, this test could only measure the automatic character of the choice of transport mode. These pitfalls led to the development of a second version of the measurement tool also based on Verplanken's work (Verplanken and Orbell, 2003): "SRHI" for Self-Reported Habit Index.

SRHI presents a series of 12 statements for each behaviour whose "habitual" character is to be assessed. The respondent must answer from a series of five attitudes ranging from "I agree" to "I disagree" (see Table 22.2). In order to go further than simply taking into account the automatic character of responses to hypothetical situations, the statements included in this test follow five dimensions that characterize habits by way of a broad definition. This test takes into account a *history of repetition* of the behaviour in question, the *lack of awareness*, but also the perceived *difficulty* of the respondent in *controlling* the habit. Two other dimensions are taken into account: the mental efficiency and the expression of self-identity which is often the last stage of a strongly ingrained habit (e.g. driving a car, that's typically "me").

Although the SHRI protocol has proved reliable when measuring the strength of a habit, the principal criticism concerns the difficulty in operationalizing it in a survey. Some of the

Table 22.2 Twelve statements in the Self-Reported Habit Index (Verplanken and Orbell 2003) and its reduced version in the SRBAI (Self-Reported Behaviour-Automaticity Index) (Gardner et al., 2012)

"Behaviour X is something …"	*Components of habit*	*SRBAI*	*SRBAI + id*
1. I do frequently.	History of repetition		
2. That belongs to my (daily, weekly, monthly) routine.			
3. I have been doing for a long time.		X	X
4. I do automatically.	Lack of awareness	X	X
5. I do without thinking.		X	X
6. I have no need to think about doing.		X	X
7. I do without having to consciously remember.	Lack of control		
8. I start doing before I realize I'm doing it.		X	X
9. That makes me feel weird if I do not do it.	Expressing self-		
10. That's typically "me".	identity		X
11. That would require effort not to do it.	Mental efficiency		
12. I would find it hard not to do.			

Source: T. Buhler

12 statements are perceived as being very similar or almost synonymous for many respondents (Gardner et al., 2012) who consequently tend to respond in the same way for all 12 assertions. This observation led to a need for a parsimonious protocol for measuring habit. This has been materialized by Gardner et al. (2012) with the SRBAI (Self-Reported Behaviour-Automaticity Index): an abridged version of the SRHI limited to five questions focusing on automaticity, history of repetition, and lack of control (see Table 22.2). The 5-item SRBAI has been successfully included in several surveys of hundreds of respondents covering very diverse questions such as active travel, snacking, or alcohol consumption. It has displayed the same level of reliability and predictive utility to future behaviour as the full 12-item SRHI.

In a recent survey in March 2018 with the contribution of the *Sciences-Po* ELIPSS national panel (Buhler and Signoret, 2018) we applied the SRBAI protocol for six different transport modes in the case of everyday travel (see Figure 22.1). As the question of identity was also of interest and as it has been proven to be the ultimate stage of habituation (Kaufmann, 2001), we also put assertion no. 10 of the SRHI ("Behaviour X is something … that's typically 'me'") to respondents. The five items of the SRBAI together with this item concerning identity composed what we can call the "SRBAI + id" test.

By way of illustration, Figure 22.1 shows the results of our protocol for the six main transport modes within this representative sample of the French Population (N = 2160 respondents). The variable analysed here in Figure 22.1 is a (simple) statistical mean between the six answers to the "SRBAI + id" items.[1] Respondents were able to answer on a scale from "*1: I totally disagree*" to "*7: I totally agree*".

This rather simple measure of the strength of habit varies greatly when applied to the six principal transport modes. Car-use is a relatively strong habit compared to all other modes. With a median of 5.5 (out of 7) and a third-quartile value of 6.5 we can easily conclude that driving is something that most of the French population do "for a long time", "without thinking about it", "automatically", and even "before (…) realizing it". The car-use habit is even stronger than the walking habit (for everyday travel).

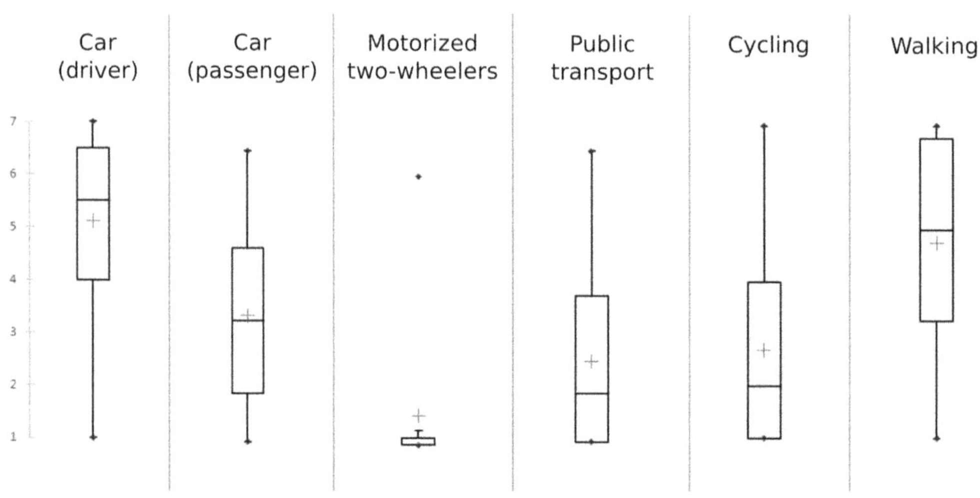

Figure 22.1 Levels of habit for the six main transport modes within the French population (N = 2160, representative sample).

Source: T. Buhler

The exclusive nature of car-use habit

Beyond the rather simple and obvious observation that car-use runs at a high level among the French population, it is important to underline the specific features of a strong car-use habit. An important observation is that it is of an exclusive nature. From the results plotted in Figure 22.1, we classified the levels of habits for each respondent and for each mode. A habit is strong (+) when it is above the third quartile for its mode (e.g. for car-as-a-driver the level of habit "+" corresponds to a habit strength of more than 6.5). Similarly, a habit is weak (–) when it is below the first quartile for its mode. An average level of habit (=) occurs in the remaining cases closer to the median. We tried to observe the potential combinations between categorized levels of habit for each mode using Multiple Correspondence Analysis[2] (MCA, see Figure 22.2). From this set-up, we can observe a principal fact: when someone has a strong habit of car driving (+), it is generally combined with very weak habits (–) for all the other modes (see Figure 22.2, *grey ellipse*). This first cluster concerns 536 people (25% of the sample). The contrary result is also true: a weak habit of car-use (–) is correlated with a strong habit for all others modes (+) (*black circle, 563 people or 26%*). This group includes people who do not or cannot use a car in their day-to-day life and have to offset this through an ability to use all the other modes. A third group is characterized by a median level of habit for all modes (*dashed circle*) and concerns 1061 people or 49% of the sample. In other words, a strong habit of driving leads to car-use becoming an exclusive way of life. Beyond a certain point, there is no scope for other transport habits.

Discussion and conclusion

Although it has long been left aside by the "classical" strands of thought in transport and mobility studies (transport economics, engineering), and more generally by the social sciences community, the concept of habit is essential for understanding the rationales of mobile individuals, their capacities to act or change, or to engage in several activities at once. Following a centuries-old definition, habit is at one and the same time a capacity to learn, to

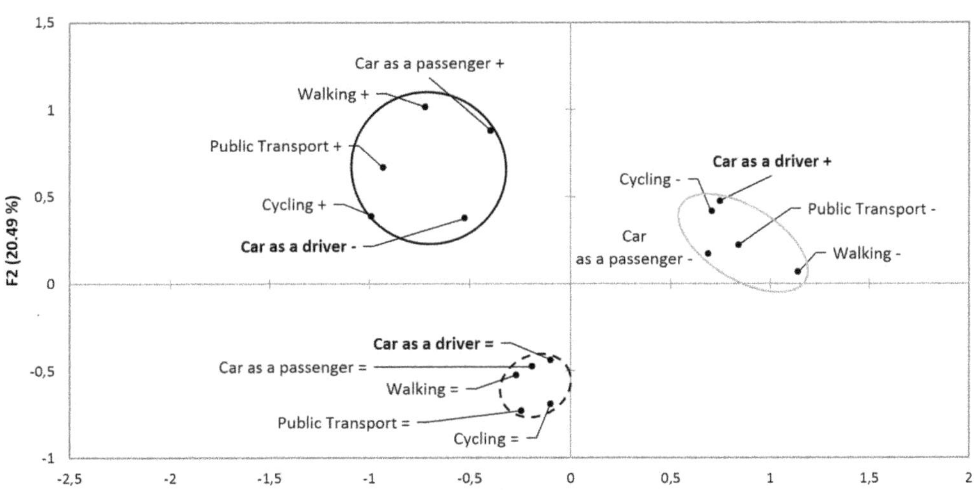

Figure 22.2 Car-use habit strength as the major variable of the combination of habits. [Multiple Correspondance Analysis (F1 + F2 = 87.66 %)] [N = 2160].

Source: T. Buhler

integrate or to embody new patterns of thinking and acting through repetition or education (*ethos*), and a propensity to act in a certain way, a power to act efficiently (*hexis*).

In the case of the car-use habit, the repetition of travel by car in a stable geographical and temporal context (*ethos*) leads, for example, to a certain mastery in the use of travel time (*hexis*). In other words, the car-use habit is correlated with the ability to detach oneself from driving and to use these new cognitive resources for texting, phoning, or talking with passengers while driving, and so for creating a "pleasant" experience of travel (Brette et al., 2014). The repetitive formation phase of the habit leads to a marked readiness to travel by car, finding it "pleasant", and in the final stage of the habit, believing that there is no viable alternative (idem). About a quarter of the French population seems to fit this profile, which is far from insignificant. Beyond the idea of a force of resistance to change, it is also worth looking to habit to explain the skills deployed in everyday travel.

It seems worth taking account of this habitual dimension of mobility behaviour in both research and in public policy design, at least to distinguish the most accustomed daily drivers (with a strong habit) from those who drive everyday but would be more amenable to switching modes. For the latter, classical public policies based on incentives and restrictions may "suffice" to get them to switch modes. For the former, recent research programmes focus on the question of "windows of opportunity" (Brette et al., 2014). Indeed, whatever the strength of a habit, it is based on stable everyday contexts in terms of geography, time, and personal life (Hodgson, 2010). The idea beyond that is to take into account the important moments of change in the person's life cycle (moving house, birth of a child, retirement, etc.) to identify potential moments to suggest new (mobility) lifestyles to people who are better able to materialize their intentions. Urban public policies could also target the creation or reinforcement of active travel habits from the earliest age. This can be done for example by providing free public transport for minors, free bike rental for students, workshops for learning how to change a bike inner tube or how to find your way around the city as a pedestrian. Again, as Aristotle asserted (e2005, p. 24) "it makes no small difference, then, whether we form habits of one kind or of another from our very youth; it makes a very great difference, or rather all the difference".

Notes

1 Dots represent the lowest and the highest values, the cross symbolizes the mean, and the three horizontal lines of the "box-plot" represent the median, the first, and the third quartiles of the series.

2 It is statistically impossible to combine more than five variables in an MCA. We eliminated the strength of habit for motorized two-wheelers as it is by far the weakest habit.

References

Aristotle. e2005, *Nicomachean Ethics*, translated by W.D. Ross, Digireads.com.

Bissell, D. 2012, 'Agitating the powers of habit: Towards a volatile politics of thought', *Theory & Event*, vol. 15, no. 1, p. 128.

Brette, O., Buhler, T., Lazaric, N., & Maréchal, K. 2014, 'Reconsidering the nature and effects of habits in urban transportation behaviour', *Journal of Institutional Economics*, vol.10, no.3, pp.399–426.

Buhler, T. 2015, *Déplacements urbains: sortir de l'orthodoxie. Plaidoyer pour une prise en compte des habitudes*. Lausanne: Presses Polytechniques et Universitaires Romandes.

Buhler, T. & Signoret, P. 2018, Panel national mobilité quotidienne – Wave 1 (February–March 2018). Paris: Fondation Nationale des Sciences Politiques (FNSP), Centre de Données Socio-Politiques (CDSP), version 1.

Collins Thesaurus. 2019, www.collinsdictionary.com/dictionary/english-thesaurus site visited on January, 7th, 2018.

Fujii, S. & Gärling, T. 2003, 'Development of script-based travel mode choice after forced change', *Transportation Research Part F: Traffic Psychology and Behaviour*, vol.6, no.2, pp.117–124.

Gardner, B. 2009, 'Modelling motivation and habit in stable travel mode contexts', *Transportation Research Part F: Traffic Psychology and Behaviour*, vol.12, no.1, pp.68–76.

Gardner, B., Abraham, C., Lally, P. & de Bruijn, G.J. 2012, 'Towards parsimony in habit measurement: Testing the convergent and predictive validity of an automaticity subscale of the Self-Report Habit Index', *International Journal of Behavioral Nutrition and Physical Activity*, vol.9, no.102, pp. 1–12.

Gärling, T. & Axhausen, K. 2003, 'Introduction: Habitual travel choice', *Transportation*, vol.30, no.1, pp.1–11.

Gärling, T., Fujii, S. & Boe, O. 2001, 'Empirical tests of a model of determinants of script-based driving choice', *Transportation Research Part F: Traffic Psychology and Behaviour*, vol.4, no.2, pp.89–102.

Goodwin, P.B., Dix, M.C. & Layzel, A.D. 1987, 'The case for heterodoxy in longitudinal analysis', *Transportation Research Part A: General*, vol.21, no.4–5, pp.363–376.

Héran, F. 1987, 'La seconde nature de l'habitus. Tradition philosophique et sens commun dans le langage sociologique', *Revue française de sociologie*, vol.28, no.3, pp.385–416.

Hodgson, G.M. 2004, 'Reclaiming habit for institutional economics', *Journal of Economic Psychology*, vol.25, no.5, pp.651–660.

Hodgson, G.M. 2010, 'Choice, habit and evolution', *Journal of Evolutionary Economics*, vol.20, no.1, pp.1–18.

Kaufmann, J.-C. 2001, *Ego: pour une sociologie de l'individu*. Paris: Nathan.

Kaufmann, V. 2000, *Mobilité quotidienne et dynamiques urbaines*. Lausanne: Presses Polytechniques Romandes.

Klöckner, C.A. & Matthies, E. 2004, 'How habits interfere with norm-directed behaviour: A normative decision-making model for travel mode choice', *Journal of Environmental Psychology*, vol.24, no.3, pp.319–327.

Lahire, B. 1998, *L'homme pluriel: les ressorts de l'action*. Paris: Nathan.

Leroi-Gourhan, A. e1993, *Gesture and Speech*, translated by A. Bostock Berger, Cambridge: MIT Press.

Maréchal, K. 2009, 'An evolutionary perspective on the economics of energy consumption: The crucial role of habits', *Journal of Economic Issues*, vol.43, no.1, pp.69–88.

Maréchal, K. 2010, 'Not irrational but habitual: The importance of "behavioural lock-in" in energy consumption', *Ecological Economics*, vol.69, no.5, pp.1104–1114.

Mauss, M. e1973, 'Techniques of the body', translated by Ben Brewster, *Economy and Society*, vol.2, no.1, pp.70–88.

Ravaisson, F. e2008, *Of Habit*. London: Continuum International Publishing Group.

Schwanen, T., Banister, D. & Anable, J. 2012, 'Rethinking habits and their role in behaviour change: The case of low-carbon mobility', *Journal of Transport Geography*, vol.24, pp.522–532.

Schwartz, S.H. 1977, 'Normative influences on altruism', in *Advances in Experimental Social Psychology*. New York: Academic Press, pp. 221–279.

Verplanken, B., & Aarts, H. 1999, 'Habit, attitude, and planned behaviour: Is habit an empty construct or an interesting case of goal-directed automaticity?', *European Review of Social Psychology*, vol.10, no.1, pp.101–134.

Verplanken, B., Aarts, H., van Knippenberg, A. & van Knippenberg, C. 1994, 'Attitude versus general habit: Antecedents of travel mode choice', *Journal of Applied Social Psychology*, vol.24, no.4, pp.285–300.

Verplanken, B., & Orbell, S. 2003, 'Reflections on past behavior: A self-report index of habit strength', *Journal of Applied Social Psychology*, vol.33, no.6, pp.1313–1330.

Verplanken, B., Walker, I., Davis, A. & Jurasek, M. 2008, 'Context change and travel mode choice: Combining the habit discontinuity and self-activation hypotheses', *Journal of Environmental Psychology*, vol.28, no.2, pp.121–127.

Watson, J.B. e1970, *Behaviorism*. New York: Norton.

23
RESIDENTIAL MOBILITY

Patrick Rérat

Introduction

Where do people live? Why do they move? How do they choose their dwelling and residential location? These questions are crucial to understand the dynamics of cities. Population flows are indeed a major driving force shaping urban regions and are at the core of processes such as suburbanization, gentrification or reurbanization.

Broadly defined, residential mobility refers to a household's change of residence over short distances (e.g. within a region or a metropolitan area). Thus, it does not usually affect the whole organization of daily life in terms of work, shopping and leisure locations. Migration, in contrast, involves longer distances across national borders (international migration) or between regions (internal or inter-regional migration). It redefines the spaces of daily life through a process of uprooting and re-rooting.

The distinction between residential mobility and migration is not always clear though. Migration implies decisions with respect to residential mobility such as selecting a neighbourhood or a dwelling. The institutional organization and size of countries differ greatly which makes the definition of internal migration variable. Moreover, the spaces of daily life are blurred due to long-distance commuting. However motivations vary according to distances: the shorter the move, the more it relies on the characteristics of the dwelling and the residential context; the longer the move, the more it refers to employment or education (Niedomysl 2011; Owen and Green 1992).

Residential mobility has been addressed by geography, sociology, demography and economics. In this chapter I first outline the classic approaches and then discuss the notion of residential choice. I also give an overview of urban phenomena where residential mobility is at stake and conclude with some final remarks.

Classic approaches[1]

Perspectives addressing residential mobility can be distributed between macro- and micro-analytical perspectives (Cadwallader 1992). The former look at aggregate residential phenomena and explain them based on the context (e.g. characteristics of spatial entities). The latter focus on individuals and their motivations and study decision-making.

A second categorization separates "deterministic" and "humanistic" perspectives (Boyle et al. 1998). The former suggest that moving is an obvious solution, given the context (and/or structures), and tend to minimize the role of individuals. The latter consider that actors consciously make decisions and have a certain freedom of choice.

Research on residential mobility has often referred to one of five traditional theoretical approaches whose principles are outlined below.

The neoclassic perspective regards individuals as economically rational (*homo economicus*). They optimize their utility based on differences of income, amenities, costs, etc. The public choice theory (Tiebout 1956) for example posits that individuals decide to move based on the combination of services provided by a local community and its price in terms of tax that best matches their preferences. Thus, by moving, individuals "vote with their feet".

The behaviourist approach highlights the importance of socio-psychological mechanisms in the decision to move. Individuals tolerate a degree of discomfort but, once a certain level of stress is reached, they seek a residential context that offers what is perceived as a better quality of life (Cadwallader 1992). Households choose from among a limited number of alternatives based on a minimum level of satisfaction ("satisficer"). For Rossi, residential mobility is a spatial process through which a family adjusts its housing consumption to its needs, notably in terms of dwelling size (Rossi 1955).

The structuralist approach highlights the social constraints that affect individuals and limit their room for manoeuvre. It explains residential phenomena on a structural level (economic and political framework, conflictual relations between classes, etc.). Authors turned to (neo-) Marxist theories to analyse residential phenomena with regard to the organization of the capitalist mode of production. Gentrification for example can be interpreted in light of neo-liberal policies and capital investment–divestment–reinvestment cycles within the built environment (Smith 1996).

The humanistic approach refers here to the human geography movement that emerged in the 1970s, which highlights action (agency). Much research – in sociology and anthropology – shares these principles, focusing on individual experiences and characteristics such as beliefs, feelings, values, emotions and attachment (Christie et al. 2008). To go back to gentrification, humanist accounts have highlighted the residential aspirations of parts of the middle class in favour of urban areas and their rejection of suburban lifestyle (Ley 1996).

The institutional approach does not have the status of an established theory. It is concerned with the roles of managers and institutions without proposing an interpretive grid (Knox and Pinch 2000; Pahl 1969). It examines the role of intermediaries in the housing market (builders, developers, real estate agents, local authorities, financial institutions, etc.). These gate keepers link available resources (e.g. land and capital) and potential clients, and structure the housing supply (Briggs et al. 2010; Teixeira 2006).

Given the variety of approaches to residential phenomena, recent developments in mobility and migration studies are characterized by a call to incorporate different points of view – each of which highlight specific mechanisms and are not necessarily incompatible – and to adopt theoretical and methodological pluralism based on the line of questioning. This is what allows the concept of residential or housing choice.

Residential choice

The term residential/housing choice, at first glance, does not seem to transcend the conflict between deterministic and humanistic approaches, given its positive connotation. However, residential choice should not be regarded solely as the result of aspirations but as a choice under constraints. More specifically, it depends on households' needs and preferences within a limited range of options defined by the opportunities and constraints of the housing market (availability of certain types of housing in a given residential context, prices, etc.) and the resources and restrictions associated with households themselves (Van Ham 2012).

Restrictions can be objective (income, location of workplaces, etc.) or subjective (schemes of perception and action arising from belonging to a social class, etc.). Residential choice thus appears to be socially constituted. Because of the existence of constraints and restrictions, residential choice is the result of trade-offs, be it between the people affected by the move or various criteria related to housing. The development of transport infrastructures and the increase in travel speed have lowered the constraints of distances and have widened the search area of households.

The concept of residential choice means that people enjoy a certain leeway, even if the range of possibilities varies greatly. This theoretical position legitimates the study of five sets of characteristics: the unit of analysis, profile, trajectories, criteria and decision-making.

Unit of analysis

Works on mobility have long focused on the individual as a unit of analysis partly due to theoretical perspectives (e.g. neoclassic) and to data availability. Rossi's research on "Why families move" (1955) paved the way to studies integrating the family or the household as a unit of analysis. Household location choices involve several members with heterogeneous preferences and influences and imply trade-offs, negotiations and compromises (Coulter et al. 2012). Differences may be observed between choices made by single individuals and those made by the same individuals when choosing collectively (Marcucci et al. 2011).

Scholars have also stressed the importance of the entourage that includes parents, children, spouses and siblings who do not share the same apartment in the residential choice (Mulder 2007). This importance may refer to the proximity of mutual aid networks such as taking care of children (Vignal 2005), the transmission of housing preferences (Lux et al. 2018) or the intergenerational financial support to access home-ownership (Hochstenbach 2019).

Profile

The second dimension, profile, refers firstly to classic variables such as position in the life course (age, type of household, etc.), socio-economic status (education, etc.) and national origin. The impact of age has been well documented (Rogers 1988): the highest mobility rate is observed among young adults when they leave their family, gain residential independence, study or enter the labour market, and start an adult life. After the twenties the mobility rate declines and residential stability becomes the norm. A rebound is hypothesized at retirement (although not observed in all countries) and a last peak takes place at the end of life due to the move to nursing homes.

To this "vertical" differentiation can be added a "horizontal" distinction based on lifestyles. This relates to individualization theories that argue that contemporary societies are characterized by the actors' "disembedding" from traditional social constraints, releasing them from traditional scripts dictating how they should live their lives (Rye 2011). The contemporary individual is characterized by a level of choice unavailable to previous generations (Beck and Beck-Gernsheim 2002) and has led to a diversification of lifestyles and residential aspirations even within social classes and age groups.

The concept of lifestyle is subject to a variety of definitions based on either practices in time and space, latent variables (opinions or values), socio-demographic variables or a combination of these (Jansen 2014). The commonality of the body of work dealing with lifestyle is its assertion that traditional demographic and socio-economic variables are insufficient to explain residential choices. It has been used to analyse the difference between urban and suburban middle classes (Lufkin et al. 2018). Some, albeit rare, studies have explored residential choice (Coolen et al. 2002; Rössel and Hoelscher 2012) through residents' values such as tradition, success, autonomy and safety (Schwartz 1994).

Profile is important as residential mobility is doubly selective. On the one hand and as said above, the propensity to move varies according to population groups. In general, the average is higher among young adults, singles and childless couples (versus married people or families), tenants (versus homeowners) and highly qualified people. Moreover, each territorial context has a specific hosting potential (Kaufmann 2011) that is more or less attractive to certain population groups. For example urban centres are characterized by an under-representation of families and an overrepresentation of singles, childless couples and flat shares (Rérat 2012a).

Trajectories

The third aspect, trajectory includes past, present and future housing locations and characteristics. It is a way to analyse a move in the perspective of the biographies of individuals. The concept of life cycle featured successive stages in the progression of a family marriages (Glick 1989). Given that the family institution has been subject to major challenges (unregistered partnership, divorces, remarriages, etc.), it has been replaced by the concept of life course. This approach is a way of structuring a complex set of events that include decisions about occupational, marital and housing careers (Mulder and Clark 2002).

The life course approach takes into account three interdependences between (1) the past, the present and the future; (2) the different spheres of action (family, education, work, leisure, etc.) and (3) individual action and contexts, "since life course patterns are embedded in macro-social structures and cultural beliefs and guided by market opportunities, institutions and social networks" (Heinz et al. 2009).

Life course approaches thus look at residential choice within a triple framework: personal (formation or separation of a couple, birth of a child, etc.), residential (location and characteristics of past and current housing) and professional (entering the job market, job changes, retirement, etc.). Other scholars have suggested reconceptualizing residential (im)mobility as relational practices that link lives through time and space while connecting people to structural conditions (Coulter et al. 2016).

Criteria

Fourth, residential choices are complex and involve a multitude of criteria. Three angles of analysis exist: residential satisfaction, aspirations and motivations which focus respectively on the current situation (factors that might encourage a move), projects in a more or less distant and defined future (stated preferences) and criteria underpinning a residential choice (revealed preferences).

Moving can be based upon different factors. Clark and Onaka (1983) distinguish forced (e.g. expropriation, home foreclosures), induced (by a change in the household structure or work location) and adjustment moves (Clark and Onaka 1983). In the latter case, moving is an attempt to improve quality of life and housing in terms of occupation status, size, comfort, accessibility, etc. As housing is a composite good, households must make a certain number of trade-offs and prioritize between these elements since they cannot meet all their aspirations. Research has addressed specifically some criteria such as the tenure choice (Andersen 2011; Lux et al. 2018), the importance of the neighbourhood (Andreotti et al. 2013), the role of school choices (Boterman 2013), mobility practices (Rérat and Lees 2011), amenities and workplaces (Frenkel et al. 2013), etc.

Residential choices refer to several logics of action: a functional logic ("using"), a social logic ("meeting") and a sensitive logic ("inhabiting") (Lufkin et al. 2018). Households are simultaneously characterized by a calculating relationship to the world (objective factors such as price and functional qualities), a socio-cognitive relationship (representations such as reputation), and a sensitive relationship (attraction and repulsion).

Decision-making process

Fifth, the decision-making process has to do with the way households find a dwelling, e.g. through the market (agencies, ads) or networks (family, friends) (Authier 1998). Households are not equal as these logics depend on their economic (income), cultural (ability to process information) and social (relations) capital (Boterman 2012).

Among other factors, time or the urgency of the situation may force a household to revisit certain criteria. Furthermore, living on site – the location-specific advantage or the ties that bind people to a place (Mulder and Wagner 2012) – imply a better knowledge of the context and reactivity when it comes to seizing opportunities, whereas migrating from another region or country requires a certain understanding of the local context (which can be done through several residential stages).

Other dimensions related to decision-making refer as mentioned above to trade-offs between members of a household and criteria. It may also imply trade-offs between residential mobility and everyday mobility and strategies to avoid moving and to preserve local moorings such as opting for longer commuting trips or multiresidentiality (Rérat 2014). Residential mobility should also be seen as a process: the focus should not be on actual moves only but take into account also the stages preceding the moves (Kley and Mulder 2010).

In addition to these dimensions other crucial issues regard what happens after a move. A first one refers to what is called post-occupancy evaluation (Preiser et al. 2016) that refers to the study of buildings once inhabited, so that lessons may be learned to guide the design of future buildings (Meir et al. 2009). Participant-observer ethnography has also been used to analyse residents' adaptation and appropriation of their residential context (Gans 2017). Another issue refers to the households' spatial practices, in other words, daily mobility downstream of residential choice in the light of the compact city debate (see below).

Research perspectives

There are several on-going debates on the urban realm where residential mobility is at stake: multiresidentiality, demographic transition, urban changes and social structure, and environmental sustainability.

Multiresidentiality

In the introduction, residential mobility was defined as a change of residence. However, a growing number of households live in more than one place (multiresidentiality).

The reasons are manifold: job or education, leisure (second-homes) or familial configurations (children of separated couples, couples living apart together, etc.). Multi-local living arrangements imply a cyclical use of more than one place of residence, include moorings and immobilities as well as the recurrent movement between these places (Schier et al. 2015).

Demographic transition

Ageing is one of the major demographic trends and the rise of senior living is an important issue for housing. One key element is the residential mobility of this (heterogeneous) age group. Studies show several practices. Ageing in place is the most common one although not geographically and socially evenly distributed. Downsizing strategies are also adopted to cope with difficulty (maintenance, stairs, etc.). This reduction of housing consumption (from owning to renting) seems much more common in the US than in Europe (Banks et al. 2012). Residential choices could become more diverse as baby boomers retire. Some authors hypothesize that the elderly today are more willing to change residence to accommodate for changing lifestyle and poorer health than earlier generations (Abramsson and Andersson 2012).

Age is becoming a core dimension of urban socio-spatial change (Hochstenbach 2019) and not only in term of ageing: some central and dense areas have been experiencing rejuvenation or youthification (Moos 2016). Young adults are making location decisions in a context of lower employment security, higher costs, new residential aspirations. Studentification (the transformation of urban areas due to the move of a growing student population) and gentrification also contribute to changing the age structure of central areas (Smith 2002; Smith and Holt 2007).

Urban change and social structure

The analysis of the selective geographies of residential (im)mobility represents a way to understand urban changes. Reurbanization for example refers to the new period of demographic growth of cities (Rérat 2019) or to a process of populating and diversifying the inner city with a variety of residential groups (Haase et al. 2010). Other cities are, on the contrary, shrinking which affects population groups differently (Pallagst et al. 2014).

Gentrification shows a diversification of residential aspirations within middle classes although important differences are observed between countries (Rérat 2018). Suburbanization or urban sprawl are another major trend. They have been interpreted among others as the result of the housing choice of families towards bigger dwellings, ownership, detached houses, children-friendly environment, etc. (suburban familism) (Fishman 1987). In many post-industrial cities a suburbanization of poverty can also be observed which limits the housing options of low-income households (Hochstenbach and Musterd 2018). Spatial segregation, social polarization and the concentration of minority ethnic groups remain a central issue, given the barriers in tightening markets (Briggs et al. 2010), the way residential mobility between neighbourhoods may bring about changes in the pattern of ethnic segregation (Bolt and Kempen 2010) and policies implemented, such as housing subsidy programmes for low-income households (Basolo and Yerena 2017). A focus is also put on the housing choice after migrating and the spatial assimilation theory (migrants would adopt the same residential practices after a few years) (Andersen 2011).

Environmental sustainability

Housing is closely related to environmental sustainability. This refers to the impact of location and of urban form upon the consumption of energy and resources, notably through mobility practices (Rérat 2012b). Urban sprawl has become the most dominant trend in urbanization since World War II (European Environment Agency 2016). Yet it has been strongly criticized as it implies a high level of land consumption and automobile dependence.

To regulate urban sprawl, the compact city has been promoted (Holden 2004). It has been argued that the densification of the built environment would slow down urban sprawl and limit resource and energy consumption by reducing the role of the car and increasing the number of trips made on foot, by bicycle or by public transport. Scholars have addressed these issues by studying residential aspirations or satisfaction between various types of environments (e.g. Mouratidis 2018) and the links between residential choice and mobility practices (Ettema and Nieuwenhuis 2017; Humphreys and Ahern 2019; Rérat and Lees 2011).

Concluding remarks

Residential mobility may be analysed through two perspectives. It is an object of study in itself as in research addressing the housing choices of population groups in a context of demographic changes and diversification of lifestyles. It constitutes also an indicator at the service of other scientific objects such as the understanding of urban dynamics. Current issues such as market (de)regulation, ecological transition, spatial justice, strengthen the importance and relevance of this field of study.

For more than a decade scholars have reconceptualized how contemporary life is configured by mobilities (Sheller and Urry 2006). While there is a burgeoning literature on international migration, far less attention has been devoted to rethinking residential (im)mobility (Coulter et al. 2016). This might be surprising as residential location is crucial in the way people (have to) deal with distance and space. A closer dialogue with the new mobilities paradigm could invigorate approaches of residential mobility regarding several issues presented above such as the experiences and meanings of both residential mobility and immobility as well as their articulation with other forms of spatial mobilities.

Note

1 This section extends a previous text on residential mobility (Rérat 2016).

References

Abramsson, M. and Andersson, E.K., 2012. Residential Mobility Patterns of Elderly—Leaving the House for an Apartment. *Housing Studies*, 27 (5), 582–604.

Andersen, H.S., 2011. Motives for Tenure Choice during the Life Cycle: The Importance of Non-Economic Factors and Other Housing Preferences. *Housing, Theory and Society*, 28 (2), 183–207.

Andreotti, A., Le Galès, P. and Javier Moreno Fuentes, F., 2013. Controlling the Urban Fabric: The Complex Game of Distance and Proximity in European Upper-Middle Class Residential Strategies. *International Journal of Urban and Regional Research*, 37 (2), 576–597.

Authier, J.-Y., 1998. Mobilités et processus de gentrification dans un quartier réhabilité du centre historique de Lyon. *In*: Y. Grafmeyer and F. Dansereau, eds. *Trajectoires familiales et espaces de vie en milieu urbain*. Lyon: Presses universitaires de Lyon, 335–352.

Banks, J., Blundell, R., Oldfield, Z. and Smith, J.P., 2012. Housing Mobility and Downsizing at Older Ages in Britain and the USA. *Economica*, 79 (313), 1–26.

Basolo, V. and Yerena, A., 2017. Residential Mobility of Low-Income, Subsidized Households: A Synthesis of Explanatory Frameworks. *Housing Studies*, 32 (6), 841–862.

Beck, U. and Beck-Gernsheim, E., 2002. *Individualization*. London: Sage.

Bolt, G. and Kempen, R.V., 2010. Ethnic Segregation and Residential Mobility: Relocations of Minority Ethnic Groups in the Netherlands. *Journal of Ethnic and Migration Studies*, 36 (2), 333–354.

Boterman, W.R., 2012. Deconstructing Coincidence: How Middle-Class Households Use Various Forms of Capital to Find a Home. *Housing, Theory and Society*, 29 (3), 321–338.

Boterman, W.R., 2013. Dealing with Diversity: Middle-class Family Households and the Issue of 'Black' and 'White' Schools in Amsterdam. *Urban Studies*, 50 (6), 1130–1147.

Boyle, P., Halfacree, K. and Robinson, V., 1998. *Exploring Contemporary Migration*. Essex: Longman.

Briggs, X.D.S., Comey, J. and Weismann, G., 2010. Struggling to Stay Out of High-Poverty Neighborhoods: Housing Choice and Locations in Moving to Opportunity's First Decade. *Housing Policy Debate*, 20 (3), 383–427.

Cadwallader, M., 1992. *Migration and Residential Mobility: Macro and Micro Analysis*. Wisconsin: The University of Wisconsin Press.

Christie, H., Smith, S.J. and Munro, M., 2008. The Emotional Economy of Housing. *Environment and Planning A*, 40, 2296–2312.

Clark, W.A.V. and Onaka, J.L., 1983. Life Cycle and Housing Adjustment as Explanations of Residential Mobility. *Urban Studies*, 20 (1), 47–57.

Coolen, H., Boelhouwer, P. and Van Driel, K., 2002. Values and Goals as Determinants of Intended Tenure. *Journal of Housing and the Built Environment*, 17, 215–236.

Coulter, R., van Ham, M. and Feijten, P., 2012. Partner (dis)Agreement on Moving Desires and the Subsequent Moving Behaviour of Couples. *Population, Space and Place*, 18, 16–30.

Coulter, R., van Ham, M. and Findlay, A., 2016. Re-thinking Residential Mobility: Linking Lives through Time and Space. *Progress in Human Geography*, 40 (3), 352–374.

Ettema, D. and Nieuwenhuis, R., 2017. Residential Self-Selection and Travel Behaviour: What are the Effects of Attitudes, Reasons for Location Choice and the Built Environment? *Journal of Transport Geography*, 59, 146–155.

European Environment Agency, 2016. *Urban Sprawl in Europe*. Luxembourg: Publications Office.

Fishman, R., 1987. *Bourgeois Utopias: The Rise and Fall of Suburbia*. New York: Basic Books.

Frenkel, A., Bendit, E. and Kaplan, S., 2013. Residential Location Choice of Knowledge-Workers: The Role of Amenities, Workplace and Lifestyle. *Cities*, 35, 33–41.

Gans, H.J., 2017. *The Levittowners: Ways of Life and Politics in a New Suburban Community*. New York: Columbia University Press.

Glick, P.C., 1989. The Family Life Cycle and Social Change. *Family Relations*, 38 (2), 123.

Haase, A., Steinführer, A., Kabisch, S., Buzar, S., Hall, R. and Ogden, P., 2010. Emergent Spaces of Reurbanisation: Exploring the Demographic Dimension of Inner-City Residential Change in a European Setting. *Population, Space and Place*, 16 (5), 443–463.

Heinz, W.R., Huinink, J. and Weymann, A., eds., 2009. *The Life Course Reader. Individuals and Societies across Time*. Frankfurt/New York: Campus Verlag.

Hochstenbach, C., 2019. The Age Dimensions of Urban Socio-Spatial Change: The Age Dimensions of Urban Change. *Population, Space and Place*, 25 (2), e2220.

Hochstenbach, C. and Musterd, S., 2018. Gentrification and the Suburbanization of Poverty: Changing Urban Geographies through Boom and Bust Periods. *Urban Geography*, 39 (1), 26–53.

Holden, E., 2004. Ecological Footprints and Sustainable Urban Form. *Journal of Housing and the Built Environment*, 19, 91–109.

Humphreys, J. and Ahern, A., March 2019. Is Travel Based Residential Self-Selection a Significant Influence in Modal Choice and Household Location Decisions? *Transport Policy*, 75, 150–160.

Jansen, S.J.T., 2014. Different Values, Different Housing? Can Underlying Value Orientations Predict Residential Preferences and Choice? *Housing Theory and Society*, 31 (3), 254–276.

Kaufmann, V., 2011. *Rethinking the City: Urban Dynamics and Motility*, 1st ed. Milton Park, Abingdon, Oxon; New York, NY: Lausanne, Switzerland: Routledge; EPFL Press.

Kley, S.A. and Mulder, C.H., 2010. Considering, Planning, and Realizing Migration in Early Adulthood. The Influence of Life-Course Events and Perceived Opportunities on Leaving the City in Germany. *Journal of Housing and the Built Environment*, 25 (1), 73–94.

Knox, P. and Pinch, S., 2000. *Urban Social Geography: An Introduction*. Harlow: Pearson Prentice Hall.

Ley, D., 1996. *The New Middle Class and the Remaking of the Central City*. Oxford: Oxford University Press.
Lufkin, S., Thomas, M.-P., Kaufmann, V. and Rey, E., 2018. Linking Spatial Characteristics to Residential Lifestyles: A Framework for Analysing the Hospitality Potential of Urban and Architectural Designs. *Articulo – revue de sciences humaines*, https://journals.openedition.org/articulo/3498.
Lux, M., Samec, T., Bartos, V., Sunega, P., Palguta, J., Boumová, I. and Kážmér, L., 2018. Who Actually Decides? Parental Influence on the Housing Tenure Choice of Their Children. *Urban Studies*, 55 (2), 406–426.
Marcucci, E., Stathopoulos, A., Rotaris, L. and Danielis, R., 2011. Comparing Single and Joint Preferences: A Choice Experiment on Residential Location in Three-Member Households. *Environment and Planning A: Economy and Space*, 43 (5), 1209–1225.
Meir, I.A., Garb, Y., Jiao, D. and Cicelsky, A., 2009. Post-Occupancy Evaluation: An Inevitable Step toward Sustainability. *Advances in Building Energy Research*, 3 (1), 189–219.
Moos, M., 2016. From Gentrification to Youthification? The Increasing Importance of Young Age in Delineating High-Density Living. *Urban Studies*, 53 (14), 2903–2920.
Mouratidis, K., 2018. Is Compact City Livable? The Impact of Compact Versus Sprawled Neighbourhoods on Neighbourhood Satisfaction. *Urban Studies*, 55 (11), 2408–2430.
Mulder, C.H., 2007. The Family Context and Residential Choice: A Challenge for New Research. *Population, Space and Place*, 13 (4), 265–278.
Mulder, C.H. and Clark, W.A.V., 2002. Leaving Home for College and Gaining Independence. *Environment and Planning A*, 34, 981–999.
Mulder, C.H. and Wagner, M., 2012. Moving after Separation: The Role of Location-Specific Capital. *Housing Studies*, 27 (6), 839–852.
Niedomysl, T., 2011. How Migration Motives Change Over Migration Distance. Evidence on Variation across Socio-Economic and Demographic Groups. *Regional Studies*, 45, 843–855.
Owen, D. and Green, A.E., 1992. Migration Patterns and Trends. *In*: A. Champion and A. Fielding, eds. *Migration Processes and Patterns. Volume 2: Research Progress and Prospects*. London: Belhaven Press, 17–38.
Pahl, R.E., 1969. Urban Social Theory and Research. *Environment and Planning*, 1 (2), 143–153.
Pallagst, K., Wiechmann, T. and Martinez-Fernandez, C., 2014. *Shrinking Cities: International Perspectives and Policy Implications*. 1. publ. New York: Routledge.
Preiser, W.F.E., Rabinowitz, H.Z. and White, E.T., 2016. *Post-Occupancy Evaluation*. Abingdon: Routledge.
Rérat, P., 2012a. The New Demographic Growth of Cities: The Case of Reurbanisation in Switzerland. *Urban Studies*, 49 (5), 1107–1125.
Rérat, P., 2012b. Housing, the Compact City and Sustainable Development: Some Insights from Recent Urban Trends in Switzerland. *International Journal of Housing Policy*, 12 (2), 115–136.
Rérat, P., 2014. Highly Qualified Rural Youth: Why Do Young Graduates Return to Their Home Region? *Children's Geographies*, 12 (1), 70–86.
Rérat, P., 2016. Residential Mobility. *Mobiles Lives Forum*, https://en.forumviesmobiles.org/marks/residential-mobility-3204.
Rérat, P., 2018. Spatial Capital and Planetary Gentrification: Residential Location, Mobility and Social Inequalities. *In*: L. Lees and M. Phillips, eds. *Handbook of Gentrification Studies*. Cheltenham, UK; Northampton, MA, USA: Edward Elgar Publishing, 103–118.
Rérat, P., 2019. The Return of Cities: The Trajectory of Swiss Cities from Demographic Loss to Reurbanization. *European Planning Studies*, 27 (2), 355–376.
Rérat, P. and Lees, L., 2011. Spatial Capital, Gentrification and Mobility: Evidence from Swiss Core Cities. *Transactions of the Institute of British Geographers*, 36 (1), 126–142.
Rogers, A., 1988. Age Patterns of Elderly Migration: An International Comparison. *Demography*, 25 (3), 355–370.
Rössel, J. and Hoelscher, M., 2012. Lebensstile und Wohnstandortwahl. *KZfSS Kölner Zeitschrift für Soziologie und Sozialpsychologie*, 64 (2), 303–327.
Rossi, P.H., 1955. *Why Families Move: A Study in the Social Psychology of Urban Residential Mobility*. Glencoe: The Free Press.
Rye, J.F., 2011. Youth Migration, Rurality and Class: A Bourdieusian Approach. *European Urban and Regional Studies*, 18 (2), 170–183.

Schier, M., Hilti, N., Schad, H., Tippel, C., Dittrich-Wesbuer, A. and Monz, A., 2015. Residential Multi-Locality Studies – The Added Value for Research on Families and Second Homes: Multi-Locality Studies – A Residential Perspective. *Tijdschrift voor economische en sociale geografie*, 106 (4), 439–452.

Schwartz, S.H., 1994. Are There Universal Aspects in the Structure and Contents of Human Values? *Journal of Social Issues*, 50 (4), 19–45.

Sheller, M. and Urry, J., 2006. The New Mobilities Paradigm. *Environment and Planning A*, 38, 207–226.

Smith, D.P., 2002. Extending the Temporal and Spatial Limits of Gentrification: A Research Agenda for Population Geographers. *International Journal of Population Geography*, 8, 385–394.

Smith, D.P. and Holt, L., 2007. Studentification and 'Apprentice' Gentrifiers within Britain's Provincial Towns and Cities: Extending the Meaning of Gentrification. *Environment and Planning A*, 39, 142–161.

Smith, N., 1996. *The New Urban Frontier: Gentrification and the Revanchist City*. London and New York: Routledge.

Teixeira, C., 2006. A Comparative Study of Portuguese Homebuyers' Suburbanization in the Toronto and Montreal Areas. *Espace, populations et sociétés*, 1, 121–135.

Tiebout, C.M., 1956. A Pure Theory of Local Expenditures. *The Journal of Political Economy*, 64 (5), 416–424.

Van Ham, M., 2012. Housing Behaviour. *In*: W.A.C. Clark, K. Gibb and D.F. Clapham, eds. *The Sage Handbook of Housing Studies*. London: Sage, 47–65.

Vignal, C., 2005. Les espaces familiaux à l'épreuve de la délocalisation de l'emploi: ancrages et mobilité des salariés de l'industrie. *Espaces et sociétés*, 120–121, 179–197.

Wolpert, J., 1965. Behavioral Aspect of the Decision to Migrate. *Papers in Regional Science*, 15 (1), 159–169.

24
URBAN MOBILITY AND MIGRATIONS

Aurore Flipo

Introduction. A city of strangers

In its first accounts of the city, sociology has paid particular attention to the figure of the stranger, defined by Simmel as one of the sociological forms the urban setting was likely to produce. Developed in reference to rural ethnology, urban sociology has initially conceptualized the city as an environment defined by mobility and social distance, as opposed to the anchorage and community-based relationships of rural life (Agier, 1996). It is, however, rather a dialectic between distance and proximity that defines the urban way of life for Simmel, as the type of social relation that is allowed by the very possibility of mobility. Indeed the stranger "has not quite abandoned its freedom to come and go" (Simmel, 1950: 404). This particular relationship with place and space is critical to understand the significance of mobility in the urban context.

Later on, the Chicago school conceptualized the city as one of immigrants, with spatial mobility being at the core of the urban process of differentiation and segregation (Grafmeyer and Joseph, 1979). Urban ecology has investigated ethnic segregation as a process of urban growth and assimilation of wide cohorts of new inhabitants coming from distant places and countries. This perspective, intrinsically linked with the historical context of 1920s' Chicago, has contributed to the birth of a sociology of immigration (Réa and Tripier, 2008). It has also contributed to defining migration as a process of uprooting, focusing on the city as a settlement area.

However, subsequent research has demonstrated that the worlds of mobility and migration are diverse and intertwined (Dureau and Hily, 2009). Practices of transnationalism, while changing in shape and intensity with the advent of the "global world", have always been part of migration systems (Waldinger, 2015). Recent research in history on the development of cities has also demonstrated that the modern European city is not a city of uprooted rural emigrants: "the nineteenth century city was a terminus for only a relatively small number of immigrants" (Winter, 2009: 2). Indeed, "all cities' life is made of this permanent flux of comings and goings", recalls Agier (2016), and migrants (whether transnational or not) having always been at the centre of this process.

This fact – that is not so distant from Simmel's account of urbanity – has gained a new visibility and importance in societies in which mobility is a cardinal value and cities are more and more dependent on global networks and flows (Sassen, 1991; Appadurai, 1996). This has led some authors to argue that cities have moved from a process of immobilization of the bodies, typical of the industrial city, to a process of mobilization, typical of the neoliberal city (Rousseau, 2008; Jaffe et al., 2012).

In this wide galaxy of mobilities, some are deemed illegitimate and undesirable, while others are encouraged and induced. International mobility in particular reveals the paradox between the simultaneous increase of liberalization and coercion of mobility practices. Indeed, the difference between migrants, refugees, asylum-seekers, undocumented migrants, international students, expatriates and so on is not so much to be found in the motives or content of their mobility, as in the way their mobility is managed, and more importantly the degree of constraint that is exerted on them. This differential treatment reflects the social and political construction of wanted and unwanted hosts in the city, and has tangible consequences on individuals' experiences of mobility in the city.

Circulatory territories and networks

In the past 15 years, research on migration has paid particular attention to the salience of transnational territories, routes and networks migrants define and sustain while moving across borders. Those transnational "social spaces" (Pries, 1999) and "communities" (Portes, 1996) challenge the borders and definitions of migration. Indeed, not all migrants settle down and cross-border mobility (which is usually defined as *migration*) can be a step preceding or following other mobility patterns, whether at the local, national or international levels. Patterns of mobility and anchorage are very diverse and happen within different timescales (Dureau and Hily, 2009).

One example of this diversity of mobility behaviours is that of transmigrants, characterized by the existence of "a chain of numerous national and urban steps" that shape a wider "circulatory territory" (Tarrius, 2010). Joining cities and villages in a transnational network, those patterns of mobility and migration are labelled "territories" because they create a dispersed yet spatialized social and cultural system. Living across borders, transmigrants draw routes and exchange goods that constitute the infra-level of globalization (Tarrius, 2002). In this context, mobility-related knowledge and social networks are particularly valued. More broadly speaking, cross-border migration creates a culture of mobility and bond that can be passed through families, groups and generations (Diminescu, 2008). However, this richness and complexity is ignored, Tarrius argues, because of the sedentary-centred view on migration that does not perceive the inclusion of migrants' livelihoods into wider spatial networks (Tarrius, 1993).

Indeed, while cities are seen and defined as social and spatial systems based on density and proximity, migration can be defined as social and spatial systems based on dispersed networks. This asymmetry leads to underestimate the importance of the various ways migrants link distant livelihoods. For example, De Villanova (1994) has shown how migrants from Portugal in France have constructed bipolar residential patterns while accessing ownership in both countries. Like for Romanian migrants' "double homes" (Benarrosh-Orsoni, 2019), the transnational anchorage of migrants is not given, but constructed by the inherent duality of the migrant figure (Sayad, 2006), who is both presence and absence (De Gourcy, 2018). Indeed, mobilities are not only about those who arrive, but also those who leave: the culture of migration also contributes to social change in the places of origin, importing new objects, practices and habits (Kandel and Massey, 2002). In this respect, the social morphology of the city, whose networks extent way across local and international borders, can be in conflict with its political morphology, which is territorially defined and limited.

The city's inner borders

As Agier (2016) recalled, cities can be defined as a nexus of different kinds of mobilities and flows. Hence, some authors have argued that cities' main function is to govern mobility (Darling, 2017), with "urbanism being a politics of movement" (Magnusson, 2000). In parallel, the process of border establishment and border control has been increasingly transferred from the fixed, geographical frontier towards mobile, diffuse, individualized "borderities" (Amilhat-Szary and Giraut, 2015). As a consequence, the control of borders is more and more conflated with the control of mobilities, which is increasingly done at a distance (Bigo et al., 2010).

Cities have played a major role in the "re-scaling" process of borders (Darling, 2017), through "the many ways in which the state and its borders occupy space in the city and seek to manage individual and collective mobilities and identities" (Young, 2011). For example, Monica Varsanyi has showed how some American cities have implemented ordinances and regulations that can be described as "immigration policies through the backdoor" (Varsanyi, 2008). Taking the example of the use of trespassing ordinances in order to arrest undocumented migrants, Monica Varsanyi argues that local authorities have developed their own local immigration agenda to control the mobility of undocumented daily labourers. Another example can be found in local policies of "hostile environment", which are based on voluntarily excluding measures such as banning undocumented migrants' children from public school meals.

Cities have been labelled by some researchers "differentiation machines" (Isin, 2002) in which "differences are labelled and perpetuated through spatial practices and technologies of governance" (Young, 2011). However, the process of social and spatial differentiation that is taking place inside global cities is not a new phenomenon. As we noted earlier, the Chicago school has highlighted how processes of incorporation, assimilation and segregation have led to the creation of "moral regions", corresponding with diverse origins, ways of life and cultural habits in the beginning of the twentieth century's Chicago. Before that, pre-Modern European cities were based on various spatialized forms of social segregation that relied on formal regulations of mobility, the most extreme form of which being the *ghetto*. Most of the political rights and fiscal advantages were limited to the category of *bourgeois*, while distinguishing them from other classes of population, based on their professional status and/or their geographical origin. Those distinct rights and duties were politics of immigration regulation that preceded and, in many respects, inspired the politics of immigration at the national level (Gotman, 2003).

The process of separating, sorting out and ultimately segregating between populations is connected with economic growth and the division of labour. In China, Laurence Roulleau-Berger and Shi Lu (2003) recount how the city of Shangaï has been growing while creating foreign concessions outside its walls. Social segregation of labour is often mixed, formally or informally, with ethnic segregation of migrant workers. Even when they are not legally restricted, the opportunities of residential location of migrant workers are frequently determined by the spatialization of productive activities and chain migration.

Migrant workers needed by the economy are often kept at the margins of the *polis*, their stay being assumed as temporary (Piore, 1979; Sayad, 2006), while those who are deemed "undesirable" are relentlessly pushed back beyond symbolic or, more and more often, actual walls. Camps represent the most visible form of social and spatial alienation of refugees and asylum-seekers that can be characterized by the fact of being outside the common law (Agier, 2008). Encamped refugees and asylum-seekers are "locked-up outside" (Kobelinsky et al., 2009), in an "outplace" that is both nowhere and everywhere: "Borderlands and camps are a moving reality in space, but also 'liquid' in substance" that belong to "a wider set of spaces of confinement and circulation" (Agier, 2008: 59). Camps are designed to remain away and separate from the city: spontaneous urban camps appear as unwanted and illegitimate forms of clustering that must be eradicated, or at least confined to the already most deprived areas (Darling, 2017).

As a consequence, migrants' mobilities can't be fully understood without taking into consideration their incorporation in labour and social control policies that have always included mobility control, especially over the poor. As Robert Castel recalls, "the repression of vagrancy has been one of the great obsessions of pre-industrial societies" (Castel, 1994: 14). On the other hand, the incarceration of those illegitimate mobile citizens has been used to provide pools of unfree labour (Agier, 2008). Indeed the repression of migrants' mobilities illustrates the entanglement of politics of mobility and mobilization of the urban poor (Jaffe et al., 2012).

Mobility or retention? The case of undocumented migrants

Though the interaction between migrations and mobilities in the city is not a new topic, the interest for undocumented migrants is rather recent (Darling, 2017; Sanyal, 2012). Recently, the so-called "refugee crisis" in Europe has shed a new light on the issue of mobility control of asylum-seekers. However, what the media have presented as a depiction of mobility, using words that suggest movement (roads of migration, spread of migrants, influx) are actually pictures of retention. Lines of people waiting at the border have become a popular depiction of globalization and migration, but what is shown is not mobility but rather the lack thereof.

Urban mobility also involves significant risks for undocumented migrants, contributing to making their presence invisible. The crossing of public spaces raises the issue of vulnerability towards police patrols. Indeed, because many people cross them every day, and because they are "hyperspaces" (Lussault, 2017) connected with distant places and extraneous lands, transportation infrastructures are risky places for undocumented migrants. Mobility patterns of undocumented migrants draw geographies of the least visible paths: peripheral byways, back roads and side roads, mountain and forest tracks, detours. As Stefan Le Courant puts it, "the city becomes a labyrinth with labile walls" for undocumented migrants. Borders and "borderities" (Amilhat-Szary and Giraut, 2015) are potentially everywhere and can take many different shapes. As a result, migrants learn and develop particular circulation skills that are based on the avoidance of risks of arrest and/or what is perceived as such, knowledge of specific landmarks such as NGOs, administrations, libraries, thus defining new circulatory territories that are marked by discontinuity (Baligand, 2015). For example, Le Courant (2016) describes how undocumented migrants' mental maps of the city are built around "white zones" where they never step foot.

This constant alert, resulting in a mobility derived from escape and avoidance, produces what Baligand (2015) calls the "diffracted city". According to the author, the lack of possession of a safe place in the city results in a "modification of the uses of the spatial environment". Not being able to move freely, asylum-seekers and undocumented migrants are objectified, resulting in processes of de-personnalization (Baligand, 2015). In addition to shattered geographies, undocumented migrants are affected by heteronomous time frames. The constant possibility of being displaced can create the feeling of being trapped in a never-ending present, waiting for the future to happen (Le Courant, 2014). The long-term temporality is blurred, while the daily rhythms are frequently disrupted. The lack of both spatial and temporal routines can cause psychological distress (Baligand, 2015).

On the contrary, the right to wander – one of Simmel's archetypal urban figures – is enjoyed with legality of status: "he's going out the streets [...] He's going nowhere in particular, he's only walking around", says a migrant about someone who obtained a residence permit (Le Courant, 2016). This right of free movement and free time contrasts with narratives of hide and expectation that characterize undocumented migrants' accounts of their experiences. Regularized legal status allows migrants to become subjects of their own mobility, rather than objects.

The differentiated time-spaces of mobility

As a result, migrants' mobilities in the city are characterized by distinct patterns, times and uses of the urban space, depending on the degree of constraint linked to their legal status, their resources, knowledge and social bonds. The possibility of appropriation of mobility resources varies greatly amongst individuals and contexts. In the contemporary city, the time–spaces of mobility are increasingly individualized.

In his study of 1970s' French industrial cities of Dunkerque and Lorient, Guy Barbichon showed that the urban practices of migrants differed importantly depending on whether movement was a result of a choice or constrained by employment (Barbichon, 1979). Places such as airports, ports, metro stations, that are associated with movement, travel and imagination for legal migrants (especially tourists), are associated with fear, angst and distress for asylum-seekers, who risk being arrested or even killed while on the move. Stefan Le Courant (2016) recounts that flyers distributed by NGOs in France to warn migrants against risks of being arrested recommends avoiding driving a car, taking public transportation without travel tickets, or not carrying a passport. Those artefacts which are means of transportation (or minor infractions, in the case of not having a travel ticket) for the average citizen are potentially means of deportation for undocumented migrants.

This contrast highlights the differentiation and potential segregation of time–spaces of mobility. While some of them are increasingly quicker, safer and connected to a growing diversity of places, others remain slow, dangerous and limited. For example, asylum-seekers are banned from the safest means of transportation, such as planes, and have to rely on inflatable boats and coyotes to cross borders. The more transportation becomes monitored and watched, the more those who are excluded from their use have to rely on smuggling to move. In the French border areas of Calais and the Roya valley for example, local inhabitants are forbidden to carry undocumented migrants into their personal cars.

While the means of transportation are increasingly numerous, their use remains differentiated. For example, Cristina del Biaggio has recounted how segregation between migrants and natives has become a routine event on the trains between Turkey and Greece (Del Biaggio, 2015) or how tourists and migrants wait for their trains in separate areas of Cosmo's railway station (Del Biaggio, 2016). While planes and personal cars are preferred by tourists, buses and vans are used by migrant workers across Europe and beyond, because they are cheaper and allow for the transportation of goods. Indeed, migrants' mobility also represents a wide market. From the opening of new low-cost lines to smuggling, the economy derived from the mobility needs generated from migration is massive (Hernandez-Leon, 2005).

Can the metropolis be hospitable?

Facing the reluctance of governments to ensure the protection of refugees at the national level, some thinkers have developed the idea that local governments can provide alternative, and sometimes confronting policies of hospitality towards migrants. One example is the concept of "refuge city" or "city of refuge", conceptualized in Derrida's claim for cosmopolitism (1997). The concept is derived from the creation of an "International Parliament of Writers" in 1993 to provide protection for threatened intellectuals around the world. Following Arendt, Derrida assumes that nation-states are intrinsically unable to perform and protect genuine cosmopolitism. The refuge city also relies on the idea that true hospitality must be based not only on a visitation right, following the Kantian definition, but on a residence right (Antonioli, 2009). This is what connects Derrida's city of refuge with Lefebvre's right to the city, as "politics of presence" (Darling, 2017).

Against the politics of bordering and borderities, refuge cities would provide the safe places asylum-seekers lack, where new forms of political membership can be enforced. They are, however, dependent on the degree of autonomy and self-government cities can have. Recently, various municipality-based initiatives have (re)emerged to face the "refugee crisis" at the local level (Furri, 2017). In 2015, the mayor of Barcelona, Ada Colau, launched the idea of a network of refuge cities in Europe. In France, a network of municipalities label themselves "welcoming cities", after the initiative of the mayor of Grande-Synthe Damien Carême, where NGOs and researchers have developed the concept of "communal hospitality" (Hanappe, 2018). A similar network exists in Great Britain under the name of "City of Sanctuary". At the European level, the network "Solidarcities" gathered in 2016 several European mayors to contest the European policy on asylum.

The original concept of Sanctuary city emerged in the early 1980s in the United States to protect undocumented migrants against federal laws. Local police forces were commanded not to investigate individuals' migration statuses. Some cities and States also implemented alternative forms of local citizenship that do not request the possession of federal documentation, such as the issuing of State driver licences (Varsanyi, 2006). Here again, the issue of allowing mobility appears as central. However, those practices have been decreasing since the implementation of anti-terror laws after 9/11 that have toughened safety regulations on international mobility in general and migrants in particular.

The timescales of mobility and migration

The city as a nexus of mobility is a crossing point in space, but also in time. The "passenger city" (Lévy-Vroelant, 2003) designates the ways by which cities have tried to organize inhabitants whose presence is assumed temporary. However, the dialectics between mobility and settlement are both unpredictable and contradictory: on one hand, local governments try to settle down and assimilate newly arrived populations, but some of them remain mobile. On the other hand, arrangements that are destined to be temporary tend to last in time and become integrated into the city, and some of the "birds of passage" (Piore, 1979) settle down.

For example, most hostels (*garnis*) in Paris, originally used to host temporarily mobile populations have become permanent housing solutions for settled migrants. They have proven to be, however, more open and integrated in the city than formal housing facilities such as social housing and shelter houses (Lévy-Vroelant, 2003). Similarly, refugees camps that were set up to face humanitarian emergencies tend to last over time and become alter-cities where the constitution of a form of autonomy and membership are possible (Agier, 2002). Finally, labour migration flows, initially perceived as temporary by both the sending and receiving states, and most of the migrants themselves, end up creating long-lasting social and cultural structures such as transnational communities and diasporas (Waldinger, 2015).

Slums have also been analysed as places of livelihood and citizenship-making through the concept of "subaltern urbanism" (Roy, 2011), establishing a parallel between urban refugee camps and the urban poor (Sanyal, 2012). While regularly destroyed or transformed to be replaced by formal accommodation such as public hostels, slums constantly mould the borders of the city and represent both an extremely mobile and extremely resilient form of urbanism. Besides local hospitality policies, informality can also be perceived as a means of resistance against the spatial alienation of the "undesirable" (Darling, 2017). For example, the "Jungle" in Calais has been described as a "potential city" in the way its inhabitants spontaneously recreated within what was originally a no-man's land public places, churches, restaurants and so on (Clément and Thiéry, 2016).

Conclusion: migrants and the politics of presence

The city is a spatial arrangement that was defined by Simmel as a complex dialectics between distance and proximity, mobility and settlement. The paths, the speeds, the geographies of those arrangements, however, vary greatly amongst different categories of dwellers, whether permanent or temporary. In this respect, migrants embody the unequal government of mobility in the city. Then, when defining migrants as being the mobile ones, there is a lack of understanding of the issue that is at stake: migrants are those whose mobility is for the least controlled, constrained and coerced, in opposition with other international travellers such as tourists, international business executives or diplomats. The varying degree of autonomy and heteronomy exerted on international mobilities is correlated to the process of social differentiation, thereby interacting with social hierarchies between and within countries (Flipo, 2018).

The examination of migrants' mobilities and experiences of the city also provides more general insights into the functions of the city and the dialectics of mobility and dwelling. Indeed, it has been argued that the city can be defined as a government of mobility (Darling, 2017) creating not only segregated spaces (such as ghettos), but more widely segregated geographies. The experiences of undocumented migrants then highlights the interactions between subjectivities and geographical spaces (Baligand, 2015). In addition, migrants' trajectories and mobilities have proven to be diverse, dependent on spatial, economic and historical contexts. The diversity of patterns of mobility and anchorage described in the literature has challenged the homogeneous figure of the uprooted emigrant.

Secondly, the examination of migrants' mobilities underlines how social processes of differentiation are made spatially visible, whether it be results of economic, political or symbolical distinctions. From the camps, designed to remain separate from cities, to public accommodation centres, ethnic neighbourhoods or slums, migrants' spatial inscription illustrates the work of social sorting cities perform (Isin, 2002). It also illustrates the process of "borderization", whose functions are increasingly distinguished from their territorial location (Amilhat-Szary and Giraut, 2015).

Finally, migrants question the very definition of the city, both spatially and politically. As parts of wider networks, the "global city" (Sassen, 1991) expands way beyond its territorial limits. If globalization challenges the ways citizenship and membership are politically defined (Bäubock, 2003), the city may provide new models of cosmopolitism that lay on politics of presence rather than membership or even residency: "presence is about the temporary fixing of mobilities rather than their capture within a given spatial form" (Darling, 2017).

References

Agier, M., 1996. *Les savoirs urbains de l'anthropologie*. Enquête. Archives de la revue Enquête 35–58.

Agier, M., 2002. *Aux bords du monde, les réfugiés*. Flammarion, Paris.

Agier, M., 2008. *Gérer les indésirables. Des camps de réfugiés au gouvernement humanitaire*. Flammarion, Paris.

Agier, M., 2016. Ce que les villes font aux migrants, ce que les migrants font à la ville. *Le sujet dans la cité* 7, 21–31.

Amilhat-Szary, A.-L. and Giraut, F., 2015. *Borderities: The Politics of Contemporary Mobile Borders*. Palgrave Macmillan, Basingstoke.

Antonioli, M., 2009. Jacques Derrida, de la "ville-refuge" au droit de résidence, in: *Le territoire des philosophes*. La Découverte, Paris, 139–159.

Appadurai, A., 1996. *Modernity at Large: Cultural Dimensions of Globalization*. University of Minnesota Press, Minneapolis, MN.

Baligand, P., 2015. Parcours de demandeurs d'asile : diffraction spatiale et traumatisme. *Les Annales de la Recherche Urbaine* 110, 56–63.

Barbichon, G., 1979. Les migrants dans la ville. Lorient, Dunkerque. *Les Annales de la Recherche Urbaine* 3, 3–25.

Bäubock, R., 2003. Reinventing Urban Citizenship. *Citizenship Studies* 7, 139–160.

Benarrosh-Orsoni, N., 2019. *La maison double. Lieux, routes et objets d'une migration rom*. Société d'ethnologie, Paris.

Bigo, D., Scherrer, A., and Guittet, E.-P., 2010. *Mobilité(s) sous surveillance: perspectives croisées UE–Canada*. Athéna éditions, Outremont.

Castel, R., 1994. La dynamique des processus de marginalisation: de la vulnérabilité à la désaffiliation. *Cahiers de recherche sociologique* 22, 11–27.

Clément, G. and Thiéry, S., 2016. Atlas d'une cité potentielle. Calais, "New Jungle", 2015–2016. PUCA.

Darling, J., 2017. Forced Migration and the City: Irregularity, Informality, and the Politics of Presence. *Progress in Human Geography* 41, 178–198.

De Gourcy, C., 2018. L'institution de l'absence en Méditerranée. *Revue des mondes musulmans et de la Méditerranée* vol 144.

De Villanova, R., 1994. Migrants et ... propriétaires : nomadisme ou sédentarité? Une citoyenneté bipolaire. *Les Annales de la Recherche Urbaine* 65, 68–78.

Del Biaggio, C., 2015. Dans le train pour Athènes: ségrégation, mode d'emploi. *VisionsCarto.net* [online: https://visionscarto.net/dans-le-train-pour-athenes].

Del Biaggio, C., 2016. Touristes et migrants: collision en gare de Côme. *VisionsCarto.net* [online: https://visionscarto.net/touristes-et-migrants-collision-a-come].

Derrida, J., 1997. *Cosmopolites de tous les pays, encore un effort !* Editions Galilée, Paris.

Diminescu, D., 2008. The Connected Migrant: An Epistemological Manifesto. *Social Science Information* 47(4), 565–579.

Dureau, F. and Hily, M.-A., 2009. *Les mondes de la mobilité*. PU Rennes, Rennes.

Flipo, A., 2018. Entre misérabilisme et injonction à la mobilité. Dominocentrisme et dominomorphisme dans l'étude des migrations internationales, in: *Migrations, circulations, mobilités. Nouveaux enjeux épistémologiques et conceptuels à l'épreuve du terrain*. PU de Provence, Aix-en-Provence, 45–59.

Furri, F., 2017. Villes-refuge, villes rebelles et néo-municipalisme. *Plein droit* n° 115, 3–6.

Gotman, A., 2003. Barrières urbaines, politiques publiques et usages de l'hospitalité. *Les Annales de la Recherche Urbaine* 94, 6–15.

Grafmeyer, Y. and Joseph, I., 2009 [1979]. *L'École de Chicago. Naissance de l'écologie urbaine*. Folio, Paris.

Guild, E. and Bigo, D., 2003. Le visa : instrument de la mise à distance des indésirables. *Cultures & Conflits* 49, 82–95.

Hanappe, C. (dir.), 2018. *La ville accueillante : Accueillir à Grande-Synthe : questions théoriques et pratiques sur les exilés, l'architecture et la ville*. PUCA, La Défense.

Hernandez-Leon, R., 2005. The Migration Industry in the Mexico-US Migratory System. California Center for Population Research On-Line Working Papers Series 49.
Isin, E.F., 2002. *Being Political: Genealogies of Citizenship*. University of Minnesota Press, Minneapolis, MN.
Jaffe, R., Klaufus, C., and Colombijn, F., 2012. Mobilities and Mobilizations of the Urban Poor. *International Journal of Urban and Regional Research* 36, 643–654.
Kandel, W. and Massey, D.S., 2002. The Culture of Mexican Migration: A Theoretical and Empirical Analysis. *Social Forces* 80, 981–1004.
Kobelinsky, C., Makaremi, C., and Agier, M., 2009. *Enfermés dehors: enquêtes sur le confinement des étrangers*. Editions du Croquant, Bellecombe-en-Bauges.
Le Courant, S., 2014. "Être le dernier jeune": les temporalités contrariées des migrants irréguliers. *Terrain* 63, 38–53.
Le Courant, S., 2016. La ville des sans-papiers. Frontières mouvantes et gouvernement des marges. *L'Homme* 219–220, 209–232.
Lévy-Vroelant, C., 2003. Les avatars de la ville passagère. *Les Annales de la Recherche Urbaine* 94, 96–106.
Lussault, M., 2017. *Hyper-lieux. Les nouvelles géographies de la mondialisation*. Le Seuil, Paris.
Magnusson, W., 2000. Politicizing the Global City, in: Isin, E.F. (ed.) *Democracy, Citizenship and the Global City*. Routledge, London, 289–306.
Marotel, G., Péraldi, M., and Tarrius, A., 1994. Migration et citadinité. L'approche de la ville par la mobilité. *Les Annales de la Recherche Urbaine* 64, 87–90.
Piore, M.J., 1979. *Birds of Passage: Migrant Labor and Industrial Societies*. Cambridge University Press, Cambridge, MA.
Portes, A., 1996. Transnational Communties: Their Emergence and Significance in the Contemporary World-System, in: Portes, A., Korzeniewic, R. P., and Smith, W. C. (eds.) *Latin America in the World-Economy*. Praeger, Westport, CT, 151–168.
Pries, L., 1999. *Migration and Transnational Social Spaces*. Ashgate Publishing, Aldershot.
Réa, A. and Tripier, M., 2008. *Sociologie de l'immigration*. La Découverte, Paris.
Roulleau-Berger, L. and Lu, S., 2003. Les provinciaux à Shanghaï : Formes d'inscriptions urbaines et économiques des migrants dans la ville. *Les Annales de la Recherche Urbaine* 93, 48–56.
Rousseau, M., 2008. La ville comme machine à mobilité. Capitalisme, urbanisme et gouvernement des corps. Métropoles 3. [online: http://journals.openedition.org/metropoles/2562].
Roy, A., 2011. Slumdog Cities: Rethinking Subaltern Urbanism. *International Journal of Urban and Regional Research* 35, 223–238.
Sanyal, R., 2012. Refugees and the City: An Urban Discussion. *Geography Compass* 6, 633–644.
Sassen, S., 1991. *The Global City – New York, London, Tokyo*. Princeton University Press, Princeton, NJ.
Sayad, A., 2006. *L'Immigration ou les paradoxes de l'altérité. L'Illusion du provisoire – tome 1*. Raisons d'agir, Paris.
Simmel, G., 1950. *The Sociology of Georg Simmel*, Wolff, K.H. (eds.). The Free Press, New-York, NY.
Tarrius, A., 1993. Territoires circulatoires et espaces urbains : Différentiation des groupes migrants. *Les Annales de la Recherche Urbaine* 59, 51–60.
Tarrius, A., 2002. *La Mondialisation par le bas: les nouveaux nomades de l'économie souterraine*. Ed. Balland, Paris.
Tarrius, A., 2010. Territoires circulatoires et étapes urbaines des transmigrant(e)s. *Regards croisés sur l'économie* 8, 63–70.
Varsanyi, M.W., 2006. Interrogating "Urban Citizenship" Vis-à-vis Undocumented Migration. *Citizenship Studies* 10, 229–249.
Varsanyi, M.W., 2008. Immigration Policing through the Backdoor: City Ordinances, the "Right to the City," and the Exclusion of Undocumented Day Laborers. *Urban Geography* 29, 29–52.
Waldinger, R., 2015. *The Cross-Border Connection – Immigrants, Emigrants, and Their Homelands*. Harvard University Press, Cambridge, MA.
Winter, A., 2009. *Migrants and Urban Change: Newcomers to Antwerp, 1760–1860*, Routledge, London.
Young, J., 2011. 'A New Politics of the City': Locating the Limits of Hospitality and Practicing the City-as-Refuge. *ACME: An International Journal for Critical Geographies* 10, 534–563.

SECTION V

Power, conflict and social exclusion

The phenomenon of urban mobilities lies at the heart of how contemporary societies function. As most people would realise, this also means that mobilities is concerned with social interests; with decisions about access and rights to mobilities. Section V takes stock on issues related to power, conflict and social exclusion, as these matters are inherently related to urban mobilities. It opens with **Chapter 25**, Sanneke Kloppenburg's examination of mobility and stratification. In urban areas, the potential of people to be mobile is a new form of capital. As such, it represents a crucial factor in social stratification. At the same time, stratification is not simply a question of mobility for some and immobility for others. Approaching the relationship between mobility and social stratification from the perspective of people's everyday mobility practices shows how social exclusion is manifested in the differentiated ways in which people move through the city. A focus on urban mobility systems reveals how infrastructures and technologies can produce or deepen existing social inequalities in the city. The urban conflicts that manifest around mobility are addressed further by Nixon and Schwanen in **Chapter 26**, which examines the conflicted pedestrian. Mobility is produced by, and reproduces, relationships between people, groups and things. The number and types of options to move may vary greatly between people, as may the speeds at which people travel. The uneven distributions of social, economic, political or material power across these relationships; the heterogeneity of human desires and societal visions of the good life; and the finite nature of urban space mean that tensions, conflicts or compromises around urban mobility are inevitable. This chapter turns to one mode of transport – walking – and findings from qualitative fieldwork in the cities of São Paulo and Vancouver, to exemplify and vivify the connection between conflict and mobilities in the city. Under one's own power and without a protective carapace, walking possesses a degree of vulnerability otherwise avoided when moving more quickly and/or with fortification. The chapter demonstrates that, because of this, both modal segregation and shared space, despite offering some advantages, are associated with shortcomings that leave each limited in their ability to reduce conflict in ways that are equitable and just. In light of this, it argues that pedestrian conflict cannot simply be designed away without awarding to walking a fundamental and central place among transport modes, a repositioning that finds its expression in the use of urban space. Justin Spinney continues the theme of mobility and justice in **Chapter 27**.

The first part of the chapter gives a brief account of the marginalisation of lived experience in the city. Questions of economy have led to the privileging and facilitation of functional and efficient commuter mobilities to the neglect of othered mobilities. The chapter goes on to discuss the ways in which such narrow visions have become sedimented in the standardising planning tools and methodologies used to assess movement. Transport for London's Public Transport Accessibility Tool (PTAL) is investigated as a methodology used to assess urban housing, parking and mobility needs. The chapter demonstrates how its apparent neutrality masks a series of assumptions that flatten differences in motility and experience. In doing so, it is inclined to show maximal levels of accessibility that in reality only reflect the experiences of a minority of users. The implication for transitions is that the case for any kind of change or transition will seem marginal at best, precisely because the tool is only capable of representing the mobility experiences of the few. Clearly, there is an issue of justice and inequality that needs to be addressed, yet has been obfuscated by the veneer of objectivity that such methodologies import with their assumptions. The third part of the chapter then examines the ways in which the mobility experiences of children in particular have been marginalised in planning practice. Assumptions regarding what constitutes desirable mobility (functional, efficient) have served to marginalise the more playful and less directed mobility needs of children. These assumptions have been compounded by research techniques that are concerned only with the amount of mobility children have with regard to public health agendas, and circumvent them as knowledgeable citizens. The chapter concludes with a call to challenge methodological anachronisms and the assumptions that perpetuate them if we are to shift our thinking to support sustainable and just transitions. Finally, in **Chapter 28** Vincent Kaufmann explores the social implications of spatial mobilities. Numerous works consider mobility as a physical phenomenon first and foremost, and have described it in terms of flows, practices and meanings. This focus on physical movement tends to limits their analysis to spatial mobility-related social mobility (i.e. inequalities and family/professional life course), linked to the fact that physical movement is usually a way of reaching a goal. It is therefore essential to focus on the nature of this goal in order to understand the energy and motivation that underlie mobility. Authors such as Michel Bassand, Bertrand Montulet, Luc Boltanski and Eve Chiapello, and Zygmunt Bauman were instrumental in developing the concept of mobility by transcending the question of physical movement in physical space. However, their ideas remain marginal in works on the mobility turn. Based on their work, the goal of the chapter is precisely that: to provide an overview of the relationship between spatial mobility and social mobility. After a discussion of the spatial and social nature of mobility, the chapter examines mobility as an analyser of life course and as a social norm.

25

MOBILITY AND SOCIAL STRATIFICATION

Sanneke Kloppenburg

Introduction

Not everyone moves in and to the city in the same way. Some people cannot afford a car and rely on public transportation to get to work. Others may use a newly built elevated train that takes them to the shopping mall in a speedy and comfortable way. Yet others may struggle to get their daily tasks done, because bringing their children to school, commuting to work, and doing the groceries takes them hours of navigating through congested streets. These examples tell us that access to mobilities and people's experiences of being mobile are uneven. In mobilities studies, the (differentiated) potential for mobility is now considered "a crucial dimension of unequal power relations" (Hannam, Sheller and Urry, 2006, p. 3, for similar statements see Flamm and Kaufmann, 2006, p. 167; Manderscheid, 2009, p. 8). It is also widely acknowledged that places, technologies or infrastructures can enhance the mobility of some people, while heightening the immobility of others (Hannam, Sheller and Urry, 2006, p. 3; Uteng and Cresswell, 2008, p. 7).

This chapter examines the multiple relationships between (im)mobility and social stratification in urban areas through engaging with the two claims above. What does it mean if one's capacity to be mobile is a crucial force of stratification in contemporary societies? And how can technologies and infrastructures enhance the mobility of some, while reducing others' potential to become mobile? In dealing with these questions, the chapter will first explore how in the mobilities literature the relations between mobilities and stratification have been theorised with the idea of "mobility potential" as a new type of capital. Next, it examines approaches that understand the production of inequalities from two different starting points: the design of mobility systems and the everyday practices of urban people. The chapter ends with acknowledging the importance of contextual approaches for examining mobilities and stratification.

Theoretical concepts: motility and network capital

In mobilities studies, theorisations of people's potential to be mobile have evolved around two different, but related ideas: the idea of *motility*, and that of *network capital.*

Motility: the potential to be mobile

The works of Vincent Kaufmann and his colleagues deal with the relations between social and spatial mobilities (Kaufmann, 2002; Kaufmann, Bergman and Joye, 2004; Flamm and Kaufmann, 2006; Kaufmann, Dubois and Ravalet, 2018). The key insight they offer is that people's potential and actual movements are interdependent with social structures and dynamics. Flamm and Kaufmann (2006) argue that in modern society, spatial mobility is crucial for engaging in diverse types of activities such as work, leisure and meeting friends and family members. In such a society, they argue the capacity to be mobile is "a deciding factor of social integration" (p. 167).

To analyse people's mobility potential, Kaufmann introduced the concept of *motility*. At the individual level motility is defined as a "set of characteristics that enables people to be mobile, including physical capacities, social conditions of access to existing technological and transportation systems, and acquired skills (e.g. training, driver's license, and international English for travel" (Kaufmann, Dubois and Ravalet, 2018, p. 199). The key point here is that not everyone has the same degree of motility: those with low motility have relatively more difficulties accessing activities and relevant others, which means they may experience social exclusion. Motility should be seen as a form of capital in relation to the opportunities the surrounding physical and social environments offer, and Kaufmann and colleagues argue that motility "forms theoretical and empirical links with, and can be exchanged for, other types of capital" (2004, p. 752). Like other forms of capital – economic, social, cultural, etc. – people can accumulate motility over time, for example through experiences with specific means of transportation. Motility can also be exchanged for other types of capital. For example, the capacity and willingness to be highly mobile (long-distance commuting or changing residence) can be exchanged for economic capital in the form of a next career step (Kaufmann, Dubois and Ravalet, 2018). Building on the definition above, Kaufmann, Bergman and Joye (2004) distinguish three elements of motility: *access*, *competence* and *appropriation*.

Access includes access to transportation infrastructures (e.g. highways, railways, airports, etc.) and to mobile technologies (e.g. cars, bicycles, smartphones). Access depends on the availability of transportation and communication infrastructures in a particular geographical area (e.g. a city or neighbourhood) and on the services available (e.g. public transportation services, car rental, parking spaces, bicycle racks, etc.), and the conditions under which the available options can be used.

Competence or *skills* include people's physical abilities to move (e.g. to walk, to ride a bicycle), the ability to plan trips and activities and use information for that (e.g. navigation systems, public transport trip planners), but also knowledge about traffic rules, language skills, or having a driver's licence. Skills or competences are interdependent with access and appropriation. For example, if a person does not have access to a good quality bicycle infrastructure, she might not consider it an option to cycle to her appointments in the city centre, which means she will not develop the skills to cycle in high-traffic urban spaces. At the same time, people who can drive a car have greater access to highways (than people who do not have a driver's licence) which gives them the possibility to consider the option of commuting by car against commuting by train.

Appropriation refers to how people or groups evaluate and use their (perceived or real) skills and their possibilities for access. Elements such as values and habits, but also people's plans, needs and aspirations shape appropriation. Appropriation helps to explain how and why people do or do not turn their access and skills into actual movement. For example, a person with good access and skills might decide to (or feel forced to) not commute over large distances because of his caring responsibilities for young children. Another person with similar access and skills might be very willing to commute or even relocate to a different city. This also shows that in order to understand an individual's motility, simply "measuring" or describing a person's access and skills is insufficient: whether or not people make use of skills and access depends on people's subjective experiences of being hampered or facilitated to use particular means of transportation in order to organise their daily activities.

Looking at the *potential to* be mobile, and not just to people's *actual* movement, sheds light on the different possibilities and constraints people experience with regards to their movements. For example, people with low motility may have little access to transportation infrastructures and lack skills to become mobile (e.g. a driver's licence), which leads to little actual movement. Yet, that high motility does not necessarily transform into high mobility practices becomes particularly clear in a recent study by Kaufmann, Dubois and Ravalet (2017). Here, they show how different forms of motility "fuel specific social inequalities" (p. 2). Based on longitudinal research among 1735 individuals, they distinguish six groups: *the unmotile, very motile, reluctant to be mobile, willing to be mobile, reversibles* and *irreversibles.* Their findings show that those individuals that had good access and skills (the *very motile*) often do not transform their motility into high mobility. It seems that this group can choose to not use their skills and access and instead remain immobile. People who can be characterised as *unmotile* – people with poor access and skills – on the other hand, are more often forced to move for work. What the concept of motility therefore also brings to the fore is that not everyone has the same degree of choice when it comes to high mobility.

With the concept of motility, a better understanding can be gained of the agency people exert and experience with regard to their mobility. The agentic concept of appropriation is key in explaining why people do or do not turn their access and skills into movement. Hence, although Kaufmann and colleagues stress that it is important to take into account the specific contexts that enable or constrain movement, it could be argued that their concept of motility prioritises the human agent over the structural context in explaining inequalities (c.f. Doherty, 2015). In addition, the emphasis is on explaining (high) mobility rather than immobility. With regards to this latter point, there has recently been a call to complement the idea of motility with that of immotility (Ferreira, Bertolini and Næss, 2017). Immotility would refer to the potential to not move, or the "positive side of stillness" (p. 26). In societies characterised by a "burden of mobility" (Shove, 2002), one could argue that the potential to stay put is also a form of capital.

Network capital: forming and sustaining social networks

The idea of motility as a form of capital is closely linked to how Urry views the role of network capital in mobile societies. He argues that it is not movement as such that is significant in contemporary mobile lives, but the *social relations* that stem from these mobilities. Participation in society requires that people can access various *networks* of work, family, friendship and leisure. Hence, it is the "ability to form and sustain networks" (2007, p. 196) that matters for social inclusion. Urry calls this "network capital" and defines it as the "capacity to engender and sustain social relations with those people who are not necessarily proximate and which generates emotional, financial and practical benefits" (2007, p. 197). Urry distinguishes eight elements that in combination produce a distinct form of social stratification. These include

> an array of appropriate documents, visas, money, qualifications that enable safe movement; others at-a-distance that offer hospitality; movement capacities; location free information and contact points; communication devices; appropriate, safe and secure meeting places; access to multiple systems; and time and resources to manage when there is a system failure.
>
> *(Urry, 2012, p. 27)*

Urry places this notion of network capital in the context of a networked society in which networks of friends, family, leisure and work change form. Such social networks are becoming more spatially distributed and less overlapping. For many people, their relevant others – family members and friends – may not only live close by, but also increasingly at a distance. In other words, the geographical distance between nodes generally grows. Maintaining these networks requires communication and intermittent travel. Even though face-to-face meetings become less frequent when geographical distance between network members grows, Urry considers them crucial for maintaining networks. As he argues, "a network only functions if it is intermittently 'activated' through occasioned co-presence from time-to-time" (2007, p. 231). Communication technologies cannot substitute this, but rather help coordinate and plan face-to-face meetings.

Network capital should not be seen as a characteristic of individuals, but as something that arises from people's relationships with others and with the affordances of the environment. Network capital, social networks and mobility systems interrelate. For example, new mobility systems (e.g. the internet, air travel etc.) and the mobilities they generate produce new social relations, which in turn result in higher levels of travel because people need to travel (and communicate) in order to sustain their social networks. In other words, changes in mobility systems transform what is necessary for social inclusion (Urry, 2007).

In mobile societies, mobility capital becomes an important factor in social stratification. Those people with high network capital can make and sustain connections with relevant others (distanced as well as nearby) and benefit from those relations. People with low network capital, on the other hand, may face constraints with regards to networking and meetings. Yet, if people's networks are local, limited mobility is not by definition a problem for maintaining existing networks. What it does mean is that it might be more difficult for people to form new networks. To examine if and how mobilities may engender social relations, it is therefore important to analyse the spatial and temporal characteristics of people's social networks. Social exclusion needs to be understood with reference to the social networks people want to be part of and the constraints they might experience in accessing these networks (Cass, Shove and Urry, 2005; Urry, 2007).

Urry gives several illustrations of how inequalities in network capital can play out (Urry, 2012). One example is the impact of Hurricane Katrina in New Orleans in 2005. People with high mobility capital who had good access to cars, communication technologies and contacts were able to leave the area before the flooding occurred, while those with low network capital were stuck in their houses and neighbourhoods. Another example is the inequalities in network capital of car owners and users versus those who do not have access to a car and have to walk or cycle. Here Urry points out how car crashes cause death and injuries especially among those who do not own a car, i.e. pedestrians and cyclists.

Not all mobility scholars agree with the emphasis Urry places on intermittent travel as *necessary* to meet up with other people. Urry even argues that "a socially inclusive society would elaborate and extend the capabilities of co-presence to all its members. It would minimise 'coerced immobility'" (2007, p. 208). Yet, with our current carbon-based mobility systems, granting everybody equal rights of co-presence would result in major environmental problems. Urry therefore also proposes that for many groups, such an equal distribution would mean that their capabilities to meet friends, family and colleagues have to be significantly reduced. Yet, he hardly ever discusses whether networks for work and family could also become less stretched-out, how we can decide on what travel is really "necessary", or how cities and regions could be designed to afford more proximate living.

Mobilities and social inequalities in the city

Where the previous section discussed conceptualisations of the relationships between mobility and stratification, this section takes a closer look at a number of empirical studies on mobility-related inequalities and exclusion in urban areas. It discusses approaches that seek to understand how urban transit infrastructures are productive of inequalities as well as approaches that start from the perspective of the people's everyday mobility practices. In doing so, it intentionally draws on examples of urban mobilities in Latin America and Asia. Mobilities research is still predominantly focused on mobility in the European context, and much of the conceptual work originates from Europe. Yet, mobilities approaches are increasingly used to examine transport interventions and social exclusion in the South (e.g. Uteng and Lucas, 2018). Cities in the Global South face huge challenges of a growing urban population, traffic congestion and socio-economic segregation, making questions of mobilities, immobilities and social stratification extremely urgent.

The production of inequalities in urban mobility systems

Mobility systems are according to Urry "the enduring systems that provide what we might call the infrastructures of social life. Such systems *enable* the movement of people, ideas, and information from place to place, person to person, event to event" (2007, p. 12). In an urban context they include infrastructures such as car travel, bike-sharing systems, high-speed trains, and metro systems. When new mobility systems are introduced in cities, this may "intervene" in inequalities in various ways. New mobility systems are designed with particular goals for the management of urban mobilities in mind, with ideas about who the future users of such systems are, and how, where, when and with whom these "mobile subjects" (Richardson and Jensen, 2008) want to travel to and within the city. Urban mobility systems therefore always produce (new) potential and actual mobility practices for some individuals and groups, but not for others.

A study on the Bangkok Sky Train by Richardson and Jensen (2008) provides a clear example of how mobility systems can (re)produce and even exacerbate inequalities. The Bangkok Sky train consists of an elevated rail infrastructure 12 metres above the city streets. This mobility system provides a cooler (air-conditioned), more predictable and faster way of moving compared to other means of mobility at ground level, where the streets are heavily congested, noisy and polluted. The infrastructure connects sites such as the business district, shopping malls, hotels and condominiums. Richardson and Jensen (2008) argue that as this system creates new and alternative possibilities to travel to and access places and sites in the city, it thereby also creates changed conditions for certain mobile subject types to emerge (p. 218). The Bangkok Sky Train was designed for three types of mobile subjects in particular: mobile shoppers, Western-oriented business people and foreign tourists. In facilitating the mobilities of these mobile subject types, the mobility system at the same time "reinforces layering and segregation" (p. 229). It creates a new physical and socio-spatial layer for the elite, who can travel in speedier and more comfortable ways than those subjects who cannot afford a ticket and remain at ground level. As Richardson and Jensen (2008) argue:

> In Bangkok, the Sky Train creates a particular problem of 'double segregation': in spatial terms as the clean-aired infrastructure overlays the smog-filled streetscape below, and in socio-economic terms as it separates its users along the lines of income, and thus restricts the relative motility of the less well-off inhabitants, compared to Bangkok's middle class, international business elites, and tourists.
>
> *(p. 229)*

With the case of helicopter travel in Sao Paolo, Cwerner similarly illustrates how helicopters create new patterns of vertical urban mobility.[1] The system of helicopter travel enables elite travellers in this megacity to practice of form of "detached" and "frictionless" mobility (Cwerner, 2006, p. 211). In doing so, it unevenly distributes motility: creating greater motility for the urban elites, while restricting the relative motility of others. These two studies about new urban mobility systems approach mobility as an additional layer over an already differentiated society (see also Adey, 2010). They are able to show how existing social inequalities are reproduced or deepened through mobility systems and the particular types of mobilities these enable.

Everyday mobility practices and social exclusion

An alternative lens to understanding urban inequalities, is through a focus on people's everyday mobility practices. Such a perspective sees mobility practices as social activities that are embedded in everyday life: people often travel together with others (Jensen, Sheller and Wind, 2015), and with the goal of accomplishing other social projects such as getting to work or school, or shopping (Peters, Kloppenburg and Wyatt, 2010; Cass and Faulconbridge, 2016; Greene and Rau, 2018). This also means that people typically use various forms of mobility and mobility systems to plan and coordinate their everyday activities (e.g. combining walking with bus travel, or cycling with train travel).

The work of Chilean sociologists Ureta and Jirón is very insightful in examining the daily mobility practices and strategies of different socio-economic groups in the city of Santiago de Chile. Ureta (2008) focuses on low-income people who live in housing estates that are located far away from the city centre. He argues that social exclusion is not just a matter of residential segregation, but also of difficulties in accessing transport and communication networks. It appeared that Ureta's respondents had to travel long distances to get to work, because jobs were often located in other parts of the city. For their modes of transport, these people relied almost completely on public transport in the form of buses. Ureta gives the example of a mother of two who works in a supermarket in another part of the city. To get to work, she needs to take a bus, which is relatively expensive and takes about two hours to reach her work place. So, on an average day this woman spends four hours at a minimum to go to and return from work, and in the case of traffic congestion, her journey can take even longer. Because of their low motility, most of the mobility of the families in Ureta's study is limited to travelling to and from work and school. Travelling to visit family members and friends or leisure location appeared to be very limited, because people could not spend more money and time to become mobile. What this study shows, is that mobility "does not liberate [these people] from poverty, but [is] a structural part of what it means to be poor in contemporary Santiago" (2008, p. 275). Social exclusion does not necessarily mean immobility, but is also enacted in the *ways people move* through the city (p. 286).

In a similar vein, Jiron (2007) shows how everyday mobility practices are differentiated along the lines of gender, age and class. Looking at three different income groups, she analyses the difficulties people experience in accessing places, people and activities on a daily basis. To get insight into their practices and experiences, Jirón joined people on their journeys. This in-depth fieldwork allowed her to show how access to mobility is gendered: because the women in Jirón's study are responsible for household duties and childcare, they face specific temporal, financial and organisational barriers to travelling. This results in limited options to travel (e.g. only having access to public transportation, and only at rush hours (Jirón, 2009) or the feeling of being trapped at home (Jirón, 2010).

Other studies that take an everyday life perspective have examined how particular groups negotiate their access to and use of mobility infrastructures and urban spaces. This includes examination of the everyday mobilities of migrants (e.g. Buhr, 2018) Muslim migrant women (e.g. Uteng, 2009; Warren, 2017), young people and children (e.g. Barker et al., 2009; Skelton, 2013; Jensen, Sheller and Wind, 2015), women (e.g. Turdalieva and Edling, 2018), older people (e.g. Nordbakke and Schwanen, 2014; Stjernborg, Wretstrand and Tesfahuney, 2015), wheelchair users (Pyer and Tucker, 2017) and homeless people (Bourlessas, 2018). Understanding everyday mobility from the perspective of a particular subject position makes clear how different people have different degrees of access to and control over mobility, and more importantly, how such differentiated mobility practices may reflect and reinforce societal inequalities (see also Massey, 1993).

Conclusions

This chapter has discussed how mobilities scholars emphasise the importance of physical movement for social integration. Related to that is the idea that constraints on mobility may engender social inequality. If people experience restrictions and difficulties in their access to and use of urban transportation infrastructures, this may reduce their opportunities to sustain social networks and to form new ones, to access the job market, to visit the doctor, etc. Mobilities studies therefore emphasise that the potential to be mobile is a new form of capital in mobile societies, and represents a crucial factor in social stratification. At the same time, scholars have shown that social exclusion does not necessarily mean immobility, and social inclusion does not presuppose mobility. A closer look at everyday mobility practices shows that immobilities may include both forced immobilities (for example because people do not have the resources to become mobile) and the choice to remain immobile or to restrict your mobility (for example by choosing to live in an urban enclave). Similarly, high mobility may be a preferred option, resulting from a high degree of motility and hence choice over when, where, how and with whom a person moves through the city. On the other hand, the example of low-income people in Santiago de Chile showed that moving over large distances can also be a form of confined or coerced mobility if it is the only option available in order to access jobs. Stratification is thus not simply a question of mobility for some and immobility for others, but rather of the differentiated ways in which people move through the city.

This chapter has also emphasised the need to examine inequalities in the context of transformations in urban infrastructures or mobility systems. The mobility systems of the city are not fixed, but changing as a result of the implementation of new policies and technologies for urban transportation and planning. Relevant developments include the introduction of mass transit systems, increased automobility, and the use of mobile phone apps for planning and sharing urban mobility. Implied in the introduction of such new mobility systems are decisions about which mobilities are supported and which ones are not. Mobility systems thus create particular ways of travelling for particular types of travellers, and in doing so are productive of inequalities. Yet, material infrastructures do not *determine* how people use them: it is in the *interactions between* mobility systems and human (travelling) subjects that inequalities are produced (Richardson and Jensen, 2008, p. 218). In order to understand how mobilities reflect or reinforce inequalities, contextual approaches are needed that approach people as actors who are embedded in social networks and situated in particular spatial and political contexts (also see Manderscheid, 2014). This enables a better understanding of not just the possibilities and constraints people experience in accessing other people, places and activities, but also of the degree of control they experience over their actual movement.

Note

1 It should be noted that although vertical mobility systems (Graham and Hewitt, 2013) are often associated with the smooth and speedy movement through corridors and fast-lanes of the (urban) elite (Birtchnell and Caletrío, 2014), this is not necessarily the case. Several South-American cities, for example have implemented aerial cable car systems to integrate marginalised neighbourhoods with the rest of the city (see e.g. Brand and Dávila, 2011).

References

Adey, P., 2010. *Mobility*. London and New York: Routledge.

Barker, J., Kraftl, P., Horton, J. and Tucker, F., 2009. The road less travelled: New directions in children's and young people's mobility. *Mobilities*, 4(1), 1–10.

Birtchnell, T. and Caletrío, J., 2014. *Elite mobilities*. London and New York: Routledge.

Bourlessas, P., 2018. 'These people should not rest': Mobilities and frictions of the homeless geographies in Athens city centre. *Mobilities*, 13(5), 746–760.

Brand, P. and Dávila, J.D., 2011. Mobility innovation at the urban margins: Medellín's Metrocables. *City*, 15(6), 647–661.

Buhr, F., 2018. A user's guide to Lisbon: Mobilities, spatial apprenticeship and migrant urban integration. *Mobilities*, 13(3), 337–348.

Cass, N. and Faulconbridge, J., 2016. Commuting practices: New insights into modal shift from theories of social practice. *Transport Policy*, 45, 1–14.

Cass, N., Shove, E. and Urry, J., 2005. Social exclusion, mobility and access. *The Sociological Review*, 53 (3), 539–555.

Cwerner, S.B., 2006. Vertical flight and urban mobilities: The promise and reality of helicopter travel. *Mobilities*, 1(2), 191–215.

Doherty, C., 2015. Agentive motility meets structural viscosity: Australian families relocating in educational markets. *Mobilities*, 10(2), 249–266.

Ferreira, A., Bertolini, L. and Næss, P., 2017. Immotility as resilience? A key consideration for transport policy and research. *Applied Mobilities*, 2(1), 16–31.

Flamm, M. and Kaufmann, V., 2006. Operationalising the concept of motility: A qualitative study. *Mobilities*, 1(2), 167–189.

Graham, S. and Hewitt, L., 2013. Getting off the ground: On the politics of urban verticality. *Progress in Human Geography*, 37(1), 72–92.

Greene, M. and Rau, H., 2018. Moving across the life course: A biographic approach to researching dynamics of everyday mobility practices. *Journal of Consumer Culture*, 18(1), 60–82.

Hannam, K., Sheller, M. and Urry, J., 2006. Mobilities, immobilities and moorings. *Mobilities*, 1(1), 1–22.

Jensen, O.B., Sheller, M. and Wind, S., 2015. Together and apart: Affective ambiences and negotiation in families' everyday life and mobility. *Mobilities*, 10(3), 363–382.

Jiron, P., 2007. Unravelling invisible inequalities in the city through urban daily mobility. The case of Santiago de Chile. *Swiss Journal of Sociology*, 33, 1.

Jirón, P., 2009. *Mobility on the move: Examining urban daily mobility practices in Santiago de Chile*. UK: London School of Economics and Political Science.
Jirón, P., 2010. Mobile borders in urban daily mobility practices in Santiago de Chile. *International Political Sociology*, 4(1), 66–79.
Kaufmann, V., 2002. *Re-thinking mobility: Contemporary sociology*. Burlington: Ashgate.
Kaufmann, V., Bergman, M.M. and Joye, D., 2004. Motility: Mobility as capital. *International Journal of Urban and Regional Research*, 28(4), 745–756.
Kaufmann, V., Dubois, Y. and Ravalet, E., 2018. Measuring and typifying mobility using motility. *Applied Mobilities*, 3(2), 198–213.
Manderscheid, K., 2009. Integrating space and mobilities into the analysis of social inequality. *Distinktion: Scandinavian Journal of Social Theory*, 10(1), 7–27.
Manderscheid, K., 2014. Criticising the solitary mobile subject: Researching relational mobilities and reflecting on mobile methods. *Mobilities*, 9(2), 188–219.
Massey, D., 1993. Power-geometry and a progressive sense of place. In J. Bird, B. Curtis, T. Putnam, G. Robertson and L. Tickner, eds. *Mapping the futures: Local cultures, global change*. London: Routledge, 59–69.
Nordbakke, S. and Schwanen, T., 2014. Well-being and mobility: A theoretical framework and literature review focusing on older people. *Mobilities*, 9(1), 104–129.
Peters, P., Kloppenburg, S. and Wyatt, S., 2010. Co-ordinating passages: Understanding the resources needed for everyday mobility. *Mobilities*, 5(3), 349–368.
Pyer, M. and Tucker, F., 2017. 'With us, we, like, physically can't': Transport, mobility and the leisure experiences of teenage wheelchair users. *Mobilities*, 12(1), 36–52.
Richardson, T. and Jensen, O.B., 2008. How mobility systems produce inequality: Making mobile subject types on the Bangkok Sky Train. *Built Environment*, 34(2), 218–231.
Shove, E., 2002. Rushing around: Coordination, mobility and inequality. *Paper presented at the ESRC Mobile Network Meeting*, Department for Transport, London, October.
Skelton, T., 2013. Young people's urban im/mobilities: Relationality and identity formation. *Urban Studies*, 50(3), 467–483.
Stjernborg, V., Wretstrand, A. and Tesfahuney, M., 2015. Everyday life mobilities of older persons: A case study of ageing in a suburban landscape in Sweden. *Mobilities*, 10(3), 383–401.
Turdalieva, C. and Edling, C., 2018. Women's mobility and 'transport-related social exclusion' in Bishkek. *Mobilities*, 13(4), 535–550.
Ureta, S., 2008. To move or not to move? Social exclusion, accessibility and daily mobility among the low-income population in Santiago, Chile. *Mobilities*, 3(2), 269–289.
Urry, J., 2007. *Mobilities*. London: Polity.
Urry, J., 2012. Social networks, mobile lives and social inequalities. *Journal of Transport Geography*, 21, 24–30.
Uteng, T.P., 2009. Gender, ethnicity, and constrained mobility: Insights into the resultant social exclusion. *Environment and Planning A*, 41(5), 1055–1071.
Uteng, T.P. and Cresswell, T., 2008. *Gendered mobilities*. Aldershot: Ashgate.
Uteng, T.P. and Lucas, K., 2018. *Urban mobilities in the global south*. London: Routledge.
Warren, S., 2017. Pluralising the walking interview: Researching (im) mobilities with Muslim women. *Social & Cultural Geography*, 18(6), 786–807.

26

THE CONFLICTED PEDESTRIAN

Walking and mobility conflict in the city

Denver V. Nixon and Tim Schwanen

Introduction

In this chapter we start with the premise that (im)mobility is an inherently relational phenomenon. That is, mobility is produced by, and reproduces, relationships between people, groups, and things which may themselves be voluntarily or involuntarily moving, or immobile. The number and types of options to move may vary greatly between people, as may the speeds at which people travel. The uneven distributions of social, economic, political, or material power across these relationships, the heterogeneity of human desires and societal visions of the good life, and the finite nature of urban space mean that tensions, conflicts, or compromises around urban mobility are inevitable.

Mobility conflict may arise intermodally—between people using different modes of transport—or intramodally—among members of a particular mobility practice. An example of *inter*modal conflict may be a driver failing to stop for a pedestrian at a crosswalk. *Intra*modal conflict, on the other hand, may involve road rage between two drivers, or a physical assault on a pedestrian by another pedestrian, for instance. Conflict about mobility may arise in a different timespace than actual transport environments, such as courtrooms, legislatures, or social media. Various regulatory approaches—some physical and others social or legal—have been used by transport authorities to minimize conflicts. With respect to intermodal conflict, approaches may be divided into those that rely upon spatial segregation and those that trust in shared space.

This chapter turns to one mode of transport—walking—and recent findings from qualitative fieldwork in the cities of São Paulo and Vancouver, to exemplify and vivify the connection between conflict and mobilities in the city. Under one's own power and without a protective carapace, walking possesses a degree of vulnerability otherwise avoided when moving more quickly and/or with fortification. We will see that, because of this, both segregation and shared space, despite offering some advantages, are associated with shortcomings that leave each limited in their ability to reduce conflict in ways that are equitable and just. In light of this, we argue that pedestrian conflict cannot simply be designed away without awarding to walking a fundamental and central place among transport modes, a repositioning that finds its expression in the use of urban space.

The aforementioned fieldwork is drawn from two different studies: semi-structured interviews with forty-six pedestrian, cyclist, and driver commuters in Vancouver, Canada, in

2009–2010, and thirty-eight semi-structured interviews with leaders, staff, and beneficiaries of grassroots walking and cycling infrastructure initiatives in the city of São Paulo, Brazil, in 2017 (for more details on methodology, see Nixon, 2012, 2014; Nixon and Schwanen, 2018; Schwanen and Nixon, 2019).

Both studies shed light on questions of pedestrian conflict. For the Vancouver study this is owed to its focus on how particular transport mode practices shape people's understandings of their commute environments and fellow commuters, and in what ways this knowledge may reproduce particular transport systems and relationships. In contrast, the São Paulo research examined attempts by civil society organizations to prevent harm inflicted upon vulnerable pedestrians through initiatives that support walking (and cycling) among disadvantaged social groups, such as those with disabilities, women and gender variant people, low-income residents of marginalized neighbourhoods, children, and older people. It focused on initiatives such as regular group walks, the provision of street furniture, temporary street-closures, and pathway and staircase improvements. The research sought to understand the nature of these initiatives, their successes and failures, and the extent to which they contribute to more equitable and just transport systems and cities. The two studies complement each other with respect to exploration of both walking practices and infrastructure, and overlap with regard to *in situ* experiences of 'individual' and group mobility practices and social relationships in transport contexts.

The conflicted pedestrian

Where is safe and secure space for walking in the city? There are competing interests over which mobility practices should take precedence in urban environments where space is limited and costly. A complex of pre-existing and present power relations mediate who can walk in particular urban spaces (Middleton, 2018). In the transport environment, this power dynamic is established in part by the vulnerability of the walking body to moving vehicles (including bicycles), as well as malintent pedestrians. The body, and all of its senses, are directly exposed to everything near to it in the transport environment. This vulnerability may be demographically specific—for instance a woman or gender variant person's higher probability of being the victim of sexual assault, or a child's growing but still immature traffic awareness—or it may transcend personal traits—a male heavyweight boxer or bodybuilder is no match for an oncoming SUV. Pedestrians are not passive victims, however, and many find ways to reduce their vulnerabilities, such as with personal listening devices or sunglasses when on sidewalks, or through group crossings of intersections so as to leverage the visibility of a 'critical mass'.

With respect to intermodal vulnerabilities, authorities have also attempted to reduce these by either separating pedestrians from other forms of traffic, or having all modes use the same space in the hope that the 'forced' interactions will inspire 'civilized' sharing of urban space. However, the disciplining and self-disciplining that both of these approaches entail or require in a context that privileges and prioritizes drivers through concepts such as 'Level of Service' (LOS)[1] (Bonham, 2006; Vallières, 2006) where pedestrians are seen as 'obstructions' that cause 'turbulence' in the system (Patton, 2007), has meant that many pedestrian freedoms—such as the ability to pause, to move continuously, or to walk in curves or angles—have been curtailed.

Modal segregation

Before the introduction of the automobile, urban spaces in 'the West' tended to be shared by any and all extant modes—pedestrians, cyclists, horses, omnibuses, and streetcars. In some places, particularly where public sector budgets are limited, separation was never introduced; in other places, such as where mode splits favour pedestrians, as in many slums, imposed separation may never have been accepted in practice. Separating modes by segregating them physically has been the most common way that modern transport engineers attempt to mitigate conflict between pedestrians, cyclists, and drivers. However, the need for the modes to cross each other's pathways complicates what would otherwise be a tidy solution to conflict, assuming that the apportioning of space itself is not contested. To deal with this, a number of different types of crossings have been conceived and implemented, including unmarked crosswalks (an invisible pathway between sidewalk corner and sidewalk corner), marked crosswalks (with paint, such as 'zebra stripes', indicating the path where pedestrians can cross and where drivers must stop), and signalled crosswalks (usually marked crosswalks that include light signals to indicate when drivers must stop and/or pedestrians may walk).

From the perspective of pedestrianism, there are several problems with crosswalks. First, at crosswalks without signals drivers often do not comply with their requirement to stop when a pedestrian is waiting to cross (DeVeauuse et al., 1999; Stapleton et al., 2017; Trinkaus, 1997). As Ashley, a middle-aged pedestrian in Vancouver claimed:

> There could be a woman with kids and a stroller, pouring rain. … And the cars just rush to get to that red light. And there's people standing there … And they just have no clue. They pay no attention whatsoever. And they're driving down the street like they're on a freeway somewhere and [yet] it's a neighbourhood.
>
> *(Ashley, Vancouver)*

Recent studies suggest that the probability of serious injury or fatality to pedestrians from an automobile collision is approximately 10% at 20km/h and increases exponentially to over an 80% chance at 50km/h (Jurewicz et al., 2016; Kröyer, 2015). The physical power differential between a multi-tonne steel object moving at speed and an unprotected human body means that many pedestrians will not challenge passing automobiles to stop, and so they may have to wait for inordinate amounts of time before they may cross.

In São Paulo, Brazil, like many places in the world, drivers seldom stop for pedestrians at crosswalks, despite being required to do so by law. A self-styled superhero from there, 'Super Ando' (meaning Super-Walk, and inspired by Mexico's equivalent, Peatónito), has responded to this problem by donning a mask and cape and stopping cars at crosswalks so as to allow pedestrians to cross (Figure 26.1). He does this part-time at different locations in the city. Super Ando receives a large amount of mass media attention and insists that the symbolic dimension of what he does—i.e. the tangible representation of the skewed relationship between drivers and pedestrians—is the most important part of his voluntary work. In his interview he iterated his slogan, 'you shouldn't need to be a superhero to cross the street in São Paulo.'

In Vancouver, Alice, a middle-aged pedestrian commuter, described both her frustrations attempting to cross the street, as well the physical aggression she felt forced to employ in order to reclaim space to cross the street:

> [Drivers are] just, like I said, sort of more 'me' focused and so the poor pedestrian on the side of the road is, like, secondary, third, you know … I'm going to work, you're going to work. But how long do I have to wait before I get to cross the street? And I can't just plough through' cause my safety's in jeopardy. You can just plough through' cause I'm not going to hurt your car unless I stick out my umbrella and make a big scratch down the side … That's the only retaliation I get, you know.
>
> *(Alice, Vancouver)*

Figure 26.1 Super Ando: the street-crossing pedestrians' superhero of São Paulo.
Source: Super Ando, superandoape@gmail.com

Similarly, another pedestrian, Angelika, described her own use of an umbrella to cross streets as part of her 'pedestrian rage' which she said was a response to the unequal outcomes of car–pedestrian collisions.

The second problem is that where crosswalks are controlled by automated signals, their timing has traditionally favoured the throughput of automobiles. That is, pedestrians are here too obliged to wait for unreasonably long periods of time to cross so that motor vehicle arterials can achieve maximum LOS. Attempts are made to align the rhythms of signal changes with the pace of motor vehicles moving at the speed limit so as to increase 'green waves' that minimize impedance for those travelling at that speed. Pedestrians walk at different speeds, or tempos, from cars and among each other, and so the alignment of walk signals with their arrivals at the crossing seldom align, thus imposing a temporal cost and awkward arrythmia to the embodied experience of movement (Spinney, 2010).

Third, the locations and configurations of crosswalks often force pedestrians to walk in not only temporal but also spatial patterns they would not otherwise. The 'desire lines' of pedestrians (as seen in the physical traces of worn paths in grassy parks or in mud or sand) seldom take right angles lest forced to do so, and yet crossing locations, particularly within gridiron street patterns, require them. As Adams declares, 'pedestrians are natural Pythagoreans' (2004, p. 40). Similarly, sparse marked or signalled crossings in the longer blocks of suburban environments often require pedestrians to deviate from the straightest path, for example walking west 100m to cross, and then east 200m to arrive at where they originally wished to go. By crossing mid-block, or walking in diagonals relative to the street grid, pedestrians shorten their commute distance and avoid waiting at lights (Nixon, 2012). Through this tactic (de Certeau, 1984) they maintain a pace and rhythm that better suits the affordances of their mode.

However, such crossings are illegal in many jurisdictions. Norton (2007) traces the historical origins of the concept of 'jaywalking'—where 'jay' was a derogatory term roughly equivalent to 'country bumpkin'—finding it to be a product of the automobile industry designed to clear pedestrians from the streets they once occupied along with horses, carriages, and sometimes streetcars. The polyrhythms (Lefebvre, 2004 [1992]; Nixon, 2014) and polytempos of bodies moving at vastly different speeds, some in hard and heavy shells, others unprotected, were unacceptable to an emerging industry that was economically and politically supported by the elite, and that wished to maximize the speeds achievable with a petroleum motor (Norton, 2007). To rid early twentieth-century US streets of the 'slow' mingling incompatible with desires for motor driven speeds, pedestrians were relegated to the edges of the streetscape, and increasingly required to cross the space appropriated for automobiles strictly at designated locations. The accusation of jaywalking was first deployed to publicly shame errant pedestrians, but was later hardened into legal regulation in many places.

The drivers interviewed in Vancouver expressed their own frustrations with what they saw as oblivious pedestrians, regardless of the type of crossing. When asked whether significant interactions occur with other modes, Alonso, a semi-retired professional driver, responded as follows:

> No, I would say probably most likely pedestrians because they are ignorant people, more than anything. They think, oh it's a crosswalk and it's mine now. You see me or you don't see me, it doesn't matter. That's the attitude they have, which is not good. They don't look before they cross … They just walk. And that's a big, big problem. That's why people get killed—because they think they own the road at that point.
>
> *(Alonso, Vancouver)*

Readily apparent in Alonso's comments is a conflict over the implied right to space at a given time, and how, from his perspective, pedestrian mortality is owed to inattentiveness. The message here is clear: a crosswalk is not a pedestrian space unless they look and confirm that there are no cars coming. In other words, drivers are partly if not wholly *not* responsible for pedestrian safety. Two other Vancouver car commuters, Branden (an IT professional and recent immigrant in his thirties) and Joselyn (a middle-aged research manager), claimed that, to their annoyance, pedestrians walk whenever, wherever, and however they want—particularly moving slowly. This echoes a European study that found that 30% of drivers would react negatively and aggressively to slow pedestrians (Parker et al., 2002). It is also reflected in recent changes to the Brazilian national crossing-light manual. According to Vitoria, a long-time pedestrian advocate who has worked on walking issues in both the government and NGOs:

> This update is the cross-light program that is being adopted in São Paulo and it is disrespecting two articles of the traffic legislation; these are two articles that say that the conductor has to wait for the pedestrian to finish crossing when there is a shift in the traffic light. They changed the traffic light programming so that this doesn't happen, meaning, only the platoon crosses, the pedestrians in the first lines. Those who arrive a little bit late can't cross anymore.
>
> *(Vitoria, São Paulo)*

These kinds of driver (and governmental) interpretations of the 'right' ways and places that pedestrians should be walking reveal how ingrained the 'transport hierarchy', in which the automobile is sovereign, has become (Nixon, 2014). Super Ando would like to see all road users in São Paulo exercise a mutually respectful form of citizenship that leaves room for longer crossings:

> I usually say to the drivers in Portuguese, "motorista cidadão respeto o pedestre," so citizen driver respects the pedestrian ... So for me it's very important, to generate this empathy with every [person] so the driver will see, "Whoah, really it's very cool. Respect, it's better for everybody." And there isn't a real cost to wait 10 seconds to a mother with her baby cross the street. Ten seconds! Whoa, very nice.
>
> *(Super Ando, São Paulo)*

Intermodal shared space

With respect to mobility, 'shared space' means spaces in which different transport modes are mixed rather than separated. Whereas most places that were never subjected to modal separation represent unplanned shared space, planned sharing is usually deployed to revert space back to an 'open' design. Inspired by the pioneering work of Hans Monderman in the Dutch province of Fryslân in the 1980s (Moody and Melia, 2014), planners and engineers across the Western world are increasingly experimenting with conversions of city space back to shared-use in the belief that with more subtle physical cues (such as paving stones and raised tiles) and different legal regulations (such as lower speed limits or laws placing heavier responsibility on drivers for pedestrian safety) these environments will create more convivial spaces that free pedestrians from the spatiotemporal restrictions and liabilities of segregation, and yet allow for the ongoing presence of automobiles.

However, much like segregation, shared space has a number of challenges with respect to tensions and conflict. The lag in provision of physically separated facilities for cyclists relative to the resuscitation of cycling as a viable mode of transport has meant that in many places—not including the Netherlands or Denmark—pedestrians and cyclists have actually been sharing space for some time. In Simpson's (2017) investigation of the ways multiple modes co-produce affective atmospheres vis-à-vis infrastructures, cyclists in Plymouth, UK, were compelled to use pedestrian-only spaces by inhospitable automobilized environments. While this avoided tensions with motor vehicles, it created them with pedestrians. Both cyclists and pedestrians preferred painted separators and signals to open sharing or hard barriers. Pedestrians interviewed in Vancouver were also found to be ambivalent about cyclists. Azalea, a doctoral student near retirement age in Vancouver who walked daily said:

> There are some very polite and respectful cyclists out there. And there are some who just are so aggressive, and just ... I don't think they care for the safety of people.
>
> *(Azalea, Vancouver)*

Vitoria echoed these sentiments with respect to the São Paulo context:

> Everyone has been happened to get bumped by bikes. A lot of elderly people complain about it. I hear a lot of complaints, especially from elderly people that have trouble hearing bikes on the sidewalk, so yeah.
>
> *(Vitoria, São Paulo)*

Thus, despite a legal obligation to avoid colliding with pedestrians, 'strong and fearless' (Dill and McNeil, 2013) cyclists give the impression of disregard, thereby increasing feelings of vulnerability among pedestrians.

The difference in speed, weight, and vulnerability between pedestrians and cyclists is dwarfed by the differences between those on foot and those in cars. Moody and Melia (2014) found in their study of a square converted to shared space in Ashford, Kent, UK, that this intermodal power relation resulted in pedestrians diverting from their desire lines and giving way to motor vehicles. These relationships were strongest for women and parents. This dissonance is even larger when those on foot are abled in ways that do not meet the expectations of planners and engineers. For instance, Imrie (2012) argues that shared spaces are predicated upon an assumed occupation by strictly stereotypically abled bodies in possession of all sensory faculties, particularly the perfect vision required to negotiate movement with others behind windshields who may themselves not see a pedestrian. Despite the competent walking skills and agency that children develop through training with parents and teachers, the negotiation of risk perception between them with respect to the presence of even slow-moving automobiles often finds children's mobility shifted from foot to the cars creating the hazard (Kullman, 2010).

This brief overview of pedestrians in shared space shows that it may not be a reliable 'cure' for the issues that plague segregated mobility environments; rather, it may reproduce disadvantage and conflict for pedestrians.

Intramodal conflict

Tensions and conflicts do not arise only through intermodal dissonance. Conflict can also be found among users of a particular mode. Here there are two primary sources of conflict, (a) heterogeneous uses of, and practices within, pedestrian spaces, and (b) interpersonal insecurities.

Sidewalk practices are diverse, and may be said to possess the qualities described by Demerath and Levinger (2003) as breadth of experience, identity expression, pausability, and collaborative creativity. Pausability in particular, or even immobility, may constitute pedestrian practices that do not agree with the flows expected by regulatory regimes predicated on sidewalk LOS (Blomley, 2010). As described above, informal economic practices sometimes repurpose the streetscapes delineated by planners and engineers, especially in Southern cities. The use of sidewalks for selling vegetables, or sleeping, may not be seen by some as the proper use of that space, thereby leading to conflict. For example, the slow and staccato rhythms of migrant street vendors in Hanoi, Vietnam, were seen as contrary to the modern fluid pedestrian mobility envisioned by the socialist government and resulted in a ban through which the vendors further adapted their movements (Eidse et al., 2016).

Much of the writing on urban walking has focused on the experience of the *flâneur*—a term connoting a young, able-bodied man of high enough socioeconomic standing to facilitate free wandering in the city—or his equivalent (Middleton, 2011). However, flâneurs make up a small minority of those who walk in the contemporary city, and of the majority remaining many walk out of necessity rather than choice. In contrast to the flâneur, the streets can be places of interpersonal risk and fear for many of those outside the confines of an automobile (Bauman, 1994; Middleton, 2018). Experiences and concerns around personal security are thus another way that pedestrian intramodal conflict may emerge. Perceived risk varies with places and times; for instance, a park or alley during midday may not arouse the same concerns that they do at night (Koskela and Pain, 2000). The women pedestrians in Vancouver developed skills or tactics to minimize the threat, such as through constant situational awareness and the need for subtly defensive body language, as described by Rosemary, a middle-aged business analyst:

> If they're coming towards me, then I'll try to make eye contact to say, I don't know, look at them so that they know I've seen them, but not actually look at, engage them.
>
> *(Rosemary, Vancouver)*

One organization in São Paulo created a phone application for women to collocate and walk together; unfortunately, they found that local cultural barriers, such as women's distrust of other women, undermined easy resolution of the problem through digital technology. Rather, the creators began working in local schools to promote respect and equal rights for women.

It may be that the right kinds and numbers of 'eyes on the street' are one of the best ways to reduce intramodal conflicts among pedestrians (Cozens and Hillier, 2012; Jacobs, 1992 [1961]), though this requires a pedestrian-friendly environment that allows for all sorts of different tempos and rhythms of walking.

Final reflections

Mobility reflects and reproduces the social world. Therefore, the conflicts manifest, and latent, in social and material differences may emerge through the processual practices of mobility. Earlier transport rationales that favoured rapid movement from point A to point B established 'a hierarchy which not only values some travel practices (rapid, direct, uninterrupted) and some travellers (fast, orderly, single-purpose) over others but also enables their prioritization in public space' (Bonham, 2006, p. 58). Patton (2007) suggests that optimization for both pedestrians and drivers is impossible, and Norton argues that 'whether an area should be automobile oriented or pedestrian oriented is fundamentally a value-based decision. More sophisticated analytic techniques cannot take the politics out of decisions over the allocation of public goods' (2007, p. 942). Similarly, Imrie critiques architectural domination of policy processes for downgrading politics and values, and asks, 'who is entitled to what space, and in what ways can their access to it be guaranteed?' (Imrie, 2012, p. 2264). What connects all of these perspectives and the preceding discussion is how the use of urban space is allocated—not technically, but in the way that the values behind the rationales support or undermine equity and justice.

Current mobility relationships were not always configured as they are now, and with enough effort they can be changed again. This effort was apparent among the interviewees in São Paulo. Clarita, the leader of a small NGO in São Paulo that educates and networks Latin American children's mobility organizations, described the work her organization did to change the way children understand mobility and their relationship to transport and the city:

> The only regulated education [about transport] that we have here in Brazil, and in a lot of parts of the world, about city and urban space is road safety education for children. It's a bullshit. You can write it. Bullshit. Bullshit. Bullshit … Why? Because basically there's a bunch of adults explaining some rules to children and saying, "City is like this. I don't care if you don't like it. City is like this. You have to know the rules. And if you don't follow the rules and somebody kills you, it's your fault." So all the modern concepts about sharing urban space, redistribution of urban space … all the modern concepts of city that is not only for cars … they don't exist in that kind of education … So, I try to fight against that, doing things a little bit different.' Cause I think you can explain a lot of concepts about the city and you can allow the children to think by their own, just to know that, yes, the city can be changed, public space can be changed, it can be better, and you can work with that also.
>
> *(Clarita, São Paulo)*

Change that grants pedestrians a central place in the constellation of mobility, and the urban space that this entails, may step away from conflict.

Note

1 Level of Service is a qualitative classification assigned to a street segment that indicates the degree and type of 'flow' of automobile traffic. The top LOS categories indicate that motor vehicles can proceed at the posted speed limits unimpeded. LOS is a product of twentieth century rationalist transport discourses that frame movement as an attempt to overcome the friction of distance between an origin and destination. Thus recent attempts to create LOS equivalents for pedestrians and cycle facilities often prioritize the same dimensions of flow and efficiency.

References

Adams, J., 2004. Streets and the culture of risk aversion. In J. Thrift, ed. *What are we scared of? The value of risk in designing public space*. London: CABE Space, pp. 34–44.

Bauman, Z., 1994. Desert spectacular. In K. Tester, ed. *The Flaneur*. London: Routledge, pp. 138–157.

Blomley, N., 2010. *Rights of passage: Sidewalks and the regulation of public flow.* Abingdon, Oxon: Routledge.

Bonham, J., 2006. Transport: Disciplining the body that travels. *The Sociological Review*, 54(1_suppl), pp. 57–74.

Cozens, P. and Hillier, D., 2012. Revisiting Jane Jacobs's 'Eyes on the Street' for the twenty-first century: Evidence from environmental criminology. In S. Hirt with D. Zahm, eds. *The urban wisdom of Jane Jacobs*. Abingdon: Routledge, pp. 196–214.

de Certeau, M., 1984. *The practice of everyday life*. Berkeley, CA: University of California Press.

Demerath, L. and Levinger, D., 2003. The social qualities of being on foot: A theoretical analysis of pedestrian activity, community, and culture. *City & Community*, 2(3), pp. 217–237.

DeVeauuse, N., Kim, K., Peek-Asa, C., McArthur, D. and Kraus, J., 1999. Driver compliance with stop signs at pedestrian crosswalks on a university campus. *Journal of American College Health*, 47(6), pp. 269–274.

Dill, J. and McNeil, N., 2013. Four types of cyclists? Examination of typology for better understanding of bicycling behavior and potential. *Transportation Research Record: Journal of the Transportation Research Board*, 2387, pp. 129–138.

Eidse, N., Turner, S. and Oswin, N., 2016. Contesting street spaces in a socialist city: Itinerant vending-scapes and the everyday politics of mobility in Hanoi, Vietnam. *Annals of the American Association of Geographers*, 106(2), pp. 340–349.

Imrie, R., 2012. Auto-disabilities: The case of shared space environments. *Environment and Planning A*, 44 (9), pp. 2260–2277.

Jacobs, J., 1992 [1961]. *The death and life of Great American Cities*. New York: Vintage Books.

Jurewicz, C., Sobhani, A., Woolley, J., Dutschke, J. and Corben, B., 2016. Exploration of vehicle impact speed–injury severity relationships for application in safer road design. *Transportation Research Procedia*, 14, pp. 4247–4256.

Koskela, H. and Pain, R., 2000. Revisiting fear and place: Women's fear of attack and the built environment. *Geoforum*, 31(2), pp. 269–280.

Kröyer, H. R., 2015. Is 30 km/ha 'Safe' Speed? Injury severity of pedestrians struck by a vehicle and the relation to travel speed and age. *IATSS Research*, 39(1), pp. 42–50.

Kullman, K., 2010. Transitional geographies: Making mobile children. *Social & Cultural Geography*, 11(8), pp. 829–846.

Lefebvre, H., 2004 [1992]. *Rhythmanalysis: Space, time and everyday life*. Translated by S. Elden and G. Moore. London: Continuum.

Middleton, J., 2011. Walking in the City: The geographies of everyday pedestrian practices. *Geography Compass*, 5(2), pp. 90–105.

Middleton, J., 2018. The socialities of everyday urban walking and the 'Right to the City'. *Urban Studies*, 55(2), pp. 296–315.

Moody, S. and Melia, S., 2014. Shared space: Research, policy and problems. *Proceedings of the Institution of Civil Engineers-Transport*, 167(6), pp. 384–392.

Nixon, D., 2012. A sense of momentum: Mobility practices and dis/embodied landscapes of energy use. *Environment and Planning A*, 44(7), pp. 1661–1678.

Nixon, D., 2014. Speeding capsules of alienation? Social (dis) connections amongst drivers, cyclists and pedestrians in Vancouver, BC. *Geoforum*, 54, pp. 91–102.

Nixon, D. and Schwanen, T., 2018. Emergent and integrated justice: Lessons from community initiatives to improve infrastructures for walking and cycling. In: N. Cook and D. Butz, eds. *Mobilities, mobility justice and social justice*. London: Routledge, pp. 129–141.

Norton, P., 2007. Street rivals: Jaywalking and the invention of the motor age street. *Technology and Culture*, 48(2), pp. 331–359.

Parker, D., Lajunen, T. and Summala, H., 2002. Anger and aggression among drivers in three European countries. *Accident Analysis & Prevention*, 34(2), pp. 229–235.

Patton, J. W., 2007. A pedestrian world: Competing rationalities and the calculation of transportation change. *Environment and Planning, A*, 39(4), pp. 928–944.

Schwanen, T. and Nixon, D., 2019. Urban infrastructures: Four tensions and their effects. In: T. Schwanen and R. van Kempen, eds. *Handbook of urban geography*. Edward Elgar Publishing, pp. 147–162.

Simpson, P., 2017. A sense of the cycling environment: Felt experiences of infrastructure and atmospheres. *Environment and Planning A*, 49(2), pp. 426–447.

Spinney, J., 2010. Improvising rhythms: Re-reading urban time and space through everyday practices of cycling. In T. Edenser, ed. *Geographies of rhythm: Nature, place, mobilities and bodies*. Basingstoke: Ashgate, pp. 113–128.

Stapleton, S., Kirsch, T., Gates, T. and Savolainen, P., 2017. Factors affecting driver yielding compliance at uncontrolled midblock crosswalks on low-speed roadways. *Transportation Research Record (TRB)*, 2661, pp. 95–102.

Trinkaus, J., 1997. Stop sign compliance: A final look. *Perceptual and Motor Skills*, 85, pp. 217–218.

Vallières, L., 2006. *Disciplining Pedestrians: A critical analysis of traffic safety discourses*. (Doctoral dissertation, Department of Sociology and Anthropology-Simon Fraser University).

27

TRANSITIONS

Methodology and the marginalisation of experience in transport practice

Justin Spinney

Introduction

How we are mobile is a constant source of problematising activity: too mobile, of the wrong sorts; in the wrong places; by the wrong people; for the wrong reasons. Or not mobile enough; too slowly; using the wrong modes. The negative consequences are multiple: environmental catastrophe; societal polarisation; community breakdown; health crises; cultural dilution; re-borderings; spread of disease. If this were the whole story then perhaps our transition would be an easy one, yet we must also acknowledge that as much as our current ways of life are problematic, there are many things to treasure related to high levels of mobility and accessibility: the thrill of travel and discovery; of making and maintaining friendships and familial connections; the intrinsic sensory rewards, exhilaration and contentment of different mobilities and immobilities.

One solution to our dilemma is that we must transition away from our current (im)mobilities to a better (more sustainable) state; one that hopefully keeps much of the good whilst ditching the bad. Some advocate a clear break: that what we are doing must stop and that any transition must be rapid, complete and universal, entailing a wholesale rethinking of current ways of life. Others are more pragmatic, asking for a slower, piecemeal and particular transition. Quite often in such visions, technology will be the medium through which transition is performed, enabling current behaviours to remain intact but with lesser consequences.

As this suggests, what is at stake here are questions around what kinds mobility we value and how this shapes what we are trying to transition toward. Such questions of experience, difference and justice are central to a mobilities perspective yet, as Temenos et al. (2017) point out, despite the politics of mobility implicit in transition, the literature has yet to fully engage with them. Martens (2017) has eloquently argued that fair and just mobility systems should accommodate a range of different ways of life and capabilities. Accordingly, any future mobility systems and transition toward them must apprehend and accommodate a plurality of mobilities.

Questions of how transitions to more sustainable mobility will occur have most commonly been theorised through the multi-level perspective (MLP) (Geels 2006, 2014; Temenos et al. 2017:114). Evolving out of innovation studies, this approach suggests that transitions come about through interactions between niches, socio-technical regimes and socio-technical landscapes (ibid.). Whilst change on the scale required will entail large shifts across all three elements, empirical research has tended to focus on physical infrastructure and technological innovation to the neglect of issues of cultural meaning and the role of professional expertise (Temenos et al. 2017:116–17). Whilst the obduracy of systems relies in no small part on socio-technical regimes such as technical standards, laws and methodological techniques (Temenos et al. 2017:114), these have been some of the least studied elements of transition.

Accordingly, in this chapter I engage with the intersection between diversely experienced socio-technical landscapes and the socio-technical regimes that make them (in)visible. Studies in transitions often take a user perspective, arguing that change will occur as users interact with new technologies to create new ways of being (c.f. Watson 2012). In this chapter I want to take a similar approach but from the perspective of socio-technical regimes, arguing that innovations in the technologies employed to understand mobility are required. These shifts may include technological advancements (such as bio-sensing, for example Osborne 2019; Osborne & Jones 2017; Spinney 2015) but in the first instance are behavioural innovations in as much as change will only occur when we expand what we consider to be the desirable range of experiences in mobility and create methodological tools capable of apprehending and observing them.

Accordingly, the main focus of this chapter is to explore the importance of experiencing mobility and how it relates to methodological concerns. In a departure from much research on mobile methods, the focus here is not on go-along style methods (Jungnickel 2015; Laurier 2010; Parent 2016; Pink 2007), but instead looks at the ways in which the experiential – and differences in the ways in which we experience mobility – are marginalised in the tools used by practitioners to understand mobility and accessibility. This builds upon previous work by the likes of Spinney (2015); Merriman (2013); Dewsbury (2000) etc. who have all advocated for a methodological sensitivity to mobility as essential to mobilities scholarship. As such, the focus here is on mobilising method in ways that are sensitive to differential experiences of mobility.

In the first part of the chapter I give a brief account of the marginalisation of lived experience in the city. Here I argue that questions of economy have led to the privileging and facilitation of functional and efficient commuter mobilities to the neglect of othered mobilities. I then go on to discuss the ways in which such narrow visions have become sedimented in standardising planning tools and methodologies used to assess movement.

In the second part I investigate Transport for London's Public Transport Accessibility Tool (PTAL) as a methodology used to assess urban housing, parking and mobility needs. In doing so I show that its apparent neutrality masks a series of assumptions that flatten differences in motility and experience. In doing so it is inclined to show maximal levels of accessibility that in reality only reflect the experiences of a minority of users. The implication for transitions is that the case for any kind of change or transition will seem marginal at best precisely because the tool is only capable of representing the mobility experiences of the few. Clearly there is an issue of justice and inequality that needs to be addressed yet has been obfuscated by the veneer of objectivity that such methodologies import with their assumptions.

In the third part of the chapter I then look at the ways in which the mobility experiences of children in particular have been marginalised in planning practice. Here I argue that assumptions around what constitutes desirable mobility (functional, efficient) have served to marginalise the more playful and less directed mobility needs of children. These assumptions have been compounded by research techniques that are concerned only with the amount of mobility children have with regard to public health agendas, and circumvent them as knowledgeable citizens. The chapter concludes with a call to challenge methodological anachronisms and the assumptions that perpetuate them if we are to shift our thinking to support sustainable and just transitions.

Mobility in the city

The origins of the modern polis are numerous, dating back thousands of years to the Civic Republicanism of the Greek and Roman empires. However, the shape and function of modern cities as we know them is in large part a product of capitalist systems of production. Here, the healthy circulation of the city has long been ordered according to the instrumental rationalities of economy; its spaces and movements ordered scientifically and functionally to facilitate efficient production and consumption (Sennett 1994). The modernist city is thus most often theorised as the colonial city where technologies of planning and architecture come together to build new societies and indoctrinate citizens within the spatial confines of rationally planned towns.

The modern city is also characterised by unprecedented levels and diversity of mobility. Managing these to maintain economic efficiency has precipitated an increasing distinction between the spaces of roads, streets and public spaces. Public highways were once quite public spaces as the name suggests (Fyfe 1998; Rabinow 1989; Sennett 1994) but the social function of these spaces has been eroded due to an emphasis on movement (Carmona, Heath, Oc & Tiesdell 2003:79). For modernists the streets became a place for getting from A to B; a system rather than a place to live in (Fyfe 1998:1). For Corbusier, a street made for speed was a street made for success and the sacrifice of many traditional street activities such as play and socialising was a price worth paying (Corbusier quoted in Fyfe 1998:2–3).

As such, cities have privileged and reproduced particular aspects of life – those linked to economic and 'rational' command functions – over the more playful and less 'productive' aspects of social life. Of course that is not to say that the modern city does not accommodate both: because of the links between production and consumption it does. However, playful functions tend to be separated and ghettoised to ensure that the important business of mobility remains unaffected. Moreover, despite attempts at separation, humans have an unwavering ability to misuse and appropriate spaces to their own ends. However, it is fair to say that the economically productive functions of the adult world remain emphasised and spatialised at the expense of other (less productive) ways of being and experiencing.

The marginalisation of experience through method

As Cresswell (2006) has noted, certain groups have had a greater say in defining who should be able to move around and how they can do it. Numerous authors have pointed out that the tools of professionals such as engineers, planners and architects who design tools to assess movement and space have served to define ideal movements and render invisible a range of mobile experiences (29). Hill (2003) argues that such tools are rooted in Taylorist understandings of efficient and predictable movement. He goes on to say that they provide a means to marginalise user experiences that fall outside of a 'normal' range, making it easy to dismiss users and define appropriate uses (26). Indeed, Taylor lamented the 'laxity, ambiguity and unnecessary complexity' of considering multiple subjects which he claimed were an obstacle to 'clear thinking' (Taylor, 1911 in Southworth & Ben-Joseph 2003:66).

The methods that we use to measure and gather data on phenomena import and enact power relations because of the ways in which they marginalise and homogenise everyday experience. As Petts and Brooks (2006) have demonstrated, whilst expert knowledge is perceived to be derived from verifiable empirical observation and distinctive techniques, lay knowledges are perceived to be based in everyday, casual common-sense understandings (Petts & Brooks 2006:1046). However, as Relph (1976) noted some 40 years ago, in its distancing from the lifeworld, science has reconstituted it as a set of idealised scientific images where the 'subjective, transient and trivial' are marginalised. The methods that we employ to understand how and why we move are therefore central to the ontological politics of mobility.

But all is by no means lost. As Johnson (1999) goes on to point out, 'no matter how sophisticated our abstractions become, if they are to be meaningful to us, they must retain their intimate ties to our embodied modes of conceptualisation and reasoning' (Johnson 1999:81). Here Johnston makes an argument that for such tools to remain useful, and to be representative, any methodology requires a sensitivity to a range of experiences. One conclusion that Law (1999) draws from this is that to create mobility systems that reflect the needs of a wide range of users, we need to recognise the embodied pleasure and skill involved in different modes, and the ways in which these skills are differently experienced and distributed (580).

Quantitative and audit measures of experience have tended to be marginalised in mobilities scholarship (see Manderscheid 2014 for a notable exception) because of their emphasis on reducing experience to numbers and the consequent loss of detail and context. However, a case can certainly be made that such methods are still able to represent experience, not in the same level of detail and richness that some qualitative methods achieve, but in the sense that they retain the potential to represent differences in the way that mobility is experienced. Indeed, as Merriman (2013) and Spinney (2015) have argued, what is required in such contexts is a 'bodily attentiveness' to experiential differences. In this sense it is not the tools themselves that are at fault, but the assumptions and absences that constitute the tools. The episteme in question then is not just how we know what we know, but *who* we know through our methods, and more importantly, who we *don't* know.

Valuing experience as a question of justice

But why is method important in the context of transitions? The answer here lies in the link between transition and questions of justice and inequality. Martens (2017) argues that transport is a social good because of the accessibility it confers on people. For many activities, whether we participate in them frequently or not makes little difference to our quality of life. Going to the cinema, or concerts regularly are all 'nice to haves' but not wholly necessary to achieve 'the good life'. Transport, however, is fundamental to our ability to participate meaningfully in economic, civil, political and social aspects of life. Transport must therefore be considered a social good and this recognition places an additional responsibility upon us to consider how it is distributed. Accordingly, when we consider any transitions in our mobility we must consider the extent to which any new arrangements accommodate the full range of users.

Such an understanding asks us to think about what we mean by transition, and in particular transition as a political rather than neutral process. The term transition does not help us here because it carries with it the veneer of neutrality – that a transition just occurs. But of course, we cause something to transition, and we can cause something to transition from one state to many other potential states, none of which are inevitable. Transition is likely to be painful: moving from one state to another requires effort, adaptation. Any process of metamorphosis requires contortion, realignment. Where does the pain fall, upon whom? Here we get to the politics of transition with regard to what we consider to be important in the decision-making process, who is involved in deciding, and who are the winners and losers (Temenos et al. 2017).

Fundamental in answering these questions is the need to unpick taken for granted understandings of mobility and what counts and matters in our mobility practices. As Temenos et al. (2017) have noted, radical socio-technical innovation and transition is theorised to occur in 'niche' spaces of experimentation. The research and development that occurs in such spaces has the potential to change broader systems (for example mobility systems) which are comprised of, 'technologies, infrastructures, regulations, policies, values, and practices that together enable mobility in a given society' (114). What this tells us is that any successful transition relies not only on shifts in our thinking, but fundamental shifts in the methods that we use to mobilise that thinking. In particular, those who become invisible in our measurements and debates about what matters are the most likely to lose out in any transition. Any transition that seeks to be based on principles of justice must recognise the importance of tools that accommodate the lived experiences and realities of the many, not the few. Thus in the sections that follow I discuss a number of methods with a view to highlighting the user experiences that they render invisible, and what could be done to make them visible again.

(Under)valuing different experiences: accessibility measures

My first example here is a tool developed and used by Transport for London (TfL) to measure accessibility – the Public Transport Access Level (PTAL). It is a quantitative tool used to assess how well different places are connected to each other by the public transport system (TfL 2015:4). PTAL is used by TfL for a number of reasons:

- To identify places that may benefit from transport improvements
- To understand the likely impacts of plans for new routes, stations or roads
- To identify the most suitable locations for medical and other services, so that people can reach them easily
- To understand which locations are most suitable for developing more houses and offices
- To recommend whether different locations need more or less car parking.

(TfL 2015:4)

In order to calculate a PTAL value, a number of things are required. Firstly, a list of places/origins (houses, offices, shops, services); the location of all public transport stops (Service Access Points – SAPs) close to these places (bus, metro and rail); data on the routes of these services including frequency; and details of the available walking routes from the specified places to the SAPs (TfL 2015:17). There is no need to go into detail on the ways in which calculations are made – it is quite straightforward and produces an indexed figure (between 0 and 6) enabling comparison of places and visual mapping using GIS tools. What I wish to focus on here are the differences in experience that the tool dismisses through a series of assumptions.

Firstly, the tool assumes a walking speed of 4.8 km/h. Whilst this is relatively low it still assumes a fit, able-bodied and unencumbered body, and indeed a body that can walk at all. This is compounded by the assumed walk network which does not account for challenging topography or street design (such as heavily barriered and signalised street environments) that may provide a deterrent or increase journey time for users. As such the tool does not consider the very different times required to access SAPs by some user groups, or the additional physical and emotional energy required to do so (see for example Spinney & Middleton 2019).

Secondly and related to this point, the walk network utilised assumes that everyone would be equally happy to use the streets regardless of time of day. Such an assumption forgets the fact that women (and to some degree the elderly) are far more aware of social safety issues and many would be unwilling to use the prescribed routes at certain times of day (c.f Greed 2005; Xie & Spinney 2018; Yavuz & Welch 2010). This means that the PTAL effectively assumes a male subject and is gender-biased.

Thirdly, the creators of PTAL admit that, 'issues such as … step-free access are not part of the PTAL measure' (TfL 2015:10). In fact the tool does not take into account any additional access or egress time at a station, assuming an isotrophic surface that everyone can equally traverse instantaneously. As a result, the measure does not account for the additional time taken to access rail and metro stations for less able and encumbered users, or indeed whether these destinations can be accessed at all (c.f Pyer & Tucker 2014).

Fourth, PTAL does not factor in the cost of accessing public transport, assuming that if the transport is available then it will be affordable. However, it is evidently not the case that all users will be able to afford all modes on offer and, for these users, the accessibility calculation would be much lower if this were factored in (Martens 2017). As such, the tool does not accommodate those on low incomes.

Fifth and finally, PTAL calculations in terms of routes and frequency of service are based upon the morning traffic peak. This means that the accessibility index calculated will be a maximal one because frequencies are highest in the morning and evening peaks. As a result the tool assumes a commuting subject that effectively marginalises other users. If for example you work at night, wish to travel at night, or travel in the day time when frequencies are reduced, the resulting accessibility level is likely to be much lower. Users most likely to experience very different accessibility in this circumstance include those with caring responsibilities, the elderly and those on lower incomes (Hine & Mitchell 2001; Lucas & Musso 2014; Markovich & Lucas 2011).

These shortcomings are easily rectified, but the key issue is that PTAL calculates a maximal level of accessibility that is only really relevant to a young, male, working core 9–5 hours with a good income. Whilst for such a user a PTAL of 6 may well be possible, for a different user living in exactly the same location the PTAL may be much closer to 0. The implication for transitions is that, by calculating a maximal level of accessibility that appears to encompass all users and ranges of experience, the case for any kind of change or transition will seem marginal at best precisely because the tool represents high accessibility for all, when in reality it is only high accessibility for the few. Clearly there is an issue of justice and inequality that needs to be addressed yet has been obfuscated by the veneer of objectivity that such methodologies import with their assumptions.

Valuing children's experience: the value of play

A study by Ipsos-Mori (2010) highlights a central problem of the marginalisation of children's experiences in planning for mobility. In the study, when parents were asked what they deemed most important regarding cycling, 66% stated enjoyment (2010:65). When children were asked this question, 89% stated enjoyment was most important (2010:79). Conversely, 70% of adults associated cycling with health whilst only 38% of children did so (2010:65/79). The disconnects here point to a partial instrumentalisation of mobility in adult life as a vehicle to achieve external goals, contrasted with children's readings of it as (a playful) end in itself. The conclusion that children's valuations of their own mobility prioritise its ludic and social elements is backed up by a number of studies.

Jay, Mahdjoubi, Greene and Walton (2009) in their exploration of the effects of public artwork installations on young people's valuations of using cycle routes found that artworks that participants found to be 'interactive' or 'fun' gave an enriched experience of using the route and helped to, 'alleviate some of the boredom of cycling along a route' (Jay, Mahdjoubi, Greene & Walton 2009:11). Likewise Clayton and Musselwhite (2013) emphasise the benefit in designing more playful cycle routes suggesting that 'such additions might be a motivator for children to use cycle networks more frequently and independently, to use it for fun as opposed to simply for access, and even making the cycle path a destination in its own right' (12). Similarly, in 2009 the UK-based sustainable transport charity Sustrans produced the guide 'Routes to Play' aiming to increase awareness of children's active travel needs. The document clearly acknowledges the unstructured nature of children's play and the fact that children see no distinction between active travel and play: 'Walking and cycling provide excellent play experiences in themselves as well as allowing children and young people to get to play spaces and recreational facilities under their own steam' (Sustrans 2009:9).

According to Oliver et al. (2016), when walking without adults, children's mobility is less structured, more playful, slower, more exploratory and challenging (Oliver et al. 2016:2). When discussing the walking practices of 9–16 year olds in the UK, Horton, Christensen, Kraftl and Hadfield-Hill (2014) emphasise its spatial–temporal characteristics noting its 'boundedness, intensity and circuitousness' (101). They also note its 'characteristic sociality, narrativity, playfulness and taken-for-grantedness' (101). What this points to again is that, in contrast to adult valuations, the meanings of mobility for young people are defined by a relative lack of 'function' in an instrumental sense, contributing rather to processes of identity formation, social bonding, making sense of the world and sensory pleasure. Indeed as Horton, Christensen, Kraftl and Hadfield-Hill (2014) note in the case of walking, 'walking was not, for these children and young people, most importantly an instrumental means of getting "from A-to-B"', with many explicitly describing their walking practices as a form of play (104–110).

The key insight from these literatures that we want to highlight is that they shift our thinking away from seeing walking and cycle routes as movement-based 'transport' infrastructure toward seeing them as place-based mobility infrastructure central to playfulness, social formation and identity construction/negotiation that sits at odds with accounts of it that cater for it as 'just movement' and accordingly seek to enhance efficiency and safety.

This rendering invisible of non-instrumental forms of mobility is exemplified in the focus on journeys to school in the literature (Carver, Timperio, and Crawford 2012), in particular the idea that the school journey is analogous to a 'commuting' journey that should therefore be efficient, productive and subject to conduct. One example of this is the development of 'Safer Routes to School' programmes which typically focus upon 'a combination of engineering, education and training, and other initiatives to reduce dependence on the car for the journey to school' (Bradshaw, 1999 in Barker 2003:137). The forms of mobility encouraged as part of these programmes – walking, buses, car sharing and cycle trains (Barker

2003:137), are evidently more about safety and efficiency then any notion of play or leisure. Couched in such terms, the benefits of children's movement and activity are largely subordinate to public health agendas where physical activity is reduced to a form of exercise, and social connectedness is simply a function of mental wellbeing. Hence rather than being intrinsically rewarding in its own right, children's movement becomes important only because it produces an outcome deemed important by policy makers.

Very clearly, playful and non-productive forms of life are increasingly written out of public and especially movement spaces, in favour of more 'rational' and productive forms of life and comportment. This chimes with wider literatures on play which draw attention to its marginalisation in urban life (Spencer 2013; Stevens 2007). The idea of play as important in its own right as a sensory pleasure rather than something that should be undertaken for health and wellbeing is an important yet neglected one. Health and wellbeing follow if we activate our bodies for sensory pleasure, the reverse – that sensory pleasure will follow if we activate our bodies because we believe it is something we should do for our health and wellbeing – is much less certain and policy consistently fails to recognise this.

So why are we producing such mono-dimensional readings of children's mobility? One of the key issues here is the lack of child-centred research with methodologies privileging adult assumptions and policy agendas over children's valuations which may be significantly different to those reported by adults. A review of literature on children's mobility points to the fact that methodologies to elicit children's valuations of mobility are largely gathered through the testimony of adults, and a limited suite of methods (most often surveys) (Barker, Kraftl, Horton & Tucker 2009; Chaudhury, Oliver, Badland, Garrett & Witten 2017; Hillman, Adams & Whitelegg 1990; Hood 2001). In earlier studies such as 'One False Move' (Hillman, Adams & Whitelegg 1990) there is an emphasis on capturing the views of children via parents where surveys with parents were used to report on the 'mobility licences' granted to their children. Whilst children reported on aspects of their actual mobility, any differences in reporting were not examined (463). More recently, Barker (2008) reports on a study where even though parents were asked to involve their children in survey completion, there was no control over and very little idea whether children were actually involved in completing the survey (189).

There is evidence of a growing trend in children's research to use child-centred and co-design methodologies (Derr & Tarantini 2016; Wake 2011 2017; Walsh et al. 2010). Engaging with both individuals and communities who can often still see the world with an unchecked sense of imagination is an opportunity that planners have begun to recognise as offering a counter-narrative to the prevailing power-narratives (often experienced implicitly as socio-economic and political issues) that define the adult world. Thus, co-design with children can provide an un-rivalled perspective on the assumptions and failures of the modern world, by allowing outcomes to emerge from their world without being prescribed by the limitations of an adulterated perspective.

Despite the increasing uptake of co-design methods and the potential for children's voices to be heard, it is clear, however, that children's valuations are not being listened to and translated into policy outcomes. Reflecting on their own experiences of participatory working, both Barker (2003) and Kelley (2006) note the continued marginalisation of both qualitative methodologies and children in policy circles where children are 'constructed as the passive recipients of policy rather than as political actors and are rarely involved as stakeholders in decision-making processes' (Barker 2003:141). As Barker goes on to note, despite collecting in-depth data on children's mobility, it would appear that any understandings that do not fit neatly into the predetermined agendas of political and economic elites do not become policy (Barker 2008: 184). Indeed as Johnson (2017) notes, if,

> power relationships between children and young people (CYP) and adult decision-makers are not taken into account in PAR, then adults are likely to ignore CYP's evidence or use it in a tokenistic way. Decisions are then made on the basis of more quantitative evidence or from processes that are more adult centric.
>
> *(106)*

As Kraftl (2015) notes, through such exclusions certain kinds of life are rendered visible, whilst others have been rendered invisible (220).

My point here is that once again, in order to effect a just transition that accommodates the embodied experiences of children, methodologies must be developed that allow not only the playfulness and 'irrational' nature of children's mobility to come to the fore, but that create participatory spaces where local power dynamics can be addressed and children's voices can be heard and acted upon (Johnson 2017:105).

Conclusions

This chapter has sought to bring together the rich vein of methodological questioning and innovation witnessed in mobilities scholarship with transitions thinking which at best has engaged only superficially with 'the heterogeneity in needs, capabilities, and experiences of mobility system users'. (Temenos et al. 2017:116). This chapter asks us to stop and understand what we are trying to achieve and for whom: what are our motivations? As such, whose experience matters is a question of mobility justice: our motivations in any transition should be to create systems that offer favourable and wide-ranging experiences for as wide a range of users as possible. It follows that if we are to more accurately assess the range of experiences that matter to people, we need to pay attention to the tools and processes that we use to measure, assess and translate experience into goals. These tools are performative in that they don't just report on specific experiences, they actively produce these by making some present and some absent, and in so doing enable and constrain the possibilities that we might transition toward. Therefore, as much as it may seem distant to the meta-perspective taken by transition theories such as the MLP, more mundane issues of method and experience must take centre stage in ensuring just mobility transitions.

Accordingly this chapter calls for a renewed research agenda centred on a questioning of assumptions around what kinds of mobility experiences should be prioritised, for whom, and what tools and techniques are required to make these visible. I use the term renewed here because this agenda is anything but new, but is part of an ongoing 'collaborative turn' in planning (Bickerstaff, Tolley & Walker 2002; Brindley, Rydin & Stoker 1996; Fainstein 2000; Healey 1996, 1997; Tewdwr-Jones & Allmendinger 1998), which has sought to involve users in the design and materialisation of projects. In essence, the call to arms of this chapter is to embrace Hajer and Kesselring's (1999) call for more participatory governance capable of representing the interests of a wider community of practice. Methodological innovation doesn't seem radical – its practice often feels very mundane and messy (Spinney & Jungnickel 2019), but its outcomes can be transformatory in opening up what we consider important constituents of mobility systems. In such a reading, technology is only one possible route to transition. Instead we need to explore how we can move sedimented understandings and mobilise methodologies.

References

Barker, J. (2003) Passengers or political actors? children's participation in transport policy and the micro political geographies of the family. *Space and Polity*, 7(2), pp. 135–151.

Barker, J. (2008) Methodologies for change? a critique of applied research in children's geographies. *Children's Geographies*, 6(2), pp. 183–194.

Barker, J., Kraftl, P., Horton, J. & Tucker, F. (2009) The road less travelled – new directions in children's and young people's mobility. *Mobilities*, 4(1), pp. 1–10.

Bickerstaff, K., Tolley, R. & Walker, G. (2002) 'Transport planning and participation: the rhetoric and realities of public involvement. *Journal of Transport Geography*, 10, pp. 61–73.

Brindley, T., Rydin, Y. & Stoker, G. (1996) Popular planning: coin street London. In S. Campbell & S. Fainstein (eds.) (2003) *Readings in planning theory* (Blackwell, Malden, MA), pp. 296–317.

Carmona, M., Heath, T., Oc, T. & Tiesdell, S. (2003) *Public places, urban spaces: the dimensions of urban design* (Architectural Press, Oxford).

Carver, A., Timperio, A. & Crawford, D. (2012) Young and free? a study of independent mobility among urban and rural dwelling Australian children. *Journal of Science and Medicine in Sport*, 15(6), pp. 505–510.

Chaudhury, M., Oliver, M., Badland, H., Garrett, N. & Witten, K. (2017) Using the Public Open Space Attributable Index tool to assess children's public open space use and access by independent mobility. *Children's Geographies*, 15(2), pp. 193–206.

Clayton, W. & Musselwhite, C. (2013) Exploring changes to cycle infrastructure to improve the experience of cycling for families. *Journal of Transport Geography*, 33, pp. 54–61. Available from: http://eprints.uwe.ac.uk/19626 [Accessed August 2018].

Cresswell, T. (2006) *On the move* (Routledge, Oxford).

Derr, V. & Tarantini, E. (2016) 'Because we are all people': outcomes and reflections from young people's participation in the planning and design of child-friendly public spaces. *Local Environment*, 21(12), pp. 1534–1556.

Dewsbury, J. D. (2000) Performativity and the event: enacting a philosophy of difference. *Environment & Planning D: Society & Space*, 18, pp. 473–496.

Fainstein, S. (2000) New directions in planning theory. *Urban Affairs Review*, 35(4), pp. 451–478.

Fyfe, N. (1998) Reading the street. In N. Fyfe (ed.) *Images of the street: planning, identity and control in public space* (Routledge, London), pp. 1–10.

Geels, F. (2006) Multi-level perspective on system innovation: relevance of industrial transformation. In X. Olsthoorn & A. Wieczorek (eds.) *Understanding industrial transformation: views from different disciplines* (Springer, Dordrecht), pp. 163–186.

Geels, F. (2014) Energy, societal transformation, and socio-technical transitions: expanding the multi-level perspective. *Theory, Culture and Society*, 31, pp. 21–40.

Greed, C. (2005) Overcoming the factors inhibiting the mainstreaming of gender into spatial planning policy in the United Kingdom. *Urban Studies*, 42(4), pp. 719–749.

Hajer, M. & Kesselring, S. (1999) Democracy in the risk society? learning from the new politics of mobility in Munich. *Environmental Politics*, 8(3), pp. 1–23.

Healey, P. (1996) The communicative turn in planning theory and its implications for spatial strategy formation. In S. Campbell & S. Fainstein (eds.) (2003) *Readings in planning theory* (Blackwell, Malden, MA), pp. 237–255.

Healey, P. (1997) *Collaborative planning: shaping places in fragmented societies* (Macmillan, London).

Hill, J. (2003) *Actions of architecture: architects and creative users* (Routledge, London).

Hillman, M., Adams, J. & Whitelegg, J. (1990) *One false move … a study of children's independent mobility* (Policy Studies Institute, London).

Hine, J. & Mitchell, F. (2001) Better for everyone? travel experiences and transport exclusion. *Urban Studies*, 38, pp. 319–332.

Hood, S. (2001) *The State of London's children report* (Officer of the Children's Rights Commissioner for London, London).

Horton, J., Christensen, P., Kraftl, P. & Hadfield-Hill, S. (2014) 'Walking … just walking': how children and young people's everyday pedestrian practices matter. *Social & Cultural Geography*, 15(1), pp. 94–115.

Ipsos-Mori. (2010) Research to explore perceptions and experiences of bikeability training amongst parents and children. London. Available from: https://webarchive.nationalarchives.gov.uk/20110119211546/http://www.dft.gov.uk/pgr/sustainable/cycling/bikeabilitytraining/pdf/bikeabilitytraining.pdf

Jay, T., Mahdjoubi, L., Greene, L. & Walton, K. (2009) A toolkit for the evaluation of routes and public artworks with young people. Unpublished.

Johnson, M. (1999) Embodied reason. In G. Weiss & H. Haber (eds.) *Perspectives on embodiment* (Routledge, London), pp. 81–102.

Johnson, V. (2017) Moving beyond voice in children and young people's participation. *Action Research*, 15(1), pp. 104–124.

Jungnickel, K. (2015) Jumps, stutters, blurs and other failed images: using time-lapse video in cycling research. In C. Bates (ed.) *Video methods, Routledge's advances in research methods series* (Routledge, Advances in Research Methods series, London), pp. 121–141.

Kelley, N. (2006) Children's involvement in policy formation. *Children's Geographies*, 4(1), pp. 37–44.

Kraftl, P. (2015) Alter-childhoods: biopolitics and childhoods in alternative education spaces. *Annals of the Association of American Geographers*, 105(1), pp. 219–237.

Laurier, E. (2010) Being there/seeing there: recording and analyzing life in the car. In B. Fincham, M. McGuinness & L. Murray (eds.) *Mobile methodologies* (Ashgate, Aldershot), pp. 103–107.

Law, R. (1999) Beyond 'women and transport': towards new geographies of gender and daily mobility. *Progress in Human Geography*, 23(4), pp. 567–588.

Lucas, K. & Musso, A. (2014) Policies for social inclusion in transportation: an introduction to the special issue. *Case Studies in Transport Policy*, 2, pp. 37–40.

Manderscheid, K. (2014) The movement problem, the car and future mobility regimes: automobility as dispositif and mode of regulation. *Mobilities*, 9, pp. 604–626.

Markovich, J. & Lucas, M. (2011) *The social and distributional impacts of transport: a literature review* (Transport Studies Unit, Oxford).

Martens, K. (2017) *Transport justice* (Routledge, New York).

Merriman, P. (2013) Rethinking mobile methods. *Mobilities*, 9(2), pp. 167–187.

Oliver, M., McPhee, J., Carroll, P., Ikeda, E., Mavoa, S., Mackay, L., Kearns, R., Kyttä-Pirjola, M., Asiasiga, L., Garrett, N., Lin, J., Mackett, R., Zinn, C., Barnes, H.M., Egli, V., Prendergast, K. & Witten, K. (2016) Neighbourhoods for active kids: study protocol for a cross-sectional examination of neighbourhood features and children's physical activity, active travel, independent mobility and body size. *BMJ Open*, 6(8), p. e013377.

Osborne, T. (2019) *Embodying heritage: a biosocial investigation into emotion, memory and historic landscapes* (Unpublished PhD Thesis, University of Birmingham).

Osborne, T. & Jones, P. (2017) Biosensing and geography: a mixed methods approach. *Applied Geography*, 87, pp. 160–169.

Parent, L. (2016) The wheeling interview: mobile methods and disability. *Mobilities*, 11, pp. 521–532. doi: 10.1080/17450101.2016.1211820.

Petts, J. & Brooks, C. (2006) Expert conceptualisations of the role of lay knowledge in environmental decisionmaking: challenges for deliberative democracy. *Environment and Planning A*, 38, pp. 1045–1059.

Pink, S. (2007) Walking with video. *Visual Studies*, 22, pp. 240–252.

Pyer, M. & Tucker, F. (2014) 'With us, we like physically can't': transport mobility and the leisure experiences of teenage wheelchair users. *Mobilities*, 12, pp. 36–52.

Rabinow, P. (1989) *French modern: norms and forms of the social environment* (MIT Press, Cambridge, MA).

Relph, E. (1976) *Place and placelessness* (Pion, London).

Sennett, R. (1994) *Flesh & stone* (Faber & Faber, London).

Southworth, M. & Ben-Joseph, E. (2003) *Streets and the shaping of towns and cities* (Island Press, Washington, DC).

Spencer, B. (2013) Playful public places for later life: how can neighbourhood public open space provide opportunities for improving older people's quality of life by enabling play? Unpublished PhD Thesis, Faculty of Environment and Technology University of the West of England, Bristol.

Spinney, J. (2015) Close encounters? mobile methods, (post)phenomenology and affect. *Cultural Geographies*, 22, pp. 231–246.

Spinney, J. & Jungnickel, K. (2019) *Studying mobilities* (Sage Research Methods Foundations, London). DOI: 10.4135/9781526421036.

Spinney, J. & Middleton, J. (2019) Social inclusion accessibility and emotional work. In I. Docherty & J. Shaw (eds.) *Transport matters* (Policy Press, Bristol), pp. 83–106.

Stevens, Q. (2007) *The Ludic City: exploring the potential of public spaces* (Routledge, Oxon).

Sustrans. (2009) *Routes to play: a guide for local authorities* (Sustrans, Bristol).

Temenos, C., Nikolaeva, A., Schwanen, T., Cresswell, T., Sengers, F., Watson, M. & Sheller, M. (2017) Ideas in motion: theorizing mobility transitions, an interdisciplinary conversation. *Transfers*, 7(1), pp. 113–119.

Tewdwr-Jones, M. & Allmendinger, P. (1998) Deconstructing communicative rationality: a critique of Habermasian collaborative planning. *Environment and Planning A*, 30, pp. 1975–1989.

Transport for London. (2015) *Assessing transport connectivity in London* (Transport for London, London).

Wake, S. (2011) Using principles of education to drive practice in sustainable architectural co-design with children, *Paper presented at 45th Annual Conference of the Architectural Science Association, ANZAScA 2011*, University of Sydney.

Wake, S. (2017) Looking ahead investigating performance art with schoolchildren as a catalyst for urban redesign. In M. A. Schnabel (ed.) *Back to the future: the next 50 years, (51st international conference of the Architectural Science Association (ANZAScA))* © 2017, Architectural Science Association (ANZAScA), pp. 65–73.

Walsh, G., Druin, A., Guha, M., Foss, E., Golub, E., Hatley, L., Bonsignore, E. & Franckel, S. (2010) Layered elaboration: a new technique for co-design with children, CHI, pp. 1237–1240.

Watson, M. (2012) How theories of practice can inform transition to a decarbonised transport system. *Journal of Transport Geography*, 24, pp. 488–496.

Xie, L. & Spinney, J. (2018) 'I won't cycle on a route like this; I don't think I fully understood what isolation meant': a critical evaluation of the safety principles in Cycling Level of Service (CLoS) tools from a gender perspective. *Travel Behaviour and Society*, 13, pp. 197–213.

Yavuz, N. & Welch, E. (2010) Addressing fear of crime in public space: gender differences in reaction to safety measures in train transit. *Urban Studies*, 47(12), pp. 2491–2515.

28
SOCIAL IMPLICATIONS OF SPATIAL MOBILITIES

Vincent Kaufmann

Introduction

A great deal of reflection on mobilities has been developed under the banner of the mobility turn since the early 2000s. This includes theoretical apprehension of the phenomenon (Adey 2010; Cresswell 2006; Urry 2007), how mobilities are experienced (Merriman 2012), the role mobilities play in the constitution of the contemporary individual (Kellerman 2006), how they have evolved over time, etc. Numerous works consider mobility as a physical phenomenon first and foremost, and have described it in terms of flows, practices and meanings. This focus on physical movement tends to limit their analysis to spatial mobility-related social mobility (i.e. inequalities and family/professional life course) linked to the fact that physical movement is usually a way of reaching a goal. It is therefore essential to focus on the nature of this goal in order to understand the energy and motivation that underlie mobility.

Bassand and Brulhardt (1980), Montulet (1998), Boltanski and Chiapello (1999) and Bauman (2000) were instrumental in developing the concept of mobility by transcending the question of physical movement in physical space. Yet, their ideas remain marginal in works on the mobility turn. Based on their work, my goal in this chapter will be precisely that: to provide an overview of the relationship between spatial mobility and social mobility. After a discussion of the spatial and social nature of mobility, this chapter will look at mobility as an analyzer of life course and as a social norm.

Mobility as a result of social and spatial change

"Being mobile" refers to the dual faculty of moving or changing locations and of adapting to new situations, changing statuses, position or even skills. In contemporary societies characterized by the globalization of exchanges, mobility is often presented as an important condition for social integration, including access to employment or professional careers. This dual faculty means that mobility is a two-sided phenomenon involving both social change and physical movement (Gallez & Kaufmann 2009). These two sides systematically characterize the various forms of mobility, which can be described in terms of their ability to generate either little or marked social change and physical movement. That is why, for example, upward career mobility within a company (notably characterized by an increase in salary) is likely to affect the choice of holiday destinations and leisure mobilities more generally. Another example is a job change involving a change in workplace locations and thus in commute patterns (number of kilometers traveled per day, travel time, transportation modes used, etc.).

This conception of mobility originated in Chicago in the 1920s. The members of the Chicago School became interested in the urban dimension of mobility, considering it above all as a social phenomenon with spatial implications. Though interactions between the city, its morphology and social relationships were at the heart of their work, space was neither an explanatory factor nor the main object of analysis for the researchers. Instead, they focused on the social system and its functioning, organization and evolution (Grafmeyer & Joseph 1979). For them, urban space was a mirror of processes and mechanisms that were social in nature—a kind of privileged indicator or revealer of phenomena that are difficult to measure. Urban space was also a distinctive contextual factor in that it could favor the development of social relationships or, on the contrary, hinder change. Be it residential or daily, mobility in physical space is a feature of urban lifestyles. For Burgess, mobility is a good "indicator of the state of the urban metabolism" (Burgess 1925: 38). In 1925, Park wrote that

> transportation and communication, tramways and telephones, newspapers and advertising, steel construction and elevators—all things, in fact, which tend to bring about at once a greater and a greater concentration of the urban populations —are primary factors in the ecological organization of the city.
>
> *(Park 1925: 2)*

Based on these considerations, Chicago School researchers quickly came to the conclusion that not all spatial and social change is mobility. Roderick McKenzie thus developed the idea that some change—like going out to buy a newspaper or cigarettes—is too insignificant to consider as mobility (McKenzie 1927). In my opinion, this consideration remains valid; considering any physical movement or activity change in daily life as mobility would broaden the notion of the latter so much that it would become all-encompassing and thus lose its analytical capacities.

For example, a business trip wherein the employee remains in his or her role of senior manager, going from the airport to the hotel and never leaving his or her small professional world, does not really count as social mobility as there is no socially significant change to speak of (Montulet 1998). Similarly, the spatially complex daily activity schedules of many parents—shuttling children around, buying household supplies, appointments, etc.—cannot be described as having high social mobility because they all stem from the same role and same activity sphere (Jurczyk 1998). Therefore, I propose considering mobility as any socially significant change, meaning any change that modifies an individual's identity. This definition of mobility excludes changes of position and movement within the same activity sphere and instead focuses on changes between the private sphere and professional spheres and between personal and professional schedules.

Mobility as a life course analyzer

Considering mobility as a social and spatial phenomenon makes it a powerful analyzer for life courses and career paths. In modern societies, social integration mainly happens through work (Cuin 1983; Erikson & Goldthorpe 1992). In fact, professional careers and the inequalities that characterize them have been major themes of sociology since its inception. Changing jobs and how socio-professional positions are maintained, acquired or lost have been the subject of extensive debate around the notion of mobility.

In 1927, Russian-American researcher Pitirim Alexandrovich Sorokin published a book entitled *Social Mobility* in the United States, in which he lays the foundations for the analysis of social mobility, which hitherto were understood essentially based on scattered empirical works. He distinguishes between two types of movement: vertical mobility, which implies a change of position on the social ladder (ascending or descending) and horizontal mobility, which refers to a change of status or social category but involving no change in the relative position on the social ladder (changes in the family situation following a divorce or marriage, change of religion or political group, change of job in terms of qualifications but with no change of salary). Social mobility can be analyzed at the generational level (intra-generational mobility) or in reference to the parents' position (intergenerational mobility).

Since Pitirim Sorokin's early work on the American society of the 1920s, this question has developed markedly, particularly in terms of the analysis of intergenerational socio-professional mobility tables and their structure, notably with the studies of Bendix and Lipset (1966), Blau and Duncan (1967). The latter focus on the distinction between structural mobility and net mobility (see Cuin 1983; Wright et al. 1982).

In this work, changes in socio-professional category relative to social origin is considered a key indicator of a society's meritocratic nature. Industrial society has valued social mobility since its inception because it allows us to establish collective development dynamics based on individuals' desire to improve their personal socio-economic position. In this conception, everyone is involved in production in hopes of improving their living conditions and social status based on merit. This idea is founded on two principles. The first affirms individual freedom in the definition and implementation of the statutory project. The second calls for the principle of individual equality such that an individual's social origin is no longer an obstacle to their desired social ascension. Paradoxically, this is an egalitarian discourse in a competition for statuses that are inherently unequal. The paradox is generally raised by the implementation of procedures that attempt to guarantee baseline equality among the various actors according to their social origin.

While these considerations do not include space explicitly, starting in the 1980s, certain authors began highlighting the face that spatial mobility was becoming increasingly important—even central—to socio-professional paths. Castells (1996) and Bauman (2000) thus noted that individuals are more or less attached to the local world in which they live depending on their professional position, and that this degree of attachment was an essential characteristic of their social mobility potential. Bassand and Brulhardt (1980) show that, depending on the area of origin—i.e. center or periphery—upward professional mobility requires specific spatial mobility. Montulet (1998) also evokes the existence of a requirement to be spatially mobile in the professional world. Boltanski and Chiapello (1999) summarize these observations in *The New Spirit of Capitalism*, asserting that status hierarchies are being challenged and that social mobility is now expressed through the constant renewal of projects. The stakes involved in upward professional career paths have changed: it is no longer a matter of achieving a status within a hierarchical structure, but of being able to "bounce back", to move from one project to another, "surfing" from one enviable position to another in an ever changing environment. As a result, social criticism has also evolved. Nowadays, it is less a question of inequalities with regard to high-power positions than of inequalities in terms of flexibility potential:

> In a connectionist world, mobility, the ability to move autonomously not only in geographical space but also between people, in mental spaces or between ideas is an essential quality of big league players, whereas minor league ones are characterized first by their fixity (their rigidity). Still, it is important not to give too much importance to the difference between strictly geographical or spatial mobility and other forms of mobility.
>
> *(Boltanski & Chiapello 1999: 445–446)*

In today's world, the acquisition or loss of socio-professional status often involves spatial change. Professional careers (and life courses and spatial mobility more generally) thus overlap. What does the research tell us about this?

To begin, while correspondence between the degree of spatial change and the quality of social change when achieving mobility existed in the past, this is no longer necessarily the case in a contemporary, globalized, interconnected world marked by the speed potential of rapid transportation modes and the immediacy of remote communication. This is how long distance commuting—which is characterized by travelling long and lengthy distances to work and returning home each evening—developed (Viry & Kaufmann 2015). It is also about the role changes that occur without actually moving in physical space (like when employees work from home) that daily life now allows for. These forms of mobility are characterized by reversibility, meaning the quality of quickly returning to the initial state, both socially and spatially.

Research also indicates that mechanical systemic relationships can be identified between the social and spatial dimensions of mobility (Courgeau 1993; Gobillon 2001). The latter are forged notably based on reversible mobility, with mobility involving considerable spatial change and little social change, or conversely little spatial change and considerable social change (see Table 28.1). Several authors consider that these asymmetrical mobilities are characteristic of contemporary societies, given their spatial and social importance (Kaufmann 2011; Montulet 1998). The speed of transportation and connected objects allows for forms of ubiquity that have indeed transformed the formerly systematic relationship between the distance traveled and the degree of change brought about by this movement.

Table 28.1 Typification of social and spatial changes

Type of mobility	*Change*	*Effects on identity*
Little spatial or social change	Change of position at the same hierarchical level within a company	Change of professional mission
	The birth of a second child	Intensification of parents' role
Considerable spatial change, little social change	Relocation of a company	New status of long-distance commuter
	Moving to a remote periurban area	New status of long-distance commuter
Little spatial change, considerable social change	Professional advancement within a company	New socio-professional position with managerial tasks
	Marital separation with no move	New marital status
Considerable social and spatial change	Migration	Recomposing of social relationships tied to the move

Finally, research shows that a combination of social and spatial change in mobility largely developed before movement itself based on motility and its various pillars (i.e. access, skills and mobility projects, see Chapter 4) (Kaufmann 2011).

The relationships between social and spatial change primarily concern access. A socio-professional position and given residential location offer specific mobility possibilities. Being the daughter or son of a manager or qualified employee or a farmer do not offer the same mobility opportunities (Wright 1992). Similarly, living in the outskirts, a regional center or a large city do not offer the same mobility opportunities (Bassand & Brulhardt 1980). Thus do we stumble into the classical sociological issue of social and spatial inequalities.

Patterns of change also have to do with mobility skills. Depending on an actor's skills, some things are feasible while others are not, both spatially and socially speaking (Vignal 2005). More specifically, a number of studies show that the ability to uproot and create new social ties is an essential skill associated with mobility (Viry & Kaufmann 2015). Similarly, the ability to use one's travel time efficiently, especially during regular long commutes, is a key mobility skill (Ravalet et al. 2015).

Finally, the links between spatial and social change have to do with projects. For example, an individual can contemplate a professional career and use spatial change to this end (i.e. move to a big city to have better job opportunities). Conversely, an individual might decide to invest in family life and, consequently, look for a secure job closer to home, even if it is redundant and leaves little chance for career advancement.

Mobility as a social norm

In all of the aforementioned scenarios, mobility was understood as a basic necessity (Gallez 2014) in the sense of Article 13 of the Universal Declaration of Human Rights on freedom of movement (see Chapter 13). In contemporary Western societies, mobility has also become a dominant social norm (Mincke & Montulet 2019) that notably stems from the imaginary that equates rapid, long-distance travel—and reversible mobility more generally—with democratization. Thanks to such travel possibilities, individuals are free to establish the desired contacts unhindered by spatial or temporal barriers. This rhetoric suggests that the individuals most likely to hold enviable social statuses are also those willing to merge into a logic of unfettered flexibility. The peculiarity of this ideology of mobility is to assume that mobility in physical space implicitly promotes the fair distribution of individuals within societies. Were this the case, simply promoting spatial accessibility would be sufficient to create social equality. In this interpretation, the massive development of reversible, rapid, long-distance mobilities demonstrates that societies throughout the world are experiencing a shift towards a more just world.

Regarding the current, dominant social norm of mobility as a requirement of flexibility and reversibility, two observations should be made:

The first is that this norm has not always been associated with positive values (Gallez & Kaufmann 2009). In France, for example, the term was first used in the abstract, psychological sense of "inconstancy" and "instability" (Dictionnaire historique de la langue française 2016). The social culture of the 17th century was based on assumptions of permanence and stability. Thus, as historian Roche (2003a) reminds us, for Pascal, "man's unhappiness stems from the fact that he does not know how to stay in one place"(Roche 2003b: 2) and for Kant, the literary journey is the equivalent to a real journey.

The second is that several social norms of mobility coexist in contemporary Western societies. Obviously, there is the dominant requirement for flexibility and reversibility, which is sometimes difficult to avoid in the work world. However, the latter coexists with other mobility cultures. Maksim (2011), for instance, shows that people with low incomes develop mobility practices that are specific and specialized enough to compensate for this economic handicap but do not correspond to the dominant model of the mobile person. These mobility cultures notably are built on mutual aid and the ability to uproot and re-root in another region.

The plurality of the social norms of mobility, and the confusion between mobility as a staple good and mobility as a norm are a source of tension, as illustrated by the intense controversies surrounding mobility policies within the European Union. The EU is a paradigmatic example of a model of economic and political integration that is built on mobility as a social norm. As Favell (2008) points out, the economic and political integration of Europe is, on the one hand, largely based on the idea that it is by promoting various mobilities (and the resulting increase in exchanges and associated meetings) that a model of regulation and common identity can emerge. On the other hand, for Favell, mobility appears both as an individual coping mechanism to access resources that are unequally distributed across continents and as an essential component for developing other types of capital:

> The theory about the positive effects of internal mobility continues to be held, especially among advocates of European enlargement. In a monetary union, mobile labour is seen as the security valve that will cushion asymmetric shocks to the system. Moreover, less negatively, increased intra-European mobility is still seen as a proactive source of dynamism and growth, in an otherwise dangerously sclerotic Europe, especially if it encourages the professional and the highly skilled to move.
>
> *(Favell 2008: 17)*

In this general context, there are three types of initiatives that promote mobility: those designed to regulate mobility conditions within the European space, including the development of the four freedoms (goods, capital, services and persons) (Kaufmann & Audikana 2017); those aimed at equipping and configuring the European continent from a material and territorial standpoint (for example, through the creation of a trans-European transportation network); and, finally, those designed to encourage and stimulate mobility between different spaces. These three types of initiatives affect the three dimensions of motility (social conditions of access, knowledge/skills and mobility projects) and not only seek to guarantee access, but often have the more ambitious purpose of developing knowledge and mobility skills at the European level and of stimulating mobility projects at this scale (Kaufmann & Audikana 2017).

An examination of these three types of interventions in European Union policies highlights the fact that the latter quite often promote flexibility and reversible mobility. In this respect, the European Union helps spread and reinforce the social norm of dominant mobility and the ideology that goes with it by playing on words, confusing mobility as a basic need and mobility as a social norm (Gallez 2014). This is how the principle of freedom of movement—a basic need provided for in the Universal Declaration of Human Rights—is reflected in the proactive promotion of flexibility and reversible mobility inside the European Union in all its forms. These different translations of the great founding principles of European construction are strongly opposed politically for different reasons, precisely because they translate mobility as a basic need through an ideology of the mobility that is only shared by a portion of the elites.

References

Adey P. (2010) *Aerial Life: Spaces, Mobilities, Affects*. Oxford: Wiley-Blackwell.

Bassand M. & Brulhardt M.-C. (1980) *Mobilité Spatiale*. St-Saphorin: Georgi.

Bauman Z. (2000) *Liquid Modernity*. Cambridge: Polity Press.

Bendix R. & Lipset S. M. (eds.). (1966) *Class, Status and Power*. New York: Free Press.

Blau P. & Duncan D. (1967) *The American Occupational Structure*. New York: John Wiley & Sons.

Boltanski L. & Chiapello E. (1999) *Le nouvel esprit du capitalisme*. Paris: Gallimard.

Burgess E. W. (1925) "The Growth of the City: An Introduction to a Research Project", in Park R., Burgess E. W. & McKenzie R. D. (eds.) *The City*. Chicago, IL: University of Chicago Press, pp. 47–62.

Castells M. (1996) *The Rise of the Network Society – The Information Age*. Oxford: Blackwell.

Courgeau D. (1993) "Nouvelle approche statistique des liens entre mobilité du travail et mobilité géographique", *Revue économique*, 44(4), pp. 791–808.

Coutard O., Dupuy G. & Fol S. (2004) "Mobility of the Poor in Two European Metropolises: Car Dependence versus Locality Dependence", *Built Environment*, 30(2), pp. 138–145.

Cresswell T. (2006) *On the Move. Mobility in the Modern Western World*. London: Routledge.

Cuin C.-H. (1983) *Les sociologues et la mobilité sociale*. Paris: PUF.

Dictionnaire historique de la langue française. (2016) Paris: Le Robert.

Erikson R. & Goldthorpe J. H. (1992) *The Constant Flux: A Study of Class Mobility in Industrial Societies*. Oxford: Clarendon Press, 429 p.

Favell A. (2008) *Eurostars and Eurocities: Free Movement and Mobility in an Integrating Europe*. Malden, MA: Blackwell.

Gallez C. (2014) "La mobilité: bien premier, nuisance ou norme sociale ? Controverses autour de la régulation des mobilités quotidiennes. La mobilité: concepts et valeurs", Mars 2014, Paris, France.

Gallez C. & Kaufmann V. (2009) "Aux racines de la mobilité en sciences sociales. : Contribution au cadre d'analyse socio-historique de la mobilité urbaine", in Flonneau M. & Guigueno V. (eds.) *De l'histoire des transports à l'histoire de la mobilité? état des lieux, enjeux et perspectives de recherche*. Paris: Presses universitaires de Rennes, pp. 41–55.

Gobillon L. (2001) "Emploi, logement et mobilité résidentielle", in *Economie et statistique*. Paris: INSEE, pp. 349–350.

Grafmeyer Y. & Joseph I. (1979) *L'école de Chicago. Naissance de l'écologie urbaine*. Paris: Champs Flammarion.

Jurczyk K. (1998) "Time in Women's Everyday Lives – Between Self-Determination and Conflicting Demands", *Time & Society*, 7(2), pp. 283–308.

Kaufmann V. (2002) *Re-thinking Mobility*. Burlington: Ashgate.

Kaufmann V. (2011) *Re-thinking the City*. London and Lausanne: Routledge/EPFL Press.

Kaufmann V. & Audikana A. (2017) *Mobilité et libre circulation en Europe. Un regard Suisse*. Paris: Economica.

Kellerman A. (2006) *Personal Mobilities*. London and New York: Routledge, Taylor & Francis Group.

McKenzie R. D. (1927) "Spatial Distance and Community Organization Pattern", *Social Forces*, 5(4), pp. 623–627.

Maksim H. (2011) "Potentiels de mobilité et inégalités sociales: La matérialisation des politiques publiques dans quatre agglomérations en Suisse et en France", Doctoral thesis, EPFL, no 4922.

Merriman P. (2012) *Mobility, Space, and Culture*. New York and London: Routledge.

Mincke C. & Montulet B. (2019) *La société sans répit – Sociologie de la mobilité contrainte*. Paris: Editions de la Sorbonne.

Montulet B. (1998) *Les enjeux spatio-temporels du social – mobilités*. Paris: L'Harmattan.

Offner J.-M. & Pumain D. (1996) éds. *Réseaux et territoires – significations croisées*. La Tour d'Aigues: L'aube.

Park R. E. (1925) "The City: Suggestions for the Investigation of Human Behaviour in the Urban Environment", in Park R., Burgess E. W. & McKenzie R. D. (eds.) *The City*. Chicago, IL: University of Chicago Press, pp. 1–46.

Ravalet E., Vincent-Geslin S., Kaufmann V., Viry G. & Dubois Y. (2015) *Grandes mobilités liées au travail: perspective européenne*. Paris: Economica.

Roche D. (2003a) *Humeurs vagabondes*. Paris: Fayard.

Roche D. (2003b) Cafés Géographiques de Paris, Daniel Roche, Alexandra Monot, Paris, 25 novembre 2003.

Sorokin P. (1927) *Social Mobility*. New York: Harper & Brothers.

Urry J. (2000) *Sociology beyond Societies, Mobilities for the Twenty First Century*. London: Routledge.

Urry J. (2007) *Mobilities*. Oxford: Polity Press.

Vignal C. (2005) "Injonctions à la mobilité, arbitrages résidentiels et délocalisation de l'emploi", *Cahiers internationaux de sociologie*, CVIII, pp. 101–117.

Viry G. & Kaufmann V. (eds.). (2015) *High Mobility in Europe: Work and Personal Life*. London: Palgrave McMillan.

Wright E. O. (1992) "The American Class Structure", *American Sociological Review*, 6/47, pp. 709–726.

Wright E. O., Costello C., Hachen D. & Sprague J. (1982) "The American Class Structure", *American Sociological Review*, 47(6), pp. 709–726.

SECTION VI

Urban planning, design and governance

The importance of urban mobilities for economies, cultures and life directs attention to how these elements are decided upon. Systems of regulation and cultures of planning and governance conditions the frames for urban mobilities – as does architecture and design decisions and interventions. Section VI turns to these elements of urban planning, design and governance. In **Chapter 29**, Malene Freudendal-Pedersen investigates urban mobilities planning and everyday life. The chapter discusses this dilemma between the technocratic outlook that still dominates the ideas of future cities and how it frequently does not relate to the lived everyday life. It is suggested that instead of only having a focus on the technological feasibility and legitimacy of measures, based on existing data, models and calculations, storytelling in urban mobilities planning can facilitate the following questions – why, for what and for whom? These are essential questions when planning urban living. Ole B. Jensen discusses the notion of mobilities design in **Chapter 30**. Recent research into the nexus between urban mobilities and urban design has made it clear that there is a new field of research emerging. Such a new field of 'mobilities design' draws upon a number of classic urban design discussions such as the design and materialisation of public spaces and streetscapes. Furthermore, it is argued that from the perspective of urban design thinking, transit spaces, infrastructural landscapes and places of mobilities often have underutilised potential to become more interesting and better accommodating to contemporary citizenship. The influence from urban design theory and practice on mobilities thinking is thus important. However, the reverse direction of influence (i.e. from mobilities research to design) is equally important. By drawing on the now well established 'mobilities turn' as well as newer theories within materialities, actor-networks, assemblages and non-human agencies, the research field of mobilities design offers new theoretical and conceptual insights into the design of mundane urban mobility. The chapter exemplifies this interaction between design and mobilities through several empirical touch points. In **Chapter 31**, Shelley Smith analyses the movement of public space. The act of movement itself is the point of departure for the re-examination of public space as both a physical and social space, as well as how viewing public space through the lens of movement contributes to the design and experience of public space as a feature of contemporary cities for their inhabitants. The urban cannot be separated from the mobilities that support and develop modern lives within it. Today's cities are composed of complex settings of physical and virtual mobilities and they have changing rhythm, speed and reach. Understanding the flow in cities purely as individual choices, technological transformations or economic forces overlooks the fact that practices and networks are culturally assembled when producing and performing city space. Contemporary planning paradigms and practices are still 'technocentric' and the planning of urban movements focuses on traffic as an unchallenged principle for the efficient organisation of societies. However, the question of how to structure existing and future cities and the *scapes* of cities is also a question of how to 'design' the social layout and human interactions. Storytelling in planning and politics offers a different perspective and creates the opportunity to ask the question regarding why, what and for whom. It can put the emphasis on what matters to people and also enables a focus on everyday life; its hopes, dreams and expectations; and its interconnectedness with urban mobilities. In **Chapter 32**, Huijbens and Johannesson address urban tourism. The chapter explains mobilities in terms of the infrastructure developed for urban tourism in Iceland spurred by the rise of the creative city discourse. Access to and access within the city are fundamental to a city's status when it comes to globalised urban tourism. The development of airports, high-speed mass transits and attractions such as revamped downtowns and waterfronts shows how tourism is playing an increasing role in urban governance, affecting planning and thus impacting the city fabric and attitudes and livelihoods of its residences. The chapter explores the vagaries thereof in Reykjavik and Iceland, and through the particular example of Airbnb. **Chapter 33** concerns Lassen and Larsen's discussion on airport cities. Questions such as what is an airport city, why do we need this concept and whether the concept can be contested are discussed. The chapter firstly introduces aeromobilities research as a theoretical foundation for understanding and analysing airport cities. Secondly, the changing historical role of airports within contemporary society is explored. Thirdly, a number of conceptual notions of the airport city are analysed, including some of the more problematic perspectives of the concept. Here, the chapter argues for a holistic, interdisciplinary and future-orientated approach to airport city development. Section VI concludes with **Chapter 34**, where Francisco Klauser turns to surveillance and urban mobility. The chapter explores the complex interactions between urban surveillance and mobility, highlighting how, today, digital technologies, whether mobile themselves or fixed in place, control and regulate movement within and between cities. This discussion is structured into three main parts, relating to (1) separation and access control, (2) the management of humans and non-humans on the move, and (3) the internal organisation of interconnected places. From this perspective, the city is portrayed as a complex system of separations and connections, in which differing spatial logics of surveillance support, modify, limit and indeed co-produce each other.

29

PLANNING FOR URBAN MOBILITIES AND EVERYDAY LIFE

Malene Freudendal-Pedersen

Mobilities and cities

Cities are nodes in global networks (Castells 1996; Graham and Marvin 2001), and through immense physical and virtual mobilities they are shaped, produced and reproduced (Freudendal-Pedersen and Cuzzocrea 2015). This goes along with a rapidly increasing urbanization, and the often-used term "urban age" describes how today's urban spaces are connected in global patterns and structures providing new opportunities as well as challenges. This entails positive economic and social effects, such as wealth, international collaboration and exchange. But simultaneously this is followed by increased inequality, climate change, urban sprawl and mobile lifestyles highly dependent on oil and other fossil fuel resources. The extension and speeding-up of mobilities systems has also led to the rise of mobile forms of working and living (Freudendal-Pedersen 2009; Kesselring 2008; Urry 2007) and today mobilities are an intrinsic component of everyday life – an everyday life that is characterized by these global processes. Global politics and economies have impacted on the price of food we buy in the supermarket, the cost of land our home is built upon, and the materials it is built and renovated with (Beck 2016). More and more of our everyday practicalities, such as banking, shopping and health issues, are digitized and handled online. The city and its mobilities are controlled by traffic lights while our own movement is guided by real time apps on cell phones or navigation systems. These global processes infiltrate all aspects of living and enhance an on-going reflexive individualization process. It creates an everyday life where an immense amount of opportunities are considered when designing the best everyday life possible. This results in a complex constant juggling of everyday life mobilities routines, where mobilities become the tool to create lifestyles entailing the "right" components (Freudendal-Pedersen 2009; Giddens 1991). Thus, in the Urban Age, the future of urban mobilities has become a key topic when mobilities have made possible an immense growth of cities, but at the same time have significant influence on the form and the planning of scapes and places (Jensen 2014). Throughout history, transportation has contained the idea and promise of frictionless speed (Jensen and Freudendal-Pedersen 2012; Urry 2007), as that which would lead to a better and happier life. Today we see that the realization of the vision of "seamless mobility" and a "zero-friction society" (Hajer 1999) also means unforeseen amounts of congestion, noise and environmental problems (Adey et al.

2013; Urry 2011) while at the same time ideas of the liveable city play a significant role in the international city competition. These things do not go hand in hand very well, and the acknowledgement that the question of how to structure the existing and future cities, and the *scapes* of cities, is also a question of how to "design" the social layout and human interactions, needs to have a greater influence on planning (Bertolini 2017; Freudendal-Pedersen and Kesselring 2016). In this chapter I will discuss this dilemma between the technocratic outlook that still dominates ideas of future cities and how it often does not relate to the lived everyday life. I suggest that instead of only having a focus on the technological feasibility and legitimacy of measures, based on existing data, models and calculations, storytelling in urban mobilities planning can facilitate the questions – why, for what, and for whom? A question, I argue, that is essential when planning urban living.

Planning cities with a technocratic outlook

The Industrial Revolution produced a number of inventions which fundamentally changed the conditions for transportation. The invention of the steam engine, and thus rail transportation, made land transportation covering long distances in a short time possible. Both speed and capacity increased rapidly and led to increased industrial specialization through manufacturing located independently of natural resources (Wolf 1996). Food production moved out of the cities and cities grew accordingly, facilitated by new transportation systems (Steel 2008). The development of the combustion engine, the automobile, the first highways constructed during the 19th century, as well as the airplane that, after World War I, increasingly became a fast way to transport people and goods over long distances, further speeded up this process. In the 1950s, the introduction of containers meant massive efficiency gains in freight transportation, facilitating and accelerating globalization. These inventions needed infrastructure to enable this new and faster movement, and the construction of infrastructure has been a major project during the 20th century (Graham and Marvin 2001). These "traffic systems" consist of roads or other pathways for movement, terminals for conducting and organizing movements, traffic steering systems, service and supply networks and, most importantly, modes of transportation adjusted to these systems. Traffic represents a quantifiable figure of these modes of transportation. Its numbers show how many cars, trucks, bikes, pedestrians, buses, skateboarders, etc., there are on a given street crossing, in a given time frame, or a given trip, for a given purpose. This short historical outline shows, in a simple manner, how traffic is being measured and, thus, conceptually captivated, and provides a good framework to understand why the technocratic outlook still dominates transportation planning, politics and research (Jensen and Richardson 2003). The transdisciplinary mobilities research opens up new understandings of the interconnectedness among the city, its mobilities and the people using it. Mobilities research regards cities as inseperable from the mobilities that support and develop modern lives within them. Based on existing data, models and calculations, planners and engineers mostly envision what, and how, decisions on spatial development, technology implementations and other forms of regulations impact on CO_2 emissions, congestion, land use, the densification of cities and the consequences for ecosystems, etc. This is not because planners and engineers do not want all the other elements of mobilities to be part of the decision-making process; mainly it is because the relevant data does not show up in data sets, models or simulations. In this sense many aspects of why things matter to people (Sayer 2011) stays invisible. As one of my engineering colleagues once asked me: "How do you put numbers on birds singing, or the embodiment of cycling through the city, on a summer evening?" I could only answer: "I am not sure you can, but

I know for sure that it is important." Therefore, it is essential when thinking about future cities to acknowledge that social innovations are needed as much as technological ones (Hajer 2016).

This invisibility of many aspects of urban life means that, despite the acknowledgement of the interconnections between social and technological innovations, modern planning paradigms are still "technocentric" (Hajer 1999). The consequence is that the idea of "seamless mobility" becomes the unchallenged principle for an efficient organization of the urban and it plays along with a neoliberal concept of an economy based on global flows of trade and workforce (Larner 2000; Tickell and Peck 2002). Understanding the flow in cities purely as individual choices, technological transformations, or economic forces, overlooks the fact that practices and networks are culturally assembled when producing and performing city space (Jensen, Sheller and Wind 2014). In the following part, I will elaborate on this through a focus on everyday life and its communities that stands as an invertible part of the city but also as an important resource (Radywyl and Bigg 2013; Rosa 2013).

The flows of everyday life in the city

Different mobilities produce and reproduce social life and cultural forms, and it is in these mobilities that cultural patterns and identities are shaped and reshaped (Shove, Pantzar and Watson 2012; Spinney 2010). The people we meet and socialize with – friends, family, colleagues and acquaintances – are, in one way or another, reached by at least one other form of mobilities. This can be, and often is, very local, but we increasingly get used to having close connections to people living far away. Sometimes we move physically to meet them, but we also maintain relationships and create communities online. Today we see close relationships in online communities (Wellman and Gulia 1999) and "living together apart" has become the reality for an increasingly number of families. In this way mobilities enable an individual's composition of the many fragments and moments that comprise time and make the late modern individual's autobiographical narrative possible. (Freudendal-Pedersen 2009; Kesselring 2006). Mobilities research provides new perspectives on the idea of mobility as "hidden" or "introvert" and moves away from sedentary concepts and methodology. This entails both an ontological and methodological counterpoint, introducing a new call for analysis that investigates the interconnectivity between people and places (Sheller and Urry 2006; Urry and Buscher 2009), and recognizing that people, places and institutions are connected in many ways and that these connections are decisive for the way cities and society develop. Thus, it becomes important to uncover why these mobilities take place, how they are carried out and which consequences the mobile practices have for both people and places. This means understanding mobility not just as an event or connection between people and places, but also as a force that creates new ways of organizing society and shaping identities in the process. Individuals master everyday life mobilities in ways that give meaning to themselves and their loved ones. This mastering draws patterns and imprints that individuals share in modern life. The changes and breaks in daily rhythms, routines and actions are not merely matter-of-fact; they lay imprints in the way meaning and place is constructed (Freudendal-Pedersen 2015; Murray 2009). Some years ago, I did research on "Why Copenhageners bike" and this certainly had a lot to do with the practicalities of moving from one place to the other; but it also had a large amount of embodied emotions related to physical and mental health, to smelling, hearing and feeling the city and being part of the city's organism (Freudendal-Pedersen 2015, 2018). Cycling is a big part of creating Copenhagen and its daily rhythms, which, to a high degree, influence the urban and its mobilities in a very specific way.

The urban everyday life is lived in a "mobile risk society" (Kesselring 2008). Risks are not increased in modern societies, but our knowledge of risks is increased due to increased mobilities (Beck 1992). Instant communication means an overload of information about global events and new knowledge (time–space compression); therefore, everyday life today means living with a constant component of reflexivity and time pressure, and thus a lot of choices made possible by mobilities (Beck 1992; Giddens 1991). One way of handling this everyday life is through the guidance of actions provided by lifestyles (Giddens 1991) which most often implies being part of different communities. Traditionally the understanding of community was related to a place – a locality; but, as already mentioned, today's increased mobilities make communities appear in many places, both physically and virtually. In *Sociology Beyond Society* (2000, 134) Urry refers to Hetherington (1997, 185–9) who uses the metaphor of a ship:

> Places are about relationships, about the placing of materials and the system of difference that they perform. Place should be thought of as being placed in relation to a set of objects rather than being fixed through subjects and their uniquely human meaning and interaction.

This underlines how increased physical and digital mobilities have changed the form of communities as something only place-specific. In pre-modern place-bound societies with low pace, the mobilities of communities are illustrated by Simmel (1972) as concentric circles, with the centre being that of the individual's identity and belonging. In modern societies, mobilities' increasing speed and extent have caused these circles to intersect, illustrating multiple identities in many communities. This in no way undermines the significance of place (Freudendal-Pedersen and Kesselring 2018). Everyday life stabilities and routines with kids, homes, friends, leisure activities and all the mobilities this entails are still, in most cases, place-bound. Modern, living communities still rely upon the need for exchanging everyday experiences and sharing responsibilities. Different forms of presence, commitment and intentionality are simultaneously localized, and continued at a distance, through physical and virtual mobilities (Amin and Thrift 2002). Individuals are continuously seeking new communities to emerge around the very same dynamics that reshaped or eroded previous ones. I will argue that the dynamics Tonnies' (1957) saw in relation to traditional communities under the pressure of urbanization; Durkheim's (1997) discussions on the move from mechanical to organic solidarity, placing human life into spheres of functionalized institutional set-ups based on contracts; Weber's (1978) studies of how instrumental and bureaucratic rationality is overtaking the rules of social organization; or Bauman's (2000) concept of liquid modernity, where the rate of change brings a sense of common ground into constant flux, are still relevant. Nevertheless, they all underestimate the need for ontological security that is still pertinent. We still need communities to create meaning, shared responsibility and validation in everyday life, to be assured that lifestyle choices make sense. Globalization, urbanization and mobilities have created opportunities for individualization and changed communities but have not eroded the need for communities. Today, communities exist both in and through places as well as their intersecting routes (Ingold 2007; Jensen 2013). This is what creates the urban as a place of attraction. If you look at the pictures from the most liveable cities in the world (this goes for all the different lists) they show people meeting and dwelling in urban space. It is the practice of "individuals in relation" that constitutes the city and mobilities makes this possible. The question then becomes how to make this matter in mobilities planning when it fits very badly into mathematical models.

The argumentative turn in planning

The "argumentative turn" in policy analysis (Fischer and Forester 1993) addresses this through the analysis of shifts in society's discursive patterns and structures. It understands planning as a form of storytelling and stands in the tradition of "communicative action planning" (Sandercock 1998) influenced by Habermas' work on communication and action planning. The argumentative turn considers discourse and talk as one of the main social activities for initiating social change. It enhances what Hajer (2016) names as "ontological expansion", which is the transformative capacity of planners to create things that don't exist. This is a step away from economic models based on the idea that people are persuaded by facts, and instead focuses on how change is motivated by perspective. Also, by examining and thus showing the articulation of discourses on planning, it is revealed how facts and future predictions on "facts" (fiction) blur. Within argumentative planning, changing practice is thus always seen to include the change of perceptions, problem definitions and the social construction of solution strategies. It also demands we ask the right questions. When we discuss how the future city should develop we might think about rephrasing the question, "How do we make smart cities liveable?" into: "How do we make liveable cities smart?" This creates a different outset where technology is not determining urban life, but urban life is what prescribes which technologies are integrated into the urban. Today the idea of automation and the smart city has a significant role in politics and business (and I would argue much less in planning). On a political, strategic level, there is also a lot of discussion of MaaS (mobility as a service) as an integral part of the development of smart cities. This demands a new type of planning where transforming the car-dependent city and the "system of automobility" (Urry 2004), is in the centre. This requires imaginary spaces, where thinking about the interconnection of future mobilities and the city can move into new directions where other alternatives are examined and can be utilized to generate new policies. This outlook necessitates a subject-oriented approach in urban planning where sustainability and socially cohesive cities are essential, rather than "nice-to-have" features of a utopian, post-materialist world. Communicative planning and storytelling offer the tools to understand the significance of creating new "utopias" about everyday life rhythms, hopes, dreams and expectations (Fischer and Gottweis 2012; Freudendal-Pedersen and Kesselring 2016). As I mentioned previously with regards to the research on cycling in Copenhagen, these hopes and dreams are not the first thing to pop up, but by showing an interest in people's everyday life stories, many new orientations towards possible different futures open up (Freudendal-Pedersen 2009; Kaplan and Ross 1987).

With an outset in the forecasting of what automation, and the technologies of automation, will do to the city, its everyday life and its mobilities in the future, they have a lot in common; almost all of them describe a future where emotions are erased and irrelevant. It reminds me of futuristic movies (*Mad Max*; *WALL-E*), and novels (*2001: A Space Odyssey*; *1984*) and also different futurologists extrapolating upon future lives. A good example is a clip from 1960 on YouTube where Arthur C. Clarke (author of *2001: A Space Odyssey*) talks about the future city. He is basically able to describe quite accurately the mobile world of today with unforeseen (or maybe not) technological opportunities. Where he falters, however, is on the emotional part of modern lives. The basic idea is that the significance of trusting, face-to-face relationships and physical presence disappear, or become less significant, because we have technologies that can make us escape this. This is also pointed out by Castells when he says that:

> the predictions of futurologists over the last 20 years have not come to pass. The death of the city has been announced a thousand times, for example. The reasoning behind this declaration was that the modern communication technologies and the internet would render the city redundant, as individuals could remain connected regardless of where they lived.
>
> *(2008, VI)*

Concluding remarks

Within city planning, the legacy of Marinetti/Futurism and Le Corbusier/modernism has made visible imprints, not least in relation to mobilities. The idea was to create separated spaces for moving and living, big traffic corridors were leading cars into close proximity with all everyday functions. Today the neighbourhoods, or cities created from this idea, are seen as examples of misanthropic city planning, but at the time they were yet another example of a utopian basis which, through a rational understanding of humans, attempted to create the good life in the city; a rational outlook where emotions are understood as an "irrelevant accompaniment". However, "emotions are not merely an irrelevant accompaniment to what we are doing, like muzak in a supermarket, but a kind of bodily commentary on how we, and our concerns, are faring" (Sayer 2011, 39). Sayer also introduces the story of the anthropologist, Renato Rosaldo, to further explain the significance of taking emotions seriously. Renato Rosaldo describes how he needed to experience the unfortunate death of his wife in an accident to understand the relationship between headhunting and grief, which was the focus of his study a decade earlier. Rosaldo, through this experience, understands that

> anthropologists writing about the ways in which cultures deal with death did so 'under the rubric of ritual rather than bereavement', so that the emotional force of the experience – the thing that matters most to the people themselves – was edited out.
>
> *(Sayer 2011, 3)*

Sayer's argument helps to unfold how we fail to comprehend and understand what goes on not only in virtual, but also physical, mobilities. Paying attention to emotions does not equate to neglecting rationalities or systemic thinking; as Sayer (Sayer 2011, 4) puts it: "I have no truck with a romanticism that attempts to deflate reason or rationality. Rather I argue that, properly understood, reason is involved with all these things." Through a shift in focus away from individuals as rational economic beings when deciding which mobilities to use in everyday life, perspectives on the emotionally driven practice of everyday life mobilities become visible. This can help planning ask the right questions and, through this, create storytelling that responds to the needs and aspirations of citizens and politicians when suggesting alternative mobilities futures in the city.

References

Adey, Peter, David Bissell, Kevin Hannam, Peter Merriman, and Mimi Sheller. 2013. *The Routledge Handbook of Mobilities*, edited by Peter Adey, David Bissel, Kevin Hannam, Peter Merriman, and Mimi Sheller. London: Routledge.

Amin, Ash, and Nigel Thrift. 2002. *Cities: Reimagining the Urban*. Cambridge: Polity Press.

Bauman, Zygmunt. 2000. *Liquid Modernity*. Cambridge: Polity Press.

Beck, Ulrich. 1992. *Risk Society: Towards a New Modernity*. London: SAGE Publications Ltd.

———. 2016. *The Metamorphosis of the World*. Cambridge: Polity Press.

Bertolini, Luca. 2017. *Planning the Mobile Metropolis: Transport for People, Places and the Planet*. London: Palgrave Macmillan.

Castells, Manuel. 1996. *The Rise of the Network Society: The Information Age: Economy, Society, and Culture Volume I*. New York: Wiley-Blackwell.

———. 2008. "The Networked City." In *The Social Fabric of the Networked City*, edited by Géraldine Pflieger, Luca Pattaroni, Christophe Jemelin, and Vincent Kaufmann, v–xiii. Oxford: Routledge.

Durkheim, Emile, and Lewis A. Coser. 1997. *The Division of Labor in Society*. New York: Free Press.

Fischer, Frank, and John Forester. 1993. *The Argumentative Turn in Policy Analysis and Planning*, edited by Frank Fischer and John Forester. Durham, NC: Duke University Press Books.

Fischer, Frank, and Herbert Gottweis. 2012. "The Argumentative Turn Revistited." In *The Argumentative Turn Revisited*, edited by Frank Fischer and Herbert Gottweis, 1–27. Durham, NC: Duke University Press.

Freudendal-Pedersen, Malene. 2009. *Mobility in Daily Life: Between Freedom and Unfreedom*, vol. 2012. Farnham: Ashgate.

———. 2015. "Cyclists as Part of the City's Organism - Structural Stories on Cycling in Copenhagen." *City & Society* 27 (1): 30–50.

Freudendal-Pedersen, Malene, and Valentina Cuzzocrea. 2015. "Cities and Mobilities." *City and Society* 27 (1). doi:10.1111/ciso.12050.

Freudendal-Pedersen, Malene, and Sven Kesselring. 2016. "Mobilities, Futures and the City. Repositioning Discourses - Changing Perspectives - Rethinking Policies." *Mobilities* 11 (4): 573–584.

Freudendal-Pedersen, Malene, and Sven Kesselring. 2018. "Networked Urban Mobilities: An Introduction." In *Exploring Networked Urban Mobilities*, edited by Malene Freudendal-Pedersen and Sven Kesselring, 1–19. London and New York: Routledge.

Giddens, Anthony. 1991. *Modernity and Self-Identity: Self and Society in the Late Modern Age*. Stanford, CA: Stanford University Press.

Graham, Steve, and Simon Marvin. 2001. *Splintering Urbanism, Networked Infrastructures, Technological Mobilities and the Urban Condition*, 1st ed. London: Taylor and Francis.

Hajer, Maarten. 1999. "Zero-Friction Society." 1999. www.rudi.net/books/11454.

———. 2016. "On Being Smart about Cities: Seven Considerations for a New Urban Planning and Design." In *Untamed Urbanism*, edited by Adriana Allen, Andrea Lampis, and Mark Swilling, 50–63. Oxon and New York: Routledge.

Ingold, Tim. 2007. *Lines: A Brief History*. Oxon and New York: Routledge.

Jensen, Ole B. 2013. *Staging Mobilities (International Library of Sociology)*. London: Routledge.

———. 2014. *Designing Mobilities*. Aalborg: Aalborg Universitetsforlag.

Jensen, Ole B., and Malene Freudendal-Pedersen. 2012. *Utopias of Mobilities. Utopia: Social Theory and the Future*. Farnham: Ashgate.

Jensen, Ole B., and Tim Richardson. 2003. *Making European Space: Mobility, Power and Territorial Identity*. London: Routledge.

Jensen, Ole B., Mimi Sheller, and Simon Wind. 2014. "Together and Apart: Affective Ambiences and Negotiation in Families' Everyday Life and Mobility." *Mobilities* 10 (3): 363–382.

Kaplan, Alice, and Kristin Ross. 1987. *Everyday Life, Yale French Studies*. New Haven, CT: Yale University Press.

Kesselring, Sven. 2006. "Pioneering Mobilities: New Patterns of Movement and Motility in a Mobile World." *Environment and Planning A* 38 (2): 269–279.

———. 2008. "The Mobile Risk Society." In *Tracing Mobilities*, edited by Weert Canzler, Vincent Kaufmann, and Sven Kesselring, 77–102. Aldershot and Burlington: Ashgate.

Larner, Wendy. 2000. "Neoliberalism: Policy, Ideology, Governmentality." *Studies in Political Economy* 63 (1): 5–25.

Murray, Lesley. 2009. "Making the Journey to School: The Gendered and Generational Aspects of Risk in Constructing Everyday Mobility." *Health, Risk & Society* 11 (5): 471–486. http://eprints.brighton.ac.uk/7364/1/Murray_Making_the_journey_to_sch_refereed.docx.

Radywyl, Natalia, and Che Bigg. 2013. "Reclaiming the Commons for Urban Transformation." *Journal of Cleaner Production* 50: 159–170. doi:10.1016/j.jclepro.2012.12.020.

Rosa, Hartmut. 2013. *Social Acceleration: A New Theory of Modernity (New Directions in Critical Theory)*. New York and Chicester, West Sussex: Columbia University Press.

Sandercock, Leonie. 1998. *Towards Cosmopolis. Planning for Multicultural Cities*. Chichester u.a.: Wiley.

Sayer, Andrew. 2011. *Why Things Matter to People: Social Science, Values and Ethical Life*. Cambridge: Cambridge University Press.

Sheller, Mimi, and John Urry. 2006. "The New Mobilities Paradigm." *Environment and Planning A* 38 (2): 207–226.

Shove, Elizabeth, Mika Pantzar, and Matt Watson. 2012. "The Dynamics of Social Practice." In *The Dynamics of Social Practice: Everyday Life and How It Changes*, 1–19. doi:10.4135/9781446250655.n1.

Simmel, Georg. 1972. *Georg Simmel on Individuality and Social Forms (Heritage of Sociology Series)*. Chicago, IL: University of Chicago Press.

Spinney, Justin. 2010. "Improvising Rhythms: Re-Reading Urban Time and Space through Everyday Practices of Cycling." In *Geographies of Rhythm: Nature, Place, Mobility and Bodies*, edited by Tim Edensor, 113–129. Farnham: Ashgate.

Steel, Carolyn. 2008. *Hungry City: How Food Shapes Our Lives*. London: Chatto & Windus.

Tickell, Adam, and Jamie Peck. 2002. "Neoliberalizing Space." *Antipode* 34 (3): 380–404. doi:10.1111/1467-8330.00247.

Tonnies, Ferdinand. 1957. *Community and Society*. Mineola and New York: Courier Dover Publications.

Urry, John. 2000. *Sociology beyond Societies: Mobilities for the Twenty-First Century (International Library of Sociology)*. London: Routledge.

———. 2004. "The 'System' of Automobility." *Theory, Culture & Society* 21: 25–39.

———. 2007. *Mobilities*. Cambridge: Polity Press.

———. 2011. *Climate Change and Society*. Cambridge: Polity Press.

Urry, John, and M Buscher. 2009. "Mobile Methods and the Empirical." *European Journal of Social Theory* 12 (1): 99–116.

Weber, Max. 1978. *Economy and Society: An Outline of Interpretive Sociology, Band 1*, edited by University of California Press. Berkeley, CA.

Wellman, Barry, and Milena Gulia. 1999. "Virtual Communities as Communities: Net Surfers Don't Ride Alone." In *Communities in Cyberspace*, edited by Marc A Smith and Peter Kollock, vol. 8, 167–194. New York: Routledge.

Wolf, Winfried. 1996. *Car Mania: A Critical History of Transport, 1770–1990*. Chicago, IL: Pluto Press.

30

MOBILITIES DESIGN

Cities, movements, and materialities

Ole B. Jensen

Introduction

Cities are dynamic, complex, and networked artefacts. The history of urban analysis is full of discussions about the relationship between what we may term the 'armatures' and the 'enclaves' of the city (Shane 2005). The 'circulation system' (Lynch & Hack 1984) of the city in its widest sense is a network of pipes, tubes, paths, lines, roads etc. These are the 'armatures' in which people, information, matter, goods etc. move and thus sustain urban life as we know it (Graham & Marvin 2001; Graham & McFarlane 2015). The movement, however, goes to locations. In the city, these are housing estates, business quarters, offices, factories, shopping malls, sports venues etc. These are the 'enclaves' of the city and they are the more or less bounded spaces where urban activities take place. The relative openness of urban enclaves is a key feature to observe since no house is a house without its connections (or disconnections) to other houses, other places (Lefebvre 1974/91:92–93).

The 'turn to Mobilities' (Jensen, Kesselring & Sheller 2019) within urban studies is therefore important. Regardless if we are looking at a two-bedroom flat or a city quarter, its relational interdependency with other enclaves is vital. Without sewers, power, people, light, goods, information etc., flowing 'in and out' in complex circuits, a place would not be a place. Many different scholars have argued for such a relational and mobility-oriented notion of places and cities (e.g. Amin & Thrift 2002; Graham & Marvin 2001; Jensen 2013; Lynch 1981; Massey 2005; Sloterdijk 2016). One might argue that the study of cities is the study of Mobilities. However, it is also the study of immobility since not rarely is one person's or group's mobility connected to another's immobility (e.g. through so-called 'barrier effects' of new roads hindering people in practising Mobilities as they have been doing up until a new urban freeway comes into place). As Urry reminds us it is the dialectic relationship between mobility and 'mooring' that we should focus on (Urry 2003). There are many ways in which one can acknowledge the importance of the mobile/immobile relationship and its connections to armatures and enclaves. In this chapter, I point to the meeting of Mobilities research and urban design as one important way of enhancing our analytical understanding of cities as mobile phenomena. Recent research into the nexus between urban Mobilities and urban design has made it clear that there is a new field of research emerging. The emerging research field of 'mobilities design' (Jensen 2017) draws upon a number of classic urban design discussions such as the design and materialization of public spaces and streetscapes. It is to this emerging field that this chapter is dedicated.

The chapter has the following structure: After the introduction, section two engages with the traditional way in which planning, transportation, and architecture are 'living separate lives' both within city government organization but also in the professions and their ways of engaging the city. Section three, then, is dedicated to unfolding the new perspective of mobilities design as a critical creative comment on the established 'pillarization' of urban disciplines. After this critique, I turn to examples. Because of the limited space in this context, I can only in brief illustrate the key dimensions of some already researched cases. Detailed references will be provided so the reader will be able him or herself to engage these cases further. The chapter ends with a concluding section fleshing out the contours of a future research agenda for urban Mobilities design.

Urban mobilities: between planning, transportation, and architecture

The problem of a disconnection between urban planners, architects, and traffic engineering might be said to be one of the prevailing issues from where to address the question of urban mobilities. It is not a new mismatch. Already Sir Colin Buchanan turned to the lack of constructive communication between the professionals within the realms of architecture and traffic (Buchanan 1964). In his now seminal report 'Traffic in Towns' Buchanan not only targeted the problem of car-dependent urbanism, he also pointed out that the lack of dialogue between traffic engineering and architecture might be part of the problem. To remedy this, Buchanan boldly proposed the development of a new perspective, that of 'traffic architecture' as the new cross-disciplinary language:

> There is a new and largely unexplored field of design here, but it involves abandoning the idea that urban areas must consist of buildings set alongside vehicular streets, with one design for the buildings and another for the streets. This is only a convention. If buildings and access ways are thought of together, as constituting the basic material of cities, then they can be moulded and combined in a variety of ways many of which are more advantageous than the conventional street. A useful term to describe this process is '*traffic architecture*'.
>
> *(Buchanan 1964:67–68, italics in original)*

Buchanan's glorious efforts neither solved the problem of the car-based city, not did it make way for a new set of cross-disciplinary understandings. Rather, urban development over the last many decades has been marked by increased professional specialization and disciplinary compartmentalization.

The value of looking at cities through the lens of Mobilities as well as understanding Mobilities in the light of urban design (Jensen 2013) seems underestimated and worth exploring. This is precisely what is done under the heading of 'mobilities design' (Jensen & Lanng 2017). Partly inspired by Buchanan's call for an urbanism sensitive to a more integrated understanding, research into Mobilities design has explored how a theoretical framing and a new cross-disciplinary vocabulary is needed in order to propose new insights and solutions. For lack of a better term, we may call this a more 'holistic' understanding of the city and its mobile practices. Different scholars may represent such more 'comprehensive' understandings. One is Tim Ingold who speaks of a 'meshwork' of people, spaces, technologies, artefacts, and atmospheres (Ingold 2011). Another way to escape the disciplinary stray jackets would be to take point of departure in the actual mobile situation (Jensen 2013, 2017; Jensen & Lanng 2017). By 'starting in situ' we put neither academic nor municipal organizations and borders in the spotlight (unless of course these are part of what we wish to understand). Rather, we explore the pragmatic question: 'What makes this mobile situation possible?'

To exemplify, we may look at a pedestrian street crossing. Such are vital 'critical points of contacts' (Jensen & Morelli 2011) within a networked city and they deal with the negotiation between e.g. car drivers and pedestrians by proxy of the traffic light algorithms deciding when and for how long each mode will have green light or red light. The point is that neither planning nor architecture alone is sufficient to understand the mobile situation of a person crossing the street in a traffic light controlled zebra crossing. We need a more holistic focus on anything that affects the person's mobility. Sometimes we must look to architecture, sometimes to computer science (as when identifying traffic light algorithms as the key to which flows that are given priority). I admit this example may be easily constructed in all its simplicity. However, the principle is the same if we aggregate up to the level of an urban highway system. If we ask what are affording the mobile practices within such a system, then we need to follow whatever relevant trace of technology, means of transportation, organizational layout etc. It is actually just a case of taking the point of departure in the passenger, driver etc. and his or her mobile experience as a holistic event. Sir Norman Foster coins this well when he speaks of how to design metro systems:

> A metro system is an excellent demonstration of how the built environment influences the quality of our lives. The building of tunnels of trains is usually seen in isolation from the provision of spaces for people – even though they are part of a continuous experience for the traveller, starting and ending at street level.
>
> *(Foster 2007:484)*

The 'continuous experience' is what should be at the centre of the analysis and this is precisely what Mobilities design aims to do. It can easily sound as if I advocate the abolition of knowledge harvested over centuries within planning, architecture, or transportation engineering. This is certainly not the case. What the argument points at is, however, that those frameworks and disciplines need to become better integrated. This has been the case since Buchanan's bold proposition in 1964. However, the increasing complexity of infrastructural landscapes and the new digital networked media seems to suggest that we actually are even more in need of an approach to look more holistically at urban Mobilities. One exploration into how this may be changed is precisely the emergent field of 'Mobilities design' to which I now turn.

What is Mobilities design, and what is the problem it solves?

By drawing on the now well-established 'mobilities turn' as well as newer theories within materialities, actor-networks, assemblages, and non-human agencies the research field of Mobilities design offers new theoretical and conceptual insights to the design of mundane urban mobility. The Mobilities design research agenda put a number of things in the foreground (see Jensen 2017; Jensen & Lanng 2017 for elaborations). In this chapter, I focus on two key themes: (1) The actual mobile situation; and (2) The materialities, multi-sensorial, and embodied dimension.

The mobile situation

The framework of 'Staging Mobilities' (Jensen 2013) is the outset, pointing to a particular interest in mobilities 'in situ' or 'mobile situations'. According to this framing, any mobile situation can be (analytically) dissected into three mutually interdependent dimensions: materiality (e.g. space, infrastructure, technology), sociality (mobilities as social interaction), and embodied performances (i.e. the fact that we are 'doing mobility' with our bodies and involving multiple sensations). Imagine yourself being in a specific mobile situation such as driving in the city looking for a parking space, you are materially situated in a car moving in a street system. Moreover, you are negotiating with other mobile people (pedestrians, cyclists, lorry drivers, other car drivers etc.). In addition, obviously you are driving the car. This simple operation is more complex than it seems but, for the purpose of this example, let us simply focus on the fact that you are seated behind a steering wheel and commanding an amount of horsepower through your physical and cognitive capacities. Thus framed and 'situated' you also will notice that some of the mobile situation's properties are given from 'outside' of the situation. The law specifies rules of the road, the city planners may define the streetscape, and the car producer has made the technology you rely on. Seen this way we could say that the mobile situation has been staged 'from above' as it were. However, the situation is also staged 'from below' through the choices, feelings, and sentiments influencing the particular way you are moving in the actual situation. The 'staging mobilities' framework thus put focus on the actual situation and enables a pragmatic enquiry pivoting around the key question: 'what enables this particular situation?' Mobile situations 'make up the city' on a large scale and across vast spaces and complex intertwining temporalities (just think about the timetabling of a city). The city is an aggregate and complex artefact; however, the staging Mobilities framework suggests that we also learn to see the city and its mobility as being made by manifold mobile situations. Put differently, the complex and aggregated city is but the sum of its multiple situated practices.

Materialities and the multi-sensorial, embodied dimension

As already mentioned, one of the three key themes of the Staging Mobilities framework is dedicated to embodied Mobilities. The way in which we perform Mobilities in specific situations is, as mentioned, also a function of our bodily capacities. Moreover, we engage with the mobility systems of the city through a multi-sensorial engagement (Jensen 2016). The city and its infrastructures enable us to 'become mobile' in complex processes where we feel, hear, sense, see, and smell the materialities around us (e.g. ticket machines, sloping pavements, other people's bodies etc.). We are cultivating a multi-sensorial sensitivity to materials:

> textures and densities, liquidities and radiances, thus act as sets of imperatives within and through which movement and sensation are inspired and performed … materiality, in this reading, is multiple: the term connotes forces and processes that exceed any one state (solid, liquid, gas), and are defined ultimately in terms of movement and processes rather than stasis.
>
> *(Anderson & Wylie 2009:326)*

The situational and pragmatic focus described here lends itself particularly well to an analysis of design and architecture. The close zoom as it were on the situation and the questioning of what kinds of 'affordances' we find in the situation are questions closely related to the designed systems and technologies as well as to the materialities and architectures of the spaces 'hosting' urban mobile situations. In the words of Adey et al: 'Indeed one of the defining characteristics of mobilities research is its attention to the mobilities of multiple materialities, both human and non-human' (Adey et al. 2014:265). The argument is thus that the emerging research perspectives of Mobilities design contain a new sensitivity to the 'material surfaces, the tactile engagements with technologies, the spatial volumes shaped by architectural intervention, the socio-technical geographies of complex networks and so on' (Jensen & Lanng 2017:40). By applying, the Mobilities design approach we are trying to fix the destructive effects of disciplinary compartmentalized gazes as one thing. However, more importantly, we are trying to get as close as possible to the key question we should ask any mobile subject: how does it feel to move/be moved? The emergent perspective of Mobilities design connects to the traditional question within phenomenology. However, given the influence of new perspectives on materialities and the problematization of 'human exceptionalism', we are facing a 'post-phenomenological' research agenda (Ihde 1993; Verbeek 2005). There is a rich literature addressing cities and Mobilities (of all sorts) from the vantage point of science and technology studies (Hommels 2006), infrastructural geographies (Graham & McFarlane 2015), actor-network theory (Latour 2005; Yaneva 2017), post-phenomenology (Ihde 1993), philosophy of technology (Verbeek 2005), and assemblage theory (Farias & Bender 2010). The list could be extended, but suffice to say that the 'socio-technical' perspective and the critique of 'human exceptionalism' all point towards re-thinking the city. We need what Amin and Thrift termed an 'ontology of encounter or togetherness based on the principles of connection, extension and continuous novelty' (Amin & Thrift 2002:27). It may be difficult to summarize and synthetize all these important influences into one coherent approach. However, the emerging articulation of 'mobilities design' (Jensen 2017; Jensen & Lanng 2017) is one dimension of this work.

Mobilities design is a pragmatic enquiry into what enables (or prevents) particular mobile situations. The link to design thinking and materialities is close, as we need an expanded material vocabulary to enhance our analysis of actual urban Mobilities. The theoretical scaffolding for this task can only be hinted at as above in this context. However, the real importance lies in the empirical application of Mobilities design thinking so to these we now turn.

Two examples of urban mobilities design

This section brings examples of this interaction between design and Mobilities through a few empirical touch points. Obviously, the multitude of Mobilities within cities makes the choice of illustrative cases a bit arbitrary. Here I have chosen to use cases from my earlier research into urban Mobilities. I want to draw on two examples of how to apply Mobilities design thinking. In the first case, the focus is on the process and the co-creating dimension where different professions and perspectives are brought together by the Mobilities design frame. The second case is one of more analytical value where I briefly illustrate the critical potential of Mobilities design analysis in understanding the materialization of injustice seen in the 'urban warfare' against homeless people.

The first example of Mobilities design illustrates its inborn cross-disciplinary gaze and potential for breaking down (some of) the organizational, disciplinary, and mental barriers between urban planners, engineers, architects, and urban designers. A few years ago, a Mobilities design workshop was organized in the municipality of Vejle, Denmark (Jensen, Lanng & Wind 2016; Lanng, Wind & Jensen 2015). Together with two colleagues, I co-organized a workshop on the question of what to do with the centrally located train station area of the Danish city Vejle. The train station served as a 'hinge' between the waterfront and the city centre and was not exactly a welcoming gate to the city. In the context of this chapter the end proposal is less interesting than the fact that after presenting traffic engineers, architects, planners, and urban designers from different organizational areas of the Municipality with this 'new field' of Mobilities design, they engaged in a design workshop where much of their mutual distrust and the classic territorial behaviour was left outside. This may of course be due to the good job done by the invited architectural companies who participated (and who were very receptive to our brief for the workshop). However, it is also one of the conclusions from this workshop that the cross-disciplinary and more 'holistic' Mobilities design approach offered the participants a chance to 'play on neutral ground' as it were. We experienced engineers voicing about pavement materials and aesthetics (the traditional domain of the architects) whilst the urban designers and architects did not shy away from voicing about traffic models, traffic data, and other often 'black boxed' dimensions of the traffic engineering world. I will try not to stretch the conclusions from this, but the point is that on a very practical level asking the simple situational and pragmatic key question in mobilities design we opened up for a very different and nuanced discussion of Mobilities in the city of Vejle. We may put it like this – the positive engagement with the mobile situation and the key pragmatic question led us to conclude not only that this is a useful method and epistemological framing for different urban stakeholders, it is also an ontological testament to the fact that cities are messy, unruly, and mobile places whose workings need different analytical vocabularies than those we have left from Modernist city planning and urbanism (Shane 2005). The engagement with practice in workshops such as the one described thus feed into the way we conceptualize and theorize the city and its mobile practices. In this short chapter it will not be possible to expand this discussion but it has led to the coining and articulation of 'mobilities design' as both an approach and a field of study (Jensen 2014; Jensen & Lanng 2017) under the label of 'material pragmatism' (Jensen 2017). The point of material pragmatism is that we should ask the simple pragmatic and situational question related to what enables or prevents something or someone to move. Next to this we should then insert it into a situational framework paying attention not just to the holistic experience of the mobile subject but also to the material conditions and the multi-sensorial engagement herewith. Mobilities design solves the problem of understanding the city and its movements in a 'monochrome' frame of mind. It enables us to draw in design competencies and designerly ways of thinking. It enables us to connect the creative and utopian 'what if …?' question with the critical and constructive 'what now …?' in a double movement of opening up to 'wild ideas' and closing in on hard choices, ethics, and pragmatic consequences.

The second example is reported in more detail elsewhere (Jensen 2019). Under the label of 'Dark Design', I used the Mobilities design framework to critically deconstruct material acts of injustice in the city. The background is the situation found in many contemporary cities where we see urban furniture designed to prevent people from lingering, and in general to materialize the exclusion of 'unwanted subjects' such as homeless people. Let me start with a quote from a homeless person offering a striking first-person account of the theme:

> From ubiquitous protrusions on window ledges to bus-shelter seats that pivot forward, from water sprinklers and loud muzak to hard tubular rests, from metal park benches with solid dividers to forests of pointed cement bollards under bridges, urban spaces are aggressively rejecting soft, human bodies. We see these measures all the time within our urban environments, whether in London or Tokyo, but we fail to process their true intent. I hardly noticed them before I became homeless in 2009. An economic crisis, a death in the family, a sudden breakup and an even

> more sudden breakdown were all it took to go from a six-figure income to sleeping rough in the space of a year. It was only then that I started scanning my surroundings with the distinct purpose of finding shelter, and the city's barbed cruelty became clear
>
> *(Andreau 2015)*

When the sensitivity to materials and multi-sensorial experiences is connected to the critical potential of Mobilities design we get a critical-creative framework for deconstructing what many people in the mundane everyday life seem to take for granted (e.g. if you aren't looking to sleep on a bench by the bus stop, you hardly contemplate the subtle leaning that prevents bodies from taking horizontal rest). The analytical potential of Mobilities design thus enforces our attention to materials, designs, and the 'stuff' that makes up our everyday mobility landscape. Furthermore, we are able to 'connect the dots' enabling us to understand the complex and important relationship between materiality and mobility. Put differently:

> Some of the interventions and designs directly orchestrate flows and movements by rendering benches, doorways and grass lawns uninhabitable. This is the subtle play between mobility and immobility in which the embedded rationalities of dark design may work to produce spaces for lingering and occupancy or push bodies away to elsewhere. Urban no-go areas and design blockings force movement to 'free zones', areas not yet imprinted with dark design. Therefore, while bum-proof benches and metal spikes nested into concrete are stationary and sedentary interventions and devices, they afford and enforce movement to other places, establishing an urban mosaic of 'go/no-go' areas. Places of forbidden access exist alongside places of access, creating an urban jigsaw puzzle constituted through corridors of movement/access and immobility/exclusion.
>
> *(Jensen 2019:123–124)*

What looks like static artefacts randomly inserted into the urban fabric by mysterious agencies (who puts the spikes where they are and why?) actually proves to be part of an inconspicuous system of Mobilities. This is not a grand infrastructural system of roads or bus lines, but rather the subtle mechanism of moving and pushing 'unwanted bodies' around the city. Furthermore, there are important lessons to take from the before-mentioned 'socio-technical' perspective. One is the fruitful notion of 'delegation' (Yaneva 2017). Designed artefacts may work through delegation enabling them to modify people's acts (e.g. preventing people from sitting or sleeping through sloping surfaces or the placement of spikes). When applied to the way in which mobility injustice becomes materialized, the connection of design, materials, power, and Mobilities becomes very clear:

> Spikes, for example, substitute for the acts of human operators, such as private guards, police officers or janitors, who otherwise would have had to perform the socially exclusionary acts of dispelling the unwanted from the chosen sites of intervention. The capacity to make a difference in the material networks of dark design makes the chosen artifacts perform tasks in a subtle interplay between humans and nonhumans.
>
> *(Jensen 2019:125)*

Much more needs to be said to do justice to the potential of connecting Mobilities research with urban design, but the point should hopefully be clear by now. More work needs to be undertaken to create a more coherent analytical framing as well as more empirical work must be carried out to further explore the potential of Mobilities design research.

Concluding remarks

An argument has been made for connecting Mobilities research with urban design. This goes both in terms of cross-fertilizing concepts and theories, but also from the way designers 'look at things'. Furthermore, the openness towards socio-technical perspectives that already has taken place within the Mobilities research community suggests that we stand with a strong analytical tool for making sense of urban Mobilities. The research I am undertaking explores how this may work under headings such as 'mobilities design' and 'material pragmatism'. From the two short cases presented in this chapter, we see that Mobilities design might work at (at least) two levels. First, we find a potential in its holistic ambition to transcend the disciplinary boundaries and taken-for-granted perspectives that for too long have troubled the city. This is the 'lesson from Vejle' where we saw that, even though workshop participants might not have had a full academic overview, they reacted positively to the spanning out of a 'neutral ground' and a level 'playing field' full of new ways of thinking about the city. Mobilities design is on this level offering a potential to work with procedures transgressing unproductive disciplinary fixities. The second level where we see the potential of Mobilities design is related to the materialities of Mobilities. By engaging with architectural and design-oriented understandings of spaces, materials, and design, we learn about the 'distributed agencies' of materials and design interventions. Nowhere is this more painfully evident than in the artefacts and objects utilized to keep 'unwanted subjects' from taking rest in the urban spaces of the contemporary city. There is already a body of literature dealing with some of these questions (e.g. Architecture for Humanity 2006; Weizman 2017). The future of urban Mobilities research points to exploring more cases and examples, all based on a simple pragmatic enquiry into the way cities work. Cities are complex artefacts and they are the habitats of an increasing number of people. Approaching the city through the perspective of Mobilities design is one way of getting closer to understanding this fascinating phenomenon.

References

Adey, P., Bissell, D., Hannam, K., Merriman, P. & Sheller, M. (eds.). (2014) *The Routledge Handbook of Mobilities*, London: Routledge.

Amin, A. & Thrift, N. (2002) *Cities. Reimagining the Urban*, Oxford: Polity.

Anderson, B. & Wylie, J. (2009) 'On Geography and Materiality', *Environment & Planning A*, 41, pp. 318–335.

Andreau, A. (2015, February 19) 'Anti-homeless Spikes: "Sleeping Rough Opened My Eyes to the City's Barbed Cruelty"', The Guardian. Retrieved from www.theguardian.com/society/2015/feb/18/defensive-architecture-keeps-poverty-undeen-and-makes-us-more-hostile?CMP=share_btn_link

Architecture for Humanity. (2006) *Design like You Give a Damn: Architectural Responses to Humanitarian Crises*, London: Thames & Hudson.

Buchanan, C. (1964) *Traffic in Towns*, Harmondsworth: Penguin.

Farias, I. & Bender, T. (eds.). (2010) *Urban Assemblages. How Actor-Network Theory Changes Urban Studies*, London: Routledge.

Foster, N. (2007) *Norman Foster Works 3*, Munich: Prestel.

Graham, S. & Marvin, S. (2001) *Splintering Urbanism. Networked Infrastructures, Technological Mobilities and the Urban Condition*, London: Routledge.

Graham, S. & McFarlane, C. (eds.). (2015) *Infrastructural Lives. Urban Infrastructure in Context*, London: Routledge.

Hommels, A. (2006) *Unbuilding Cities. Obduracy in Urban Socio-Technical Change*, Cambridge, MA: MIT Press.

Ihde, D. (1993) *Postphenomenology: Essays in the Postmodern Context*, Evanston, IL: Northwestern University.

Ingold, T. (2011) *Being Alive. Essays on Movement, Knowledge and Description*, London: Routledge.

Jensen, O. B. (2013) *Staging Mobilities*, London: Routledge.

Jensen, O. B. (2014) *Designing Mobilities*, Aalborg: Aalborg University Press.

Jensen, O. B. (2016) 'Of "Other" Materialities: Why (Mobilities) Design Is Central to the Future of Mobilities Research', *Mobilities*, vol. 11, no. 4, pp. 587–597.

Jensen, O. B. (2017) 'Urban Design for Mobilities – Towards Material Pragmatism', *Urban Development Issues*, vol. 56, pp. 5–11. doi:10.2478/udi-2018-0012.

Jensen, O. B. (2019) 'Dark Design. Mobility Injustice Materialized', in N. Cook & D. Butz (eds.) *Mobilities, Mobility Justice and Social Justice*, London: Routledge, pp. 116–128.

Jensen, O. B., Kesselring, S. & Sheller, M. (eds.). (2019) *Mobilities and Complexities*, London: Routledge.

Jensen, O. B. & Lanng, D. B. (2017) *Mobilities Design. Urban Designs for Mobile Situations*, London: Routledge.

Jensen, O. B., Lanng, D. B. & Wind, S. (2016) 'Mobilities Design – Towards a Research Agenda for Applied Mobilities Research', *Applied Mobilities*, vol. 1, no. 1, pp. 26–42.

Jensen, O. B. & Morelli, N. (2011) 'Critical Points of Contact – Exploring Networked Relations in Urban Mobility and Service Design', *Danish Journal of Geoinformatics and Land Management*, vol. 46, no. 1, pp. 36–49.

Lanng, D. B., Wind, S. & Jensen, O. B. (2015) 'Meget Mere End Et Parkeringshus. Mobilitetsdesign I Praksis – Et Eksempel Fra Vejle Stationsområde', *BYPLAN*, 4/15, pp. 40–46.

Latour, B. & Yaneva, A. (2008) 'Give Me a Gun and I Will Make All Buildings Move: An ANT's View of Architecture', in R. Geiser (ed.) *Explorations in Architecture: Teaching, Design, Research*, Basel, CH: Birkhäuser, pp. 80–89.

Latour, B. (2005) *Reassembling the Social*, Oxford: Oxford University Press.

Lefebvre, H. (1974/91) *The Production of Space*, Oxford: Blackwell.

Lynch, K. (1981) *Good City Form*, Cambridge, MA: MIT Press.

Lynch, K. & Hack, G. (1984) *Site Planning*, Cambridge, MA: MIT Press.

Massey, D. (2005) *For Space*, London: Sage.

Shane, D. G. (2005) *Recombinant Urbanism. Conceptual Modelling in Architecture, Urban Design, and City Theory*, Chichester: Wiley.

Sloterdijk, P. (2016) *Foams. Speheres III*, Cambridge, MA: MIT Press.

Urry, J. (2003) *Global Complexity*, Oxford: Polity.

Verbeek, P. (2005) *What Things Do. Philosophical Reflections on Technology, Agency, and Design*, University Park, PA: The Pennsylvania State University Press.

Weizman, E. (2017) *Forensic Architecture. Violence at the Threshold of Detectability*, Cambridge, MA: MIT Press.

Yaneva, A. (2017) *Five Ways to Make Architecture Political: An Introduction to the Politics of Design Practice*, London: Bloomsbury.

31

THE MOVEMENT OF PUBLIC SPACE

Shelley Smith

Where and when we are: contemporary urbanity, mobility and public space

In the 20th century a number of inventions forever changed the world. The car, the telephone, the television, the airplane, the computer all impacted the physical spaces and social aspects of individual lives lived in collective urban situations. Here and now, in the 21st century, the impact of transportation and information technologies is clearly seen – and experienced – on a daily basis, and the ways in which we perceive the urban environment of the city is seen through a prevalence of mobile, technical and informational modalities.

The contemporary urban situation is characterised by factors such as: globalisation, bigness, shrinking cities, urban sprawl, and consumerism, and all of these are directly linked to an increase in mobility. Perhaps the singular most influential factor in contemporary urbanism, then, has been mobility. It has shaped the form of cities and affected urban life in such an irrepressible way that not only cities and the societies that they mirror, but also the relationships between cities and cities, between people and people and between cities and people have been irrevocably altered.

Within cities, public spaces have been the physical places in which society was most visible. However, as cities have become larger and more extended, as we have utilised technology to traverse distance via communication and travel, the spaces of the public and the people in the publics have changed. We have become more mobile in our lives and through movement more diverse, less place-fixed and place-dependent. This in turn poses a number of challenges for public space in both a physical and social capacity. This not least of all regarding how quality of life, and notions such as belonging and ownership are defined, experienced and on a more conscious level, provided and designed for. When both urban form and urban life have been radically changed by increased mobility, the form that public spaces take and the role that public spaces play needs to be re-examined. In order to ensure that public space reflects contemporary society and provides spaces that serve these publics, they must respond to a set of parameters that are defined within the contexts of mobility.

As the title of the paper signals, it is the act of movement itself in mobility, which is regarded as a key factor, and in this chapter it is the point of departure for the re-examination of public space as both a physical and social space. And further, how viewing public space through the lens of movement contributes to the design and experience of public space as a feature of contemporary cities for their inhabitants.

Traditional public space

Public space is a primary, and necessary, result of urban settlement. Historically, public space was the forum for people meeting and exchanging news, gossip, goods and services. It was also the place of seeing and being seen. It was an open space, accessible – and with an overarching intention of being inclusive and available to all in the community. Public space was though literally the space of the public – stemming from the Latin, *publicus* (from *poplicus*), meaning 'of the people' and 'common' – it provided a physical place for the unfolding of social acts by members of the urban settlement. It was a place of politics and play, of ritual and celebration. It performed an important function as a forum for gathering and meeting. Public space, as a spatial physical entity, was framed and fixed – clearly defined. It was a common space with known and recognisable borders, and it *contained* the publics that entered it, as well their activities. Historically, the spatiality of public space was that of fixivity and stasis.

The fixed nature of traditional public space can perhaps be linked to the very nature of its location – in that of urban *settlement.* The settling notion lies at the base of urbanity. Urbanity is defined as the locus of human settlement. Both as a verb and as a noun, *settle* is derived from the root *setlan* (old English), meaning 'cause to sit, place, put', from *setl* 'a seat'. Derived from the Latin *sella,* also meaning 'chair'. The etymology of *settle* means 'come to rest' (Oxford Dictionary 2018). The pause of movement. Settlements signalled the end of wandering, the end of nomad existence – a choice to come together. These were societies defined by the choice to stop, rather than to move. Traditional public spaces reflected this. Today, however, people are on the move; to urban centres, from troubled areas, to other places – for short and long term. Increased mobility has become a factor of contemporary society and one that also directly affects the role and shaping of public space.

Contemporary public space

The idea of public spaces has remained somewhat unchanged, despite the fact that the conditions for its existence have changed radically. We can easily imagine public space – and most often do so as a kind of space that more fittingly falls into the category of traditional public space. The public spaces we imagine are fixed in space and clearly defined by the surrounding built environment and with a focus on the relationship between the user and the public space.

Our mental imaging of public spaces has been to a very large degree one of fixivity. Being remembered almost as tourist snap shots. And although this type of public space is still to be found as traditional squares and historic places, the reality of contemporary public space does not conform to this picturing. There are many other types of public space that burst the borders of the snap shot, that are identified by a different set of criteria and that perform very different functions in a temporality and a spatiality that challenges traditional ways of understanding public space. This is, to a very large degree, due to the increase of mobility in contemporary societies. Today there is a focus not only on the relationship between the users and the public space, but increasingly between the users themselves in these spaces. Introducing the concept of movement and understanding public space 'in motion' and 'through motion' requires a different mindset and a different way of seeing it.

Mobilities as an urban paradigm

The increase of mobility that had started in the 20th century and continued to increase into the 21st, created radically new situations that the majority of people were impacted by. This new mobilities paradigm, most famously developed by sociologist John Urry (Urry 2000, 2007), and coined as the 'mobilities turn', was primarily a sociological paradigm. It aptly described a contemporary urban situation that was largely created and defined by mobility, but one that could be viewed through the lens of cross-disciplinarity in which not only sociology, but also other disciplines such as geography, anthropology, urban design and architecture, as well as studies relating to urbanity and culture generally, were seen as contributors to the discourse. The focus of the mobilities turn was to define mobility as more than just the journey from A to B, i.e., to also understand the potential experiential qualities of the journeys (Urry 2000, 2007; Jensen 2014). With an increase in mobility, an increase in the desire for seamless and smooth journeys has occurred. But journeys are filled with pauses and waiting – with points of non-moving in the line of flow. This new paradigm also redefined urban spaces through the prevalence of both the journey and the journeyer, i.e., spaces that have come into being solely as spaces of transit and people that are in movement and transit as a result of a number of types of increased mobility. These spaces of congregation are different than traditional public spaces, but in the sheer numbers of people they assemble, they can be considered a new type of public space that is characterised not only by the link to mobility and movement, but also because they are not public in the traditional sense. They are semi-public/semi-private spaces, often owned by companies and therefore subject to their rules for access, behaviour and potential belonging. There is a need for understanding who the users and stakeholders are, and how public space – and its users – are defined in a situation where *public* and *space* have become distanced from each other through new definitions of privatised 'public' space, such as semi-public and semi-private space.

There has been a fundamental change in how public space is described spatially and socially, in great part due to mobilities. This opens the need for reassessing public space in the mobilities turn, and examining the kinds, qualities and locations of meetings that this introduces and allows.

> We must rekindle the lost relationship between the social and the physical space, between form and meaning, with an eye to differences and relationships, as well as bear in mind the demands of a mass culture in flux.
>
> *(Hajers and Reijndorf 2001:227)*

Moving people: migrating populations and the role of public space

There are more people living in urban settlements at this time in history than at any other. Over one half of the world's population now lives in an urban setting and in 2020, it is estimated that 68% of the world's population will be living in urban areas (UN DESA 2019). In addition to this, one of the elements of the mobility turn, migrating populations – the movement of people from one country and culture to another – has increased significantly and greatly impacts the role of public space. Whether fleeing from war, terror or poverty, or exchanging one country for another based on leisure, work or retirement reasons, mobility as a means and condition is extremely present in the make-up of the city. The enormous amounts of refugees, expats, visitors and new citizens of host countries change the population of the countries they arrive at and move into, creating a diverse and multicultural society – at times in countries that have not previously experienced a great influx of 'foreign elements'. This can place a strain – real or perceived – on the economic, physical and social infrastructure, creating a situation stemming from fear of the unknown – the stranger – and ending potentially in a lack of acceptance, and in worst case scenarios in resentment, hate and violence. Here, public space as a visible and inclusive container for all elements of the societies that they should serve, can provide a moderating and socially educating function. An incredibly important function in contemporary societies of diverse peoples.

With the role of educating multicultural societies about themselves, public space is the space in which we see and 'get to know' 'the other' – or are at least, where we are exposed to 'the other', 'the stranger', i.e., to those who are not us. Making the stranger visible and accessible begins to break down the barriers that fear sets up. Sociologist Lyn Lofland uses the term 'public realm' and defines this as 'the city's quintessential social territory'. She sees it as a 'teaching space' where we learn 'cosmopolitanism' – the ability of (inter)acting together from a point of difference. We can be together and exchange without being the same. (Lofland 1998)

Lofland also describes the nature and types of the spaces that could function as a public realm, indicating that in-transit space, small-scale segregation and formation of 'hard edges', which allow 'meaningful' contact with others who are different from us, are ways in which interaction as cosmopolitans could be encouraged (Lofland 1998).

Lofland's description explicitly names spaces of mobility and the spatialities described coincide in more detail to the spaces of transit. The spatial devices Lofland indicates that could affect and encourage social practices, i.e., the breaking down of barriers between diverse groups, are characteristic of spaces of transit in which the possibility for contact with an extensive group of others is made possible, safe, natural and perhaps also fleeting enough to allow a comfortable, but present form of contact with 'the other'. The in-transit experience, and the spaces facilitating this, contain the conditions for the manifestation of the public realm.

Urbanist Ali Madanipour echoes many of Lyn Lofland's views in the following statement about public spaces in general and emphasises the importance of ways in which the presence of 'the other' is facilitated:

> Improvements in public spaces would inevitably improve the quality of urban life for more citizens. It could therefore be a route to a social integration in increasingly fragmented urban societies … Public spaces that allow the symbols and self-expressions of different groups to be displayed in public, to allow diverse groups use the same space and mingle with one another, and so to become aware of themselves and others are essential for the health of a city. In the absence of such channels for exploring the self and others, the fragmented social groups can remain separate and alienated from each other, with possible explosive results.
>
> *(Madanipour 2005:15)*

Public space has the potential to instigate contact between diverse societal groups by being a physical forum for the enactment of contact. Sociologist Maarten Hajer and urban planner

Arnold Rejndorp, also indicate the need for devices in public space that allow for exchange between diverse groups stipulating that where the notion of 'meeting' exemplified the function of traditional forms of public space, 'exchange' is a more fitting term to describe the kind of contact between public space users happening today.

> The core of successful public space thus lies not so much in the shared use of space with others, let alone in the 'meeting', but rather in the opportunities that urban proximity offers for a 'shift' of perspective: through the experience of otherness one's own casual view of reality gets some competition from other views and lifestyles.
>
> *(Hajers and Reijndorf 2001:89)*

With many diverse groups, a less committal and more fleeting exchange is sufficient and perhaps even desired. It may not even involve physical contact or being physically close, but may be simply a matter of seeing and/or hearing – becoming sensorially aware of other groups or individuals that are different from ourselves. Public space provides a relatively safe forum for the observance of the other who may be not only unknown, but also perceived as the 'dangerous' other. Such a forum is instrumental in breaking down barriers that sanction and encourage exclusion because of fear (Lofland 1998).

People moving: urban lifestyles and changing views

In contemporary societies, people are also on the move by choice. One of the factors of increased mobility, is increased lifetime and lifestyle mobility. French urbanist, François Ascher, described the relational differences between pre-mobile cultures and post-mobile cultures through the analogy of weaves of fabric. At the turn of the past century, urban social fabric was comprised of few but thick threads, typically that of work and family; these relations were constant and stable throughout the span of a lifetime. In addition, location was also largely unchanged with few, if any, moves of home within a lifetime, or even generations. Contemporary urban social relations, on the other hand, are comprised of many more, but thinner threads. The number and variety of personal relations, including family, friendship, memberships, associations etc., has increased substantially, and although often of shorter duration, they can change many times throughout a lifetime, are not restrained by physical boundaries and occur over much larger geographical areas. (Ascher 1997:28)

While mobility has been key in changing the weaves of social patterns in contemporary urbanity, it has also profoundly affected our understanding of space and of place. In his seminal work, *Space and Place: The Perspective of Experience* (Tuan 1977), Yi Fu Tuan maintains that space and place are co-dependent and argues that what defines space is the ability to move from one place to another while to exist, place requires space. Put even more clearly as: '[p]lace is security, space is freedom' (Tuan 1977:3), Tuan's definition of space and place are interesting in terms of contemporary public *space* in that the spatial aspect referred to implies an implicit 'free' and moveable aspect, rather than the more (spatially and conceptually) fixed nature of *place*. Movement is identified as the defining factor of space. Understanding space and the spaces of publics through Tuan's definition allows for greater 'freedom' – a definition that is in sync with a society defined by movement and mobility. Herein the spatial essence of contemporary public space can be seen as a moveable, dynamic and flexible entity – a spatial concept that coincides with Ascher's description of the woven fabric of social relations.

From an anthropological perspective, these changing weaves and moveable spaces characterise a radically different social and spatial situation in what had been deemed 'known' territory, i.e., contemporary Western urban societies. Anthropologist Marc Augé called this *supermodernity* and his focus was on the changing social relations of people and their attachment – or not – to physical place (Augé 1995). He coined the term *non-place* to describe spaces that we increasingly find ourselves in, but with which we have no traditional anthropological connection to in terms of socio-cultural and historical relatedness. Highways, airports and shopping malls exemplify non-place. They are spatial products of increased mobility and can be placed within the mobilities turn paradigm not least of all through their exemplification. Non-place defined a new kind of space that did not conform to traditional notions. The main experience of these spaces is related to being in transit – being in between – and Augé indicates that the space of the traveller may indeed be the archetypical non-place (Augé 1995). In this sense there is a direct correlation to the mobilities turn and the notion that the in-transit experience – the journey – is more than the space between departure and arrival.

As a result of the frequency with which we find ourselves in these in-transit situations and the sheer numbers of diverse users in the spaces of mobility, a whole range of spaces that would previously not be considered public spaces, are now deemed so: transit nodes such as stations and airports, the modes of transit themselves such as trains (commuters) and the changing locus of cars from car park to road. Many of these spaces are huge, characterised by movement and often have blurred boundaries. The concept of space itself has changed from that of fixivity and stasis, of space defined via clear borders and demarcation. In contemporary cities, space and time together have created a dance of movement that is fluid. Mobility and the in-transit condition of contemporary societies has changed the *spatiality* of public space. It is no longer only fixed and settled, it is no longer the 'chair' for pause, but must also be considered as a dynamic place of exchange and flow.

Designing public spaces of mobility: rethinking requirements for public space and people

In a situation defined to a large degree by mobility and in which new social and spatial situations exist, it is crucial that an understanding of how the concept of space itself is perceived and experienced in this new situation is addressed. Through the mobilities turn, the perception of space changed from that of settled, fixed or pause(d) space to that of spaces that afforded movement, and were in fact defined by it. Movement occurs at different tempi and particularly, increased speeds have prompted new perceptions. Moving through space quickly emphasises the visual sense and negates attention to detail. In speed, we *see* the big picture, the broad strokes, the large lettering, the bright colours. A primacy of the visual sense occurs and use of our other senses is reduced. The premise of speed as a factor in the experience of space is also taken up in the notion of 'zoomscape', a term introduced by architectural historian, Mitchel Schwarzer. Schwarzer indicates that in movement, neither place as a fixed entity nor the sensorial experience of it are readily accessible. And in addition to this, we have, through the prevalence of media and transit, developed a distanced relationship to space.

> The experience of the built environment emerges on the go or in distant places, and, in either case, more and more through visual observation. The influence of place on our understanding of architecture is less pronounced than ever. The nonvisual senses,

> especially touch, play a diminishing role. Today, sight moves with the swiftness of vehicles or camera edits. It shifts off-site with camera images or, with vehicles, to a state one could call passing-site.
>
> *(Schwarzer 2004:20)*

François Ascher echoed the same message in his description of the need for signage on the Paris highways. Drivers are simply going too fast to visually make the connection between where they are on the highway and the part of the city where they want to get off. The intended destination is seen in passing. Instructions for how to get there – while you are (almost) there – are required (Ascher 1997). There is a link to be made here between the tempi of the spaces of mobility and the connection to, or distance from, them that we experience.

However, not all mobilities space is the space of speed. Public space also needs to be considered in other aspects of temporality: fast and slow public space, fast and slow users. That contemporary public space is related to mobilities does not necessarily mean that it is always fast space and always distant space. As indicated earlier in this chapter, mobility contains wait time and pause time and stop-over time and information regarding this is not only available via temporality, it is also present in materials. The spaces of fast mobilities afford fast movement through their materials and their qualities, e.g., asphalt, tile and smoothness, while carpeting, cobblestone and textured surfaces make speed more challenging and signal slowing down or even stopping. As embodied spaces of mobility, public spaces are spaces of experiencing, and their materiality makes different types and velocities of use possible (Smith 2015). In addition, it also gives clues as to how to use space by extending invitations to the users, promoting new ways of being together and providing more intimate and relational sensorial experiences.

New spatial lenses

The consideration of public space through mobilities is a new way of looking at public space – a consideration of how space is experienced and contact can occur in a world defined and created by mobilities. In a discussion of place and space, Tim Ingold in his book, *Being Alive: Essays on Movement, Knowledge and Description*, utilises the drawing of a circle to indicate the pathway along which movement is made. He states,

> My contention is that lives are led not inside places but through, around, to and from them, from and to places elsewhere (Ingold 2000a:229). I use the term *wayfaring* to describe the embodied experience of this perambulatory movement. It is as wayfarers then that human beings inhabit the earth (Ingold 2001a:75–84).
>
> *(Ingold 2011:148)*

For Ingold it is not *in* places, but *along* paths leading to and from them that there is the greatest possibility for experience.

> It [human existence] unfolds not in places but along paths. Proceeding along a path, every inhabitant lays a trail. Where inhabitants meet, trails are entwined, as the life of each becomes bound up with that other. Every entwining is a knot, and the more that lifelines are entwined, the greater the density of the knot.
>
> *(Ingold 2011:148)*

Drawing on the mobilities turn and applying this to contemporary public space and the unfolding of human existence that takes place there, the analogy of wayfinding is particularly significant and a useful image for the imagining of movement *and* connection, and in fact, new methods through which to comprehend. The imagery is implicitly mobile – addressing the movement of humans and indicating a current situation in 'mobile times', but also addressing human contact and the density of meaning that can occur through the crossing of paths. Continuing perhaps from the weaves of Ascher's example, Ingold's knots – again a technique for joining and connecting the threads of individual lives, now becomes groups, and addresses even more profoundly and fluidly, the notion of new types of connections and exchanges between people. 'Places, in short, are delineated by movement, not by the outer limits to movement' (Ingold 2011:149). In terms of the mobility turn, this perspective introduces a way of looking at the paths, or the spaces of movement, with a focus on the movement itself – the trails left by the 'movers' – the wayfarers – and points to the potential for contact – the density of experience made by the crossing and intertwining these trails. Interestingly, the notion that space is fluid and unbounded comes clearly to the fore when the borders defining traditional public spaces are dissolved and this emphasises the relationships between users and the experiential potential this holds.

Changing minds and tools: reflections and conclusions

The professions creating public spaces have been hard pressed to keep up with the changes that mobility has brought about to the social and spatial aspects of public space. Planners, designers and policymakers involved in the creation of contemporary public space require new ways and new tools with which to work with these new spaces, to understand new perceptions of them and to create experientially rich environments for the unfolding of human activity in urban settings deeply influenced by mobility. The design of public space has historically been undertaken by a variety of specific professionals: architects, urban planners, landscape architects, urban designers; however, the traditional borders between these professions have to a large extent disintegrated today. Add to this the multitude of adjacent professions that can be linked to public space: sociology, ethnography, geography, anthropology, psychology, philosophy to name a few, and the potential for cross-pollination of ideas and methods between professions is enormous. Inviting professional diversity in the groups designing public space is one of the most available ways of addressing and working with these new forms of public space and the societal roles they can perform. An example of this is Maarten Hajer and Arnold Reijndorp's partnership in penning *In Search of New Public Domain*, a book which maintained that the radical changes in contemporary urbanity had not been followed up with just as radical changes in the tools with which to design public space. Both professional backgrounds of sociology and urban planning were activated in a common 'call to action' providing a socially focussed intention and a practical toolbox consisting of three planning strategies and five spatial interventions – hands-on methods to be applied in reality to encourage real exchanges between diverse users of public space. While the strategies provide overriding ideas, the interventions provide more specific spatial actions on a smaller scale. These strategies and interventions activate inclusivity in diversity and identify spatial ways to encourage exchanges between different groups to take place.

> Public domain centres around experiencing cultural mobility: for the opportunity to see things differently the presentation of new perspectives, as much as the confrontation with one's own time-worn patterns. Being coerced to conform does not tally with this perspective of a properly functioning public domain. Being challenged to relate to others does.
>
> *(Hajers and Reijndorf 2001:116)*

The use of intervention as a device links to a more lived and embodied experience of public space – one that addresses the diversity of contemporary society and the notion that public space can – and should – provide an educating function – a way of telling and showing about each other. Intervention in itself indicates a kind of temporality, a less permanent, perhaps, approach which could provide more robust, flexible and inclusive public spaces. And it also leans towards a smaller-scale, more sensorially relatable entity in the large scale of space that activates materiality to give clues to use and ways of connecting.

The unfolding of stories, materialities and outside/inside views are examples of the potential connections that could be made. This borrows from ethnographic techniques and here the notions presented by Ingold are a case in point. Seeing the *publics* of public space through Ingold's optic as the movement, as the line, and requiring of the line maker, the wayfarer, an engagement with space and contact with the other wayfarers creating knots – densities of experience. The movement of people in space with a view to their connectedness in a new way is applicable to understanding public space in the mobility turn. A shift is made between the focus on *publics* and *space* to *publics* and *publics* with an understanding that publics *are* movement and that their movement creates densities of spatial and social experience.

Public space in a contemporary situation must provide a different and expanded function than that of meeting. Issues of mobility of moving people – creating new publics and new mobilities creating new spaces that lie outside of the traditional understanding of space and public space. In contemporary urbanity, space is more fluid, faster paced, less permanent – it is characterised by movement – comprised of movement – of moving publics and moving spaces – of the mobility of people and the spaces of mobility. And the design of public space must respond by being more flexible, inclusive and open in its offerings to the publics it may server rather than merely being a container. It must move rather than stand still.

References

Ascher, Francois. Modernities, Discontinuities and Urbanities: What Issues for European Towns? In *The European 5 Charter*, Paris, 1997.

Augé, Marc. *Non-Places: Introduction to an Anthropology of Supermodernity*. Verso, London, 1995.

Hajers, Maarten and Reijndorf, Arnold. *In Search of New Public Domain*. NAi Publishers, Rotterdam, 2001.

Ingold, Tim. *Being Alive: Essays on Movement, Knowledge and Description*. Routledge, London, 2011.

Jensen, Ole B. *Designing Mobilities*. Art and Urbanism Series, No. 4, Aalborg University Press, Aalborg, 2014.

John, Urry. *Mobilities*. Polity, Cambridge, 2007.

Leach, Neil. *The Anaesthetics of Architecture*. MIT Press, Cambridge, MA, 1999.

Lefebvre, Henri. Right to the City. In *Writings on Cities*, Elenore Kofman and Elizabeth Lebas (eds.), Blackwell, Oxford, 1996.

Lofland, Lyn. *The Public Realm: Exploring the City's Quintessential Social Territory*. Aldine de Gruyter, New York, 1998.

Madanipour, Ali. Public Spaces of European Cities. In *Nordic Journal of Architectural Research 2005: Volume 1*, 7–16. Aarhus, Denmark, 2005.
Schwarzer, Mitchell. *Zoomscape: Architecture in Motion and Media*. Princeton Architectural Press, New York, 2004.
Smith, Shelley. A Hop, Skip and a Jump: Examining the Experiential Potential in Contemporary Urban Public Space. In *Architecture and Stages of the Experience City*, Hans Kiib (ed.), 347–353. Aalborg University Skriftserie, Aalborg, 30, 2009.
Smith, Shelley. Discovering Urban Voids and Vertical Spaces. In *Performative Urban Design*, Hans Kiib (ed.), 146–154. Art and Urbanism Series, No. 2, Aalborg University Press, Aalborg, 2010.
Smith, Shelley. Exploring the Art of Urban Design as Sensorial Experience, in the Art of Urban Design – Part II, Vol. 168, Issue 6, *Proceedings of the Institution of Civil Engineers: Urban Design and Planning*, ICE Publishing, London, 2015.
Smith, Shelley and Steinø, Nicolai. The Appropriation of Public Space. In *Intervening Spaces: Respatialisation and the Body*, Nycole Prouse (ed.), 99–117. Brill Rodopi, Leiden, 2018.
Tuan, Yi-Fu. *Space and Place: The Perspective of Experience*. University of Minnesota Press, Minneapolis, MN, 1977.
Urry, John. *Sociologies beyond Societies: Mobilities for the 21st Century*. Routledge, London, 2000.

Online sources

https://en.oxforddictionaries.com/definition/settle last accessed 29.09.18 and www.etymonline.com/word/settle last accessed 29.09.18.
UN DESA – United Nations Department of Economic and Social Affairs www.un.org/development/desa/en last accessed 15.03.19.

32
URBAN TOURISM

Edward H. Huijbens and Gunnar Thór Jóhannesson

Introduction

As transport and communication technologies continue to progress and information becomes all-pervading it would seem as if our world is constituted purely by flows of which rapidly growing international tourism is one. Indeed, now dominant production factors, time, information and capital, remain highly mobile. This mobility is expressed in the ways people "invest" their time in travel and how online information presence is becoming vital to secure a place's seat in the minds of the globetrotting masses. At the same time growing awareness of local culture, place attachments and valuation are clearly observable. Thus, friction is manifest, e.g. in the search for meaningful and authentic place experiences by the very same masses. But whilst factors of production seem to continue to be borderless and the world of tourism booms, these only become so through grounded processes in particular places (Amin, 2002). This chapter will focus on urban tourism and the ways in which it manifests global mobility in Iceland.

Tourism plays a role in urban governance, affecting planning and thus impacting the city fabric and attitudes and livelihoods of its residents in diverse ways. Indeed, changes in the urban caused by rapid development of tourism are not always positive and expected by residents (Waddell, 2002) as for instance increasing concerns about overtourism in cities manifest (Milano, Cheer and Novelli, 2019). In this sense tourism is a "vehicle for transition, an integrated part of transitions, and a consequence of transition" (Müller, 2018, p.2). A key task remains to challenge the seemingly footloose nature of tourism flows and tourism investment in the city to sustain residents' place attachments. With emphasis on the role of infrastructure development in Iceland, this chapter shows how the creative city discourse is to facilitate mobilities and related infrastructure development, thereby tapping and stoking "the creative furnace inside every human being" (Florida, 2005, p.5). How can urban tourism become a platform for a host of mobilities sustaining the city in a global urban tourism hierarchy, where creativity is the norm, and at the same time introduce a multitude of new mobilities and ways of being within the city itself? From international airport development and ideas of Aerotropolis (Kasarda and Lindsay, 2011) to city-bikes and fusion restaurants, tourism adds flavour to our modern cities but not without potential detrimental side effects.

This chapter will proceed in two parts before a summary conclusion. Firstly, we will present an understanding of globalised tourism and how tourism mobility ties together localities and globalised flows. Our case is the city of Reykjavík, Iceland where we will explore how the city is linked with global tourism flows through the infrastructure developed to receive it and how that channels the flow of tourism through the city and to the attractions of the island at large. Secondly, we will present our take on the urban and how it is constituted and maintained through a dense network of mobilities, augmented by tourism; again, through our case study in Reykjavík and its emphasis on being a global cultural attraction and cauldron of creativity. A particular example thereof and a consequence of tourism mobilities is Airbnb.

The globetrotting masses

Through the democratization of travel, transport technology developments, ubiquitous internet use and growing global affluence, international tourism is booming. Globetrotting tourists are seen as part of the frictionless flow of capital, information, culture and goods, and cities competing in a global marketplace are meant to entice these. Indeed, some of the key attractions in global tourism are cities such as Paris, Bangkok and New York, but apart from these iconic urban nuclei, small cities vie for a share in the flow of international tourists (Richards and Duif, 2019). For them to catch their share of the flow, friction needs to be created, i.e. the flows need to be grounded and funnelled towards meaningful experiences offered in specific destinations.

Tourists to Iceland almost all arrive on medium to long-haul flights. Boundless open skies, which came to be with international aviation agreements post 1992, allow any airline to connect almost any places in the world. The concomitant growth in aviation and lower fares worldwide facilitate tourism mobilities globally. But each trip is meant to take you somewhere and this somewhere has travelled to the minds of the potential visitors through a myriad of wirings facilitating the mobilities of images, ideas, advertisements and social media communication. Once physically travelling to the place in question, these mobilities become palpable and move from an image in the mind to actual presence, which requires a different set of infrastructures. This infrastructure is the first tangible sign of friction for the mobile individual. In the case of Iceland this would be Keflavík International airport (KEF airport), adjacent to Reykjavík city. The airport is the hub which almost all inbound tourism to Iceland will go through and it is the only functioning year round gateway to the country. In 2018 just under 10 million passengers passed through there, up from under 3 million in 2012 (see Figure 32.1). As is made clear from the masterplan of the airport, outlining investment plans till 2040, this airport is to treble in size and capacity.

Figure 32.1 shows the post 2010 boom in inbound tourism to Iceland to date. The lines show foreign overnights in all accommodation establishments, passengers through KEF airport and foreign departure counts at KEF airport by the Icelandic Tourist Board (ITB). All three give a comprehensive idea of inbound tourism in Iceland. In addition, cruise ships also visit the island. In 2017, 135 cruise vessels berthed in Reykjavík carrying in total 128,000 passengers plus crew. Almost all passengers from these vessels are day trippers (excursionists) and thus not counted as inbound tourists. The only alternative means to aviation and cruising in getting to Iceland is by ferry from Denmark to the port of Seyðisfjörður. This ferry carried just over 22,000 passengers to Iceland in 2017 (Icelandic Tourist Board, 2019).

As Figure 32.1 clearly shows, inbound tourism takes off after 2010. KEF airport is a 45-minute drive from Reykjavík city. The airport is serviced by shuttle buses to the

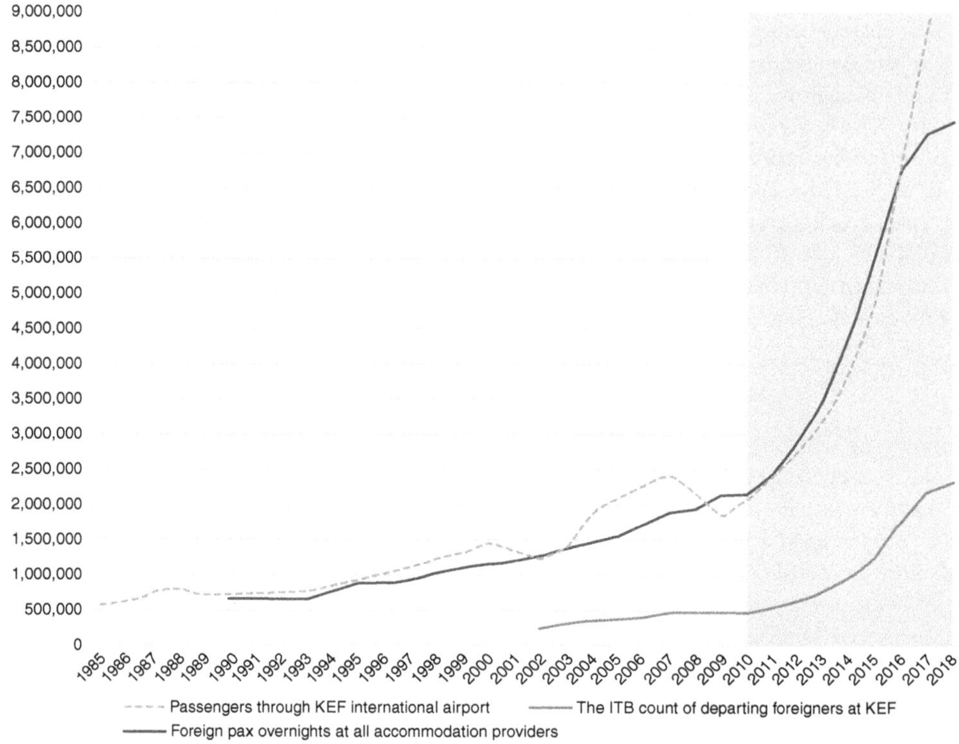

Figure 32.1 Inbound tourism in Iceland.
Source: Icelandic Tourist Board, 2019

city centre which are not linked up with the local system of public transport in the city or to the rest of the island. The only road from the airport will take these buses, tour buses and others, opting for e.g. rental cars, through the city. In addition, anyone wanting a domestic flight will need to visit the Reykjavík downtown airport. So in effect, almost all those visiting Iceland will use the city as their base of mobilities around the country. With year-round inbound tourism traffic growing every month, seasonality has all but disappeared in the city. Yet it persists throughout the country, demonstrating how access to the rest of the country is orchestrated by this disarticulated system of transport from the city heartland. To get to the rest of the country a rental car is necessary if fuzz free travel is the aim. In winter, with shorter holiday periods and more challenging weather and road conditions, visitors become city bound. Reykjavík thus sees year round tourism, transforming its service infrastructure, tourism amenities, accommodation and attractions. To more effectively funnel the inbound tourist into the country a proposed high-speed rail link is now being negotiated with the relevant municipalities en-route from the airport to Reykjavík. This would be the first ever railway of any capacity built in Iceland (Fluglestin, 2018), but again it will place all those coming from the airport squarely in the city centre of Reykjavík as of 2023, maintaining it as the actual platform for a tourist encountering Iceland.

The positive economic effects of tourism in cities have been recognised for long, leading to the rise of the city as a tourist destination (Crouch and Ritchie, 1999; Judd and

Fainstein, 1999). Visitors' movement patterns affect the scope of infrastructure, transportation development, product development, destination planning and the design of new attractions. To date, the city planning of Reykjavík and its link to Keflavík has been focused on accommodating the private car. This is part of a pan-European legacy of automotive hegemony "in which functionalism and notions of 'the modern city' inspired urban planning" (Gössling, 2017, p.3). This can most readily be seen through the fact that car rental in Keflavík is the most expensive in Europe as the only means to get from the airport, apart from the shuttle, is by private car (see Figure 32.2). In years 2017 and 2018 some signs of improvement to this can be noted, but not transforming the prevailing state to any degree.

Figure 32.3 shows the results of winter and summer surveys performed by the Icelandic Tourist Board amongst departing tourists from 1999 till 2016 when queried on what means of transport they used during their stay in Iceland. As can be seen, rented cars comprehensively overtook organised coach travel in the year 2016. The number of registered rental cars has grown by 20% year on year since 2012, whilst inbound tourism numbers grew by 24% in the same period. At the same time the number of coaches in the country grew by 8.1%, partly attributable to the growth in cruise ship arrivals, but the dominant mode of travel from these are half or whole day excursions on a coach from and to the port of call.

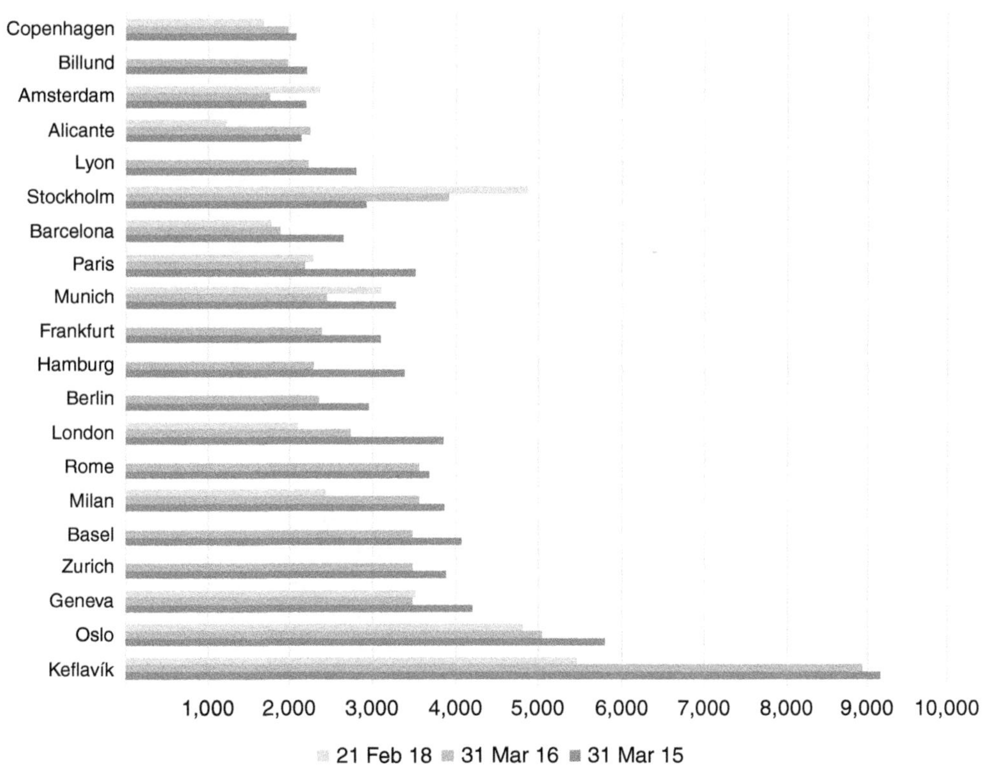

Figure 32.2 Rental car prices from Keflavík airport, compared to select European cities.
Source: touristi.is

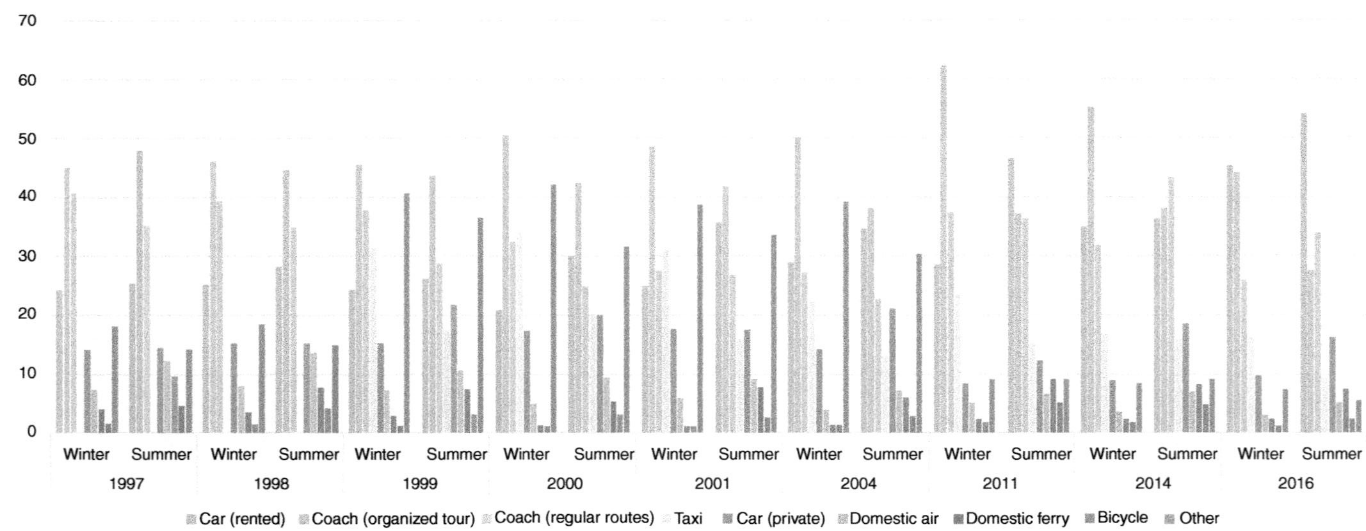

Figure 32.3 What means of transport did you use during your stay in Iceland?, 1996–2016.

Source: Icelandic Tourist Board, 2018

As Gössling (2017) will remind us;

> Movement, in contemporary society, is 'at the center of constellations of power, the creation of identities and the microgeographies of everyday life'. Consequently, movement is no longer a choice, rather it is a necessity; opportunities to be mobile shape society as much as mobility patterns are now shaped by social norms and expectations.
>
> *(p.34, citing Cresswell, 2011, p.551)*

The expectations of many visitors to Iceland, not least during the summer season, is mainly to complete the ring-road, a 1,332 km road that circles the island. For that, a rental car is imperative. The ring-road is not wholly serviced with public transport, so doing a bus trip around is impossible unless bought as a tour package. What Arnason (2015) calls "the lure of the ring" shapes the travel patterns on the island and it starts in KEF. Being mobile in the rental car allows you to explore and visit places on the ring according to your own schedule, laid out on your individual smart device, but with consequences.

> This "self-generated" evolution of the Ring Road from basic transport venue to scenic route has arguably had serious consequences for tourism management in Iceland, e.g. by undermining efforts to disperse the ever-growing numbers of visitors more evenly around the country. Paradoxically, the increased mobility offered by self-driving is offset by the urge to "tick off" all the main scenic attractions dotted alongside the Ring, which in turn leads to ever more congestion in these already over-loaded areas.
>
> *(Arnason, 2015, p.199)*

Indeed the democratisation of travel, decisions pertaining to KEF airport development and ubiquitous internet use are impacting modes of mobilities in Iceland. However, transport technology and infrastructure developments lag considerably behind, leading to several "wicked problems" in the context of tourism as identified by Urry (2016, p.132). The main "wicked problem" faced in the city of Reykjavík and the sites on the ring-road to be "ticked off" is increasing concern that the city is becoming a victim to overtourism, manifest in congestion, with rental cars and tourist coaches crowding destinations and city streets, in particular in the downtown area and the hollowing out of the city centre as a residential area. The response from the city of Reykjavík can be seen in initiatives such as the ban on coaches entering the downtown area, effective from 15 July 2017 (see: https://busstop.is/) and the introduction of city-bikes to rent by one of the two major Icelandic airlines; Wow air (see: https://wowcitybike.com/), sadly now bankrupt. Elsewhere, the only response has been to pile gravel on grassland to extend parking lots as close as possible to the attraction itself.

The urban

The urban is constituted and maintained through a dense network of mobilities, tourism being an integral part thereof. Cities generally provide a dense amalgam of tourism services and a base from which to constitute personal networks of mobility as we have demonstrated in the case of Reykjavík above. In this respect, access to and access within the city prove to be fundamental to a city's status when it comes to globalised urban tourism. But moreover, attractions are needed and a particular vibe (Judd and Fainstein, 1999; Page, 1995; Vanhove,

2017). Tapping the perceived benefits of tourism, many cities have developed access through airports and high-speed mass transits and attractions by reconstructing their downtowns and waterfronts to accommodate cultural vibrancy and creativity.

The "authority of infrastructure" which Easterling (2016, p.71) explains as undeclared in the dominant stories portraying it, is where the balance of flow and friction becomes manifest in the urban landscape; that which facilitates mobility and yet sustains the urban fabric. Cities have promoted various initiatives and campaigns to support and expand the growth of tourism, mobilising the cultural, social and physical capital present in the process. The expansion and prioritisation of the tourism sector in urban development strategies and the manifold impacts of the growing presence and prevalence of tourism on urban spaces and on the life of residents have generated new contestations of, and conflicts over, the visitor economy and tourism development in cities. These conflicts often seem to revolve around the negative effects tourism has, or is believed to have, on neighbourhoods (Colomb and Novy, 2014; Milano, Cheer and Novelli, 2019).

The pervasiveness of the private car and problems of congestion when it comes to tourism mobilities in Iceland and the city of Reykjavík result from the emphasis on road infrastructure manifesting to date the undeclared pan-European legacy of automotive hegemony. The road connecting Keflavík with Reykjavík has been expanded and improved several times, but always to accommodate private cars, not any other means of mobility. Above and beyond ease of access, the city also strives to promote why people should come. The redevelopment of the city's waterfront and downtown area is emblematic thereof and they are couched in stories of creativity and cultural vibrancy. Harpa, the culture and event centre completed 2011, despite Iceland's financial meltdown in 2008, proudly proclaims in its online adverts; "Experience award winning architecture. Discover the wonders of Harpa, Reykjavik" (see: https://en.harpa.is/). Adjacent, the Edition hotels chain of Marriott is opening one of its "luxury boutique" hotels in 2019. This particular brand of hotels are located in gateway cities worldwide, such as Bangkok, Abu Dhabi, Bodrun and Shanghai, and they are meant to make you "feel different because they make you feel something". Feeling something here refers to you being able to "rub shoulders" with locals and experience "authentic design" (see: www.editionhotels.com/the-idea/). These emblematic figures of a "placeless aesthetic" are symbolic rather than anything else (Frampton, 2007). They are meant to convey local connectedness for the globally footloose, i.e. tourists, grounding a globalised culture of commerce and design (Eldemery, 2009). The Harpa/Edition complex is a large-scale building project meant to secure Reykjavík a place in a global city competition for both residents and tourists. It is part of a process of spectacularisation of the urban environment often with the paradoxical effect of infrastructure becoming more homogeneous (Ponzini, Fotev and Mavaracchio, 2016).

A running theme of this spectacularisation in infrastructure development is the notion of "creativity", which also is closely related to urban tourism development. What is to spur your mobility to a certain place is the sense that it is a bustling cauldron of creativity. City planners and administrators adopt the notion claiming that the developed urban infrastructure will accommodate creative people and foster creativity (Gibson, 2010; Landry, 2000; Richards, 2014). This creativity is grounded in a wider discussion on the role of culture for innovation and economic development, most often emphasising how the creative capacity of individuals is a decisive factor for the prosperity and competitiveness of regions and places (Du Gay and Pryke, 2002; Florida, 2005; Gibson, 2010; Gregson, Simonsen and Vaiou, 2001). Tanggaard (2013) points out that this innovation conception of creativity casts individuals as entrepreneurs and responsive to societal changes. Accordingly, some seem to be more creative than others, for instance the "creative classes" (Florida, 2005), which are set to attract businesses and create jobs as well as giving places a necessary flair of "coolness". Also tourism is being used to vouch for a city's creative credentials. If tourism is booming and people are coming in droves, surely the city is a creative cool place? Thereby even the smallest and most remote cities are projected onto the global tourism map (Richards and Duif, 2019; Romão, Kourtit, Neuts and Nijkamp, 2018). However, creative cities strategies are often controversial. According to Borén and Young (2013, p.1801) "Certain forms of creativity become valued by urban elites and enjoy support from public funding, often with an international audience in mind". Thus, creative city programmes may be boiled down to particular gentrification projects, promoting a certain set of values of what is a positive development and who are part of that process, sustaining present class segregation or even inventing new lines of inequality in the city.

Most city developers abhor "mass" tourism and have adopted a positive disposition to creativity and coolness. City marketing and branding is about enticing the culturally savvy and enlightened tourist, making the city into an eventful and creative entrepôt to their respective countries or hinterland (Alberti and Giusti, 2012; Alvarez-Sousa, 2018). Beyond the above outlined classic waterfront development in the city of Reykjavík, the concept of creativity is put forth in tourism policy and city planning documents in various contexts, albeit loosely defined. In concrete terms, the development of Reykjavík as a creative city is most evident on the one hand in efforts to extend the city centre along the harbour area, creating a more vibrant city environment with a mix of traditional harbour facilities, tourist shops, art galleries and restaurants and on the other in investing in various cultural events such as book fairs and concerts. The increase of tourism is both a reason and an outcome of these developments. More particularly Airbnb is a creative city ideal where the savvy urban tourist, motivated by "deals" and "ideals" assemble their own experiences through their mobile online devices, motivated by word of mouth. This particular example of grounded tourism mobilities is certainly impacting the urban fabric in Reykjavík.

Airbnb is an example of a direct platform of communication and commerce between the consumer and producer, allowing for variety and creativity by both. This platform is most certainly disrupting the role of the established accommodation provider (Guttentag, 2015), but moreover it is disrupting urban fabrics worldwide. Adamiak (2018) reports that Reykjavík ranks eight among European cities when it comes to number of Airbnb listings per 1,000 inhabitants. The only capital city to rank higher is Lisboa (#6) whilst the others in the top ten are tourist cities such as Venice (#4), Split (#2), Batumi (#1) and Marbella (#3). Landsbankinn Economic Research obtained data in 2017 on Airbnb operations in Reykjavík through the analyst firm Airdna, the same source as Adamiak (2018) used, which gathers current information from Airbnb's website. They estimate that Airbnb stays numbered over 1.1 million in Reykjavík alone in 2016, exceeding 40% market share in accommodation there in the summer of 2016 and with an estimated 46 million EUR turnover in the city (Landsbankinn, 2017). Compared to the accommodation nights represented in Figure 32.1, the 1.1 million represent 16% of the total inbound tourism overnights at established accommodation providers. According to the Directorate of Internal Revenue, 4% of registered properties in Iceland are listed on Airbnb, although this has to be an overestimate as individual rooms for let are considered as whole property (Reykjavíkurborg, 2017, p.20). Arguably Airbnb has helped people deal with the aftermath of the economic crisis in 2008 and created value for property owners, but as another bank, Íslandsbanki (2017), argue in their annual analysis of tourism, Airbnb has "played a large role in hiking property prices in Reykjavík" (p.23) as the number of apartments rented far

Table 32.1 Rental price index in Reykjavík, year on year (Jan–Jan) increase in %.

Year	*% growth from Jan previous year*
2012	11
2013	7
2014	8.5
2015	8
2016	5
2017	11.5
2018	9

Source: Registers Iceland, 2018

exceeds the number of apartments built in the same period. In the same way rental prices have steadily gone up in the city, also attributed to Airbnb (see Capacent, 2016, p.80). Table 32.1 shows the development of the rental price index of the capital region as compiled by Registers Iceland (2018).

As Table 32.1 clearly demonstrates, rental prices have gone up fast. The consequence is a serious housing shortage for low-income households and concomitant social problems, such as serial relocations of single parent households. Tackling the housing issue in Reykjavík was the key debate in the spring 2018 municipal elections and the role of tourism in contributing to the problem was recognised. So another transformation wrought for the benefit of friction is one where housing becomes a commodity for tourism rental, pushing locals out, in particular those socio-economically disadvantaged.

With this critical analysis of the infrastructure formulae manifest when it comes to grounding tourism mobilities, we have shown how Reykjavík's waterfront and downtown have become ubiquitous markers of a placeless aesthetic firmly aligned with the jargon of creativity in city planning. At the same time, housing is marketed online for tourists allowing them to tailor their experiences and capitalise on the best deals. Both these processses have ramifications for the urban fabric and society at large. Urban tourism is talked about in terms of guilt free mobilities, creativity and coolness but in reality it is based on business as usual, turning a profit for those privy to the global world of flows.

Conclusion

In the opening of this chapter we asked how urban tourism can become a platform for a host of mobilities sustaining the city in a global urban tourism hierarchy and at the same time introducing a multitude of new mobilities and ways of being within the city itself? We have shown how through facilitating mobilities for tourism through airport master planning and expanding road networks the urban fabric of Reykjavík is undergoing transformation. The city vies for a spot in the global urban tourism hierarchy, introducing new developments, but set within the prevailing paradigm of automobile hegemony, the consequences are manifest i.e. in congestion and housing market upsets.

"[C]onnectivity is the platform for fuller societal development" (Khanna, 2016, p.341) and "[m]obility thus ought to be one of the paramount human rights of the twenty-first century" (p.359, see also Jensen, 2005, p.144). The individual traveller is at the heart of our ever more interconnected world, but which mobilities are provided to the visitor and how do they become of paramount importance when it comes to the future of our cities. According to a McKinsey (2016) report on the future of mobility, electrification, shared mobility, and autonomy are the key mobility trends. According to the report, effective mobility relies on efficient public transit, ease of cycling and walking and shared modes of transport facilitated by ubiquitous internet and smartphone use. This will limit congestion and pollution in the urban, but requires coherent public sector interventions to develop the necessary infrastructure and modes of transport. What is readily visible in Iceland, however, is the fact that infrastructure development is lodged within the pan-European legacy of automotive hegemony (Gössling, 2017, p.3). Rental cars have become the dominant mode of transport and although ideas are on the table for rail links and integrated public transport, none have materialised. In addition, as Urry (2016) will remind us, "But this fast-mobility city will only develop if a new post-carbon energy system is innovated and implemented around the world" (p.140). Beyond the contentious issues of tourism's carbon footprint, questions remain whether growth in mobility is good for happiness, wealth and quality of life (Whitelegg, 2015)? But this we leave for another chapter.

We claim that for urban tourism to be of benefit to the cities it needs to be delivered from the ground up, from the citizens themselves hosting those visiting and catering to their needs. The urban as emergent through global flows is a political project that needs to be grounded amidst the residents and their visitors. Thereby urban tourism can contribute to healthier globalised communities, and the mobilities outlined by McKinsey (2016) are more likely to cater to this.

References

Adamiak, C. 2018. Mapping Airbnb supply in European cities. *Annals of Tourism Research*, 71(C), pp.67–71.

Alberti, F. G. and Giusti, J. D. 2012. Cultural heritage, tourism and regional competitiveness: The Motor Valley cluster. *City, Culture and Society*, 3(4), pp.261–273.

Alvarez-Sousa, A. 2018. The problems of tourist sustainability in cultural cities: Socio-political perceptions and interests management. *Sustainability*, 10(2), pp.2–30.

Amin, A. 2002. Spatialities of globalisation. *Environment and Planning A*, 34(3), pp.385–399.

Arnason, T. 2015. The lure of the ring: Merry-go round turned vicious circle? In: Icelandic Tourism Research Centre ed., *Responsible Tourism? Book of Abstracts. The 24th Nordic Symposium in Tourism and Hospitality Research* [online]. Available at: www.rmf.is/static/research/files/ns-book-of-abstracts-websmallpdf [Accessed 7 June 2018].

Borén, T. and Young, C. 2013. Getting creative with the 'creative city'? Towards new perspectives on creativity in urban policy. *International Journal of Urban and Regional Research*, 37(5), pp.1799–1815.

Capacent. 2016. Greining á fasteignamarkaði. Reykjavíkurborg 2016. [online]. Available at: https://reykjavik.is/sites/default/files/ymis_skjol/skipogbygg/skjol/2016_10_12_capacent_fasteignagreining_fyrir_reykjavik.pdf [Accessed 30 April 2018].

Colomb, C. and Novy, J. 2014. Protest and resistance in the tourist city. [online]. Available at: www.geschundkunstgesch.tu-berlin.de/fileadmin/fg95/Veranstaltungen/2014/Program___Protest_and_Resistance_in_the_Tourist_City__.pdf [Accessed 7 June 2018].

Cresswell, T. 2011. Mobilities I: Catching up. *Progress in Human Geography*, 35(4), pp.550–558.

Crouch, G. I. and Ritchie, J. B. 1999. Tourism, competitiveness, and societal prosperity. *Journal of Business Research*, 44(3), pp.137–152.

Du Gay, P. and Pryke, M. eds., 2002. *Cultural Economy: Cultural Analysis and Commercial Life*. London, Thousand Oaks and New Delhi: Sage.

Easterling, K. 2016. *Extrastatecraft. The Power of Infrastructure Space*. London: Verso.

Eldemery, I. M. 2009. Globalization challenges in architecture. *Journal of Architectural and Planning Research*, 26(4), pp.343–354.

Florida, R. 2005. *Cities and the Creative Class*. New York: Routledge.

Fluglestin. 2018. Fluglestin. [online]. Available at: www.fluglestin.is/ [Accessed 27 April 2018].

Frampton, K. 2007. *Modern Architecture. A Critical History*. London: Thames & Hudson.

Gibson, C. 2010. Guest editorial – Creative geographies: Tales from the 'margins'. *Australian Geographer*, 41(1), pp.1–10.

Gössling, S. 2017. *The Psychology of the Car Automobile Admiration, Attachment, and Addiction*. Amsterdam: Elsevier.

Gregson, N., Simonsen, K. and Vaiou, D. 2001. Whose economy for whose culture? Moving beyond oppositional talk in European debate about economy and culture. *Antipode*, 33(4), pp.616–646.

Guttentag, D. 2015. Airbnb: Disruptive innovation and the rise of an informal tourism accommodation sector. *Current Issues in Tourism*, 18(12), pp.1192–1217.

Icelandic Tourist Board. 2018. Ferðavenjur erlendra ferðamanna. [online]. Available at: www.ferdamalastofa.is/is/tolur-og-utgafur/kannanir-og-rannsoknir/ferdavenjur-erlendra-ferdamanna [Accessed 7 June 2018].

Icelandic Tourist Board. 2019. Tourism in Iceland in figures. [online]. Available at: www.ferdamalastofa.is/en/recearch-and-statistics/tourism-in-iceland-in-figures [Accessed 8 April 2019].

Íslandsbanki. 2017. *Íslensk ferðaþjónusta*. Reykjavík: Íslandsbanki.

Jensen, A. 2005. The institutionalisation of European transport policy from a mobility perspective. In: T. U. Thomsen, L. D. Nielsen and H. Gudmundsson eds., *Social Perspectives on Mobility*. Aldershot: Ashgate, pp.127–154.

Judd, D. R. and Fainstein, S. S. 1999. *The Tourist City*. New Haven: Yale University Press.

Kasarda, J. and Lindsay, G. 2011. *Aerotropolis: The Way We'll Live Next*. New York: Penguin Books Limited.

Khanna, P. 2016. *Connectography: Mapping the Global Network Revolution*. London: Weidenfeld & Nicolson.

Landry, C. 2000. *The Creative City: A Toolkit for Urban Innovators*. London: Earthscan.

Landsbankinn. 2017. Hvað er efstá baugi í ferðaþjónustunni? Mörg brýn verkefni framundan. [online]. Available at: https://umraedan.landsbankinn.is/umraedan/samfelagid/ferdathjonusta-2017/folk-ur-ferdathjonustunni/ [Accessed 28 April 2018].

McKinsey. 2016. *An Integrated Perspective on the Future of Mobility*. New York: McKinsey & Company and Bloomberg.

Milano, C., Cheer, J. M. and Novelli, M. eds., 2019. *Overtourism: Excesses, Discontents and Measures in Travel and Tourism*. Abingdon: CABI.

Müller, D. K. 2018. Tourism and transition. In: D. K. Müller and M. Więckowski eds., *Tourism in Transitions*. Cham: Springer, pp.1–20.

Page, S. J. 1995. *Urban Tourism*. London: Routledge.

Ponzini, D., Fotev, S. and Mavaracchio, F. 2016. Place making or place faking? The paradoxical effects of transnational circulation of architectural and urban development projects. In: A. P. Russo and G. Richards eds., *Reinventing the Local in Tourism: Producing, Consuming and Negotiating Place*. Bristol: Channel View, pp.153–170.

Registers Iceland. 2018. Vísitala íbúða- og leiguverðs á höfuðborgarsvæðinu. [online]. Available at: www.skra.is/markadurinn/talnaefni/visitolur-kaups-og-leiguverds/ [Accessed 28 April 2018].

Reykjavíkurborg. 2017. *Skýrsla starfshóps um heima- og íbúðagistingu*. Reykjavík: Reykjavíkurborg.

Richards, G. 2014. Creativity and tourism in the city. *Current Issues in Tourism*, 17(2), pp.119–144.

Richards, G. and Duif, L. 2019. *Small Cities with Big Dreams: Creative Placemaking and Branding Strategies*. London: Routledge.

Romão, J., Kourtit, K., Neuts, B. and Nijkamp, P. 2018. The smart city as a common place for tourists and residents: A structural analysis of the determinants of urban attractiveness. *Cities*, 78, pp.67–75.

Tanggaard, L. 2013. The sociomateriality of creativity in everyday life. *Culture and Psychology*, 19(1), pp.20–32.

Urry, J. 2016. *What Is the Future*. Cambridge: Polity Press.

Vanhove, N. 2017. *The Economics of Tourism Destinations: Theory and Practice*. London: Routledge.

Waddell, P. 2002. UrbanSim: Modeling urban development for land use, transportation, and environmental planning. *Journal of the American Planning Association*, 68(3), pp.297–314.

Whitelegg, J. 2015. *Mobility: A New Urban Design and Transport Planning Philosophy for a Sustainable Future*. New York: Straw-Barnes Press.

33
THE AIRPORT CITY

Claus Lassen and Gunvor Riber Larsen

Introduction

The airport's influence on cities has been growing since the Second World War. Today

> airports are developing into small-scale global cities in their own right, places to meet and do business, to sustain family life and friendship, and to act as a site for liminal consumption less constrained by prescribed household income and expenditure patterns.
>
> *(Urry, 2007:138)*

More and more, cities have also started to become like airports, organised and designed around various forms of mobilities (Kasarda & Lindsay, 2011; Urry, 2007). Moreover, air travellers themselves make 'a city in the sky'. According to FlightAware, there are an average of 9,728 planes with 1,270,406 people in the sky at any given time (www.travelandleisure.com/airlines-airports/number-of-planes-in-air). In various ways, airports seem to have created new forms of urbanism (Hirsh, 2016) that cannot be ignored or overlooked in social science.

Advocates of the airport city concept have argued that the airport is amongst the most characteristic elements of metropolitan areas and therefore needs to move into the core of future urban development (Güller & Güller, 2002:9; see also Hirsh, 2016; Kasarda & Lindsay, 2011). However, the notion of the airport city is relatively new, and is yet to be fully defined, while its uses have yet to be fully explored. And additionally, it also raises a number of new dilemmas and problems in relation to noise, environmental and land-use implications related to such an airport city driven development (see Lassen & Galland, 2014). Questions such as what an airport city is, why we need this concept and whether the concept can be contested are discussed in this chapter. The ambition of the chapter is not to give a complete definition of the airport city concept. The goal is rather to identify and review a number of the various concepts and ideas that recently have contributed to the frame of such a perspective on urban development in different ways.

The chapter commences with a brief introduction to the foundations of aeromobilities research, as this is the theoretical and methodological basis for understanding and analysing airport cities. Secondly, the changing historical role of airports within the contemporary society is explored. Thirdly, a number of conceptual notions of the airport city are analysed, including some of the more problematic perspectives of the concept. Here we argue for a holistic, interdisciplinary and future-orientated approach to airport city development. Finally, the chapter's concluding remarks are made.

Airport development through the lens of aeromobilities research

Conventional mainstream air transport research is founded on a rationality of 'predict and provide' (Bloch, 2018; Lassen & Galland, 2014). Thus, traditionally, research has focused on estimating air transport growth to provide the required infrastructure and to determine aircraft supply and airport capacity. Conversely, Cwerner et al. (2009) formulated an agenda for a new aeromobilities research (see also Lassen, 2006; 2009) as a part of the 'mobilities turn' (Cresswell, 2006; Jensen, 2013; Urry, 2007). This new vein of aeromobilities research focuses particularly on how aeromobilities are produced, reproduced, conducted and regulated in relation to various spaces, networks, systems and environments, as a way of 'opening' the 'black box' of flying. This theme in aeromobilities research also indicates a need to bridge multiple scales connecting international air systems to particular local urban transformation processes and their consequences (Jensen & Lassen, 2011). In this line of aeromobilities thinking, the airport offers 'laboratory conditions' (Fuller & Harley, 2004) for analysing and understanding moving societies (Urry, 2007) and can be seen as almost 'paradigmatic localities' (Kesselring, 2009). Elsewhere, Lassen (forthcoming) has identified the following characteristics of large international airports:

- global transfer points in the global economy and key features of infrastructure in an increasingly cosmopolitan world (Fuller & Harley, 2004; Kesselring, 2009);
- international borders, an interface between the territory of nation-state and the international zone (Kesselring, 2009; Kloppenburg, 2013);
- often run by the logic of private organisations. Increased commercialisation of the airport is reflected in the changing role of the airport from a mostly passive (often) state-governed traffic hub to 'a mobile actor' itself, trying to influence the production and consumption of aviation (Bloch, 2018);
- not an ahistorical and interstitial non-place, even though it may seem so on the surface for the air traveller or the first-time observer (Adey, 2006; Cresswell, 2006; Urry, 2007);
- a highly divided and stratified space. Airports consist of a front state visible for global air travellers, but they also involve a complex, highly material and invisible backstage (Goffman, 1959; Jensen et al., 2019; Urry, 2007);
- social, economic and geographical outreach is difficult to precisely define. Even though airports do have relatively fixed coordinates, their vectoring ensures that it is hard to know where an airport ends and something else begins (Merriman, 2007);
- not just passive players in globalisation processes and infrastructural extension, but historically influential in terms of governance, airline service and regional economics. They also mediate the effects of the processes of globalisation on individual places and shape global processes (Adey, 2006; Cidell, 2006).

From this perspective, contemporary airports are much more than a way to move from place A to place B (Jensen, 2013). Therefore, they need to be explored as an integrated part of modern society. Aeromobilities research is then an attempt to move away from 'seeing airports as simply evidence of standardization brought by globalization and aims instead to re-localize airports in their specific cities' histories … as a way to re-contextualise the airport topic both in a temporal and spatial framing' (Roseau, 2012:35 in Bloch, 2018:61). Therefore, in the following pages, we will explore the airport's social development from the early pioneer phase to the present emergence of the airport city as a framework for future urban development.

The history of airports and cities

The following section will mainly draw on the work of Roseau, who identifies six urban narratives related to historical airport development (2012:36–48). The first is what she terms 'the spectacle of aeromobility'. This time period covers the 1900s, from the beginnings of human flight, when the airport consisted of temporary fields for spectacular events of flying over cities and their peripheries (Roseau, 2012:36). Simonsen (2012) talks here about 'the aviation neck', which indicates that flying in this early stage was primarily defined socially by meeting and watching aircraft up in the sky together. These 'flying meetings' were highly publicised by the mass media, which gave status to aeromobility and built up a collective understanding and acceptance of flying across society (Roseau, 2012:37). The second narrative is labelled 'the metropolis of the future', in which small airfields and aerodromes sprouted up (the 1910s and 1920s). Debate started on the manner in which the city would be recast through emerging forms of mobility. Aeromobility was at this stage considered the transport mode of the future, and future buildings were imagined with lift-garages for 'aerial motor cars' (Roseau, 2012:38). The third narrative is 'the flagship monument', in the 1930s, when airports became a core component of countries' and large cities' modernisation strategies, which could also be seen in the design of international airports (Roseau, 2012:41). The fourth narrative begins after the Second World War, when airports started to develop into 'air cities' outside of traditional city walls (Roseau, 2012:44). This stage was labelled by Roseau as the 'urban showcase', in which 'not just the planes, but the airport as a whole … was getting media attention and being turned into a suburban attraction' (Roseau, 2012:44). Generally, in the first stages of airport history, the airport was mono-modal and focused primarily on transporting people and goods from one place to another (Urry, 2007:137). The fifth narrative can be described as the 'jet age', a period of time largely characterised by oil price shocks, environmental degradation, and terrorism that all in various ways undermined the model of the perfect airport showcase and made uncertainty a key component in airport development. Moreover, the rise of the hub airport combined with the growing complexity of passenger transport formalities, safety issues and expectations undermined the principle of splitting up airports into smaller terminal buildings to 'recentralize' facilities (Roseau, 2012:44). The development of airports also meant increasing interconnection between different modes of travel (planes, trains, metro and cars), creating a situation where 'the airport was no longer isolated and specialised but developed into a multimodal hub' (Urry, 2007:137). The final narrative can be labelled the 'metropolitan archetype'. Today, the airport no longer lies outside the city, but inside a constantly changing metropolis; thus, it has become one of the dominant institutions in global cities (Roseau, 2012:46) and acts as a 'multifunctional commercialised global hub' (Urry, 2007:139):

> Airport Cities emerg[ed] at the turn of the 1990s as new huge metropolitan fragments. Whether located in, on the top of, near, outside, or far from its 'host city', the airport always develops in a symbiotic relation to the urban environment in which it is situated, which it serves and on which it depends.
>
> *(Roseau, 2012:50)*

These developments signified a double movement towards a 'civilization of airports' and an 'aeroportization of cities' (Roseau, 2012:48), where a new awareness of the airport's urban influence and characteristics can be seen amongst important urban actors, with significant implications for airport development, including the increasing tendency of airports defining themselves as airport cities in their development strategies.

Towards the airport city

The notion of the airport city extends the classical understanding of an airport as a system of runways and buildings for take-off and landing that provides facilities for passengers and civil aircraft maintenance towards a new airport business model that also includes a strong focus on non-aeronautical commercial facilities and activities. As Boucsein et al. pointed out, Güller and Güller (2002) were some of the first to articulate an 'airport driven' or 'airport-related' development through an urban model which they conceptually termed an 'airport city' (Boucsein et al., 2017:12):

> Airports are not just airports any more. Forgoing their status as simple traffic machines, airports can rightly be considered as decisive for the transformation of the metropolitan area. Propelled by a series of strategic investments, they have assumed a key position in high speed train and railway networks, a position until recently reserved for central stations alone. Being the undisputed interfaces of entire European regions, airports become centres of activity within them, new regional development poles, or simply airport cities.
>
> *(Güller & Güller, 2002:5)*

In this way, the movement towards the development of airports as airport cities relates to the transformations that have taken place over the last few decades; airports can no longer be seen as 'just' airports, but have instead 'become centres on their own, in the most radical cases pacemakers of entirely new cities' (Güller & Güller, 2002:29; see also above). One of the main ideas evident in this movement is exploiting new business opportunities that the position as a multimodal interchange hub of the region creates for many of the growing airports around the world; specifically, the development of transport infrastructure and real estate cannot be seen in terms of separate elements anymore. However, as Güller and Güller pointed out, the problem seems to be that 'the making of' airport cities (and airports) is still done from a relatively limited airport point of view (2002:5). A core element of the airport city concept is the 'outstanding urban accessibility' that airports create for the city and urban region, in which they can be seen as the 'central stations of the 21st century'. In this view, landside access is indispensable for sustaining airport growth, and therefore airport cities need similar access standards to those of cities. They need to be looked at as new regional and international interchange nodes that contribute to contemporary metropolitan conditions (Güller & Güller, 2002:10). This implies that the development of airport cities can be seen as an urban planning task. The development of airport cities necessitates that planners and scholars move away from pure 'technical airport planning' towards an urban planning approach, with a particular focus on integrated land use and transport planning (a shift that also more generally can be identified in the field of mobility planning). This also means that airport cities can no longer remain white spots in the development plans of municipalities and regions (Güller & Güller, 2002:11). This concept has, from Güller and Güller's perspective, the potential to create a high-quality accommodation of airport spin-offs, and can be seen as an attempt to move to the forefront of the discussion on future urban strategies (Güller & Güller, 2002:9). The concept of the airport city has developed as a version of future airport development, and a number of airports around the world have in parallel started to implement the idea of an airport city as a core element in their development strategies and business models. The most notable example is Amsterdam Airport Schiphol (see www.schiphol.nl/en/schiphol-group/page/amsterdam-airport-schiphol), but others, including Dubai International Airport and Singapore Changi Airport can also be interpreted as airport cities.

Aerotropolis – an urban airport region

Notably, Kasarda and Lindsay conceptualised the airport city as a new business model, which they termed 'aerotropolis' (2011:176). They argued that the future relationship between cities and airports will change the historical configuration that has placed the city in the centre of a metropolis and the airport on the periphery: the airport now will be at the centre, and the city will be built around it (Kasarda & Lindsay, 2011). Kasarda and Lindsay defined the aerotropolis as an 'airport-integrated region, extending as far as sixty miles from the inner clusters of hotels, offices, distribution and logistics facilities' (Kasarda & Lindsay, 2011:174). The social context of aerotropolis is increasing globalisation and inter-city and inter-regional competition organised around networks and driven by the internet (Kasarda & Lindsay, 2011:175–176). In such a globalised world, 'cities now connect more easily to each other than to the towns and villages that lie just beyond their borders or to the national capitals' (Kasarda & Lindsay, 2011:175). This also means that airports can no longer be seen as only a pure piece of transportation infrastructure, or as accessories of a municipality. The aerotropolis is a city itself:

> An aerotropolis isn't an airport ... and building one isn't a matter of having the longest runways or the largest landmass. Frictionlessness is the product of a whole host of attributes, many of which are invisible: tariff-free trade zones, faster customs clearance, fewer and faster permits, and a right-to-work workforce that knows what it's doing. It's the way you reduce time, the way you reduce space .... The aerotropolis is where the elastic mile, the friction of space, community without propinquity, and trade routes all come together.
>
> *(Kasarda & Lindsay, 2011:168)*

This illustrates how friction of space is a key element in Kasarda and Lindsay's thinking. They saw the aerotropolis as the nucleus of a range of new economic functions, 'with the ultimate aim of bolstering the city's competitiveness, job creation, and quality of life' (Kasarda & Lindsay, 2011:174) in a world of increased globalisation. Speed can kill places, Kasarda and Lindsay argued; however, speed may also come with the potential to resurrect places. Therefore, from Kasarda and Lindsay's point of view, cities need to be faster than their rivals, and, 'with the right infrastructure, a hinterland can become a hub' (2011:175). According to Kasarda and Lindsay, cities need to be accelerated to the pace of globalisation:

> The real question is not whether or not aerotropolises will evolve around airports (they surely will). It is whether they will form and grow in an intelligent manner, minimizing problems and bringing about the greatest returns to the airport, its users, businesses, the surrounding communities, and the larger region it serves.
>
> *(2011:169)*

Thus, according to Kasarda and Lindsay, the aerotropolis is an 'optimal way' to tackle the rising challenges of global competition that cities and airports face today. However, whether the concept of aerotropolis is the best way for cities and airports to tackle such global challenges has also, as we will show below, been questioned, and alternatives have been presented.

Airport urbanism: a people-focused approach

Hirsh (2016) addressed what he termed 'airport urbanism' by focusing on infrastructure and mobility in Asian cities. The context of airport urbanism is the growing importance of non-aeronautical revenue for the airport business model and the increasing importance of airports for regional urban economies. According to Hirsh, this means that it has become difficult to determine exactly where the airport ends and the city begins (2019). From this point of departure, Hirsh explored international air traffic and its implications for planning and designing contemporary cities. The concept of airport urbanism has, according to Hirsh, a more people-orientated focus on the design of airports, and the urban developments around the airport than, for example, the aerotropolis (2019, 2016). From Hirsh's perspective, concepts like the aerotropolis have a much stronger one-size-fits-all approach to every individual airport (2019:3).

Two fundamental questions drive Hirsh's people-focused approach: (1) How do what he terms as the 'nouveaux globalisés' travel? This growing group of air travellers are not traditional members of cosmopolites 'kinetic elite' consisting of global business people, cultural leaders or academics (Bauman, 1998); instead, they represent a new semi-privileged group that is more diverse in terms of their socioeconomic backgrounds, countries of citizenship, trajectories, trip purposes and age groups. From an Asian perspective, this group consists of four passenger types: Chinese tourists, foreign students, expatriate retirees and migrant workers (Hirsh, 2016:5). (2) How have architects and planners redesigned airports and cities to accommodate these significantly larger and more diverse flows of global passengers? Hirsch is critical of the fact that urban infrastructure studies mainly interpret urban space through visual images produced by architects, developers and urban investors (2016:7), who imagine airport infrastructure as clean, efficient, and exclusively designed for wealthy business travellers (2016:9), and who design it to accelerate the 'seamless' movement of the jet elite. This transport network, he points out, is both socioeconomically and spatially segregated from the surrounding city:

> the infrastructure systems used by the nouveaux globalisés are largely absent from official maps, ads, and brochures, and thus they remain unaccounted for in the urban scholarship on the physical dimensions of international mobility. But it is precisely these non-traditional travellers who have driven the exponential increase in global air traffic, and the expansion of airport infrastructure – over the past thirty years.
>
> *(2016:9)*

Instead of focusing on iconic megaprojects and business orientated transport modes, Hirsh investigated the parallel transport systems that are designed to 'plug less-privileged people and places into the infrastructure of international aviation' (Hirsh, 2016:11). The movement across international borders of this group of travellers is limited by income, citizenship, literacy, and birthplace–all elements that influence the urban form. Therefore, Hirsh argued that scholarship of airport urbanism needs to focus on the needs and desires of customers, and customers are not only travellers, but also the people who live, work, and own businesses at the airport and in nearby communities. This means that, from Hirsh's perspective, if airports want to be successful, they need to coordinate airside, landside and off-airport development in a holistic and mutually beneficial manner, because airports and cities grow best together (Hirsh, 2016:2). Therefore, the development of airport urbanism needs to focus on the following elements (Hirsh, 2019):

- each airport's unique mix of passengers;
- the people who work at the airport every day;
- the expansion of the airport customer base to include residents of local communities;
- attention to the needs of the local business community.

In general, the airport urbanism approach presented above has a more context-specific and people-orientated focus, that aims to develop guidelines for the needs of the individual airport, compared to the more 'universal' concept of aerotropolis. However, we will also argue that the discussion of the airport city concept needs to factor in more of the potential downsides related to airport city development.

The dark side of airport cities

One of the most significant challenges for the airport city is the impact of the climate crisis on aviation. According to the International Air Transport Association (IATA), global civil aviation accounts for 2.5% of all energy-related CO_2 emissions (IATA, 2017). Accordingly, IATA has created a plan to achieve a 50% net reduction of aviation-generated CO_2 emissions by 2050 (for more details, see www.iata.org/policy/environment/Pages/climate-change.aspx). This is also why the Airports Council International Europa (ACI-Europa) together with its sister organisations in Asia, the Pacific, Africa, North and Latin America and the Caribbean have formed the 'Airport Carbon Accreditation' system. This system covers 274 airports in 71 countries worldwide and tracks almost 44% of total global air passenger traffic. The programme plans to reduce its member airports' carbon emissions with the ultimate goal of becoming carbon neutral (see www.airportcarbonaccreditation.org/about/what-is-it.html for more details on this programme). Airport cities are key players in tackling the air sector's contributions to the climate crisis.

In addition to the climate crisis, major airports also have other impacts. In a study of Mexico City International Airport, Lassen and Galland (2014) investigated land use conflicts, noise and health problems, local air pollution, decreased urban quality, and affected liveability as some of the impacts and consequences associated with a global city airport. They argued that a much more 'integrated focus that brings such different impacts and perspectives together is needed in order to widen the understanding of the existing relationship between socio-spatial and environmental effects, increased aeromobility, airport siting conflicts, airport urban surroundings and globalization' (Lassen & Galland, 2014:132). Similarly, Ferrulli (2016) stressed that concepts like the 'aerotropolis', the 'airport city', the 'airport corridor' and the 'airport region' have a number of spatial implications on various levels and scales that need to be carefully considered (2016:3782). She argued that 'the development of sustainable airport infrastructure depends on achieving a correct balance between social and economic objectives within the limits imposed by the environment' (2016:3785). Therefore, Ferrulli proposed what she termed as a 'Green Airport Design Evaluation' method with the goal of sustainably developing airport infrastructure through a framework that can measure and monitor environmental performance of airport city development (2016:3789). Similarly, from a noise perspective, Boucsein et al. (2017) argued that when planning city airport models we need to take 'a more fine-tuned look at the different spatial/urban conditions of the airport area' (2017:21). They therefore suggested five effect types to evaluate and understand the airport and urban development: territorial, aviation, flows, allocation and urbanisation (2017:21).

The future of airport cities

We will argue that a multidisciplinary, contextual, holistic, multi-scale and multi-method approach is necessary for finding the right balance of economic and social values and adapting to the social and environmental challenges associated with aviation growth. In a 2017–2021 research project hosted by the Centre for Mobilities and Urban Studies at Aalborg University, the concept of the airport city is being empirically explored based on a comprehensive case study of the regional hub airport in Copenhagen, Denmark. With a hands-on vision of producing an aeromobilities decision support model for the Danish aviation industry and stakeholders, the research aims to conceptually explore what an airport city is, or can be, in a local context and not just as a theoretical imagination detached from an empirical reality. In six interrelated pieces of research, AirCiF has zoomed in on global aviation trends, passenger experience and airport design, aeromobilities management for business air travel, airport catchment area impact and expansion, airport governance and strategic development, and future aviation scenarios. Moreover, a number of the challenges described above are also included. Through this holistic and interdisciplinary examination, the research establishes an empirical representation of what an airport city can be, set in the context of the wider Copenhagen area and the Danish geographical, societal and administrative framework.

The idea of the airport city is at the heart of AirCiF, with the guiding hypothesis that hub airports as we know them are the result of 20th century aeromobilities, created in the industrial society, and are therefore not able to efficiently facilitate aeromobilities in the 21st century, with new demands for the airport city as described above. Specifically, this is done by focusing on mobility and the continuous travel experience in relation to managing, planning and designing airport cities. Conceptually, this research draws on the various airport city visions presented above, and encompasses a holistic understanding of the airport, aiming to explore the wider interactions between the airport and the cities and regions that surround it. Viewing Copenhagen Airport conceptually as an airport city thus allows this research to widen the empirical scope, allowing global investigation beyond the immediate airport site and its attached infrastructure, and to go past the explicitly aviatic and aeronautic services and activities associated with the airport. This results in an exploration that goes into greater detail on a range of implicit, but, as it emerges, highly important activities that are linked to the airport as a central piece of national infrastructure, both physically and immaterially, including aspects not traditionally included in 20th century airport studies. Airport development takes time and requires significant financial investments, so due diligence is needed when making decisions about the airports of the future. The old conceptualisations of airports will not suffice when planning 21st century airports and airport cities. Therefore, the airport city concept must also be strongly future-orientated (Urry, 2016).

Concluding remarks

This chapter has explored the notion of the airport city. We have shown that this notion moves beyond the conventional understanding of the airport as a purely technical question limited to the airport itself towards a more 'airport driven' urban development perspective. The airport city approach is both a practice that has already been implemented by a number of global airports and a vison and model for how airports and cities can take advantage of the growth of air traffic and increasing urban globalisation in the best possible way. The airport city concept has a stronger focus on non-aeronautical commercial facilities and the social relations between the airport, the host city and the surrounding region than previous airport development concepts have had. In this way, the airport has historically moved away from being an exclusive activity on a bare field outside the city and towards being the core of societal development. However, this chapter has also shown that this relatively new concept is already a contested field that has been presented in different versions, from the aerotropolis's top-down and universal approach, to airport urbanism, which has a more people-orientated and contextual focus. In this chapter, we have also highlighted some the potential more negative impacts of airport development that airport cities also need to address in the future. We have therefore also argued for a more holistic, interdisciplinary and future-orientated approach to airport city development. This approach aims to adopt and integrate a number of different economic, social, spatial, environmental and mobilities-related impacts, that are highly relevant for airport city development. However, the discussion on the airport city has just begun; and much more research and normative discussions regarding how the prospects of the airport city can be best utilised in future urban development will be necessary in the coming years.

References

Adey, P. (2006). Airports and air-mindedness: Spacing, timing and using the Liverpool Airport, 1929–1939. *Social and Cultural Geography*, 7(3), 343–363.

Bauman, Z. (1998). *Globalization: The human consequences*. New York: Columbia University Press.

Bloch, J. H. (2018). Making of hub airports: A cross analytical approach based on aeromobilities. Ph.D.-Thesis. Department of Architecture, Design and Media Technology. The Technical Faculty of IT and Design. Aalborg University.

Boucsein, B., Christiaanse, K., Kasioumi, E., & Salewaski, C. (2017). *Noise landscape: A spatial exploration of airports and cities*. Rotterdam: nai010 publishers.

Cidell, J. (2006). Air transportation, airports and the discourses and practices of globalization. *Urban Geography*, 27(7), 651–663.

Cresswell, T. (2006). *On the move: Mobility in the modern western world*. London: Routledge.

Cwerner, S., Kesselring, S., & Urry, J. (eds.) (2009). *Aeromobilities*. London: Routledge.

Ferrulli, P. (2016). Green Airport Design Evaluation (GrADE) – Methods and tools improving infrastructure planning. *Transportation Research Procedia*, 14, 3781–3790.

Fuller, G., & Harley, R. (2004). *Aviopolis: A book about airports*. London: Black Dog Publishing.

Goffman, E. (1959). *The presentation of self in everyday life*. New York: The Overlook Press.

Güller, M., & Güller, M. (2002). *From airport to airport city*. Barcelona: Editorial Gustavo Gill.

Hirsh, M. (2016). *Airport urbanism: Infrastructure and mobility in Asia*. Minneapolis: University of Minnesota Press.

Hirsh, M. (2019). *What is airport urbanism?* Executive Summary. Retrieved from: https://airporturbanism.com/content/1-home/airport-urbanism_max-hirsh.pdf (Access Marts 2019).

IATA. (2017). *Fact sheet climate change and CORSIA*. Retrieved from: www.iata.org/pressroom/facts_figures/fact_sheets/Documents/fact-sheet-climate-change.pdf

Jensen, O. B. (2013). *Staging mobilities*. London: Routledge.

Jensen, O. B., & Lassen, C. (2011). Mobility challenges. *Danish Journal of Geoinformatics and Land Management, 46*(1), 9–21.

Jensen, O. B., Lassen, C., & Lange, I. S. G. (2019). *Material mobilities*. London: Routledge.

Kasarda, J. D., & Lindsay, G. (2011). *Aerotropolis: The way we'll live next*. New York: Farrar, Straus & Giroux.

Kesselring, S. (2009). Global transfer points: The making of airports in the mobile risk society. In Cwerner, S., Kesselring, S., & Urry, J. (eds.) *Aeromobilities*. London: Routledge, 39–60.

Kloppenburg, S. (2013). *Tracing mobilities regimes. The regulation of drug smuggling and labour migration at two airports in the Netherlands and Indonesia*. Enschede: Gildeprint Drukkerijen.

Lassen, C. (2006). Work and aeromobility. *Environment and Planning A*, 38(2), 301–312.

Lassen, C. (2009). A life in corridors: Social perspectives on aeromobility and work in knowledge organisations. In Cwerner, S., et al. (eds.) *Aeromobilities*. London: Routledge, 177–193.

Lassen, C. (forthcoming). Airports as a mobile method. In Büscher, M., Kesselring, S., & Freudendal-Pedersen, M. (eds.) *Handbook of methods and applications for mobilities research*. London: Routledge.

Lassen, C., & Galland, D. (2014). The dark side of aeromobilities: Unplanned airport planning in Mexico City. *International Planning Studies*, 19(2), 132–153. Doi: 10.1080/13563475.2013.876913.

Merriman, P. (2007). *Driving spaces: A cultural-historical geography of England's M1 motorway*. Oxford: Wiley-Blackwell.

Roseau, N. (2012). Airports as urban narratives. Toward a cultural history of the global infrastructures. *Transfers*, 2(1), 32–54.

Simonsen, D. G. (2012). *The aviation neck: Aerial and grounded bodies in early powered flight*. Paper presented at The Copenhagen Meeting, 4–7 October, Copenhagen: University of Copenhagen.

Urry, J. (2007). *Mobilities*. Cambridge: Polity.

Urry, J. (2016). *What is the future?* Cambridge: Polity.

34
SURVEILLANCE AND URBAN MOBILITY

Francisco Klauser

Today, computerised systems that act as conduits for multiple, cross-cutting forms of data gathering, transfer and analysis control, protect and manage everyday life on many levels, and for manifold purposes. Think of the rapidly expanding use of RFID chips in tickets and goods, the increasing number of surveillance cameras in public places, or of smart traffic and navigation systems. The digital age has spawned a range of novel techno-mediated forms and formats of surveillance, understood here as the ensemble of focused, systematic and routine practices and techniques of attention, relating to human or non-human objects, for purposes of influence, management, protection or direction (inspired by Lyon, 2007: p.14).

From this evolution result novel techno-mediated ways of ordering and managing presences and flows of human and non-human objects. For example, smartphones and other self-tracking devices work though the continuous localisation of mobile people and objects. Many of these devices then offer place-, user- and practice-specific information and services that act on the ways in which we access and pass through differing urban spaces (De Souza e Silva & Frith, 2012; Graham, Zook & Boulton, 2013; Widmer, 2015). In the field of smart urban infrastructure, a similar spatial dynamics can be found, responding to the need to manage the city as an interconnected, digitised and 'technologically empowered' (IBM, 2010) system of connections, processes and flows.

Connecting with this problematic, this chapter discusses and problematises contemporary surveillance from a specific urban and mobility-related viewpoint, highlighting how digital technologies, whether mobile themselves or fixed in place, control and regulate movement within and in-between cities. Drawing upon existing literatures that explore the relationships between surveillance and space (Graham, 1998, 2005; Koskela, 2000; Franzén, 2001; Coaffee, 2004; Duarte & Firmino, 2009; Zurawski, 2013; Klauser, 2013, 2017), this discussion is structured into three main parts. They outline three complementary ways to approach and problematise the imbrications of urban surveillance and mobility, relating to (1) separation and access control, (2) the management of humans and non-humans on the move, and (3) the internal organisation of interconnected places.

Separation and access control

The first direction of research from which to addresses issues of urban surveillance and mobility revolves around the 'splintering urbanism problematics' (Graham & Marvin, 2001). Relevant literatures highlight and problematise current trends towards the fragmentation of cities into a patchwork of 'more or less purified insides, separated from more or less dangerous outsides' (Franzén, 2001: p.207). Hereby, separation and enclosure are seen not only on a physical level, but also related to the creation of more or less rigid and exclusive forms of togetherness, thus structuring the city into a patchwork of secluded spherical conglomerates of co-isolation (Klauser, 2010). Resonating with Peter Sloterdijk's (2004) analysis of the 'foam city', studied examples range from inner city zones (Coaffee, 2004), gated communities (Connell, 1999) and shopping malls (Helten & Fischer, 2004; Benton-Short, 2007), to recreational facilities, leisure spaces and bunkered private homes (Flusty, 1994).

In these studies, surveillance is approached as a combined problematic of enclosure and accessibility, and set in relation to the monitoring and regulation of flows of people and objects, crossing different kinds of borderlines at particular points in space. The key issue is access control, implying a spatial logic of power that encloses and keeps places, people, objects and functions apart (Bauman, 2000: p.115; Boyne, 2000). Yet importantly, access control is not only about fixing, demarcating and fragmenting space, but also about allowing and facilitating entrance or exit, and thus flow. Enclosure lays the fixed conditions and frame within and through which circulations are allowed to develop, implying a range of more or less technology-driven processes of differentiation and categorisation of 'good' and 'bad' flows, and relying on differing efforts of data gathering, analysis and exchange. This resonates with Foucault's (Foucault, 2007 [2004]) understanding of the rationalities of power that characterise apparatuses of 'security', as the economy of power, starting in the eighteenth century, that characterises liberalism. Following Foucault (2007 [2004]):

> An important problem for towns in the eighteenth century was allowing for surveillance, since the suppression of city walls made necessary by economic development meant that one could no longer close towns in the evening or closely supervise daily comings and goings. [...] In other words, it was a matter of organizing circulation, eliminating its dangerous elements, making a division between good and bad circulation, and maximizing the good circulation by diminishing the bad.
>
> *(p.18)*

Host cities of sport mega events provide a powerful contemporary example of this problematic. Leading to the installation of dozens of kilometres of fences throughout the urban environment, major sporting events imply a wide range of more or less hermetically enclosed and tightly controlled security perimeters that are supported by advanced surveillance technologies and increased numbers of security personnel (Klauser, 2013). Examples range from stadium security rings to the referee headquarters, from team hotels to fan zones and from private camping grounds to official fan villages (offering accommodation and attractions for fans). Hereby, different levels of accessibility are not merely conditioned by a rigid binary opposition between those permitted and those prohibited to enter, but adapted meticulously to the specific profiles (needs, risks, etc.) of entrants. This exemplifies the contemporary evolution towards an ever-more sophisticated and complex analytics of access management that depends on ever more sustained efforts in gathering, analysing and exchanging data, in the aim of developing codes that can be used to automatically assess and manage people's level of admittance to a whole range of services, activities and spaces.

Such processes of ordering (i.e. privileging and restricting) presences and flows raise a series of power issues that are of particular importance, especially if the differential treatment of people and objects not only results from risk assessments and security considerations, but also responds to private interests and commercial rationales. Also, the study of access control is crucial not only to an understanding of how surveillance affects secluded spaces themselves, but also to highlight the resulting implications for the remaining outside space. For Franzén (2001: p.206) 'the urban order in a particular place is determined, at least partially, by the unintended, and cumulative, consequence of all border controls'. One particularly important question arising here for future research is how access control relates to current trends of polarisation and hierarchisation of space, reinforcing, for example, the opposition between economically attractive zones and areas of concentrated social disadvantage.

Importantly, a crucial lesson derived from this line of research is that the distinction between inter-state border control and the monitoring of everyday urban borders and access points is increasingly blurred from a spatial, functional, technological and organisational viewpoint (Albert & Jacobsen, 2001; Lyon, 2005; Salter, 2005; Amoore, Marmura & Salter, 2008: p.96). As Graham (2010) puts it:

> Borders cease to be geographical lines and filters between states (always an over-simplified idea) and emerge instead as increasingly interoperable assemblages of control technologies strung out across the world's infrastructures, circulations, cities and bodies.
>
> *(p.132)*

Again, consider the example of sport mega events. Here, 'risky' foreign fans encounter access control and denial somewhere in between their private home (through requests to report to their national police during the tournament, or by unsuccessfully trying to book an aeroplane or tournament ticket online), at the national border of the host nation or at the entrance gates of stadiums, training grounds or fan zones (through police spot checks and biometric fingerprint identification devices). Thus access control in urban space is part of a larger, multi-scalar system of threat filters situated both within and outside the host nation's territory, implying an increased dissociation of surveillance-as-border-control from the territorial frontiers (Albert & Brock, 1996: p.62). This 'de-bordering process' (Rumford, 2006; Côté-Boucher, 2008) is at the very core of the splintering urbanism problematic. As such, it is not exclusive to mega-event security, but the special conditions of, and measures at, sport mega events add further importance to it.

In turn, this exemplifies that if access control is shaped by the search for the right balance between enclosure and openness, fixity and fluidity, this does not mean that these rationales are necessarily to be seen as antagonistic. Rather, they embody and nourish each other (Foucault, 2007 [2004]: p.107). With the management of circulations and openness, as in the example of sport mega events, the problem of fixity and enclosure is not eliminated, but made more acute. Access control highlights that surveillance often combines in both complementary and conflictual ways elements of fixity and flow. Indeed, it is in the very balancing hence implied that lies the key problematic of the surveillance of movement, in terms of power, social justice, etc., as I will discuss in more detail below.

Management of humans and non-humans on the move

A second direction of research into the problematic of urban surveillance and mobility revolves around the question of how – and to what effect – multi-layered surveillant assemblages coalesce around mobile people and objects themselves. The key concern here lies in the continuous monitoring and management of people and objects on the move (Buhr, 2003; Côté-Boucher, 2008; Cowen, 2010; Firmino, Duarte & Ultramari, 2011; Martin, 2012; Salter, 2013; Klauser & Albrechtslund, 2014). There are two complementary aspects to highlight.

Firstly, think of the infrastructural dimension of the surveillance of (human and non-human) objects on the move. Relevant research is concerned with how digital technologies today permeate the key infrastructural networks underpinning everyday urban life (Debrix, 2001; Wekerle & Jackson, 2005). Examples range from computerised motorways and energy grids, to the digitisation of water pipelines and public transport systems. Thus, from this perspective, the contemporary 'smart city' appears as a vast 'programme of government of movement' (Côté-Boucher, 2008), aimed at establishing the routes along which movement happens, and that allow the channelling, monitoring and restriction but also facilitation and speeding up of various types of circulation, from passage point to passage point, from separation to separation, from enclosure to enclosure. The emerging system of 'conductive lines' thus stands for a spatial logic of surveillance that maps to, without being synonymous with, enclosure and access control.

Secondly, increased attention has been paid in recent years to the surveillant capacities of increasingly mobile, ubiquitous and smart information and communication technologies, with a particular interest in how such devices embrace and manage circulations. This applies for example to smartphones and other self-tracking devices, which work through the continuous geo-localisation of mobile people and objects (Dodge & Kitchin, 2007). The place-, user- and practice-specific information and services, offered by such devices, organise, guide and regulate flows and presences of people and objects as they navigate through urban space (Widmer & Klauser, 2013). What matters is the regulation and management of openness and fluidity, rather than the fixing and enclosing of particular places, people, objects and functions (Farman, 2011; Monahan & Mokos, 2013).

This, again, implies important implications in terms of urban organisation and polarisation. Namely, as shown by De Souza e Silva and Frith (2012), location-aware mobile applications enable, but also push, their users to target specific relationships with the spaces and people around them. Addressing this problematic through the lens of socio-spatial media such as Foursquare, Widmer (2015) concludes:

> The visibilities and invisibilities of places are ... shaped by Foursquare's personalisation algorithms, which prioritise two specific alterities: users' friends on the platform and people whose behaviours are similar to theirs. By highlighting recommendations on places frequented by these two alterities, Foursquare enables its users to make informed decisions and go to places approved by people who have tastes similar to their own. It has been argued that, by reducing alterity to these two figures ('my friends' and 'people like me'), Foursquare could potentially accentuate homophily patterns and strengthen specific forms of togetherness. The 'bubble' metaphor, proposed by Pariser (2011), has proved to be useful in describing those homophilous communities mediated by software sorting operations.
>
> *(p. 72)*

Taken together, the two lines of enquiry provide much needed accounts of how surveillance works to align the circulation of mobile bodies, data, objects and services with localisation, identification, verification and authentication controls, and of how the practices and techniques of surveillance engage with the key infrastructural networks that aim to filter and manage movements within and between cities (Debrix, 2001; Wekerle & Jackson, 2005).

Managing movement through internal organisation

Whilst both aforementioned research directions offer important insights into the control and management of spatially articulated separations and connections on the urban scale as a whole, attention should also be given to how surveillance relates to, and permeates, monitored places and buildings themselves if we are to understand the management of movement across and within urban space. For example, transport nodes such as airports, railway and metro stations are not only to be studied as filtering points for the procession of urban mobility as a whole (Castells, 1996; Fuller, 2002), but also as security zones in their own right, i.e. as carefully planned worlds of spatially articulated limits, channels and filters that respond to the need to control and regulate flows of people and objects, processed from point to point, from zone to zone (Klauser, 2017).

In scholarly research, attention has also to be given to how surveillance monitors and affects movements within particular spaces of surveillance and security, thus setting surveillance in relation to the 'internal organisation' of particular geographical locales, from buildings to public squares and larger urban areas. Of particular note here is Andrzejewski's (2008) study of the imbrications of architecture and surveillance in factories, post offices, prisons, religious camps and private homes in Victorian America, which shows in detail how specific forms of spatial organisation enable, and result from, the monitoring of indoor micro-movements by servants, workers, believers, etc. Others have investigated, for example, the spatial articulation of surveillance relating to the management of micro-flows in airports (Klauser, 2009), shopping malls (Helten & Fischer, 2004) and football stadia (Bale, 2005; Hagemann, 2007), as carefully organised and subdivided spaces of surveillance.

This brings us back to the example of sport mega events (Klauser, 2013, 2017). For example, at the European Football Championships 2008 in Switzerland and Austria, stadium perimeters were subdivided into 13 different zones, each of which was matched with different routes, control points and modalities of access for differing fan communities, dignitaries, commercial partners etc. In the case of fan zones, internal structuring consisted of the use of 'wave breakers' (physical barriers and obstacles (Republik Österreich, 2008: p.48) and of the erection of stands and platforms related to special activities and access criteria. Internal zonal surveillance ranged from patrolling security agents and CCTV cameras to the Scanning Infrared Gas Imaging System for the detection of hazardous chemicals deployed in Bern and Basel (Projektorganisation Öffentliche Hand, 2008: p.47). In both cases, internal monitoring and organisation aimed for the planned, hierarchical and functional distribution of people and objects across the enclosed inside, bound up with differential access control. Unlike in open public space, security in fan zones and stadiums was delegated to Euro 2008 SA, which accomplished the task by contracting private security companies. Constructional, technical, organisational and operational security aspects in both zones thus fell under the authority of UEFA, with national and international police ready to intervene if need be. Thus surveillance differed from adjoining space not only in intensity, spatiality and internal organisation, but also in substance and regulative principle.

In sum, if we are to understand the regulation of movement within and across cities, it also matters to study how particular spaces are internally organised around surveillance and how, in turn, surveillance is shaped by the specific characteristics of particular places. Yet in such micro-geographical studies, the wider picture regarding surveillance and intra- or inter-urban mobilities should not be lost. Therefore, one central objection is that such studies often ignore the broader networks within which the studied micro spaces of surveillance are positioned and monitored. In this sense, this research direction not only contributes to, but also depends on the previously outlined perspectives.

Conclusion

The three directions of research discussed above are not mutually exclusive. Indeed, many studies touch on various spatial logics and scales of surveillance. For example, Jones's (2009) analysis of 'checkpoint security' and Graham's (2010) work on 'passage point urbanism' powerfully combine the separation/access-control and 'mobility management on the move dimensions' of surveillance. However, both Jones and Graham largely overlook how exactly spatial enclaves are organised and monitored internally through everyday surveillance practices. In turn, many studies dealing with urban enclosures and fragmentation emphasise both the access control to, and internal surveillance of, secluded zones, but do not consider how flows of people and objects are monitored, channelled and filtered in between fortified places.

The present chapter invites further development of the surveillance–mobility problematics, in approaching urban surveillance simultaneously in its enclosing and separating dynamics, and in relation to the opening up of cities and to the management of flows of people and objects on the move. From this perspective, the city has been portrayed as a complex system of separations and connections, in which differing spatial logics of surveillance call on each other, support each other, modify and shape each other.

This discussion was mostly programmatic and exploratory in aim and style. Today, little is known about the dissonances and resonances between surveillance practices and techniques relating to separation, access control, circulation and internal organisation. The ways in which surveillance focuses on, embraces and in turn results from urban mobilities should be explored in much more empirical detail. The ambition thereby should be to study and to conceptualise surveillance as an ensemble of heterogeneous techniques and practices of control and power that are intrinsically bound up with mobility, through multiple processes and relationships, on different scales and for numerous reasons. To conclude, I here want to reiterate two key issues that should guide such a research agenda.

Firstly, the collection, classification and analysis of data is never neutral, but depends on a range of choices made and interests conveyed by the involved actors, whether aimed at greater efficiency, convenience or security. Information management, and management through information, thus work through and depend on novel forms and possibilities of differentiation and prioritisation, used to orchestrate everyday life and affecting the life chances of individuals or social groups in ways that are often unknown by the public. Critical debates about the implications of surveillance on everyday urban mobilities should therefore move beyond privacy, data protection and accountability issues, to challenge, more generally, the functioning of surveillance as a technique of social sorting that restricts, facilitates and organises flows and presences in and through urban space in often problematic ways (Graham, 1998, 2005; Lyon, 2003).

Secondly, if surveillance orchestrates urban space by sorting access, filtering flows and organising presences, it does so in ever more automated ways (Hinchliffe, 1996; Thrift & French, 2002; Dodge, Kitchin & Zook, 2009; Kitchin & Dodge, 2011; Amoore, 2011). Thus the key issue here relates to the agency and power dynamics implied by software. How do computerised and increasingly automated techniques of surveillance mediate the organisation and production of particular forms of mobility within and across cities? What are the wider societal implications thereof? Hereby, it will be of crucial importance to avoid the trap of overstating the agency of software in a deterministic way. In recent years, an ever more impressive body of research has shown that the effects of surveillance are much more contingent and complex than often expected. Drawing particularly upon Michel de Certeau's (1984) work, scholars have stressed the micro tactics and strategies deployed by individuals and social groups to resist, bypass and subvert surveillance (Marx, 2003). It is in this spirit that the question of how software-mediated surveillance orchestrates space must be approached. Future research should further pursue this line of investigation, so as to provide more detailed accounts of how exactly emerging geographies of regulation-at-a-distance work to align the circulation of mobile bodies, data, objects and services with localisation, identification, verification and authentication controls, and of how the practices and techniques of surveillance engage with the key infrastructural networks that aim to stop and accelerate, channel and filter movements within and between cities.

Acknowledgement

This chapter draws upon two previously published pieces of work, relating to surveillance and sport mega events (Klauser, 2013), and to the interactions between surveillance and space more generally (Klauser, 2017).

References

Albert, M., & Brock, L. (1996) Debordering the world of states: New spaces in international relations. *New Political Science*. 35(1), 69–106.

Albert, M., & Jacobsen, D. (2001) *Identities, Borders, Orders: Rethinking International Relations Theory*. University of Minnesota Press, Minneapolis.

Amoore, L. (2011) Data derivatives: On the emergence of a security risk calculus for our times. *Theory, Culture & Society*. 28(6), 24–43.

Amoore, L., Marmura, S., & Salter, M. (2008) Editorial: Smart borders and mobilities: Spaces, zones, enclosures. *Surveillance & Society*. 5(2), 96–101.

Andrzejewski, A. V. (2008) *Building Power: Architecture and Surveillance in Victorian America*. University of Tennessee Press, Knoxville.

Bale, J. (2005) Stadien als Grenzen und Überwachungsräume. In: Marschik, M., & Zinganel, M. (eds.) *Stadion: Geschichte, Architektur, Politik*. Turia + Kant, Wien, pp. 49–88.

Bauman, Z. (2000) *Liquid Modernity*. Polity Press, Cambridge.

Benton-Short, L. (2007) Bollards, bunkers and barriers: Securing the national mall in Washington DC. *Environment and Planning D: Society and Space*. 25(3), 424–446.

Boyne, R. (2000) Post-panopticism. *Economy and Society*. 29(2), 285–307.

Buhr, B. L. (2003) Traceability and information technology in the meat supply chain: Implications for firm organization and market structure. *Journal of Food Distribution Research*. 34(3), 13–26.

Castells, M. (1996) *The Information Age: Economy, Society and Culture Volume 1: The Rise of the Network Society*. Blackwell, Oxford.

Coaffee, J. (2004) Rings of steel, rings of concrete and rings of confidence: Designing out terrorism in central London pre and post September 11th. *International Journal of Urban and Regional Research*. 28(1), 201–211.

Connell, J. (1999) Beyond Manila: Walls, malls and private spaces. *Environment and Planning A*. 31(3), 417–439.
Côté-Boucher, K. (2008) The diffuse border: Intelligence-sharing, control and confinement along Canada's smart border. *Surveillance & Society*. 5(2), 142–165.
Cowen, D. (2010) A geography of logistics: Market authority and the security of supply chains. *Annals of the Association of American Geographers*. 100(3), 600–620.
Debrix, F. (2001) Cyberterror and media-induced fears: The production of emergency culture. *Strategies*. 14(1), 149–167.
de Certeau, M. (1984) *Practice of Everyday Life*. University of California Press, Berkeley.
De Souza e Silva, A., & Frith, J. (2012) *Mobile Interfaces in Public Spaces*. Routledge, London.
Dodge, M., & Kitchin, R. (2007) The automatic management of drivers and driving spaces. *Geoforum*. 38(2), 264–275.
Dodge, M., Kitchin, R., & Zook, M. (2009) How does software make space? Exploring some geographical dimensions of pervasive computing and software studies. *Environment and Planning A*. 41(6), 1283–1293.
Duarte, F., & Firmino, R. J. (2009) Infiltrated city, augmented space: Information and communication technologies, and representations of contemporary spatialities. *The Journal of Architecture*. 14(5), 545–565.
Farman, J. (2011) *Mobile Interface Theory, Embodied Space and Locative Media*. Routledge, London.
Firmino, R. J., Duarte, F., & Ultramari, C. (2011) The rising of the ubiquitous city: Global networks, locative media and surveillance technologies. In: Firmino, R. J., Duarte, F., & Ultramari, C. (eds.) *ICTs for Mobile and Ubiquitous Urban Infrastructures: Surveillance, Locative Media and Global Networks. Information Science Reference*, Hersey, pp. 1–13.
Flusty, S. (1994) *Building Paranoia: The Proliferation of Interdictory Space and the Erosion of Spatial Justice*. Los Angeles Forum for Architecture and Urban Design, Los Angeles.
Foucault, M. (2007 [2004]) *Security, Territory, Population*. Palgrave Macmillan, London.
Franzén, M. (2001) Urban order and the preventive restructuring of space: The operation of border controls in micro space. *Sociological Review*. 49(2), 202–218.
Fuller, G. (2002) The arrow – Directional demiotics: Wayfinding in transit. *Social Semiotics*. 12(3), 131–144.
Graham, S. (1998) Spaces of surveillant-simulation: New technologies, digital representations, and material geographies. *Environment and Planning D: Society and Space*. 16(4), 483–504.
Graham, S. (2005) Software-sorted geographies. *Progress in Human Geography*. 29(5), 562–580.
Graham, S. (2010) *Cities under Siege. New Military Urbanism*. Verso, London.
Graham, S., & Marvin, S. (2001) *Splintering Urbanism*. Routledge, London.
Graham, M., Zook, M., & Boulton, A. (2013) Augmented reality in urban places: Contested content and the duplicity of code. *Transactions of the Institute of British Geographers*. 38(3), 464–479.
Hagemann, A. (2007) Filter, Ventile und Schleusen: Die Architektur der Zugangsregulierung. In: Eick, V., Sambale, J., & Töpfer, E. (eds.) *Kontrollierte Urbanität*. Transcript Verlag, Bielefeld, pp. 301–328.
Helten, F., & Fischer, B. (2004) Reactive attention: Video surveillance in Berlin shopping malls. *Surveillance & Society*. 2(2/3), 323–345.
Hinchliffe, S. (1996) Technology, power, and space: The means and ends of geographies of technology. *Environment and Planning D: Society and Space*. 14(6), 659–682.
IBM. (2010) *Smarter Cities with IBM Software Solutions*. Available from: http://public.dhe.ibm.com/software/ch/de/multimedia/pdf/transcript-smarter-cities-with-ibm-software-solutions-eng.pdf [Accessed 30th May 2016].
Jones, R. (2009) Checkpoint security: Gateways, airports and the architecture of security. In: Aas, K. F., Gundhus, H. M., & Lomell, H. M. (eds.) *Technologies of Insecurity*. Routledge, London, pp. 81–101.
Kitchin, R., & Dodge, M. (2011) *Code/Space. Software and Everyday Life*. MIT Press, Cambridge.
Klauser, F. (2009) Interacting forms of expertise in security governance: The example of CCTV surveillance at Geneva International Airport. *British Journal of Sociology*. 60(2), 279–297.
Klauser, F. (2010) Splintering spheres of security: Peter Sloterdijk and the contemporary fortress city. *Environment and Planning D: Society and Space*. 28(2), 326–340.
Klauser, F. (2013) Spatialities of security and surveillance: Managing spaces, separations and circulations at sport mega events. *Geoforum*. 49, 289–298.
Klauser, F. (2017) *Surveillance and Space*. Sage, London.

Klauser, F., & Albrechtslund, A. (2014) From self-tracking to smart urban infrastructures: Towards an interdisciplinary research agenda on big data. *Surveillance and Society*. 18(3), 273–286.

Koskela, H. (2000) The gaze without eyes: Video-surveillance and the changing nature of urban space. *Progress in Human Geography*. 24(2), 243–265.

Lyon, D. (ed.) (2003) *Surveillance as Social Sorting*. Routledge, London.

Lyon, D. (2005) The border is everywhere: ID cards, surveillance and the other. In: Zureik, E., & Salter, M. B. (eds.) *Global Surveillance and Policing*. Willan, London, pp. 66–82.

Lyon, D. (2007) *Surveillance Studies. An Overview*. Polity Press, Cambridge.

Martin, C. (2012) Desperate mobilities: Logistics, security and the extra-logistical knowledge of 'appropriation'. *Geopolitics*. 17(2), 355–376.

Marx, G. T. (2003) A tack in the shoe: Neutralizing and resisting the new surveillance. *Journal of Social Issues*. 59(2), 369–390.

Monahan, T., & Mokos, J. T. (2013) Crowdsourcing urban surveillance: The development of homeland security markets. *Geoforum*. (early view).

Pariser, E. (2011) *The Filter Bubble: What the Internet Is Hiding from You*. Penguin Press, New York.

Projektorganisation Öffentliche Hand. (2008) *Schlussbericht EURO 2008*. Schweizerische Eidgenossenschaft, Bern.

Republik Österreich. (2008) *Euro 2008: Final Report*. Bundesministerium für Inneres, Wien.

Rumford, C. (2006) Introduction: Theorizing borders. *European Journal of Social Theory*. 9(2), 155–169.

Salter, M. B. (2005) At the threshold of security: A theory of international borders. In: Zureik, E., & Salter, M. B. (eds.) *Global Surveillance and Policing*. Willan, London, pp. 36–50.

Salter, M. B. (2013) To make move and let stop: Mobility and the assemblage of circulation. *Mobilities*. 8(1), 7–19.

Sloterdijk, P. (2004) *Sphären III, Schäume*. Suhrkamp, Frankfurt.

Thrift, N., & French, S. (2002) The automated production of space. *Transactions in Human Geography*. 27(4), 309–335.

Wekerle, G. R., & Jackson, P. S. B. (2005) Urbanizing the security agenda. Anti-terrorism, urban sprawl and social movements. *City*. 9(1), 33–49.

Widmer, S. (2015) Experiencing a personalised augmented reality: Users of Foursquare in urban space. In: Amoore, L., & Piotukh, V. (eds.) *Algorithmic Life: Calculative Devices in the Age of Big Data*. Routledge, London, pp. 59–74.

Widmer, S., & Klauser, F. (2013) Mobilité surveillée: Rôle et responsabilité des développeurs d'applications smartphone. *Espace, Populations, Sociétés*. 3, 63–77.

Zurawski, N. (2013) *Raum – Weltbild – Kontrolle: Die Bedeutung räumlicher Diskurse für Überwachung, Kontrolle und gesellschaftliche Ordnung*. Habilitation, Darmstadt.

SECTION VII

Infrastructures, technologies and sustainable development

Cities have grown into immense, complex socio-technical systems. As they represent some of the most complex human-made artefacts, the need for understanding the role of infrastructures and technologies is key. This is particularly so at the current moment in history, where human imprints on the world threaten to jeopardise development for future generations (as well as for the globe itself). Sustainable development must be addressed in any serious attempt to account for urban mobilities, and hence the theme of section seven is infrastructures, technologies and sustainable development. It opens with **Chapter 35**, Thomas Birtchnell on 3D printing and its possible implications for city logistics. The chapter challenges the necessity of global freight by suggesting that 3D printing is a future major player in the system of logistics. In pursing this line of inquiry on 3D printing, the chapter emphasises that there is a need to exercise criticality regarding technological determinism and the prescription of technological 'fixes' to societal, environmental or infrastructural woes, given that, in many instances, these introduce further uncertainties and unforeseen consequences. However, with the advent of decentralised digital methods of production (popularly termed '3D printing') and promising material objects appearing in homes, stores, libraries and places near to consumers, could the city become a space for making again? In the nineteenth century, at the time of the Industrial Revolution in the global North, urban centres were powerhouses of industry. They remained so until the late-twentieth century where major cities – that is, London, Paris, Copenhagen, New York and so on – in the global North underwent deindustrialisation. The suburban dream where the urban core was a sole preserve of work and leisure activities receded in favour of a post-industrial configuration of inner city residential living in proximity to cultural and social amenities and central business districts in tune with the global knowledge economy and financial markets. There was paltry space in this urban model for factories of any size. The small-to-medium sized manufacturing precincts that once peppered the inner cities of the global North surrendered to the vast quantities of products flowing into cities via the complex reconfiguring of off-shored production in standardised factories, benefitting from low labour costs in the global South. In order to realise economies of scale, transoceanic containerised distribution through automated ports and standardised consumption practices centring on the mall combined to meet aspirations for a rise in 'living standards'. Here, quality of life equated to the ongoing access of citizens to growing and affordable quantities of material objects regardless of per capita waste or energy-use. The chapter draws conclusions on the last mile and a shift to a system where distance is no longer a chief component in the consumption of objects by urban citizens. In **Chapter 36**, Chiara Vitrano and Matteo Colleoni investigate the phenomenon of 'Mobility as a Service' (MaaS) and how this may affect urban mobility. MaaS is a mobility distribution model that, through a single interface from which trips are planned, reserved and paid for, allows the user to combine different transport modes in a tailored mobility package. It seems to have emerged as a response to the personalised mobility needs of more and more de-synchronised urban populations, while simultaneously promoting a sustainable transport system. The chapter examines the core characteristics of MaaS and reflects upon the emerging criticalities of the model: MaaS can exacerbate mobility-related social exclusion. Here, the chapter argues that not only can MaaS exacerbate inequalities leading to insufficient mobility, but it may also strengthen the 'assumption of high mobility', the 'duty to be mobile' which lies at the basis of mobility-related inequalities. Providing door-to-door, 24/7, bundled mobility options, MaaS takes mobility for granted and immobility merely as an individual choice. **Chapter 37** is Guillaume Drevon and Alexis Gumy's discussion of multimodality seen through the rhythm of life. The car dominance in the urban transport system may contribute to health and environmental troubles. In recent years, multimodality has appeared as a central issue in public policy as it tries to encourage and develop the modal shift in favour of walking, cycling and public transportation. In the scientific debate, the interest in multimodality is relatively recent. The large majority of studies aim at identifying the predictors of multimodality behaviours depending on geographical context, household characteristics and activities patterns. Other scientific contributions show that having a multimodal behaviour is largely associated with urban lifestyles. For example, urban areas offer good conditions for promoting this kind of behaviour. However, as Drevon and Gumy reveal, few studies have investigated how good lifestyles may be at predicting a multimodal behaviour. The development of high commuting and bi-residential behaviours illustrates the diversification and complexification of lifestyles due to better accessibility and the democratisation of ICT. These technical evolutions involve a social acceleration which can be measured by the number of mobility sequences within a given period of time. The hypothesis of the chapter is that this densification of life rhythms could foster the use of a car. Based on the swiss microcensus 'Mobility and Transport', the chapter highlights the existing relations between life rhythms and mode choice on the one hand and sociodemographic characteristics on the other. **Chapter 38** is Jian Zhuo's exploration of large-scale mobility infrastructure projects from a sustainable smart city perspective. Large-scale mobility infrastructures are a physical presence in modern cities and cannot therefore be ignored. They have a powerful influence on the socioeconomic organisation of cities. In the context of globalised competition, the mega-projects as a strategic instrument for urban development remain high on local governments' agendas. The chapter conducts a brief review of the evolution of transport infrastructure in urban history and investigates the main challenges raised by these projects. Drawing on the existing research findings and planning practices experiences, it advocates a paradigm shift to the people-centred approach with global thinking and presents a new perspective for the future development of mobility infrastructure projects. Cugurullo and Acheampong devote **Chapter 39** to the phenomenon of smart cities. The smart city is now a prominent model of urban development. Among its many promises is the optimisation of urban transport and mobility. Cugurullo and Acheampong first illustrate the concept of a smart city and its many global incarnations. They then discuss smart transport as a core element of the smart-city agenda. Here, the focus is on automated and autonomous systems of traffic management, and on autonomous self-driving cars. The chapter concludes by critically evaluating the potential benefits and dangers of smart urban transport from a sustainability perspective. In **Chapter 40**, Petter Næss explores sustainable mobility and argues that three main strategies should be combined to obtain ecological sustainability: Making each separate mode of transport more energy-efficient and environmentally friendly, promoting a shift to more environmentally friendly modes of transport, and reducing the movement of persons and goods. Social sustainability requires that ecologically sustainable mobility must be achieved while securing everyone accessibility to the facilities and people they need to visit in their daily lives. Land use planning can play an important role in providing accessibility within ecologically sustainable mobility volumes. Section seven ends with **Chapter 41**, Ida Sofie Gøtzsche Lange's chapter on Terminal Towns. Lange proposes the concept of terminal towns as a way of thinking about towns or cities, which also act as transport hubs. From the etymology of the word 'terminal' the concept enables researchers to study places that are understood both as physical sites for transit, often described in a progressive sense, and as sites that are simultaneously challenged by the somehow contrasting meaning of being terminal. The chapter provides an exemplary case (Hirtshals, Denmark) and elaborates further on port cities as a special type of Terminal Town.

35

3D PRINTING AND THE CHANGING LOGISTICS OF CITIES

Thomas Birtchnell

Introduction

This chapter will challenge the necessity of global freight by suggesting 3D printing as a future major player in the system of logistics. Here utopic ideas about consumers becoming closer to manufacturing, and practices of making more broadly, appear surmountable. The word 'manufacture', *manu factum*, means 'made by hand'. Consumers of goods sold in major retailers around the world might forget that the majority of the objects sold on their shelves continue to be made by people's hands, at a distance. Whether it is American company Apple, whose laptop computers sport the words 'Designed in California assembled in China', or Chinese company Lenovo, whose laptops (although only a fraction of their total manufactures) now sport 'Made in the US', humans are still a crucial element in the global systems of production and consumption. What is different about the present state of things is that a third factor, distribution, is now a link in the chain between where objects are manufactured and where they are bought and used.

In pursuing this line of inquiry on 3D printing, I emphasize that there is a need to exercise criticality about technological determinism and prescribing technological 'fixes' (Nye, 2014) to societal, environmental or infrastructural woes given that, in many instances, these introduce further uncertainties and unforeseen consequences (Geels and Smit, 2000). However, with the advent of decentralized digital methods of production (popularly termed '3D printing') promising material objects appearing in homes, stores, libraries and places near to consumers, could the city become a space for making again? In the nineteenth century, at the time of the Industrial Revolution in the global North, urban centres were powerhouses of industry. They remained so until the late twentieth century where major cities – that is, London, Paris, Copenhagen, New York and so on – in the global North underwent deindustrialization. The suburban dream where the urban core was a sole preserve of work and leisure activities was receding in favour of a post-industrial configuration of inner city residential living in proximity to cultural and social amenities and central business districts in tune with the global knowledge economy and financial market. There was paltry space in this urban model for factories of any size. The small-to-medium sized manufacturing precincts that once peppered the inner cities of the global North surrendered to vast quantities of products flowing into cities via the complex reconfiguring of off-shored production in standardized factories benefitting from low labour costs in the global South. In order to realize economies of scale, transoceanic containerized distribution through automated ports and standardized consumption practices centring on the mall combined to meet aspirations for a rise in 'living standards'. Here, quality of life equated to the ongoing access of citizens to growing and affordable quantities of material objects regardless of per capita waste or energy use.

The next section of this handbook entry on the mobilities of cargo reviews the importance of the mobilities framing to understanding the future logistics of cities. The third section introduces a socio-technical transition to 3D printing and the fourth section attends to the last mile in light of the potential ubiquity of this innovative ecosystem. The fifth section draws conclusions about the last mile and a shift to a system where distance is no longer a chief component in the consumption of objects by urban citizens.

The mobilities of cargo

It is not surprising that a majority of media attention towards 3D printing rests on the potential for criminal and illicit uses, such as using them to make firearms in the home or to bypass border importation and exportation controls (Little, 2014). Historically, ports have been tightly integrated with criminal networks; for instance, the Sicilian Mafia acted as intermediaries between producers and exporters in the citrus trade following the nineteenth-century boom in demand due to the discovery of the link between diet and scurvy (Dimico et al., 2017). A major reason for the shift to automation in the mid-twentieth century was the continuing links between crime, port labour (stevedores) and shipping and into the twenty-first century the 'violence' of global trade remains a notable hurdle for logistical planning (Cowen, 2014). Control over the flows of commodities is a key issue for the decentralization of manufacturing through 3D printers with digital flows of products being far more predictable and open to scrutiny utilizing the benefits of cyptography, databases, online repositories and so on.

The mobilities paradigm understands global flows of objects and information in a horizontal way to people as part and parcel of the same phenomenon; namely, an irrevocable quickening of movement after the Second World War of objects of a distinct nature to be desired, rapidly replaced, and enmeshed with personal identities: 'immutable in their western-ness' (Urry, 2000: 187). As cities in the global North developed infrastructure enabling them to export and import through the automated containerization system, their citizens benefitted from easy access to cargo (Cudahy, 2006). The lion's share of this cargo is made in the 'world's factory', China (Tyfield, 2018), able to leverage its vast and youthful population of manual workers and the comparatively lower conditions and salaries that are paid to them. Here the idea of networks made ground in conceiving of cities as socio-technical constructions supporting the departure and arrival of information and goods through multi-modal logistics enclaves dedicated to freight (Graham, 2001).

While a great deal of research has opened up solutions to the challenge of congestion in cities, these are invariably from the perspectives of supply-chain, logistics and freight management expertise. While certainly there is much to gain in considering freight and road infrastructure, operational efficiency, traffic and flow management and toll and congestion charging, there is also value in mulling over behavioural and systemic solutions within the late modern capitalist society (Golob and Regan, 2000).

There are important considerations such as space compression, wherein developments in transport and technology shrink the world; time compression, wherein the time to cross distances is also reduced; and the flow society, wherein 'relations to the flow create the dynamics of the development in production and distribution' (Drewes Nielsen et al., 2003: 298). Beyond the abstract constructions of the 'space of action' between production and consumption are insightful studies of freight frictions: inefficiencies and the challenges associated with heterogenous packaging and the disconnects between and disputes around data flows between the world's ports within geopolitical entities (Gregson et al., 2017). Indeed, there are many instances of flows and pauses due to nested hierarchies with logistics systems, for instance at the municipal level (Cidell, 2012).

3D printing and mobilities

Here the idea of 'digital cargo' enters the picture. Could objects travel virtually over the Internet to consumers, instead of physically, via transoceanic shipping? Early 3D printers were modest affairs: a technical solution to a specific niche in industrial design, that is, model-making, usually in clay (Rael and San Fratello, 2017). Industrial artists would enlist their training in technical drawing and sculpting or assembly to create lifelike models at different scales to test the prototypes of industrial designers. In the late twentieth century 3D printers moved from the pre-production to production process to create parts in limited numbers of objects that were challenging, if not impossible, to make in any other way (Baumers et al., 2015). 3D printing, or digital fabrication, is able to compete with traditional industrial processes in the production of low-volume manufacturing components. So-called 'additive manufacturing', involves building up an object from nothing in contrast to traditional 'subtractive' manufacturing, involving the reduction of a volume of material to an object, found applications in aerospace and transport applications (Agapovichev et al., 2015). Notwithstanding that polymers remain a material of choice, 3D printers able to utilize metals and other more exotic materials intimate applications in end-use product creation. There are distinct opportunities and limitations for 3D printers (Table 35.1) that suggest they will not simply render the present system obsolete, but rather be a facet of a much wider transition in how consumers engage with the creation, movement, destruction and reuse of objects and materials (Weller et al., 2015).

Table 35.1 AM technology's opportunities and limitations from a technological pespective

Opportunities	*Limitations*
+ Direct digital manufacturing of 3D product designs without the need for tools or molds + Change of product designs without cost penalty in manufacturing + Increase of design complexity (e.g., lightweight designs or integrated cooling chambers) without cost penalty in manufacturing + High manufacturing flexibility: objects can be produced in any random order without cost penalty + Production of functionally integrated designs in one-step Less scrap and fewer raw materials required + Less scrap and fewer raw materials required	- Solution space limited to 'printable' materials (e.g., no combined materials) and by size of build space - Quality issues of produced parts: limited reproducibility of parts, missing resistance to environmental influences - Significant efforts are still needed for surface finishing - Lacking design tools and guidelines to fully exploit possibilities of AM - Low production throughput speed - Skilled labour and strong experience needed

Source: adapted from Weller et al., 2015: 46

First, how much freight could be reduced by widespread 3D printing happening closer to consumers? According to Deloitte, global sales relating to 3D printing by large public companies (including enterprise 3D printers, materials and services) will at most total US$3 billion in 2020, in comparison to the global manufacturing sector's revenue as a whole approximately totalling US$12 trillion annually, that is 0.02 per cent (Stewart, 2018). To fathom the reduction of freight, analysts propose that the total transportation spent by a warehouse may be reduced by up to 85 per cent and total supply chain costs by 50–90 per cent, particularly for automotive industry spare parts inventory and slow moving and custom products (Bhasin and Bodla, 2014).

Second, reductions in the shipping of raw materials will depend on local, domestic, availability of oil, coal or gas from fossil fuels or agricultural produce (for instance corn for a polyactic acid (PLA) plastic substitute) for conversion to polymers. Metals will require mineral reserves of iron (steel), bauxite (aluminium) and ilmenite and rutile (titanium) ores. In terms of recycling waste from 3D printing, there is potential for circular economies to emerge that are unfeasible in the present system, given that more than 95 per cent of metal powders can be reclaimed or recycled for further 3D printing processes (Despeisse et al., 2017). Plastics can be recycled, although they will degrade over time and each iteration will be less pure, although viable for other uses, such as packaging.

Third, what will be the nature of these technologies in terms of size, capability, complexity and so forth? While a summary of 3D printing is beyond the scope of this handbook entry, and the preserve of publications by engineering experts (see Baumers et al., 2017), it is obvious that 3D printers come in all shapes and sizes with no single forerunner in either industrial or consumer domains. The emergence of 3D printers as consumer items in their own right hinges so far on the expiry of key patents allowing start-up companies to progress the innovation in response to market pressures and windows of opportunity. In the 1980s, light-curing technologies involving computer controlled lasers that transform a liquid resin into a solid object became common in the process of rapid prototyping: creating models. In the early twenty-first century the process of fused deposition modelling (FDM) wherein a computer-controlled heated nozzle extrudes melted plastic onto a build tray became the mainstream form of industrial and consumer 3D printing.

The surge of interest in 3D printing is not only technical in origin; indeed, commentators recognize the social aspects both from media influence and in the aspirations of consumers for input into object creation (Cano, 2011). The possibilities of using rapid prototyping technologies in homes for end-use products is now a marketable attribute of many 3D printers available to consumers to buy (Rayna and Striukova, 2016). The possibilities of 3D printing for end-user products and parts is an intrinsic aspect of the hype and backlash that arose in the early 2010s, as witnessed in industry tradeshows that championed home 3D printers, such as the Makerbot Replicator (Birtchnell et al., 2017). Even by 2030 there is a probability that objects will continue to be hybrid, that is, admixtures of 3D printed and traditionally manufactured parts, since most objects at present are multi-material and contain embedded electronics (Jiang et al., 2017). In future, consumers might not even realize that 3D printing has altered the business model or distribution network behind the objects they consume. Behind the scenes there could be sustainability through improved resource efficiency in production, extending product lifespans through modular part replacement and consumer repair, and reconfiguring value chains, making them shorter and simpler (Ford and Despeisse, 2016).

The last mile

In this section, I zero in on one key challenge that disputes the notion of an ever-expanding and more efficient global containerization system in the future. The last mile is a term that conflates multiple cultural, political and logistical issues around flows of freight into and out of cities. These include, insurmountable transport congestion; competition for commercial space; strict regulations against urban noise and land-use; rising property costs and NIMBYs (not in my backyard) in gentrifying inner-city suburbs; and air, ground and water pollution from road, air and rail diesel transport. In the twenty-first century a number of these issues pre-empt a further reconfiguration that opens a window of opportunity for 3D printing to become ubiquitous. Citizens shifting to online purchasing require logistics that arrive at their homes in a timely fashion and are unwilling to travel to centralized depots or stores to collect mis-deliveries. In response, small-scale yet intensive delivery systems are required that clog transport arteries and run counter to urban visions of liveability. 3D printing would offer a significant gain in 'freights revenues' in the last mile, namely, the transportation costs from the warehouse to the individual or store (Bhasin and Bodla, 2014).

Overcoming this 'last mile' in transport logistics is now a key challenge. Solutions such as 'anywhere work' are feasible to reduce congestion through alternative labour practices, notably working from home, despite some resistance from employers to adopt a more flexible work culture (Hopkins and McKay, 2018). Alongside the dissolution of the office could be a similar trend in shopping. The adoption by consumers of digital shopping is creating opportunities for strictly virtual suppliers that foresee a print-to-demand business model – that is, manufacturing after an order has been made, rather than bringing objects to market – extending to material objects. If citizens could simply click-to-buy and then click-to-print an object in proximity to where they live there would be momentous pressures for the existing complex of production, distribution and consumption to change. Reviewing the recent history of the mobilities of objects and their digitalization, this handbook entry introduces the social scientist to potentially radical near-future socio-technical transitions in how cities feature in the global flows of material objects.

The 'last mile' in cities appears insurmountable, yet solutions are perhaps on the horizon, notably with the emergence of decentralized forms of manufacturing happening near to consumers through digital technologies termed 3D printers and the associated innovations of digital file repositories, digital scanners, design software and smart materials. To present 3D printing as a technological fix, or techno-fix, for the last mile or any other barrier in the present logistics system would be inapt. Rather, 3D printing would be a feature in the future of a socio-technical transition to a system that undervalues the present one through progress across multiple dimensions that, after Geels (2012), includes science, industry, markets and user preferences, culture and policy.

3D printers negate the need for freight tubes and other grand infrastructure projects and require radically revising the movement of objects (Birtchnell and Urry, 2012). Yet, the potential introduction of digital making technologies into people's homes and communities is not a clear-cut process of shifting to a specific future, set in stone; instead, it involves thinking the unthinkable, for instance, car-free cities (Banister and Hickman, 2013). A socio-technical transition to a digital industrial revolution (Potstada and Zybura, 2014) requires foresight on the present mobilities of cargo (Birtchnell et al., 2015) and multiple, even conflicting, visions of the future (Birtchnell and Urry, 2017). The handbook entry adopts a perspective combining both the mobilities and post-mobilities of cargo (Birtchnell and Urry, 2015).

The return of urban making with 3D printing?

A *Popular Science* feature in the early twentieth century by architect Harvey Wiley Corbett titled 'The Wonder City You May Live to See'(Popular Science Monthly, 1925) foresaw a solution to city congestion in 1950 in stratification: shifting congestion either above or below the street pedestrian level to modally distinct sections. Most beguilingly in this hierarchy is the notion of 'freight tubes', specialized tunnels able to ferry cargo from source to customer without the need for short-term distribution by road, rail or air. Fast forward to 2018 and the news of the installation of the world's first freight tubes in Toulouse France, known as the Hyperloop, offering a tangible example of how cargo and passengers could travel in the future with similar projects being planned for California between airports and city hubs (Voltes-Dorta and Becker, 2018).

Whether the Hyperloop system will ever see the light of day is a moot point. The enduring problem it aims to tackle is the 'last mile' in logistics systems, a problem that has haunted cities ever since the emergence of the automobile as a form of ubiquitous transport. The majority of cargo entering the cities of the global North in the twenty-first century arrive at container terminals or airports set apart from urban centres. Once debarked, goods are then freighted by rail or road to inventory warehouses to be distributed to places where consumers are able to procure their individual items, either in stores or more commonly, with the rise of online procurement, in homes. The last mile denotes the relatively short distance between warehouse and the ultimate destination, in light of the distances transoceanic freight has travelled overall. Where road is the mode of delivery, vans, trucks, motorbikes, bicycles and even foot couriers compete with all manner of other forms of demand on urban transport infrastructure. Most significant is the morning and afternoon peak times as commuters travel to and from the urban core and entire transport networks face near to full capacity.

Instead of adopting a narrow, last-ditchery, view of the present system of containerization as inviolate, this handbook entry contemplates a future where data flows that encompass both information about objects' movements and the objects themselves transform the physical mobilities of cargo, notably, shifting it towards the mobilities of bulk resources rather than end-user products with consequences for handling and transportation (Birtchnell, 2016).

A systematic review in combination with two rounds of expert consultation made three findings useful here (Boon and van Wee, 2018). Cities will be the most plausible location for 3D printers because they co-locate both technical expertise and material flows; individual consumers with interests in personalizing and customizing objects will derive demand for 3D printers; and, raw materials will still require shipping of polymers, metals, resins and other inputs. Taking these findings into consideration, I am able to draw three main points. First, the withdrawal of consumers from shopping precincts in the urban core as a result of the spread of 3D printing will require offsetting by some other economic or social activity beyond simply residential concerns. The cultural revitalization of post-industrial clusters offers a pathway for the mass adoption of 3D printing wherein warehouses (Sasson and Johnson, 2016) and other industrial infrastructures are repurposed for alternative uses, for instance community craft clusters or farmers' markets.

Second, if 3D printers are widely adopted by consumers directly or indirectly, via intermediaries, there will still need to be a core group of 'lead users' with some engineering training and an advanced competence in design principles. Stewardship of design will be a disruption of 3D printing as consumers are made aware of how objects are constructed and move through stages of testing and development until entering the manufacturing process (Stein, 2017). Current studies on the backgrounds of users in a popular 3D printing online repository, Thingiverse, shows the lion's share of objects on the site that are utilized for customization and download have an origin in the labour of a small number of core designers (Özkil, 2017). As designs begin to be distributed across societies for 3D printing there will need to be clarification, both legally and ethically, about intellectual property ownership and fair use, if core designers are to continue to participate in cultures of sharing (Moilanen et al., 2015). 3D printing could be complementary to traditional manufacturing technology and would overlap as it has done in pre-production since its inception in rapid prototyping, restructuring, upgrading and distributing benefits along manufacturing global value chains. Or it would be a substitution, superseding traditional manufacturing fully or significantly, through bringing production closer to the consumer and involving buyers in the processes of making. An enticement to consumers would be to offer custom options and personalization; however, substitution could destabilize the existing system detrimentally (Rehnberg and Ponte, 2018).

Third, geopolitics will not necessarily cease with the transformation of the transoceanic containerization system to solely handle bulk cargo in the form of materials for 3D printing. One possibility is the expansion of the global 'fossil-fuel' petroleum and gas pipeline system to include facilities to process and convert raw fossil fuels into suitable materials for 3D printing: polymer and metal powders, filament, or liquids. Such retrofitting and growth in order to flow to both transportation fuels and object production would entail dramatic geopolitics (Bradshaw, 2009).

In sum, 3D printing offers consumers the ability to exercise their own agency and participate in the production and distribution of the objects they use. Here, there could be a transition in the stewardship of things that would render the present system obsolete, inefficient, environmentally damaging and a historical nuance tied to, what is in hindsight, a regrettable spike in greenhouse gas emissions (Lane and Watson, 2012). Certainly, such a utopic vision (Morgan, 2015) is not the only future possible; indeed, 3D printing could also underpin a dystopic future of increasing consumption at the expense of the environment and human wellbeing. The fact that there are many possible futures open to humanity should be seen as a boon rather than a curse (Urry, 2016).

As a concluding remark, it is imperative that, given the technology of 3D printing is nascent, we do not presume to suggest the future is certain or simply a result of the present unfolding. The global system extant in the present where a lion's share of manufacturing takes place in China and other countries in Asia, where objects are mostly shipped via transoceanic containerization, and where consumer choice does not extend beyond an object's price, size and colour, is only a recent phenomenon with historic roots in the politics and cultures of the post-Second World War period. It is indeed possible to imagine multiple futures where this system no longer prevails and where the materialization of digital information sets societies on a different path. Whether this trajectory is towards low-carbon, frugal and just ways of living is far from clear, although surely demanding of our critical foresight, given the global challenges of climate change, mass extinction, over-population and income inequalities that frame the twenty-first century.

References

Agapovichev, A.V., A.V. Balaykin, V.G. Smelov and A.V. Agapovichev (2015). "Application of additive technologies in the production of aircraft engine parts." *Modern Applied Science* **9**(4): 151–159.

Banister, D. and R. Hickman (2013). "Transport futures: Thinking the unthinkable." *Transport Policy* **29**: 283–293.

Baumers, M., L. Beltrametti, A. Gasparre and R. Hague (2017). "Informing additive manufacturing technology adoption: Total cost and the impact of capacity utilisation." *International Journal of Production Research* **55**(23): 6957–6970.
Baumers, M., P. Dickens, C. Tuck and R. Hague (2015). "The cost of additive manufacturing: Machine productivity, economies of scale and technology-push." *Technological Forecasting and Social Change.*
Bhasin, V. and M.R. Bodla (2014). *Impact of 3D Printing on Global Supply Chains by 2020.* Cambridge, MA: M. Eng in Logistics, MIT.
Birtchnell, T. (2016). "The missing mobility: Friction and freedom in the movement and digitization of cargo." *Applied Mobilities* **1**(1): 85–101.
Birtchnell, T., S. Savitzky and J. Urry, Eds. (2015). *Cargomobilities: Moving Materials in a Global Age.* London: Routledge.
Birtchnell, T. and J. Urry (2012). "Fabricating futures and the movement of objects." *Mobilities* **8**(3): 388–405.
Birtchnell, T. and J. Urry (2015). "The mobilities and post-mobilities of cargo." *Consumption Markets & Culture* **18**(1): 25–38.
Birtchnell, T. and J. Urry (2017). *A New Industrial Future? 3D Printing and the Reconfiguring of Production, Distribution, and Consumption.* Abingdon: Routledge.
Birtchnell, T., J. Urry and J. Westgate (2017). Design Mobilities via 3D Printing. In: *Mobilising Design: Intersections, Affordances, Relations.* Eds. J. Spinney, S. Reimer and P. Pinch. Abingdon: Routledge: 76–86.
Boon, W. and B. van Wee (2018). "Influence of 3D printing on transport: A theory and experts judgment based conceptual model." *Transport Reviews* **38**(5): 556–575.
Bradshaw, M.J. (2009). "The geopolitics of global energy security." *Geography Compass* **3**(5): 1920–1937.
Cano, J.L.C. (2011). "The Cambrian explosion of popular 3D printing." *International Journal of Artificial Intelligence and Interactive Multimedia* **1**(4): 31–33.
Cidell, J. (2012). "Flows and pauses in the urban logistics landscape: The municipal regulation of shipping container mobilities." *Mobilities* **7**(2): 233–245.
Cowen, D. (2014). *The Deadly Life of Logistics: Mapping Violence in Global Trade.* MN: University of Minnesota Press.
Cudahy, B.J. (2006). *Box Boats: How Container Ships Changed the World.* New York: Fordham University Press.
Despeisse, M., M. Baumers, P. Brown, F. Charnley, S.J. Ford, A. Garmulewicz, S. Knowles, T.H.W. Minshall, L. Mortara, F.P. Reed-Tsochas and J. Rowley (2017). "Unlocking value for a circular economy through 3D printing: A research agenda." *Technological Forecasting and Social Change* **115**: 75–84.
Dimico, A., A. Isopi and O. Olsson (2017). "Origins of the Sicilian Mafia: The market for lemons." *The Journal of Economic History* **77**(4): 1083–1115.
Drewes Nielsen, L., P. Homann Jespersen, T. Petersen and L. Gjesing Hansen (2003). "Freight transport growth: A theoretical and methodological framework." *European Journal of Operational Research* **144**(2): 295–305.
Ford, S. and M. Despeisse (2016). "Additive manufacturing and sustainability: An exploratory study of the advantages and challenges." *Journal of Cleaner Production* **137**: 1573–1587.
Geels, F.W. (2012). "A socio-technical analysis of low-carbon transitions: Introducing the multi-level perspective into transport studies." *Journal of Transport Geography* **24**(0): 471–482.
Geels, F.W. and W.A. Smit (2000). "Failed technology futures: Pitfalls and lessons from a historical survey." *Futures* **32**(9–10): 867–885.
Golob, T.F. and A.C. Regan (2000). "Freight industry attitudes towards policies to reduce congestion." *Transportation Research Part E: Logistics and Transportation Review* **36**(1): 55–77.
Graham, S. (2001). "FlowCity." *disP – The Planning Review* **37**(144): 4–11.
Gregson, N., M. Crang and C.N. Antonopoulos (2017). "Holding together logistical worlds: Friction, seams and circulation in the emerging 'global warehouse'." *Environment and Planning D: Society and Space* **35**(3): 381–398.
Hopkins, J.L. and J. McKay (2018). "Investigating 'anywhere working' as a mechanism for alleviating traffic congestion in smart cities." *Technological Forecasting and Social Change* **142**: 258–272.
Jiang, R., R. Kleer and F.T. Piller (2017). "Predicting the future of additive manufacturing: A Delphi study on economic and societal implications of 3D printing for 2030." *Technological Forecasting and Social Change* **117**(Supplement C): 84–97.

Lane, R. and M. Watson (2012). "Stewardship of things: The radical potential of product stewardship for re-framing responsibilities and relationships to products and materials." *Geoforum* **43**(6): 1254–1265.
Little, R.K. (2014). "Guns don't kill people, 3D printing does? Why the technology is a distraction from effective gun controls." *Hastings Law Journal* **65**(6): 1505–1514.
Moilanen, J., A. Daly, R. Lobato and D. Allen (2015). "Cultures of sharing in 3D printing: What can we learn from the licence choices of Thingiverse users?" *Journal of Peer Production* (6): 1–9.
Morgan, D.R. (2015). "The dialectic of utopian images of the future within the idea of progress." *Futures* **66**: 106–119.
Nye, D.E. (2014). "The United States and alternative energies since 1980: Technological fix or regime change?" *Theory, Culture & Society* **31**(5): 103–125.
Özkil, A.G. (2017). "Collective design in 3D printing: A large scale empirical study of designs, designers and evolution." *Design Studies* **51**: 66–89.
Popular Science Monthly. (1925). *Wonder City You May Live to See May Solve Congestion Problems*. Popular Science Monthly. New York: Popular Science Publishing Company: 40–41.
Potstada, M. and J. Zybura (2014). "The role of context in science fiction prototyping: The digital industrial revolution." *Technological Forecasting and Social Change* **84**: 101–114.
Rael, R. and V. San Fratello (2017). "Clay bodies: Crafting the future with 3D printing." *Architectural Design* **87**(6): 92–97.
Rayna, T. and L. Striukova (2016). "From rapid prototyping to home fabrication: How 3D printing is changing business model innovation." *Technological Forecasting and Social Change* **102**: 214–224.
Rehnberg, M. and S. Ponte (2018). "From smiling to smirking? 3D printing, upgrading and the restructuring of global value chains." *Global Networks* **18**(1): 57–80.
Sasson, A. and J.C. Johnson (2016). "The 3D printing order: Variability, supercenters and supply chain reconfigurations." *International Journal of Physical Distribution & Logistics Management* **46**(1): 82–94.
Stein, J.A. (2017). "The political imaginaries of 3D printing: Prompting mainstream awareness of design and making." *Design and Culture* **9**(1): 3–27.
Stewart, D. (2018). "3D printing growth accelerates again." Retrieved 3 April, 2019, from www2.deloitte.com/insights/us/en/industry/technology/technology-media-and-telecom-predictions/3d-printing-market.html.
Tyfield, D. (2018). "Innovating innovation—Disruptive innovation in China and the low-carbon transition of capitalism." *Energy Research & Social Science* **37**: 266–274.
Urry, J. (2000). "Mobile sociology1." *The British Journal of Sociology* **51**(1): 185–203.
Urry, J. (2016). *What is the Future?* Cambridge: Polity.
Voltes-Dorta, A. and E. Becker (2018). "The potential short-term impact of a Hyperloop service between San Francisco and Los Angeles on airport competition in California." *Transport Policy* **71**: 45–56.
Weller, C., R. Kleer and F.T. Piller (2015). "Economic implications of 3D printing: Market structure models in light of additive manufacturing revisited." *International Journal of Production Economics* **164**: 43–56.

36

MOBILITY AS A SERVICE

Moving in the de-synchronized city

Chiara Vitrano and Matteo Colleoni

Introduction

In the contemporary city, the ability to be mobile is a crucial prerequisite to access urban opportunities (jobs, private and public services, social relations). This is due to phenomena shaping the spatio-temporal configuration of urban contexts. On the one side, the morphological changes of the city have reduced the spatial proximity of opportunities and of dwellings, leading to more intense mobilities. On the other side, temporal processes connected to the production and consumption systems lead to a polyrhythmic city, in which the socio-temporal coordination is made difficult by the de-synchronized and continuous activities of urban populations, creating the need for unsystematic and less predictable mobilities (Camarero and Oliva, 2008). These new rhythms impose the mastery of a higher amount of resources (financial, physical, cognitive, temporal) to access urban opportunities. The possession of an adequate mobility capital, or "motility" (Kaufmann, Bergman and Joye, 2004), becomes an essential precondition for the construction and maintenance of relations and social connections in the contemporary, de-synchronized city (Pflieger et al., 2008; Shove, 2002; Urry, 2007).

These processes, together with the spread of ITCs (especially smartphones) and the diffusion of the idea of *usership* instead of ownership (a pillar of the growing "sharing economy"), are having a profound impact on contemporary cities. Mobility as a Service (MaaS) seems to emerge as a product of these dynamics for what concerns urban mobility. The term "Mobility as a Service" indicates the purchase of "mobility services as packages based on consumers' needs instead of buying the means of transport" (Kamargianni et al., 2016, p. 3295). MaaS has been described as a mobility distribution model that, through a single interface for planning, reserving and paying for trips, allows the user to combine different transport modes in a tailored mobility package. Presented as "one of the promising tools for decarbonisation of the transport system" (Holmberg et al., 2016, p. 10), it aims at fostering sustainable urban transport, offering door-to-door mobility solutions and thereby discouraging private car ownership; at the same time, it aims at meeting users' needs with a flexible mobility supply.

The chapter provides an overview of the definitions, the intended aims and the criticalities of MaaS. The first part presents a review of the literature about MaaS, focusing on both scientific articles and non-academic documents and reports. It examines the core characteristics of the phenomenon and outlines the MaaS ecosystem, identifying its key actors.

The second part proposes some reflections on the intended aims of the model and stresses the potential criticalities, especially for what concerns environmental sustainability and transport-related inequalities.

What is MaaS

Gliding on the concept

The term Mobility as a Service indicates a new transport paradigm (Hietanen, 2014), which still lacks a shared definition. MaaS

> can be thought of as a concept (a new idea for conceiving mobility), a phenomenon (occurring with the emergence of new behaviours and technologies) or as a new transport solution (which merges the different available transport modes and mobility services).
>
> *(Jittrapirom et al., 2017, p. 14)*

There is still no unanimous understanding of which mobility services should fall under the label of MaaS. In the broadest sense, MaaS includes those mobility solutions, which aim at meeting the individuals' mobility needs by guaranteeing the usage of transport means without the ownership of any of those. The lack of a rigorous definition appears to some as a condition for the conceptual development of MaaS in these early and transition stages of the phenomenon (Holmberg et al., 2016; Sochor et al., 2017). According to this approach, MaaS includes peer transport services (e.g. Uber, where the provider does not own the vehicles but only the platform to pair users and drivers), car/bike/scooter-sharing (where the provider owns the vehicles), extended multimodal planners (which help users plan the best route to a destination), combined mobility services (e.g. UbiGo and Whim, which offer subscription-based combined mobility options), integrated public transport (e.g. Helsinki's Mobility on Demand, where public transport is designed to integrate other mobility options such as car/bike-sharing and taxis) and mobility brokers (which offer mobility subscriptions as part of the rent in specific city areas) (Holmberg et al., 2016). In these conceptualizations, MaaS appears as an umbrella term, which basically includes all those mobility services characterized by some degree of integration aimed at guaranteeing seamless intermodal journeys (e.g. the integration of tickets and payments, of mobility services into a single package, of ICTs) (Kamargianni, Li, Matyas and Schäfer, 2016).

Notwithstanding the contribution of these conceptualizations to the comprehension of the emerging mobility services, in its most appropriate – nevertheless quite loose – meaning, MaaS indicates a mobility distribution model that, through a single interface for planning, reserving and pay for trips, allows the user to combine different transport modes in a tailored mobility package (Hietanen, 2014; Jittrapirom et al., 2017). Even if the idea of seamless door-to-door mobility offered in transport packages was not new to the Intelligent Transport Systems research, the expression "Mobility as a Service" made its first appearance in Finland in 2014. Sampo Hietanen,[1] at that time CEO of ITS Finland, has been one of the first advocates of the adoption of the MaaS framework and has contributed to its diffusion among academics and policy makers (Heikkilä, 2014; Hietanen, 2014; Smith et al., 2018). According to Hietanen, MaaS emerges from a vision of "the whole transport sector as a co-operative, interconnected ecosystem", which "consists of transport infrastructure, transportation services, transport information and payment services" (Hietanen, 2014, p. 27).

Through an extensive review of the literature, Jittrapirom and colleagues (Jittrapirom et al., 2017) identified the core characteristics contained in the definitions of MaaS: (1) The integration of transport modes, allowing the users to choose and navigate their intermodal trips; (2) The possibility to choose between two main tariff options: the subscription to a pre-paid mobility package or a "pay-as-you-go" charge; (3) A single digital platform (app or webpage) for planning and booking the trips, paying for the tickets and getting real-time information; (4) An ecosystem made up of multiple actors; (5) The use of different technologies, such as devices (e.g. smartphones), a reliable mobile internet network, GPS, e-payment systems, database management systems; (6) A strong demand orientation, i.e. the supply of mobility solutions is centred on the user's needs and benefits; (7) The requirement of registration to the platform to access the available services; (8) The personalization of the suggested solutions on the basis of the user's profile; (9) The possibility to customize the offered service on the basis of the preferred transport modes, the CO_2 footprint or one's available time.[2]

The MaaS ecosystem

The integration required in order to develop MaaS schemes implies the interaction and cooperation of a multitude of actors. The public and private actors constituting the MaaS ecosystem can be traced back to four main and sometimes overlapping levels (Eckhardt et al., 2017):

1) The public and regulatory level, which encompasses both the national and the local authorities. The public sector's role in MaaS appears crucial, especially for its tasks to "manage negative externalities from commercial markets" (Holmberg et al., 2016, p. 27) and "ensure an equitable and sustainable transport system" (Hoadley, 2017). More specifically, the role of the public sector will depend on the trajectory followed by MaaS's business models in each context.[3]
2) The transport and mobility service providers' level, e.g. PT agencies, taxi, bike and car-sharing, car-rental and ride-sharing companies. The responsibilities of these actors include the provision of a quality service and the opening of data and APIs. Furthermore, MaaS can involve Logistics Service Providers, interested in managing the flow of goods in a seamless way. A further actor is the Mobile Service Provider (MSP), the company offering a well-functioning ICT infrastructure. Since "MaaS is about integrating transport modes through the internet" (Finger et al., 2015, p. 8), MSPs are considered to be the enablers of its development (Eckhardt et al., 2017),
3) The MaaS Operator (MO) level. The MO is a new player (Eckhardt et al., 2017) and its role is to "combine the existing transport services into a single mobile application on the 'one-stop-shop' principle and provide personalized transport plans tailored to customer needs" (Eckhardt et al., 2017, p. 28). The responsibilities of the MO include the user's experience and the development and testing of new services, business models, and combinations of additional non-transport services.
4) The end-user level. The users' role is to influence the service development on the basis of their needs, requirements, evaluation processes and acceptance of the model. They also contribute through *prosuming* (e.g. for car-pooling).

A more accurate look at the MaaS ecosystem shows a higher number of levels and stakeholders with different roles and responsibilities, which change according to the country and to the business model (Eckhardt et al., 2017).[4] Therefore, MaaS is based on new forms of partnerships, "in which private actors play a larger role in the creation of public value" (Smith et al., 2018). The choice of the business model for MaaS – market driven, public controlled or based on a public–private partnership (Smith et al., 2018) – will be paramount also to determine the inclusiveness and sustainability of MaaS. In fact, according to Pangbourne, Stead, Mladenović and Milakis (2018, p. 35), "the commodification of access to mobility by commercial intermediaries" may put at risk a sustainable transport system, since "non-motorized transport modes [...] may be side-lined as mobility options because they do not generate substantial income for MaaS" (Pangbourne et al., 2018, p. 40).

For what concerns actual implementations, some authors have noted a scarce engagement of public authorities in the first MaaS projects (Hoadley, 2017; Li, 2017), stressing at the same time their cardinal role in promoting the use of sustainable means of transport – a goal which could be not profitable for MaaS operators. It seems therefore crucial to implement MaaS in the framework of a business and governance model that pursues public priorities (e.g. reducing the environmental impact of transport and protecting the ability to be mobile of the population) and, at the same time, allows for the development of market services sharing these goals (Hoadley, 2017). In relation to this, one of the main challenges of MaaS will be guaranteeing environmental and social sustainability while, at the same time, meeting the users' needs. The criticalities emerging from these challenges will be the object of the next section.

Personalized, yet sustainable: intended aims and criticalities of the MaaS model

A user-centric service

Since MaaS aims at providing tailor-made mobility solutions, the users' needs, attitudes towards consumption and mobility, and their actual travel behaviours are among the main drivers of the service development. The main changes in broader consumption attitudes and behaviours emerge from a more and more widespread environmental consciousness, trends towards joint/shared ownership or no ownership at all,[5] the widespread dissemination of ICTs, the loss of private cars' appeal towards the youngest generations (Dant, 2014; Kuemmerling et al., 2013; Sochor et al., 2015). At the same time, travel demand is described as characterized by a greater request for flexibility and on-demand (instant) services (Dotter, 2016), according to an "everything at any time" logic pervading all current consumption practices. Therefore, end-users may expect MaaS to be "user centric, easy to plan, book and pay, as well as seamless during the actual trip, integrating all transport means and systems, using real-time data, and responding to a broad range of individual user priorities" (Giesecke et al., 2016, p. 1).

A development of the MaaS service based on users' individual requests and aimed at increasing its customer base may result in unsustainable mobility practices. Moreover, the same capacity of MaaS to meet every user's *whim* at any time – as the telecommunication or streaming services to which MaaS is likened by its promoters – is put in question and considered a "false promise of freedom" due to the risks of congestions of the transport network, which is essentially different from data networks (Pangbourne et al., 2018). According to Mulley (2017), initiatives from the public actor, such as public health messaging and abandonment of minimum car-park spaces in urban planning, can help the development and dissemination of MaaS, increasing its user base, while preserving its sustainability goals.

MaaS's intended sustainability aims, as expressed by its promoters, show a series of criticalities, which will be examined in the next paragraph. We'll focus on MaaS's impacts on emissions and congestion, on public transport, and on physical, financial, spatial and digital inequalities.

From ownership to access: do fewer private cars mean fewer trips?

MaaS promotes a "shift away from the existing ownership-based transport system toward an access-based one" (Jittrapirom et al., 2017, p. 13). It aims at making car ownership pointless, providing seamless, easy to plan and book, door-to-door mobility solutions, which solve the last mile problem including also services as ride-hailing, car-sharing and car-rental. In order to make its services more attractive than privately owned vehicles, MaaS has to facilitate multimodal and intermodal transfers through the integration of a payment and ticketing operation on a single account and platform, on a one-stop shop principle, even better if offered in a pre-paid package (Eckhardt et al., 2017; Kamargianni et al., 2016). Furthermore, the user's decision to give up their car will also depend both on the new service's price and comfort, and on their willingness to change travel behaviour (Hoadley, 2017).[6]

According to MaaS promoters, the shift from the private car in favour of easily accessible mobility services would contribute to environmental sustainability leading to a reduction of emissions and congestion, a decrease in the number of car trips and an increase in road and land use efficiency. The latter aim would be reached, reducing the need for parking spaces (due to the increase in the use of shared vehicles) and directing the traffic to less congested routes, thanks to real-time data (Karlsson et al., 2017).

Nevertheless, according to some observers, MaaS's contribution to environmental sustainability has been taken for granted and the lack of supporting evidence to this assumption has been overlooked (Li, 2017). It seems rather likely that MaaS implementation may bring an increase of car-based trips (even if shared or e-hailed), and hence result in more emissions and congestions (Eckhardt et al., 2017). First of all, as underlined by Karlsson et al. (2017), the easiest access to car-sharing may result in an increase in car use for those who would not have used a car for their trips. Secondly, the supply of mobility packages would encourage so-called "induced trips", based on the regret of not having used all the taxi/ride hailing/car-sharing trips provided by the monthly subscription (Pangbourne et al., 2018, p. 40).

Maas and transport-related disadvantage

Given the centrality of mobility in current social practices, emerging transport models as MaaS have to be examined also in relation to inclusion and transport-related disadvantage. Mobility-related exclusion indicates

> the process by which people are prevented from participating in the economic, political and social life of the community because of reduced accessibility to opportunities, services and social networks, due in whole or in part to insufficient mobility in a society and environment built around the assumption of high mobility.
>
> *(Kenyon et al., 2002, pp. 210–211)*

We argue that a MaaS-based transport system could contribute to mobility-related social exclusion by exacerbating the insufficient mobility of some categories of users. Speaking of risks of insufficient mobility in such a transport network seems to be counterintuitive. At a first glance, MaaS seems to offer a valid solution to inaccessibility to opportunities, enhancing the mobility potential of those who don't own a car. A closer look at the first implementations may confirm MaaS's potential in overcoming forms of physical exclusion – if subsidized and directed by public authorities.[7] Nevertheless, criticalities emerge in relation to MaaS's impacts on public transport, whose appropriate provision is crucial to tackle social exclusion (Lucas et al., 2008), and on spatial, financial, and digital inequalities.

According to its promoters, MaaS can benefit public transport by enlarging its potential user base and allowing better access to public transport from/to underserved locations by alleviating the first/last mile problem. Nevertheless, not all the services included in the MaaS system share the mandate of public transport and pursue equity goals, rather targeting the most profitable users (Amin, 2018). Indeed, the impact of MaaS on public transport may be disrupting. Recent figures related to the impact of ride-hailing (one of the pillar services of MaaS) in US cities show how services such as Uber have attracted users away from bus and light rail services (Clewlow and Mishra, 2017). Furthermore, researchers have found that "a majority (61%) of [ride-hailing] trips would have not been made at all, or by walking, biking, or transit" (Clewlow and Mishra, 2017, p. 26). Furthermore, in 2016, Uber tested UberPlus, a ride-hailing service, which, through monthly subscription packages, offered rides at a cheaper price than the PT tickets (Lindsay, 2018). A lack of public supervision of MaaS services could therefore lead to a cannibalization of the public transport share (Smith et al., 2018).

For what concerns spatial inequalities, a MaaS-based transport system may provide better access to those residing in already well-served areas, such as central urban areas. The last few years have seen a growing interest by private companies in running mass transit (Wilding, 2018), which can severely jeopardize public transport. If these companies were to run their bus service in the most profitable areas, attracting usual public transport users with faster, more comfortable bus services, public transport could be left without the money to cover expenses for the least utilized, often more peripheral, routes. Underserved neighbourhoods would face inaccessibility and those living on the profitable routes would experience higher congestion.[8] Even if its promoters suggest that MaaS could help overcome transport exclusion in isolated areas (Burrows et al., 2015), the same characteristics of the service (e.g. its Multi-Sided Platform nature, which increases benefits of both users and providers when actors on both sides grow (Pangbourne et al., 2018)) make it hard to offer MaaS solutions in less dense areas, such as rural areas (Karlsson et al., 2017).

With regard to financial inequalities, MaaS offers an alternative to owning a car, but the prepaid subscriptions (and the necessity to pay large sums in advance) may be equally hard to face by low-income people (Karlsson et al., 2017). In addition, MaaS may help to exacerbate inequalities in access to mobility services where it offers "premium levels of service to those who pay more (such as priority seats and boarding, or faster and safer connections)" (Hoadley, 2017, pp. 7–8). Lastly, financial inequalities may emerge with the promotion of a *cashless city*, where the diffusion of services based on cashless payments excludes those who are not registered in any banking service.

Lastly, mobility inequality may increase due to digital divide, where digital competences become a crucial resource to access transport solutions. Pangbourne argues that "MaaS's reliance on registration and digitalization [...] creates additional barriers for those who are already experiencing exclusion" (Pangbourne et al., 2018, p. 39) and deepens inequality, offering discounts to the owners of smartcards and app users.

Hence, MaaS may emerge as a transport model leading to uneven mobilities (Sheller, 2018), where those already equipped with financial, spatial and digital resources will not only be the ones benefitting from the new mobility services, but they might do it also to the detriment of the most disadvantaged.

Conclusions

MaaS has been defined as "the Spotify of Transport" or has been compared to a monthly mobile phone contract that, as any other on-demand subscription-based service, makes access to its contents (in this case, transport) extremely easy and taken for granted. We can therefore wonder whether an "all you can ride" mobility model will make the individuals' burden of mobility, "the opposite side of motility" (Urry, 2007, p. 52), heavier or lighter.

As shown in the previous paragraph, MaaS can exacerbate mobility-related social exclusion. Not only may MaaS exacerbate inequalities leading to insufficient mobility, it may also strengthen the assumption of high mobility, which lies at the basis of mobility related inequalities. Providing door-to-door, 24/7, bundled mobility options, MaaS takes mobility for granted and immobility merely as an individual choice. This appears connected to the process leading to a "self-privatised city" (Camarero and Oliva, 2008), where the physical and social environment of the city is built on the basis of individual requests. Personalization, a process characterizing the contemporary de-synchronized city, forces individuals to ceaselessly take decisions in a wider range of possibilities, holding them responsible for their choices (Mulíček et al., 2016).

As specified by Reisch (2001), the acceleration and de-synchronization of life rhythms is one of the factors at the basis of socially and environmentally unsustainable lifestyles and forms of consumption. The gaps between MaaS's intended aims and potential effects on environmental sustainability and social inclusion, shown in this chapter, seem to confirm this thesis.

Furthermore, the chapter has shown the importance of the role of public authorities to oversee MaaS development, especially to guarantee access to mobility and environmental quality. Nevertheless, public authorities seem to be inadequately prepared to play an active role in MaaS and assess its impacts, since research and implementations are still limited and much information about MaaS is produced by providers and may be driven by commercial interest. City administrations and public transport agencies will have to play a crucial role to guarantee that new mobility services aim at fulfilling not only individual needs, but also broader societal goals.

Notes

1 He is currently the founder and CEO of Maas Global, the company behind Whim, "the first all-inclusive MaaS solution" which was launched in the Helsinki region in 2016. Source: https://whimapp.com/about-us/[Accessed 3.09.208].

2 For a useful review of the current operational and pilot MaaS projects see Jittrapirom et al. (2017) and Kamargianni (2016).

3 See Mukhtar-Landgren and Smith (2018) for an analysis of the public sector's activities in the development of MaaS.

4 Further stakeholders include funding agencies, the academia (providing knowledge and evaluation), and OEMs and resellers (producing the electric vehicles fleet, implementing new technologies and creating new business models) (Eckhardt et al., 2017). Other actors, such as unions, media, marketing and advertising companies might join the MaaS ecosystem as it evolves (Kamargianni and Matyas 2017).

5 The worldwide and continuous growth of car-sharing usage is a sign of these changing attitudes and behaviours (Shaheen and Cohen, 2016).

6 Since the uses of private cars are not limited to the mere mobility of individuals or groups, the difficulty to replicate the automobiles' multiple uses in the framework of MaaS could act as an obstacle to its success. Not only are cars, for many users, privacy spaces, mobile homes, storages or offices (Laurier, 2004; Urry, 2006), they also transcend instrumental reasons, fulfilling strong symbolic and affective functions (Steg, 2005).

7 MaaS might help overcome physical constraints to mobility by offering an efficient support to public authorities in providing subsidized door-to-door on-demand transport to persons with reduced mobility, e.g. the elderly and people with disabilities (Hoadley, 2017), on condition that public actors take on an active role in providing such service in partnership with the new mobility providers (Tsay et al., 2016).

8 These aspects were the core of the limits posed by Transport for London to CityMapper's and Ford's proposals (Hern, 2018; Kobie, 2018).

References

Amin, R., 2018. Why We Can't Leave Transportation Apps to the Private Sector. *Spur*, [online] 25 April. Available at: www.spur.org/news/2018-04-25/why-we-can-t-leave-transportation-apps-private-sector [Accessed 24.09.2018].

Burrows, A., Bradburn, J. and Cohen, T., 2015. Journeys of the Future. *Introducing Mobility as a Service*. Available at: www.atkinsglobal.com/~/media/Files/A/Atkins-Corporate/uk-and-europe/uk-thought-leadership/reports/Journeys%20of%20the%20future_300315.pdf [Accessed 25/09/2018].

Camarero, L.A. and Oliva, J., 2008. Exploring the Social Face of Urban Mobility: Daily Mobility as Part of the Social Structure in Spain. *International Journal of Urban and Regional Research*, 32(2), pp. 344–362.

Clewlow, R.R. and Mishra, G.S., 2017. *Disruptive Transportation: The Adoption, Utilization, and Impacts of Ride-Hailing in the United States*. [pdf] Davis: University of California. Available at: https://itspubs.ucdavis.edu/wp-content/themes/ucdavis/pubs/download_pdf.php?id=2752 [Accessed: 10.09.2018].

Dant, T., 2014. Drivers and Passengers. In: P. Adey, D. Bissell, K. Hannam, P. Merriman, M. Sheller, eds. 2014. *The Routledge Handbook of Mobilities*. London: Routledge. pp. 367–375. Ch.35.

Dotter, F., 2016. Mobility-as-a-Service: A New Transport Model. *CIVITAS Insight No.18*. https://civitas.eu/sites/default/files/civitas_insight_18_mobility-as-a-service_a_new_transport_model.pdf

Eckhardt, J., Aapaoja, A., Nykänen, L., Sochor, J., Karlsson, M. and König, D., 2017. Deliverable 2: European MaaS Roadmap 2025. MAASiFiE project funded by CEDR. [pdf]. Available at: www.vtt.fi/sites/maasifie/PublishingImages/results/cedr_mobility_MAASiFiE_deliverable_2_revised_final.pdf [Accessed: 10.09.2018].

Finger, M., Bert, N. and Kupfer, D., 2015. Mobility-as-a-Service: from the Helsinki experiment to a European model? 3rd European Intermodal Transport Regulation Summary. Technical Report, European Transport Regulation Observer. No. 2015/01. https://cadmus.eui.eu/handle/1814/38841

Giesecke, R., Surakka, T. and Hakonen, M., 2016. Conceptualising Mobility as a Service. A User Centric View on Key Issues of Mobility Services. In: EVER, *Eleventh International Conference on Ecological Vehicles and Renewable Energies*. Monte Carlo, Monaco, 6–8 April 2016.

Heikkilä, S., 2014. *Mobility as a Service: A Proposal for Action for the Public Administration, Case Helsinki*. MSc. Aalto University. Available at: https://aaltodoc.aalto.fi/handle/123456789/13133 [Accessed 20.09.2018].

Hern, A., 2018. Citymapper Launches Bus-Taxi Hybrid Smart Ride in London. *The Guardian*. [online] 21 February. Available at: www.theguardian.com/technology/2018/feb/21/citymapper-launches-bus-taxi-hybrid-smart-ride-london-transit-app [Accessed 24.09.2018].

Hietanen, S., 2014. "Mobility as a Service": The New Transport Model? *Eurotransport*, 12(2), pp. 26–28.

Hoadley, S., 2017. Mobility as a Service: Implications for Urban and Regional Transport. Discussionpaper Offering the Perspective of Polis Member Cities and Regions on Mobility as a Service (MaaS). *Polis Traffic Efficiency and Mobility Working Group*. Retrieved at: www.polisnetwork.eu/wp-content/uploads/2017/10/polis-maas-discussion-paper-2017-final_pdf

Holmberg, P.-E., Collado, M., Sarasini, S. and Williander, M., 2016. Mobility as a Service - maas. Describing the Framework. [pdf] Göteborg: Viktoria Swedish ICT. Available at: www.viktoria.se/publications/mobility-as-a-service-maas-describing-the-framework [Accessed 12.09.2018].

Jittrapirom, P., Caiati, V., Feneri, A.-M., Ebrahimigharehbaghi, S., González, M.J.A. and Narayan, J., 2017. Mobility as a Service: A Critical Review of Definitions, Assessments of Schemes, and Key Challenges. *Urban Planning*, 2(2), pp. 13–25.

Kamargianni, M. and Matyas, M. 2017. The business ecosystem of mobility as a service. 96th Transportation Research Board (TRB) Annual Meeting, Washington, DC.

Kamargianni, M., Li, W., Matyas, M. and Schäfer, A., 2016. A Critical Review of New Mobility Services for Urban Transport. *Transportation Research Procedia*, 14, pp. 3294–3303.

Karlsson, M., Sochor, J., Aapaoja, A., Eckhardt, J. and König, D., 2017. Deliverable 4: Impact Assessment. MAASiFiE project funded by CEDR. [pdf] Available at: www.vtt.fi/sites/maasifie/PublishingImages/results/CEDR_Mobility_MAASIFIE_Deliverable_4_Revised_Final.pdf [Accessed 10.09.2018].

Kaufmann, V., Bergman, M. M., & Joye, D. (2004). Motility: Mobility as capital. *International Journal of Urban and Regional Research*, *28*(4), 745–756.

Kenyon, S., Lyons, G. and Rafferty, J., 2002. Transport and Social Exclusion: Investigating the Possibility of Promoting Inclusion through Virtual Mobility. *Journal of Transport Geography*, 10 (3), pp. 207–219.

Kobie, N., 2018, The rules refuse to bend as Citymapper moves to disrupt London Transport. *Wired.* [online] 22 February. Available at: www.wired.co.uk/article/citymapper-smart-ride-ceo-smart-bus-make-money [Accessed 24.09.2018].

Kuemmerling, M., Heilmann, C. and Meixner, G., 2013. Towards Seamless Mobility: individual Mobility Profiles to Ease the Use of Shared Vehicles. In: IFAC, *12th IFAC Symposium on Analysis, Design, and Evaluation of Human–Machine Systems.* Las Vegas, NV, USA, August 11-15, 2013.

Laurier, E., 2004. Doing Office Work on the Motorway. *Theory, Culture & Society*, 21(4-5), pp. 261–277.

Li, Y., 2017. Future Roles of Public Authorities in Mobility as a Service (MaaS). *Workshop report. SPICE project.* [pdf] Available at: http://spice-project.eu/wp-content/uploads/sites/14/2017/04/Report-from-workshop-3.pdf [Accessed 10.09.2018].

Lindsay, G., 2018. The State of Play: Connected Mobility + U.S. Cities. How next Generation Transportation is Shaping Cities. [pdf]. *CityLab Insights*. Available at: https://cdn.theatlantic.com/assets/media/files/citylab_insights_connected_mobility.pdf [Accessed 20.09.2018].

Lucas, K., Tyler, S. and Christodoulou, G., 2008. *The value of new transport in deprived areas. Who benefits, how and why?* New York: Joseph Rowntree Foundation.

Mukhtar-Landgren, D. and Smith, G., 2018. Perceived Action Spaces for Public Actors in the Development of Mobility as a Service. In: *7th Transport Research Arena*, Vienna, Austria, 16–19 April 2018.

Muliček, O., Osman, R. and Seidenglanz, D., 2016. Time-Space Rhythms of the City: The Industrial and Postindustrial Brno. *Environment and Planning A*, 48(1), pp. 115–131.

Mulley, C., 2017. Mobility as a Services (MaaS): Does it have Critical Mass? *Transport Reviews*, 37(3), pp. 247–251.

Pangbourne, K., Stead, D., Mladenović, M. and Milakis, D., 2018. The Case of Mobility as a Service: A Critical Reflection on Challenges for Urban Transport and Mobility Governance. In: G. Marsden and L. Reardon, eds. 2018. *Governance of the Smart Mobility Transition.* Bingley: Emerald Publishing Limited. pp. 33–48.

Pflieger, G., Pattaroni, L., Jemelin, C. and Kaufmann, V., 2008. Introduction: Urban Forms, Experience and Power. In: G. Pflieger, L. Pattaroni, C. Jemelin, V. Kaufmann, eds. 2008. *The Social Fabric of the Networked City.* Lausanne: EPFL Press/Routledge. pp. 1–15.

Reisch, L.A., 2001. Time and Wealth: The Role of Time and Temporalities for Sustainable Patterns of Consumption. *Time & Society*, 10(2–3), pp. 367–385.

Shaheen, S. and Cohen, A., 2016. *Innovative Mobility Carsharing Outlook. Carsharing Market Overview, Analysis, and Trends.* [pdf] Berkeley: Transportation Sustainability Research Center, University Of California. Available at: http://innovativemobility.org/wp-content/uploads/2016/02/Innovative-Mobility-Industry-Outlook_World-2016-Final.pdf [Accessed 2.09.2018].

Sheller, M., 2018. Theorizing Mobility Justice. *Tempo social, revista de sociologia da USP*, 30(2), pp. 17–34.

Shove, E., 2002. *Rushing around: coordination, mobility and inequality.* Draft paper for the Mobile Network meeting, October 2002.

Smith, G., Sochor, J. and Karlsson, I.C.M., 2018. Mobility as a Service: Development Scenarios and Implications for Public Transport. *Research in Transportation Economics*, 69, pp. 592–599.

Sochor, J., Arby, H., Karlsson, M. and Sarasini, S., 2017. A Topological Approach to Mobility as a Service: A Proposed Tool for Understanding Requirements and Effects, and for Aiding the Integration of Societal Goals. In: *1st International Conference on Mobility as a Service (ICoMaaS)*. Tampere, Finland, 28-29 November 2017.

Sochor, J., Strömberg, H. and Karlsson, I.C.M., 2015. An Innovative Mobility Service to Facilitate Changes in Travel Behavior and Mode Choice. In: *22nd ITS World Congress*. Bordeaux, France, 5–9 October 2015.

Steg, L., 2005. Car Use: Lust and must. Instrumental, Symbolic and Affective Motives for Car use. *Transportation Research Part A: Policy and Practice*, 39(2–3), pp. 147–162.

Tsay, S., Accuardi, Z. and Schaller, B., 2016. Private Mobility, Public Interest – How Public Agencies can Work with Emerging Mobility Providers. [pdf] New York: Transit Center. Available at: http://transitcenter.org/wp-content/uploads/2016/10/TC-Private-Mobility-Public-Interest-20160909.pdf [Accessed 3.09.2018].

Urry, J., 2006. Inhabiting the car. *The Sociological Review*, 54, pp. 17–31.

Urry, J., 2007. *Mobilities*. Cambridge: Polity Press.

Wilding, M., 2018, Private Companies want to Replace Public Transport. Should we let them? *The Guardian – Cities*, [online] 29 March. Available at: www.theguardian.com/cities/2018/mar/29/public-transport-transit-private-companies-citymapper-uber-whim-smart-buses [Accessed 24.09.2018].

37

UNDERSTANDING MULTIMODALITY THROUGH RHYTHM OF LIFE

Empirical evidence from the Swiss case study

Guillaume Drevon and Alexis Gumy

Background

Theoretical framework

Transports and telecommunications provide considerable speed potentials by allowing immediacy for information and idea flows. These potentials have contributed to the "shrinking" of space and time (Drevon et al., 2017). The resulting acceleration of daily life tends to shape hybrid lifestyles relying on new arrangements between daily travels and residential mobility (Ravalet et al., 2015; Vincent et al., 2016). Typical mobility patterns are thus more difficult to define. As suggested by several authors such as Hartmut Rosa (Rosa, 2010), Nicole Auber (Auber, 2009) and Robert Levine (Levine, 1997), life rhythms are deeply changing in contemporary societies. These authors share the conviction that this acceleration is relative to the conjunction between technological advances and new forms of capitalism. In this context, a rhythmic approach is relevant to understand and analyse the daily mobility of these contemporary societies (Drevon, 2019). In the field of mobility, rhythmic approaches are rather uncommon (Edensor, 2012). However, some authors like Tim Cresswell (Cresswell, 2010) argue for taking rhythms into account for mobility and public policy analyses. The situation is different with regard to geography. Based on the theoretical corpus of the Lund School (Hägerstraand, 1970), Time Geography was proposed to consider space and time together to better understand mobility. This chapter proposes a rhythmic approach of mobility based on a spatiotemporal analysis of daily mobility behaviours (Lenntrop, 1976). In particular, the empirical analysis is based on the case of Switzerland.

Developing a rhythmic approach for a better understanding of daily mobility

The concept of rhythm appears relatively implicitly in two particular main approaches. The first corresponds to the Time Geography and the second to the activities-based approach. These approaches were developed widely during the early 1970s.

The contributions of "Time Geography" are composed of three elements:

- The first corresponds to the simultaneous consideration of space and time from the space–time prism diagram and the concept of space–time path.
- The second element concerns trips constraints (capacity constraints, interaction constraints, authority constraints). All of these constraints underlie the concept of accessibility (Pirie, 1979) which corresponds to the space–time budget available for an individual in order to carry out his activities programme.
- The third contribution corresponds to the consideration of all trips as a mobility project and not as a series of independent activities.

Although Time Geography promoted time as a central element to its mobility theory, this approach does not consider people's relationship to time nor their motivations (Hallin, 1991). Alongside the Time Geography corpus, sociologists like Francis Stuart Chapin (Chapin, 1974) developed the concept of activity patterns, which focuses on the succession and sequencing of activities in everyday life (Ellegård and Vilhelmson, 2004). These authors highlighted the relationships between activity programmes and sociodemographic characteristics of individuals (Thévenin et al., 2007). This approach was developed by Peter Jones, who refined the analysis of the relationships between activity programmes and lifestyles (Jones et al., 1983). Peter Jones' purpose was to integrate activity programmes within behaviour prediction models for transportation planning. The temporal approaches of mobility are closely related to the notion of rhythm and provide interesting analytical tools for our empirical analysis. First of all, "Time Geography" suggests three-dimensional analytical support that considers both space and time. It considers mobility as a set of sequences that unfold within a spatiotemporal unit (the day, the year, the life course). Next, the activity patterns approach links activities with the sociodemographic characteristics of individuals and more broadly with their lifestyle. At the end, these approaches provide indicators of the rhythmicity of mobility in line with the definition proposed by Rosa which corresponds to the number of actions or experiences per unit of time.

From this definition, we propose to understand the rhythm as the number of actions per space–time unit. As shown in Figure 37.1, the approach consists of measuring daily life rhythms from three dimensions:

- The density of the activities programs and their complexity.
- The temporal anchoring of daily rhythms, related to the different periods of daily mobility.
- The spatial dimension, reflecting the dispersion of the daily activities.

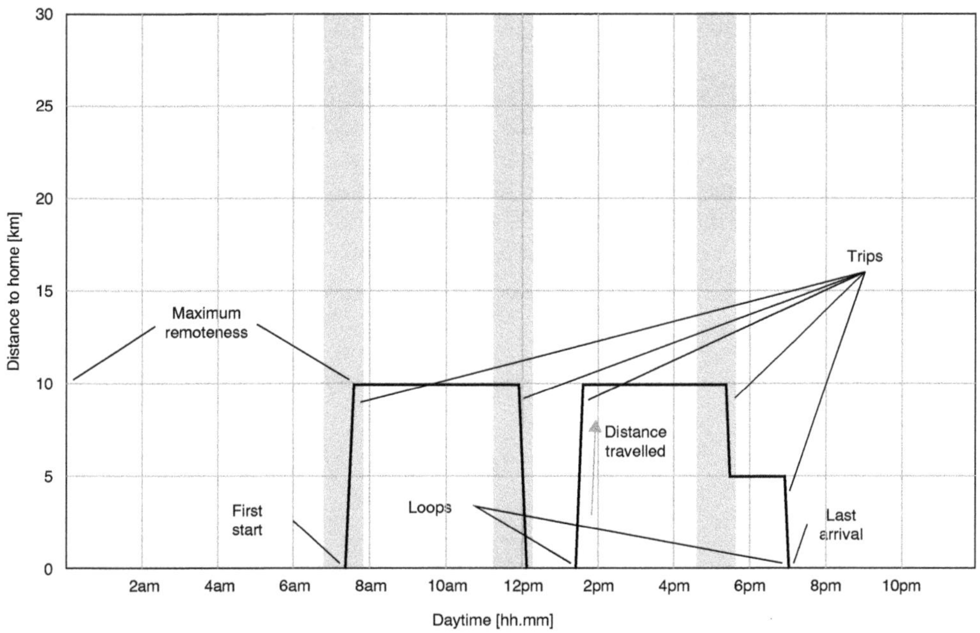

Figure 37.1 Daily rhythms indicators.
Source: A. Gumy, 2019

Data and methodology

This analysis is based on the Swiss microcensus "mobilité et transports" (2015). The survey has been carried out every five years since 1974. Based on a representative sample of the Swiss population (N=57,090), the survey collects data about activities locations, mobility practices and sociodemographic characteristics. The first step of the analysis is to select the set of variables needed to deploy our methodology.

- The first variable is the number of actions per unit of time. This first indicator measures the density and complexity of daily patterns from mobility loop types. A day could contain several simple loops or, on the contrary, only one complex loop. These two patterns have different implications in terms of rhythm.
- The second indicator corresponds to the temporal anchoring. The temporal anchoring of activities programmes is commonly associated with the notion of rush hours. They generally refer to some form of temporal constraints, for example due to a professional activity structured by predefined schedules. Rush hours are defined between 7am and 8am and between 5pm and 6pm. A third rush hour, between 11.30am and 12.30am is also included in the analysis. Two complementary variables add to the analysis of temporal anchoring that are the time of the first departure and the time of the last arrival at home. These variables define the daily time span.
- The third indicator refers to the spatial dimension of mobility. Two variables enable the integration of the spatial dimension for our rhythmic analysis. The first is the total distance travelled during the day. The second is the maximum remoteness from home. These two variables offer a measurement of the spatial dispersion of daily activities.

The analysis methodology is firstly based on techniques of descriptive and multivariate statistical analyses. In order to reduce complexity and to identify what is structuring the rhythms, the first methodological step is a Principal Component Analysis (PCA). This first level of analysis is complemented by hierarchical clustering to identify typical rhythmic profiles. The third level of the analysis corresponds to the production of an aggregated graphical indicator. This type of representation offers the possibility to visualize and better understand all the behaviours identified in the clustering phase. Finally, we use binomial regression models in order to relate rhythmic profiles to socioeconomics and mode choice.

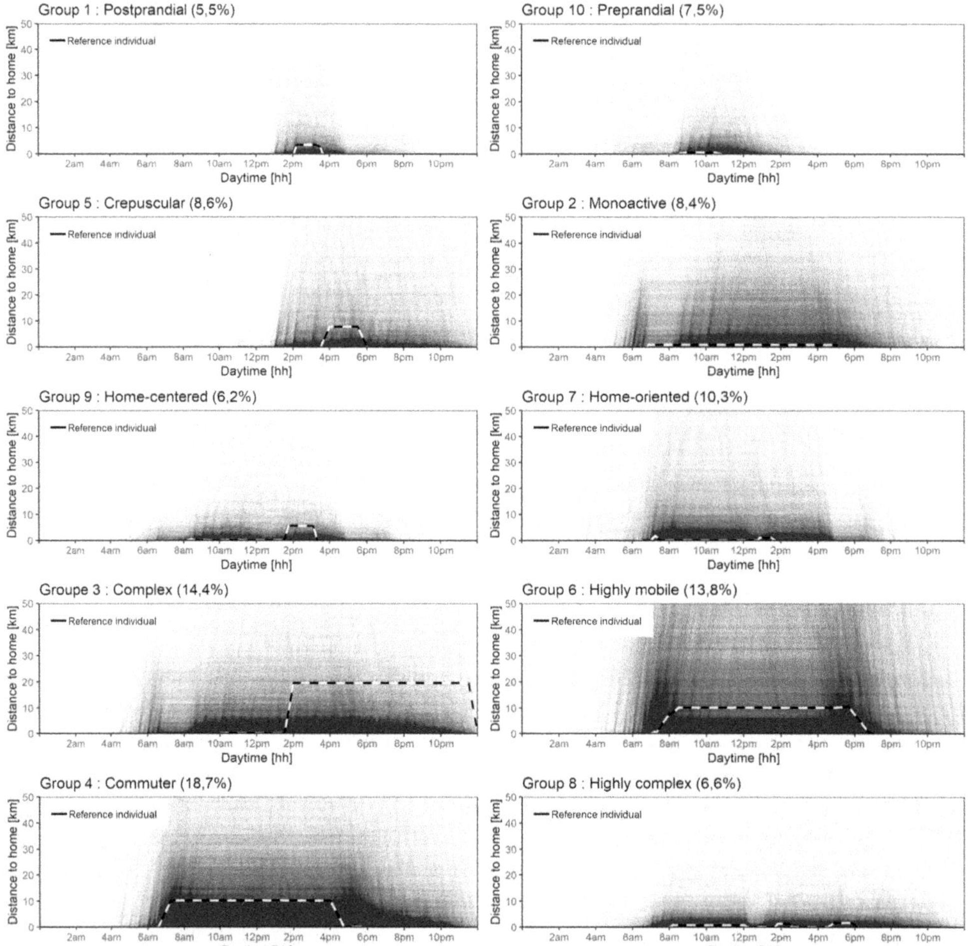

Figure 37.2 Rhythmic profiles.

Source: A. Gumy, 2019

Results

Contrasting rhythmic profiles

PCA, followed by hierarchical clustering allowed us to identify ten rhythmic profiles as shown in Figure 37.2. The dashed line indicates the median rhythm composing the group.

An important diversity between the rhythmic profiles appears. The "Commuter" profile (18.7% of the Swiss population) is the most represented among the different rhythms. It is characterized by an important temporal anchoring in the morning, with a first trip generally before 8am. The "Commuter" activities programme is relatively complex. The "Highly mobile" profile (13.8%) is singular in its spatial dimension. This second profile is characterized by a significant distance travelled during the day. The temporal anchoring shows a late return home in the evening, or even at night. The complexity of this second profile is relatively low as evidenced by the number of trips and loops performed. The "Complex" profile (14.4%) concerns individuals who make a high number of loops and trips during the day. The temporal anchoring of this third profile is mostly situated in the morning rush hour. The spatial dimension of the "Complex" profile is characterized by a dispersion of activities and a relatively small remoteness from home. The "Home-oriented" profile (10.3%) shows a low spatial dispersion of their activities and a relative proximity to home. The temporal anchoring of this profile is characterized by a first departure in the morning around 8am. The return home is relatively early in the day and spreads mainly between 1pm and 4pm. The "Home-centred" profile (5.8%) shares most of its characteristics with the "Home-oriented" profile. The difference between these two profiles concerns the level of complexity of their activities programme as the "Home-centred" profile presents a bigger concentration of trips in a close proximity, allowing for a higher amount of daily activities. The "Mono-active" profile (8.4%) concerns individuals who travel during morning and evening rush hours. The activities programmes of these individuals are not very complex. The spatial dispersion of their activities is higher than those of other profiles. The "Crepuscular" profiles (8.6%) are very different from the other profiles, especially in terms of temporal anchoring. Indeed, these individuals tend to carry out their activities only in the afternoon until the evening near their home place. The "Preprandial" profiles (7.5%) have a rhythmicity that is also temporally segmented. Indeed, these individuals rather do their activities in the morning and in a certain proximity to their home. These individuals present an activities programme that is not very dense. The "Postprandial" profile (5.4%) acts as a symmetry to the "Preprandial" profile. Individuals of "Postprandial" profile present a simple activities programme that takes place during the afternoon near the place of residence. Finally, the individuals of the "Highly complex" profile have three special features. First, they tend to carry out their first trip early in the morning and return home later compared to other profiles. Second, their activities programmes are mainly deployed in close proximity to their home. Finally, these programmes are the densest and most complex among all profiles.

The analysis of the different rhythmic profiles shows an important heterogeneity in the daily rhythmic profiles of the Swiss population. However, although the differences are striking, common rhythmic features appear between the different profiles. On the one hand, a kind of synchronization appears during the evening hours for a majority of the rhythmic profiles. On the other, common points also appear in the number of loops realized as the majority of individuals only carry out one or two loops.

Rhythmic profiles and inequalities

The analysis of the relationships between rhythmic profiles and socioeconomics shows they are based on a series of ten multivariate binary regressions presented in Table 37.1. The results are ordered according to the degree of rhythmicity measured by the sum of the factorial scores identified by the PCA. Table 37.1 shows from left to right the different profiles, from the least rhythmic ("Preprandial") to the most rhythmic ("Highly complex"). The purpose of the different models is to explain which are the factors responsible for the different rhythm types. In this perspective, the dependant variable in each model is the rhythmic profile. The reference of the model refers to all other rhythmic profiles. The reference category of the different explanatory variables is selected based on its median position (for example, the intermediate income) and/or if it is the most represented.

All regressions (Table 37.1) show a large number of significant relationships. The relation intensity is measured through odds ratios (OR) which indicates the strength of the relation. In comparison to women, men appear to be less associated with low level of rhythmicity ("Postprandial" OR = 0.57, "Prepandial" OR = 0.81, "Crepuscular" OR = 0.67). However, this relationship is also true at the other end of the scale as men are less associated with the "Highly complex" profile (OR = 0.90). They are mostly over-represented in the profiles "Monoactive" (OR = 1,22), "Highly mobile" (OR = 1,48) and "Commuter" (OR = 1,43). These results suggest gender inequalities based on both complexity of activities programs and distance travelled. Indeed, "Monoactive", "Commuter" and "Highly mobile" profiles share low complexity and high spatial dispersion as the main rhythmic characteristics. Age also appears as a factor influencing rhythmic patterns. Compared with people in the 45 to 64 years old age group, people over 65 years old tend to adopt uncomplicated, proximity-based and out of peak hours' activities programs ("Postpandial" 65 to 74 years old OR = 2.35, 75 years old and over OR = 3.73, "Prepandial" 65 to 74 years old OR = 2.37, 75 years old and over OR = 3.62 "Crepuscular" 65 to 74 years old OR = 1.59, 75 years old and over OR = 1.48). The "Highly mobile" and "Commuters" profiles rather concern the youngest part of the population ("High mobiles" 18 to 24 years old OR = 1.56, 25 to 44 years old OR = 1.30, "Commuter" 18 to 24 years old OR = 1.10, 25 to 44 years old OR = 1.11).

Looking at the type of household shows that the "Highly mobile" profile mainly concerns single-person households (OR = 1.34), while the "Highly complex" profile is, on the contrary, strongly under-represented in this category of household (OR = 0.53). Results suggest that families are associated with an increase in the number of activities carried out on a daily basis, and therefore in the complexity of their activities programme.

Income level analysis exemplifies some symmetry in the results. Indeed, profiles with the lowest life rhythms are more likely to be associated with moderate income levels ("Postprandial" moderate OR=1.67; "Preprandial" moderate OR=1.634; "Crepuscular" moderate OR=1.16) compared to an intermediate income level. The rhythmic profiles "Commuter" (comfortable OR=1.28) and "Highly mobile" (comfortable OR=1.69) are rather associated with a high level of income. This result would confirm the idea that high mobility provides access to greater economic resources, in particular through access to a spatially wider labour market. The analysis of the socioeconomics through rhythmic profiles mainly reveals effects related to gender, age and income level. The results underpin forms of gender inequality that would translate into a greater propensity for women to carry out complex activities programs focused on a closer proximity to the place of residence.

Table 37.1 Sociodemographic determinants of rhythmic profiles

	Postprandial (Ref. other rhythmic profiles)		*Preprandial (Ref. other rhythmic profiles)*		*Crepuscular (Ref. other rhythmic profiles)*		*Monoactive (Ref. other rhythmic profiles)*		*Home-centred (Ref. other rhythmic profiles)*		*Home-oriented (Ref. other rhythmic profiles)*		*Complex (Ref. other rhythmic profiles)*		*Highly mobile (Ref. other rhythmic profiles)*		*Commuter (Ref. other rhythmic profiles)*		*Highly complex (Ref. other rhythmic profiles)*	
	OR	*Sig*	*OR*	*Sig*	*OR*	*Sig*	*OR*	*Sig*	*OR*	*Sig*	*OR*	*Sig*	*OR*	*Sig*	*OR*	*Sig*	*OR*	*Sig*	*OR*	*Sig*
Constant	0.06	***	0.09	***	0.10	***	0.08	***	0.10	***	0.10	***	0.22	***	0.11	***	0.21	***	0.08	***
Gender. Ref=Woman																				
Man	0.57	***	0.81	***	0.67	***	1.22	***	0.76	***	0.95		0.90	***	1.48	***	1.43	***	0.90	***
Age. Ref=45 to 64 years old																				
18 to 24 years old	0.81		0.28	***	1.71	***	1.30	***	0.35	***	0.80	**	0.89	*	1.56	***	1.10	*	0.62	***
25 to 44 years old	0.87	*	0.62	***	1.03		1.14	**	0.71	***	0.72	***	1.07	*	1.30	***	1.11	***	0.87	**
65 to 74 years old	2.35	***	2.37	***	1.59	***	0.66	***	1.77	***	0.93		1.14	**	0.42	***	0.34	***	0.65	***
75 years old and +	3.73	***	3.62	***	1.48	***	0.60	***	1.97	***	0.73	***	0.88	*	0.18	***	0.20	***	0.31	***
Household type. Ref = Couple with children																				
Single	0.76	***	0.88		1.04		1.31	***	0.75	***	0.79	***	1.03		1.34	***	1.05		0.53	***
Couple without children	0.85	*	1.11		1.03		1.23	***	0.87	**	1.03		0.91	**	1.23	***	0.98		0.59	***
Single – parent family	0.80		0.96		0.87		0.98		0.71	***	1.02		1.03		1.20	**	1.14	*	1.00	
Other	1.05		0.82		1.00		0.95		0.77		0.71	**	0.96		1.50	***	1.19	**	0.64	**
Income level. Ref = Intermediate																				
Moderate	1.67	***	1.34	***	1.16	**	1.03		1.20	***	0.99		0.91	*	0.55	***	0.64	***	0.89	
Comfortable	0.55	***	0.65	***	0.72	***	0.90	**	0.63	***	0.99		0.83	***	1.69	***	1.28	***	1.05	

Source: G. Drevon and A. Gumy; MRMT, 2015
Method : Multivariate binary logit
Significance threshold : ***<=0.01; **<=0.05; * <=0.1

Rhythmic profiles and (multimodal) mobility practices

The following analysis aims at understanding relationships between rhythmic profiles and mobility practices. As in the previous analysis, it is based on a series of ten multivariate binary regressions.

Although the use of motorized individual vehicles (MIV) remains massive in the mobility of the Swiss population (54%), forms of multimodality are already playing an important role. Most of the time, it concerns the association of MIV with soft modes (SM) such as walking or cycling (27%) or with public transportation (PT) (8%). Combining PT and SM also concerns a significant part of the sample (16%). Therefore, it would not be appropriate to consider a single main mode identified on the basis of distance travelled or transport time budget that would reflect exclusive and partial modal behaviour. This analysis therefore focuses on identifying the propensity of the different rhythmic profiles to show forms of multimodality.

Rhythmic profiles that underlie simple activities programmes near home and outside peak hours ("Postpandial", "Prepandial", "Crepuscular") rely more on the exclusive use of SM on one hand and MIV on the other. Regarding "Postpandial" and "Prepandial" profiles, there is a tendency to combine PT and SM. The "Monoactive" rhythmic profile, which refers to programmes of activities that are not very complex but whose spatial dispersion is relatively high, appears to be rather predisposed to multimodality. Indeed, individuals who are within this group tend to use the entire modal range both exclusively and by combining several modes. The same goes for the "Highly mobile" profile where the combination between MIV and PT is particularly significant, as is the exclusive use of PT and its combination with SM. Results also suggest forms of MIV dependency. "Home-centred", "Complex" and "Highly complex" profiles have the lowest propensity to use different modes of transportation. This is especially true for PT combined with MIV.

Analysing relationships between rhythmic profiles and mobility practices shows important propensities for using different modes depending on one's life rhythm. It suggests forms of dependency, and in particular on MIV. Indeed, results would confirm that complexity fosters the use of the car. On the contrary, a smaller amount of complexity, regardless of the spatial dispersion of activities, would lead to more multimodal practices. Finally, analyses show a potential for modal shift towards simple rhythmic profiles that occur near home but for which actual use of MIV is still quite important.

Conclusion

To measure rhythms of daily life, our approach integrated density, complexity and spatial dimension of mobility. Empirical analyses based on the measurement of these different dimensions identified ten rhythmic profiles that illustrate the Swiss population's activities programmes. The profiles that are characterized by a high level of complexity are mostly realizing activities near home, thus confirming the thesis of "Time Geography" which stipulates that the transport time budget constrains the space budget. The corollary is also true as the rhythmic profiles presenting a high spatial dispersion are also those with low complexity activities programmes. This relationship between proximity and complexity is fundamental and shows the reciprocal relationship mentioned by Lefebvre between the spatial and temporal dimensions. Thus, the rhythmic approach reveals the difficulties faced by people with solving the daily spatial and temporal equation with limited temporal and spatial resources. It is also important to highlight the importance of rhythmic profiles marked by significant

Table 37.2 Analysis of the relationships between rhythmic profiles and modal practices

	Postprandial (Ref. other rhythmic profiles)		*Preprandial (Ref. other rhythmic profiles)*		*Crepuscular (Ref. other rhythmic profiles)*		*Monoactive (Ref. other rhythmic profiles)*		*Home-centred (Ref. other rhythmic profiles)*		*Home-oriented (Ref. other rhythmic profiles)*		*Complex (Ref. other rhythmic profiles)*		*Highly mobile (Ref. other rhythmic profiles)*		*Commuter (Ref. other rhythmic profiles)*		*Highly complex (Ref. other rhythmic profiles)*	
	OR	*Sig*	*OR*	*Sig*	*OR*	*Sig*	*OR*	*Sig*	*OR*	*Sig*	*OR*	*Sig*	*OR*	*Sig*	*OR*	*Sig*	*OR*	*Sig*	*OR*	*Sig*
Constant	0.02	***	0.04	***	0.08	***	0.05	***	0.10	***	0.08	***	0.30	***	0.07	***	0.32	***	0.14	***
Transportation modes used. Ref=MIV=SM																				
SM	6.25	***	4.63	***	1.34	***	1.01		0.95		2.40	***	0.44	***	0.48	***	0.41	***	0.64	***
MIV	3.09	***	3.04	***	1.59	***	3.32	***	0.46	***	1.10	*	0.38	***	3.11	***	0.55	***	0.18	***
MIV+PT	0.01		0.27		1.10		2.67	***	0.21	**	0.62		0.28	***	13.30	***	0.45	***	0.22	***
MIV+PT+SM	0.39	***	0.29	***	0.97		1.23	**	0.28	***	0.78	***	0.70	***	3.92	***	1.36	***	0.52	***
PT	1.90		2.42		1.79		5.06	***	0.00		3.03	**	0.00		4.28	***	0.42		0.00	
PT+SM	1.73	***	1.48	***	1.08		2.44	***	0.47	***	1.64	***	0.36	***	4.44	***	0.74	***	0.23	***

Source: G. Drevon and A. Gumy; MRMT, 2015
Method : Multivariate binary logit
Significance threshold : ***<=0.01; ** <=0.05; * <=0.1

desynchronization. This result would suggest forms of temporal exclusions, arrhythmias that would limit the inclusion of certain people, such as the aged and women, in the eurhythmy of a Swiss society in motion.

The different rhythm categories reflect a high diversity in daily mobility. However, it is important to remember here that this distinction only applies to the actual practices of individuals, without considering their representations towards time. Despite obvious methodological challenges, such an analysis would approach the fact that an objective increase in the rhythm of life may lead to a qualitative transformation of the subjective experience of time (Rosa, 2010).

Based on our investigations, the potential for using a rhythmic approach to understand daily mobility is obvious. Results show strong associations between rhythmic profiles and individual socioeconomics. Gender, age or income effects highlight cumulative social inequalities regarding rhythms. For example, proximity mainly concerns low incomes, while complexity is more broadly associated with women. On the one hand, these conclusions are in line with the literature on the subject (see for example Demoli, 2014), and on the other hand they reinforce the analytical power of the methodology used.

Concerning transportation planning, the rhythmic approach applied in this article reveals a modal shift potential for proximity-based and low complexity activities programmes. Analyses also highlight the rhythmic conditions of an exclusive car use. In particular, they confirm that complex activities programmes are more easily associated with the use of MIV, showing evidences that using a car stays a good candidate to avoid the spatiotemporal frictions of everyday life. The rhythmic approach shows which characteristics may lead people to combine different transportation modes, or show a multimodal behaviour. Regardless of the distance to be travelled, multimodality seems appealing when the complexity and density of activity programmes are reduced.

All these results suggest that if the acceleration and complexification of lifestyles are to continue, the modal shift from cars to other transportation modes may be compromised. Indeed, if reducing car use is an important objective of mobility policies, slowing down the rhythms of life could be an interesting way for public policies to promote modal shift. In view of the gender inequalities observed in our analyses, such desaturation policies would imply some parity considerations regarding the distribution of constrained activities between men and women. This relates to the perspective of the feminism-supported public temporal policies developed in Italy in the early 1990s aiming at rhythmic parity.

References

Auber Nicole, 2009, *Le culte de l'urgence : La société malade du temps*, Paris, Editions Flammarion.

Buehler Ralph and Hamre Andrea, 2015, "The multimodal majority? Driving, walking, cycling, and public transportation use among American adults", *Transportation*, vol. 42, no. 6, pp. 1081–1101.

Chapin Francis Stuart, 1974, *Human Activity Patterns in the City: Things People Do in Time and in Space*, New-York, Wiley-Interscience.

Cresswell Tim, 2010, "Towards a politics of mobility", *Environment and Planning D: Society and Space*, vol. 28, no. 1, pp. 17–31.

Demoli Yoann, "Les femmes prennent le volant. Diffusion du permis et usage de l'automobile auprès des femmes au cours du xxe siècle", *Travail, genre et sociétés*, vol. 32, no. 2, pp. 119–140.

Drevon Guillaume, Gwiazdzinski Luc, Klein Olivier *et al.*, 2017, *Chronotopies : Lecture et écriture des mondes en movement*, Grenoble, Elya Editions.

Drevon Guillaume, 2019, *Première proposition pour une rythmologie de la mobilité et des sociétés contemporaines*, Neuchâtel, Alphil-Presses universitaires suisses.

Edensor Tim, 2012, *Geographies of Rhythm: Nature, Place, Mobilities and Bodies*, Farnham, Ashgate.

Ellegård Kajsa and Bertil Vilhelmson, 2004, "Home as a pocket of local order: everyday activities and the friction of distance", *Geografiska Annaler: Series B, Human Geography*, vol. 86, no. 4, pp. 223–238.

Hägerstraand Tortsen, 1970, "What about people in regional science?", *Papers in regional science*, vol. 24, no. 1, pp. 7–24.

Hallin Per Olof, 1991, "News paths for time-geography?", *Geografiska Annaler: Series B, Human Geography*, vol. 73, no. 3, pp. 199–207.

Heinen Eva and Mattioli Giulio, 2017, "Does a high level of multimodality mean less car use? An exploration of multimodality trends in England", *Transportation*, pp. 1–34, https://link.springer.com/article/10.1007/s11116-017-9810-2

Heinen Eva, 2018, "Are multimodals more likely to change their travel behaviour? A cross-sectional analysis to explore the theoretical link between multimodality and the intention to change mode choice", *Transportation Research Part F: Traffic Psychology and Behaviour*, vol. 56, pp. 200–214.

Jones Peter M., Dix Martin C., Clarke Mike I. *et al.*, 1983, *Understanding Travel Behaviour*, [not printed].

Kaufmann Vincent, 2008, *Les paradoxes de la mobilité: bouger, s'enraciner*, Lausanne, Presses Polytechniques Universitaires Romandes.

Lenntrop Boe, 1976, "Paths in space-time environments: A time-geographic study of movement possibilities of individuals", *Studies in Geography Series B Human Geography*, no. 44.

Levine Robert, 1997, *A Geography of Time: The Temporal Misadventures of a Social Psychologist, or How Every Culture Keeps Time Just a Little Bit Differently*, New York, Basic Books.

Nobis Bo, 1976, "Multimodality: Facets and causes of sustainable mobility behavior", *Transportation Research Record*, no. 2010, pp. 35–44.

Nobis Claudia, 2007, "Multimodality: Facets and Causes of Sustainable Mobility Behavior", *Transportation Research Record: Journal of the Transportation Research Board*, vol. 2010, no. 1, pp. 35–44. doi:10.3141/2010-05.

Pirie Gordon H., 1979, "Measuring accessibility: A review and proposal", *Environment and Planning A: Economy and Space*, vol. 11, no. 3, pp. 299–312.

Ravalet Emmanuel, Vincent Stéphanie, Kaufmann Vincent *et al.*, 2015, *Grandes mobilités liées au travail, perspective européenne*, Paris, Edition Economica.

Rosa Hartmut, 2010, *Accélération. Une critique sociale du temps*, Paris, La Découverte.

Susilo Yusak O. and Axhausen Kay W., 2014, "Repetitions in Individual Daily Activity–Travel–Location Patterns: A Study Using the Herfindahl–Hirschman Index", *Transportation*, vol. 41, no. 5, pp. 995–1011.

Thévenin Thomas, Chardonnel Sonia and Cochey Élodie, 2007, "Explorer les temporalités urbaines de l'agglomération de Dijon. Une analyse de l'Enquête-Ménage-Déplacement par les programmes d'activités". *Espace populations sociétés. Espace populations sociétés. Space populations societies*, no. 2007/2–3, pp. 179–190.

Vincent Stéphanie, Ravalet Emmanuel and Kaufmann Vincent, 2016, "Des liens aux lieux : l'appropriation des lieux dans les grandes mobilités de travail". *Espaces et sociétés*, no. 1, pp. 179–194.

38

RETHINKING THE LARGE-SCALE MOBILITY INFRASTRUCTURE PROJECTS IN SUSTAINABLE SMART CITY PERSPECTIVE

Jian Zhuo

Introduction

Transport systems play a key role in the organization and development of the city. Occupying 15–20% of urban land area, mobility infrastructure physically supports the daily commutes of city residents and directly affects their social life. Driven by the technical progress in transport, the physical scale of mobility infrastructure increases in pace with the expansion of urban space. In the age of the carriage, boulevards were the early large-scale mobility infrastructure projects. After the emergence of steam locomotives and rail transport, some large railway stations built in the latter half of the 19th century were regarded as 'modern cathedrals'. By the early 20th century, elevated freeways and overpass junctions became the iconic elements of the era of the automobile. Today, in the context of globalization, airports, high-speed rail (HSR) lines and hubs, and deep-water ports are emblematic of national and worldwide inter-city circulation and competition. All these mega-infrastructures have not only contributed to the improvement of urban mobility but also constitute the representative urban landscape of the times.

The term 'large-scale mobility infrastructure' implies a double meaning. The physical dimensions of a project itself may be tremendous. And projects like this can generate important impacts that are very wide in spatial scope. Therefore, they can raise significant challenges to sustainability. Motorization is the main cause of a majority of these projects. There are now more than a billion motor vehicles in the world and the World Bank projects that this number will double by 2050 (World Bank, 2018). Auto use is increasing most rapidly in absolute terms in developing countries as more and more people are able to afford cars (Sumantran, Fine and Gonsalvez, 2017). Critics blame automobile dependence for producing a low density land-consuming urban form, requiring an enormous amount of precious urban land to be devoted to roads and parking, depleting non-renewable fossil fuels, creating air pollution that poses serious health risks and significantly contributing to global climate change and auto-related fatalities, injuries and property damage. The mega-infrastructures pursuing efficiency in engineering performance stimulate in return the motorization.

Today, the vision of sustainable development is transforming people's values and their conception of urban lifestyle. A city should be first of all a living place for people (Gehl, 2010). When large-scale infrastructure needs to be built in urban spaces, the design paradigm adopted should not be the same as in rural areas. Facing new demands and new challenges, city planners and engineers equipped with new technologies have opportunities to rethink mega-projects development in the cities and produce mobility systems that will be compatible to sustainable and livable environments. The first part of this chapter is a brief review of the evolution of large-scale mobility infrastructure and its importance for urban development. The second part investigates the influence of modernist functional planning principles on urban infrastructure projects from the 1920s until the 1970s and the negative impacts they caused. The third part presents how people-centered approaches have been implemented in a number of different cases in order to reduce these negative impacts. The final part of the paper states an alternative perspective for new types of mobility infrastructure for developing future healthy and smart cities.

Changing scale: from pedestrian cities to global city-regions

Before the Industrial Revolution, modern urban transport did not exist in cities. Mechanical wheeled vehicles such as horse-drawn carriages and human-powered rickshaws helped move people and goods from one place to another. A large proportion of urban trips were made by foot for short distance. These are frequently referred to as 'pedestrian cities' (Wiel, 2002). With the constraint of the low travel speed, urban functions were maintained at a walkable scale. Mobility infrastructure was built at human scale and highly integrated into the urban environment. Transport corridors served as public space as well and for movement in walking cities. Urban streets, the fundamental infrastructure for transporting people and goods, at the same time was the most important place for social life (Southworth, 2005).

Boulevards can be considered as one of the earliest types of large-scale mobility infrastructure built in cities. They are composed of multiple roadways separated by rows of border trees and designed for different users circulating at different speeds. Boulevards were first built by Baron Haussmann in the 1850s during his reconstruction of Paris. They quickly gained popularity in European cities. In the 1890s boulevards were introduced into American cities by the City Beautiful Movement (Wilson, 1994). Boulevard development in Asian and South American cities lasted till the 1940s (Jacobs et al., 2001). Usually part of big urban plans, boulevard projects have generated positive impacts on the cities where they have been built: reducing urban congestion in historical cities with medieval road systems; opening up the connection between important places; establishing a new urban spatial structure, and renewing inhabitants' recognition of the image of their cities. The construction of boulevards has also resulted in the redevelopment of roadside areas, attractive for middle class housings and recreation activities (Jacobs et al., 1994; 1995).

The invention of the Bolton-Watt steam engine in the early 19th century gave birth to the modern urban transport system. Horse-trams – continuously extended first on wooden tracks and later steel rails – were replaced briefly by cable cars in some cities and from the 1880s on by electric streetcars. Then came the autobus and automobile around the turn of the 20th century. The significant increase of urban travel speed overcame the obstructs of spatial distance and resulted in the sprawl of modern cities (Mumford, 1961; Warner, 1978). In early industrial cities, lineal urban expansion often happened along rail or tram lines. Later, in the automobile era, the development of new urban districts in suburban areas was driven by the extension of urban road networks.

Martin Wachs (1984) examined the dispersed growth of Los Angeles from 1880. Interurban electric railways encouraged residential decentralization in Los Angeles before there were many automobiles, but this pattern was reinforced after the popularization of automobiles – particularly during the 1920s. On one hand, new mobility infrastructure contributed to cities' spatial sprawl and satisfied people's preference for low-density living. On the other hand, the expanded urban area provided more space to accommodate the large-scale infrastructures.

Today, in the context of globalization, a new form of socio-economic organization of cities is clearly evident: generally called 'world cities' (Hall, 1966; 2014; Taylor and Derudder, 2015) or 'global cities' (Sassen, 2001). Urban clusters connected to a megacity core form dynamic and attractive city-regions. This so-called 'metropolitanization' phenomenon characterized the current urbanization process (Ascher, 1995). At the same time, the city networks founded beyond administrative boundaries establish a new spatial structure for inter-city interactions and upgrade local development into regional or even global scope (Beaverstock et al., 2000). Both the city-region and the city network create a huge potential for enlarging the scale of mobility infrastructures. Enhanced population concentration and multiplied development opportunities in leading world metropolitan areas have increased their financing capacity for mega-projects. Among the most remarkable ones are the Channel Tunnel between the UK and France (the Chunnel), the Øresund Denmark–Sweden link and, more recently, the Hong Kong–Zhuhai–Macao Bridge built in the south of China. The latter project connects three main cities crossing 55 miles of the South China Sea with a 33-foot wide highway, aiming to push forward the development integration of the Big Bay Areas in the Pearl Delta Region.

Mobility infrastructures play a key role in new urban dynamics. The efficient circulations of people, goods and information are crucial for local development in the post-industrial age. Robert Cervero (1998) defines a metropolitan area that has linked land use and transport as a 'transit metropolis'. London is a leading example. Transport for London controls all the major urban transport systems and plans and manages the entire transit service. Manuel Castells (2000; 2011) distinguishes between what he calls the 'space of flows' and the traditional 'space of places'. He argues that the flow of people, goods, capital and information into and out of city hubs in the networks are as important as the activities which take place within the city space itself. Flows affect the functioning and organization of the urban realm and flow analysis can play an important role in understanding the nature of cities. Derudder and Witlox (2005; 2008) mapped the world city network based on airline flows. The improvement of the accessibility of a city generally contributes to a higher ranking in the world city hierarchy (Keeling, 1995). That is why the development of mobility infrastructure remains high on local governments' agendas.

In many cases, large-scale mobility infrastructure that is highly visible serves not only as a transport project but also as a strategic tool for urban marketing (Shannon and Smets, 2016). The Golden Gate Bridge, for example, has been a San Francisco landmark for almost 80 years. In the context of the fierce competition among cities, a number of mega-projects and mega-events have been carried out to enhance cities' branding. The decision making for projects like this went far beyond technical reasoning and the scale of the infrastructure – deliberately designed to be impressive – was no longer determined only according to real functional need. The huge amount of investment for fixed assets could support a significant GDP growth. The difficulties and complexity of construction work on these types of mega-projects can demonstrate local technical strength and know-how. The superhuman scale of iconic transport infrastructures may positively reflect local dynamics and display the city's development ambitions. Successful large infrastructure projects can change local inhabitants' pride in their city and may even become a sightseeing attraction for visitors. One very successful example to illustrate this point is the Millau viaduct built over the Tran valley in the south of France.

However, large-scale mobility infrastructure projects are always an important political decision that should be made very cautiously. Many of the 20th century 'great planning disasters' that Peter Hall has documented (1982) involve transport infrastructure. These include London's motorways and airports, the San Francisco Bay Area's rapid transit system. Megaprojects require a substantial long-term public investment which has often been underestimated and has later become a heavy financial burden for local governments (Altshuler and Luberoff, 2003; Flyvbjerg et al., 2002). Construction work of big infrastructure is technically difficult and implies potential risks. The demand forecasts for large-scale mobility infrastructure projects have usually been inaccurate (Flyvbjerg et al., 2005). How to maintain the rationality to a certain degree in the decision-making process is a real political challenge (Flyvbjerg, Bruzelius and Rothengatter, 2013).

Functional modernist design and its negative impacts on cities

The scale enlargement of mobility infrastructure was not an isolated transformation. It went along with the scale change of cities. With the final purpose of serving urban development, mobility infrastructure should have been designed and built with a tight relationship with the city, placing the socio-economic interactions of people at the center of considerations. However, a lot of large-scale mobility infrastructure projects built in cities since the beginning of the motor age did not follow this fundamental principle. Their physical scale was incompatible with humanized urban space and their design did not respond to people's actual needs.

When the motorization of urban transport introduced high traffic speed into urban space, it raised additional requirements for mobility infrastructure. Some politicians at that time even claimed adapting cities to the automobiles. Progressive architects who were very active from the 1920s defined a new organizational form for urban spaces and became the founders of modernist theories of urban planning. Swiss-French architect Le Corbusier was one of the leading modernists. He famously declared that a city built for speed would be a successful city. To Le Corbusier (1924), the high speed brought by the automobile was revolutionary progress. He believed it would cure cities with traditional forms from a variety of urban dysfunctions caused by the over-concentration of people and activities after industrialization. In the famous *Charter of Athens* (1943), Le Corbusier and other modernist architects stated their vision about the ideal modern city. Two planning principles regarding mobility infrastructure were fundamental. First, with the high speed provided by modern transport, different urban functions, such as residential areas, industrial sites, public services and recreation zones, should be planned in different and separate places to avoid a chaotic mixture and mutual conflicts. Second, since high-speed transport played a key role in supporting the spatial separation of urban functions, mobility infrastructure should be free from other uses and dedicated only to traffic. Le Corbusier (1945) sharply criticized traditional multifunctional street space walled in road-side buildings.

Technical-oriented functionalism characterizes the modernist urban planning paradigm. It extensively influenced urban development and reconstruction after WWII (Hall and Tewdyr-Jones, 2010). From then on, urban road systems were planned mainly to facilitate fast circulation of motorized vehicles. Their spatial layout didn't necessarily need to coordinate with the existing urban form. As planning and design work focused essentially on technical issues, it was usually assigned to traffic engineers. A similar approach was adopted for infrastructure designed in urban and rural contexts. The specificities of urban environment were rarely taken into account. The former intimate relationship between humans and space was interrupted with the insertion of motorized vehicles. Their sizes and their velocity, taking the place of people and walking, became the reference for infrastructure design. The modern urban planning paradigm accelerated the development of large-scale mobility infrastructures in cities.

These projects have attached importance to the functionality of mobility infrastructure and paid less attention to their relationship with urban space, which generated many negative impacts on urban functions. First, large-scale infrastructures have destabilized the urban transport system since they have contributed to suburbanization (Baum-Snow, 2007). The extending commuting distance affected the modal choice of commuters generally apt to motorization. The automobile dependence turned some commuters into captured users of these infrastructures (Brueckner, 2000; Newman and Kenworthy, 1990). Most of the car drivers assumed that the circulation on the larger infrastructures would be faster. This often caused traffic congestion and high concentration of exhaust gas emission.

Second, large-scale mobility infrastructure has had a variety of negative impacts on urban physical space. Compact, continuous and integrated urban form was broken into the segmented, dispersed and segregated areas. The construction of large-scale infrastructure projects destroyed traditional urban features and characteristics. Their dominant occupation of urban space marginalized the users of non-motorized travel modes. More and more urban space was designed for the motorized vehicles and not for people. Cities gradually lost human scale (Gehl, 2010). As most of the large-scale mobility infrastructures were dedicated to traffic, they became less accessible for people without cars. To ensure high speed and road safety, large transportation infrastructures need to be closed to adjacent areas. Infrastructure initially created to facilitate connections and movements became, on the contrary, large-scale hard barriers difficult to cross. The reintegration of large transport infrastructures into the urban space was hardly possible.

Third, large-scale mobility infrastructure projects often brought about social consequences (Lefebvre, 1967). Environmental degradation along the infrastructure not only damaged the quality of space but also bred social problems. As these projects were too expensive to provide everywhere, their construction was very selective and only for some specific places. They often aggravated social segregation and development lopsidedness among different urban areas. Graham and Marvin (2001) developed an international and interdisciplinary analysis of the complex interactions between infrastructure networks and the urban built environment and pointed out a 'splintering urbanism' in the contemporary metropolis.

Rethinking large-scale mobility infrastructure projects

Modernist planning principles were generally seen as adequate for the rapid urban development and the modernization of existing cities after WWII. However, they have been widely criticized since the late 1960s because of their neglect of social life and human needs. Modern cities built according to these principles were short of livability (Jacobs, 1961; Lefebvre, 1962). The large-scale infrastructures and their negative impacts on cities were one of the most obvious problems that should be questioned and handled first. To restore human scale and to reposition people in the center of considerations became many city planners' priority. London is a notable example. In 2014, the Mayor of London launched the city's first long-term infrastructure plan. It was the first attempt to identify, prioritize and establish the cost of London's future infrastructure. The plan clearly cited 'human-centered' modes of travel as a priority (Mayor of London, 2014; 2015).

In order to deliver efficient mobility service while reducing negative impacts on the environment and preserving social cohesion, urban transport projects should be recognized and addressed as part of an overall urban development strategy, not as isolated projects. The new book of Robert Cervero (2018) and his colleagues advocated shifting the current transport paradigm into a global thinking around place, in particular to articulate transport with land use. 'Moving beyond mobility' allows the needs and aspirations of people and the creation of better communities to be prioritized. The reconstruction of the railway station in the Zuidas district of Amsterdam in the Netherlands presents a good example. This mega-project of station renewal has been integrated into the sustainable redevelopment program of the whole area by creating a new development model called 'dokmodel'. The master plan centralized all large-scale transport infrastructures into an underground 'wharf' space, which released the ground surface from the heavy traffic and allow the formation of urban districts with mixed land use, a good street plan and a traditional environment (DRO, 2009; 2012).

People-centered mobility requires better integration of city-wide transport systems. When modernist principles are replaced by the human-focused approach, the users are placed at the heart of conception. Their diversified demands for a seamless journey should be treated as individual cases, not only as part of mass traffic volume. To meet this requirement, mobility infrastructure should be designed to improve accessibility and mobility, as opposed to just being developed for modes of transport in isolation (Mitchell et al., 2016). Projects to integrate various urban travel modes have developed since the 1960s. The city of Delft in the Netherlands pioneered the concept of *Woonerfs* (Streets for living) in the late 1960s. A *Woonerf* is designed primarily as a means of traffic calming aiming to improve road safety and the cohabitation of different transport modes in the same space. The good impacts achieved in the Netherlands inspired Germany, France and other European countries to promote the 'Zone 30' districts in their cities. Within these traffic calming zones whose function is essentially residential, traffic speed is reduced to 30km/h. As lower traffic circulation requires less space, the existing mobility infrastructure is able to accommodate more other transport modes users or even provide some public space for social activities. Traffic calming zones are now frequently important components of local mobility plans in European and some North American cities. They become a useful planning tool to promote multimodal transport and to improve the quality of mobility space. In Paris, over 37% of the urban streets are already in traffic calming zones.

The people-centered approach also includes other new ideas for restoring or replacing large-scale mobility infrastructure. Allan Jacobs and Elizabeth Macdonald won the design competition for the reconstruction of one section of the Central Freeway in San Francisco which had been damaged in an earthquake in 1989. Their design turned the former elevated freeway into a surface multiway boulevard 133-feet wide, named Octavia Boulevard. The new infrastructure inaugurated in 2005 is divided into several roadways: fast traffic ways, auxiliary lanes and sidewalks separated from each other by borders of trees, designed to accommodate traffic of different speeds and also social activities (Olea, 2007). The project is reminiscent of parkway development in the United States in the late 19th century. Assessments of its impacts on traffic, neighborhood and housing price are in general very positive (Cervero, Kang and Shively, 2008).

Recently, the renovation of urban streets has become a worldwide movement. From New York to Shanghai, from Abu Dhabi to Bombay, city governments have published street design manuals based on people-centered principles. Their common goal is to return streets to urban life. In these cities, a number of roads previously dedicated to traffic have been converted into multifunctional 'completed streets'. Automobile circulation was not totally banned but canalized into narrowed lanes, which makes car drivers' behavior more respectful of other users. The space gained from the remodeling streets can be reallocated to other transport modes such as walking, cycling and public transit. Part of the street space may be turned into green areas or pedestrian zones, favorable for social activities (DOT, 2015). The complete street is no longer a functional infrastructure dedicated to traffic. It is a high-quality public space shared by different users and various transport modes.

Finally, temporal transformation of usage is an interesting practice for the reconciliation between large-scale mobility infrastructure and the urban environment. Without modifying any physical features of the infrastructure, it can be closed to traffic at specified times and turned into social space shared by different users. This transformation – without any reconstruction or remodeling work – is reversible and flexible. Wide trunk roads with heavy traffic during weekdays can be turned into car-free marketplaces on Sunday mornings. A spacious parking lot in the city center may be transformed into a children's playground during holidays. The functional temporality gives local inhabitants a fresh experience when they use the existing infrastructure. The longer the transformation lasts, the greater the impacts generated to surrounding areas. Since 2002, the motorway along the right bank of the Seine River in Paris has been closed to traffic during several weeks every summer and turned into a pedestrian resort area. The former mobility infrastructure accommodates an artificial sand beach, cool lawns, sun umbrellas, deckchairs, palm trees and cultural and sports activities, offering Parisians and visitors free places to relax and take advantage of recreational activities during hot summer days. This annual event named '*Paris Plages*' (Paris beaches) quickly gained popularity and brought new vitality to the heart of the historical city.

Towards healthy, smart and sustainable urban mobility

The current evolution of urban mobility is influenced by a variety of factors such as changing lifestyles, new values, and fast technological progress. Besides functional performance, political goals established by urban mobility plans are becoming more and more global, focusing on ecological, social and long-term targets. The *Strategic Plan 2016* published by the New York City Department of Transport proposes safe, green, smart and equitable development as the key missions for the public agency.

As the concept of sustainable development has been widely accepted by the public, the dominant position of private cars in urban transport systems is being contested and re-examined in depth. Public transit development is now widely advocated by transport plans as an important alternative to individual motorization. Large-scale mobility infrastructure previously built only for automobiles is now more and more being built or retrofitted for public transit, such as sky trains, streetcars and Bus Rapid Transit (BRT) lines. These projects evidence an important shift of local transport policies in favor of public transit and the strong political intention to develop public transport usage. Since the 1990s, the successful experience of BRT projects in South American cities has been introduced into China. More than 30 Chinese cities have built BRT lines in dense central city areas. Efficient transit corridors have resulted in the growth of transit ridership. The city of Vancouver in Canada considers the extension of its sky train network as the key urbanism catalyst. The mobility infrastructure projects are associated with Vancouver's housing, commercial space and public facilities development programs. Following Transit-Oriented

Development (TOD) principles, the areas surrounding train stations are being densified and built as pedestrian friendly, multifunctional urban cores.

It is generally accepted that large-scale mobility infrastructure implies high travel speed. Walking and cycling, the oldest non-motorized urban transport modes, have nothing to do with large scale. However, in the most modernized contemporary cities like New York, London and Paris, their modal share remains considerable and has increased in recent years. They are now recognized as 'active' travel modes whose evolution is deeply influenced by two trends.

First, it is clear that transport systems have a highly significant role to play in improving public health (Lee and Moudon, 2004). World Health Organization data estimate that physical inactivity accounts for nearly one in ten premature deaths worldwide. The dependence on automobiles has increased physical inactivity and harms healthy environments. Active travel can help cure poor health. More and more urban commuters are adopting active travel modes in place of their car for the sake of a healthier lifestyle for themselves and contributing to a cleaner world which will benefit the health of others.

Second, new information and communication technologies (NICT) have profound implications for promoting active travel. With the popularization of smartphones, shared mobility systems have experienced a rapid expansion worldwide. Since their first appearance in April 2016, dockless shared bikes in Chinese cities reached over 4 million within less than one year. Shanghai alone had over 500,000 (Gu et al., 2019). After its successful development in California, the startup company Lime is promoting electric scooter sharing systems in big cities all around the world. Abundant, cheap scooters that can get a commuter where they want to go quickly may shift modal choices so that many more people choose a combined trip using a subway or light rail system plus a scooter than would chose to take the subway or light rail if they had to walk to their destination. In some cases, electric scooters may help solve the 'last mile problem' – the dilemma that it often takes enough additional time to get from a transit hub to the commuter's destination that the commuter will choose not to make the trip. Abundant and cheap scooters that can get a commuter where they want to go quickly may encourage many more people to take a combined trip using a subway or light rail system plus a scooter for the last mile.

The changes of commuters' travel behavior and the emergence of new mobility services are increasing demand for non-motorized transport infrastructure. New York City plans to build 320 kilometers of new bike lanes by 2021 in order to double the number of cyclists. In the cities in the United Kingdom, Denmark, the Netherland and China, a number of splendid infrastructures reserved for pedestrians and cyclists have been designed and built. The elevated bicycle highway designed by Norman Forster for London is a very inspiring conceptual project. The Danish architecture firm Dissing+Weitling has built an outstanding 190-meter long Bicycle Snake Bridge in Copenhagen. With the same designer, the city of Xiamen in China opened the world's longest bicycle bridge – the 7.6-kilometer long 'Xiamen Bicycle Skyway' – and planed additionally a 20-kilometer long footpath network. With the ambition to create a large-scale network to improve conditions for cyclists and pedestrians, the city's vision is to connect the city with nature and lead residents and visitors in the exchange between bustling urban life and luscious natural landscape.

The rapid progress in information technology is making the intelligent city a prospect within reach. A mobility revolution is happening before our eyes. Smart cities innovations already imbed sensors capable of capturing transportation data in real time and delivering it to physical or digital control rooms where it can be analyzed. The push technology can send user-friendly information to cell phones or other devices to help travelers making decisions. The Internet of

Things (IOT) promises to link people, cars, public transit and traffic management to a much greater extent in the near future. Traffic signals can adjust to traffic flows altering start and stop times, not only to improve mobility, but to reduce air pollution and achieve other goals. Vehicle-to-infrastructure communication holds the promise of alerting drivers to conditions or directly controls vehicles for safety or efficiency. Vehicle-to-vehicle communication may make much more efficient use of existing infrastructure by reducing the distance needed between vehicles (Sumantran, Fine and Gonsalvez, 2017).

Automatic driving technology is redefining the automobile into a networked, movable and personal large intelligent terminal device. It will be used more and more in a sharing way. A range of autonomous vehicle prototypes (also called driverless cars, self-driving cars and robot cars) already exist with varying degrees of autonomy. Stop-start systems can stop a car at a traffic light and start it immediately afterwards. Software can slow or stop a car when the car in front slows down, the car begins to drift out of its lane, or a car in an adjacent lane gets too close. Semi-autonomous and autonomous vehicles may prevent accidents and will improve largely the road safety. Their intelligence allows a more peaceful cohabitation with other road users. Large mobility infrastructures will no longer be fixed to a given function only but may soon become more adaptive to temporal usages.

Conclusion

Mobility infrastructures play a key role in local development and urban place fabrication. As the geographic scale of urban spaces organization increases constantly, the large-scale mobility infrastructure projects will keep developing accordingly. In the context of globalized competition, these projects remaining high on local governments' agendas are regarded as a strategic development instrument. They raise significant challenges to sustainability.

The transport infrastructure in the inhabited environment is not only a functional space for automobiles. Kelly Shannon and Marcel Smets (2016) list four dimensions that need to be taken into account in the infrastructure design: as an agency of enhanced mobility, as a physical presence, as a design feature contributing to the character of a city and as a sound theoretical approach to a positive experience of collective space. However, the functionalism planning paradigm pursued the technical performance and neglected social life and human needs. In many cases, the construction of large-scale mobility infrastructures resulted in the disruption of urban space and negative social impacts.

Today, new values, changing lifestyles and fast technological progress are favorable for a paradigm shift. It becomes a wide consensus that our cities should be planned for people instead of cars. Urban designers, transport engineers and policy makers should work together and develop a global thinking on urban mobility systems. The large-scale mobility infrastructure projects should prioritize the needs and aspirations of people, create places and improve accessibility. With the human-centered design approach, they may on the one hand meet the needs of users and support the best transport services for the expanding cities, and on the other hand, deliver economic growth, health benefits and social cohesion to the places we live.

References

Altshuler A., and Luberoff, D. 2003. *Mega-projects: The Changing Politics of Urban Public Investment.* Washington, DC: Brookings Institution Press.

Ascher, F. 1995. *Métapolis ou l'avenir des villes.* Paris: Odile Jacob.

Baum-Snow, N. 2007. Did highways cause suburbanization?. *The Quarterly Journal of Economics*, 122, pp. 775–805.

Beaverstock, J. V., Smith, R. G. and Taylor, P. 2000. World city network: A new metageography?. *Annals of the Association of American Geographers*, 90(1), pp. 123–134.

Brueckner, J.K. 2000. Urban sprawl: Diagnosis and remedies. *International Regional Science Review*, 23, pp. 160–171.

Castells, M. 2000. *The Rise of The Network Society*. 2nd ed. New York: Wiley-Blackwell.

Castells, M. 2011. Space of flows, space of places: Materials for a theory of urbanism in the information age. In LeGates, R. and Stout, F., Eds., *The City Reader*, 5th ed. London: Routledge. pp. 572–582.

Cervero, R. 1998. *The Transit Metropolis: A Global Inquiry*. 4th ed. Washington: Island Press.

Cervero, R., Guera, E. and Al, S. 2018. *Beyond Mobility: Planning Cities for People and Places*. Washington: Island Press.

Cervero, R., Kang, J. and Shively, K. 2008. *From Elevated Freeways to Surface Boulevards: Neighborhood, Traffic, and Housing Price Impacts in San Francisco*. University of California: University of California Transportation Center.

Le Corbusier. 1924. *Urbanisme*. Paris: Cres.

Le Corbusier. 1943. *La charte d'Athènes*. Paris: Plon.

Le Corbusier. 1945. *Les trois établissements humains*. Paris: Denoël.

Derudder, B. and Witlox, F. 2005. An appraisal of the use of airline data in assessing the world city network: A research note on data. *Urban Studies*, 42(13), pp. 2371–2388.

Derudder, B. and Witlox F. 2008. Mapping world city networks through airline flows: Context, relevance, and problems. *Journal of Transport Geography*, 16, pp. 305–312.

DOT. 2015. *Street Design Manual*. 2nd ed. [online] Available at: www1.nyc.gov/html/dot/html/pedestrians/streetdesignmanual.shtml [Accessed 20 March 2019].

DRO. 2009. *Zuidas vision document*. [online] Available at: www.amsterdam.nl/zuidas/english/menu/zuidas-development/zuidas-vision/ [Accessed 20 December 2018].

DRO. 2012. *ZuidasDok*. [online] Available at: http://amsterdam.nl/zuidas/menu/zuidasdok/downloads/ [Accessed 20 December 2018].

Flyvbjerg, B., Holm, M. and Buhl, S. 2002. Underestimating costs in public works projects: Error or Lie?. *Journal of the American Planning Association*, 68(3), pp. 279–295.

Flyvbjerg, B., Holm, M. and Buhl, S. 2005. How (In)accurate are demand forecasts in public works projects? The case of transportation. *Journal of the American Planning Association*, 71(2), pp. 131–146.

Flyvbjerg, B., Bruzelius, Nils. and Rothengatter, W. 2013. *Megaprojects and Risk: an Anatomy of Ambition*. Cambridge: Cambridge University Press.

Gehl, J. 2010. *Cities for People*. Washington: Island Press.

Graham, S., and Marvin, S. 2001. *Splintering Urbanism: Networked Infrastructures, Technological Mobilities and the Urban Condition*. London: Routledge.

Gu, T., Kim, I. and Currie G. 2019. To be or not to be dockless: Empirical analysis of dockless bikeshare development in China. *Transportation Research Part A*, 119, pp. 122–147.

Hall, P. 1966. *The World Cities*. London: Weidenfeld and Nicolson.

Hall, P. 1982. *Great Planning Disasters*. Berkeley: University of California Press.

Hall, P. 2014. *Cities of Tomorrow: An Intellectual History of Urban Planning and Design Since 1880*. 4th ed. New York: Wiley-Blackwell.

Hall, P. and Tewdyr-Jones, M. 2010. *Urban and Regional Planning*. 5th ed. London: Routledge.

Jacobs, A., Macdonald, E. and Rofe, Y. 2001. *The Boulevard Book: History, Evolution, Design of Multiway Boulevards*. Cambridge: The MIT Press.

Jacobs, A., Rofe, Y., and Macdonald, E. 1994. *Boulevards: A Study of Safety, Behavior, and Usefulness*. Berkeley: The University of California Berkeley Transportation Center.

Jacobs, A., Rofe, Y. and Macdonald, E. 1995. *Multiple Roadway Boulevards: Case Studies, Designs, and Design Guidelines*. Berkeley: The University of California Berkeley Transportation Center.

Jacobs, J. 1961. *The Death and Life of Great American Cities*. New York: Random House.

Keeling, D. J. 1995. Transportation and the world city paradigm. In Knox, P. L. and Taylor, P., Eds., *World Cities in a World System*. Cambridge: CUP. pp. 115–131.

Lee, C. and Moudon, A. 2004. Physical activity and environment research in the health field: Implications for urban and transportation planning practice and research. *Journal of Planning Literature*, 19(2), pp. 147–180.

Lefebvre, H. 1962. *Notes on The New Town: Introduction to Modernity*. London and New York: Verlso.

Lefebvre, H. 1967. The right to the city. In Kofman, E. and Lebas, E., Eds., *Writings on Cities*. London: Blackwell.
Mayor of London, 2014. *London Infrastructure Plan 2050: a consultation*. [online] Available at: www.london.gov.uk/infrastructure [Accessed 18 May 2019].
Mayor of London, 2015. *London Infrastructure Plan 2050: update report*. [online] Available at: www.london.gov.uk/infrastructure [Accessed 18 May 2019].
Mitchell D., Claris S. and Edge D. 2016. Human-centered mobility: A new approach to designing and improving our urban transport infrastructure. *Engineering*, 2, pp. 33–36.
Mumford, L. 1961. *The City in History: Its Origins, Its Transformations, and Its Prospects*. New York: Harcourt, Brace & World.
Newman, P. and Kenworthy J. 1990. *Cities and Automobile Dependence: A Sourcebook*. Aldershot: Gower Publishing Company.
Olea, R. 2007. *San Francisco's Octavia Boulevard*. San Francisco: San Francisco Municipal Transportation Agency.
Sassen, S. 2001. *The Global City: New York, London, Tokyo*. 2nd ed. Princeton: Princeton University Press.
Shannon, K. and Smets, M. 2016. *The Landscape of Contemporary Infrastructure*. 2nd ed. Rotterdam: NAi Publishers.
Southworth, M. 2005. Designing the walkable city. *Journal of Urban Planning & Development*, 131(4), pp. 246–257.
Sumantran, V., Fine, C. and Gonsalvez, D. 2017. *Faster, Smarter, Greener: The Future of the Car and Urban Mobility*. Cambridge: MIT Press.
Taylor, P. and Derudder, B. 2015. *World City Network: A Global Urban Analysis*. 2nd ed. Oxford and New York: Routledge.
Wachs, M. 1984. Autos, Transit, and the Sprawl of Los Angeles: The 1920s. *Journal of the American Planning Association*, 50(3), pp. 297–310.
Warner, S. 1978. *Streetcar Suburbs*. Cambridge: Harvard University Press.
Wiel, M. 1999. *La transition urbaine*. Bruxelles: Mardaga Pierre.
Wiel, M. 2002. *Ville et automobile*, Paris: Descartes et Cie, p. 145.
Wilson, W. H. 1994. *The City Beautiful Movement*. Baltimore: Johns Hopkins University Press.
World Bank. 2018. *Transport overview*. [online] Available at: www.worldbank.org/en/topic/transport/overview. [Accessed 28 May 2019].

39
SMART CITIES

Federico Cugurullo and Ransford A. Acheampong

Introduction

The smart city is now one of the most popular models of urban development in the world. Among the many promises of smart urbanism, the optimization of urban transport and mobility features prominently. Alleged smart cities are expected to improve the quality, efficiency and sustainability of urban transport, by means of information and communication technology (ICT) and, more recently, artificial intelligence (AI). The aim of this chapter is to unpack the meaning and practice of smart urbanism, in relation to issues of transport and mobility. We proceed through three steps. First, we illustrate the concept of the smart city and its many incarnations across the globe. Second, we move the discussion to smart transport as a core element of smart-city agendas. In this section of the chapter, we focus (a) on automated and autonomous systems of traffic management and (b) on autonomous self-driving cars. Finally, we conclude by critically evaluating the potential benefits and dangers of smart urban transport, from a sustainability perspective.

The smart city phenomenon

A multitude of cities are currently following *smart urbanism* as a model of city-making and urban regeneration: 'the era of the smart city has arrived' (Karvonen et al., 2018a: 1). While only a couple of decades ago, the digitalization of the built environment appeared like a techno-utopian fantasy, the rhetoric is now becoming a reality. Across heterogeneous geographical spaces, emblematic examples of smart-city agendas can be found, for example, in Spain (March and Ribera-Fumaz, 2016), the United Kingdom (Cowley et al., 2018), Italy (Garau and Pavan, 2018; Grossi and Pianezzi, 2017), Africa (Odendaal, 2016), the United States (McLean et al., 2016; Wiig, 2018), the United Arab Emirates (Cugurullo, 2016a; Khan et al., 2017), Korea (Shwayri, 2013) and India (Datta, 2018). However, despite such sheer enthusiasm and popularity, smart urbanism eludes a definition. It is often unclear what makes a city a *smart* city and, as a result, alleged (and frequently self-proclaimed) smart cities, do not always manifest considerable differences from ordinary cities.

There are some common conceptual and technical pillars at the basis of smart-city initiatives, whose exploration can be useful to understand the meaning and practice of smart urbanism. In terms of key concepts, ideas and images, the smart city is inspired by an adamant faith in technology and innovation (Angelidou, 2015; Cugurullo, 2018a). In this sense, the smart city emerges as the descendant of the modernist city in which the power of the machine was believed to be the key to progress, and applied sciences (engineering in particular) were seen as a medium to control the built and the natural environment. From a technical point of view, actually-existing smart cities are filled with information and communication technology (ICT) which is used to generate large data sets (or in other words *big data*) on how the city functions, in order to improve its performance (Coletta et al., 2018; Dameri et al., 2018; Kong and Woods, 2018; Luque-Ayala and Marvin, 2015; Mohamed et al., 2018).

The many facets of smart urbanism reflect the many facts of cities and urban living. In terms of energy, for instance, smart grids can monitor energy usage to reduce energy waste and synchronize energy production with local changes in consumption. Similarly, intelligent buildings by means of a central software can automatically access and optimize lighting, heating, ventilation, safety and other key systems. When it comes to safety, holistic CCTV networks can constantly watch the city and alert the police in real time. In these terms, the evolution of the city follows the evolution of technology and, according to the smart-city rationale, when technology improves so does the city (Battarra et al., 2018; Contreras and Platania, 2018; Hopkins and McKay, 2018; Yan et al., 2018). Ultimately, smart-city believers claim that smart-city agendas can lead cities towards a condition of urban sustainability and that, with their emphasis on energy conservation and renewable energy, have the capacity to tackle pressing global challenges such as climate change, pollution and resource scarcity (De Jong et al., 2015; Glasmeier and Nebiolo, 2016; Guedes et al., 2018; Yigitcanlar et al., 2018).

Transportation being one of the fundamental elements of the metabolism of cities, it is not surprising that the movement of people across the built environment represents one of the key foci of smart urbanism. Within the logic of the smart city, a smart transport is represented by an ICT-driven infrastructure which optimizes the circulation of people, goods and services, thereby saving energy, carbon emissions and time. Empirical studies on actually-existing smart cities, show that smart transport solutions have been implemented in different ways in different cities, generating different impacts on urban mobility (Karvonen et al., 2018b). The sheer variety of smart mobilities can be understood first as a reflection of the politics of mobility. Understanding politics as the 'social relations that involve the production and distribution of power', Cresswell (2010: 21) points out that 'mobilities are both productive of such social relations and produced by them'. These relations are influenced by local contexts, and thus possess contextual qualities which might not be observable or replicable elsewhere.

Second, because smart urbanism is substantially shaped by technological innovation, the material nature of smart-city solutions depends on the specific technology that is utilized. In this sense, smart transport technologies can differ substantially across a broad spectrum of digital products and services. Examples include now common smart cards used in the context of public transit, (a) by the individual to quickly pay for the service and (b) by the provider of that service to monitor how, when and where individuals employ the service (a bus ride or a metro trip, for example). More recently, the application of artificial intelligence to optimize urban transport systems, has led to the autonomy of entire transport infrastructures and means of transport such as self-driving cars and drones. Within this heterogeneous pool of technologies, the common denominator is ICT designed to produce data on urban mobility. From a conceptual perspective, the common underpinning is the assumption that the more we know about urban mobility, the more efficient is the transport infrastructure that we can build. The chapter now turns to two representative examples of smart transport, in order to unpack how smart urban mobility is understood and implemented on the ground.

Smart traffic management: automation and artificial intelligence

A key aspect of smart transport concerns the management of urban traffic. Within this category, we can identify two sub-categories: *automated* traffic management and *autonomous* traffic management. An example of the first sub-category is the Personal Rapid Transit (PRT) in Masdar City, a new settlement currently under development in Abu Dhabi (Cugurullo, 2013a). The PRT is a transport system designed around automated driverless vehicles for up to four passengers, operating over a smart grid (see Figure 39.1).

The PRT functions in the following way. In terms of infrastructure, the streets of Masdar City are filled with micro sensors which define the tracks over which the vehicles will pass. All the tracks form an invisible grid which delineates the spaces of urban mobility: all the vehicles and, therefore, their passengers move exclusively within this smart transport grid. Passengers can enter a PRT vehicle only at a PRT station. Once inside, the passenger can select via a tablet inside the vehicle, a pre-defined destination within the grid. The moment the destination is selected, the PRT vehicle drives the passenger to the selected station. The trip takes place in an automated way, in the sense that PRT vehicles are programmed to automatically follow a specific track of the grid. They do not decide the route. Instead, they follow the route that the designers of the smart grid had programmed and

Figure 39.1 A PRT vehicle in the central station of Masdar City.
Source: Federico Cugurullo

established beforehand. In this case the control over the mobility of the vehicles and their passengers is automated, because passengers can choose the destination, knowing the route in advance, and the movement of the machine is disciplined by *a priori* decisions made in advance by a team of computer scientists, engineers and urban planners.

A different typology of smart traffic management is that exemplified by Alibaba's City Brain, a programme led by an artificial intelligence which autonomously controls urban traffic (Alibaba, 2018). City Brain generates big data by means of hundreds of cameras, and processes it, in real time, through a video-analysis software. This data, coupled with information produced by sensors installed in smart cars, tracking movement and speed, provides Alibaba's artificial intelligence with a broad-based knowledge of traffic in a given city. City Brain then takes control and operates traffic lights, thereby directing the flow of cars. Everything takes place without any inputs from computer scientists, engineers and urban planners. Decisions are made autonomously by an artificial intelligence according to its own calculations. The AI software is of course programmed in advance, but its outcomes are not. City Brains, via machine learning, learns independently about the traffic of the city through the data that it collects, and makes decisions which have not been explicitly programmed. In these terms, while Masdar City's PRT is automated (the machine follows precisely the instructions of its creators, and sticks to pre-programmed mobility patterns), Alibaba's City Brain is autonomous (the machine follows its own judgement and creates new un-programmed mobility patterns in real time).

Smart transportation systems: autonomous self-driving cars

A new form of motorized transportation in the field of autonomous mobility, driverless cars are now becoming part of the transport portfolio of smart-city initiatives (Acheampong et al., 2018). Today, partial and conditional automation technologies are available to assist the human driver with safety critical control functions such as steering, acceleration and breaking (Hashimoto et al., 2016). At the highest levels of automation, driverless cars, by using advanced sensing and communication technologies, are expected to be able to operate efficiently within a prescribed operational domain (Level-4), and to perform all driving functions under normal road conditions without any design-based restrictions (Level-5). As with most smart-city technologies, a confluence of economic and socio-environmental imperatives is being invoked to justify and propel the transition to fully automated driving in cities. From an economic point of view, major players in the automobile and transport service industries, as well as in the technology space, see artificial intelligence in general, and self-driving cars in particular, as areas where significant growth is likely to occur in the near future. Hence the interest and investments in the development of driverless cars, of companies such as Google, Uber, Apple, NuTonomy, Toyota and General Motors.

Beyond the mere economic lure, one of the main promises of automating driving is that self-driving cars can eliminate the human factor that leads to accidents, thereby significantly reducing injuries and fatalities. This assertion is based on the fact that worldwide, road traffic accident is one of the leading causes of death, claiming the lives of over 1.2 million people annually (WHO, 2015). In addition, driverless cars have the potential to make cities more inclusive, by improving accessibility across different groups of people. Problems of unequal access to urban opportunities such as employment and social services, are strongly linked to the way transportation systems are designed and operated in cities (Barnister, 2018; Jones and Lucas, 2012; Martens et al., 2012). In a future smart city where self-driving cars are ubiquitous, the constraints on motorized transport use, imposed by factors such as age-related impairments and physical and sensory disability, could no longer exist. In addition, the diffusion of self-driving cars could improve the environmental sustainability of cities (Acheampong et al., 2019). Autonomous vehicles capable of communicating with each other and moving in groups (i.e. platooning) can reduce fuel consumption and carbon emissions, while increasing road capacity. More specifically, it has been estimated that under controlled conditions, connected autonomous cars could reduce fuel consumption by 47%, and lower CO_2 emissions by up to 50% (Wadud, MacKenzie and Leiby, 2016).

Another pathway through which automated driving could help cities become more sustainable, is shared-mobility. Driverless cars can easily support free-floating car-sharing services (Guériau et al., forthcoming). Through this mode of deployment, autonomous cars could eliminate the need for fixed-stations, make service use more flexible for commuters and, above all, contribute to a reduction in car ownership (Fagnant and Kockelman, 2014; Litman, 2017). Sivak and Schoettle (2015), for example, have estimated that car-ownership could be reduced by up to 43% if individuals opted for driverless-car sharing services. In this case, a potential domino effect could be the reduction of the demand for parking spaces in cities, with considerable opportunities in terms of the redesign of the built environment (Duarte and Ratti, 2018; Zhang et al., 2015). However, whether self-driving cars can and will trigger a transition towards urban sustainability is an open question (Acheampong and Cugurullo, 2019). Several studies on autonomous vehicles, for instance, expect that 'the associated decrease in travel disutility will cause people to travel more frequently and across greater distances': a trend that might increase the demand for the energy, space and infrastructure necessary to power cars and allow their transit (Hawkins and Nurul Habib, 2019: 69). This question is connected to a broader perplexity regarding the extent to which smart urban transport in general can lead to a sustainable urban mobility, which is what the chapter will discuss in the concluding section.

The long and winding road to a sustainable smart urban mobility

As the previous sections have shown, smart transport systems are not a techno-utopian fantasy anymore (Karvonen et al., 2018c). Many cities are digitalizing their transport infrastructures with smart technologies ranging from AI-led traffic management programmes to autonomous transportation systems, which prove that ICT can make urban mobility more efficient, thereby saving energy, time and reducing carbon emissions. However, as a number of studies point out, *smart* is not necessarily equivalent to *sustainable* (Colding et al., 2018; Cugurullo, 2013b, 2016b; Kaika, 2017; Martin et al., 2018; Trencher, 2018). Building upon this premise, in this last section the chapter will address the following question: to what extent are smart transport systems actually sustainable?

From a sustainability point of view, a key problem which has been stressed particularly in the field of geography is the spatial dimension of smart urbanism. Geographers have shown that, often, smart-city agendas, despite their lofty claims of homogeneity, inclusion and large-scale master planning, actually cover only a relatively small percentage of the total area of a city. In practice, this means that smart technologies are integrated exclusively into a few buildings and infrastructures, while most of the urban fabric is left untouched. The same issue can be observed in the context of smart transport. In Masdar City, for instance, the PRT covers only 10% of the new settlement. The AI software developed by Alibaba controls only a portion of a handful of Chinese cities (Hangzhou, Suzhou and Hong Kong). Autonomous cars tend to be operational only in selected neighbourhoods. Developers claim that this geographically uneven distribution of smart transport technologies is due to their experimental nature, and that in a later stage such technologies, once fully tested and deemed safe and efficient, will be scaled up. However, empirical studies show that this is rarely the case. Smart technologies like those discussed in this chapter, are frequently developed on an individual basis without any overarching vision of urban development. Because of the lack of a cohesive urban agenda, they are disconnected from the rest of the built and the natural environment, and can therefore lead to the formation of Frankenstein cities made of incongruous elements (Cugurullo, 2016c, 2018b; Cugurullo and Ponzini, 2018). This situation is socially and environmentally unsustainable. First, when smart transport solutions are implemented only in the most affluent parts of the city, they increase social inequality and exacerbate the divide between who can afford to live in urban areas equipped with prime transport infrastructures, and who cannot. Similarly, when smart transport technologies such as self-driving cars become available only via private ownership or private transportation network companies (Uber, for example), they separate those who can pay for the car or the service from those who do not have enough economic power. Second, from an environmental perspective, the creation of smart transport technologies capable of communicating with each other, but not with local ecosystems, promotes a form of mobility which prioritizes the needs of human societies over the rest of the biophysical environment.

The above problems are interconnected to another major flaw in the practice of smart urbanism: the preponderance of the economic. Many smart-city agendas are currently being led by the private sector, often in a neoliberal fashion. The state has little or no control over the *what* and *where* of smart technology. More specifically, it is down to tech companies such as IBM, Siemens and Alibaba to choose, in line with their business agenda, the specific smart technologies that will be developed, and the parts of the city where they will be integrated into. These companies being driven and sustained by profit, the choice inevitably falls on the most remunerative option. Applied in the context of urban transport, the economic impetus of smart urbanism pushes for the development of those technologies for which developers and stakeholders see a return on investment, even with retroactive consequences. The moment the prospect of profit is not on the horizon anymore, even the development of what was promoted as a groundbreaking and efficient transport solution stops. Emblematic is the example of the PRT in Masdar City, whose implementation was stopped in 2009. Although successful in terms of energy efficiency, the PRT was ultimately abandoned by its developers because it was considered to be too expensive compared with its low potential investment return (Cugurullo, 2016a). This is when *smart* clashes against *sustainable* and when smart transport can clash against sustainable transport. There are many alternative low-tech solutions to urban mobility which can be efficient, socially just and environmentally friendly without relying on artificial intelligence and automation. These mobilities are cultivated particularly in countries with a strong welfare state, where the government, despite the rise of AI-controlled urban infrastructures, continues to invest in pedestrianized streets, cycling lanes and canals for kayaking (see, for instance, Byoghavn, 2018). In these terms, urban mobility can be sustainable without being smart. However, under the hegemony of neoliberalism, smartness has become the reality of the present, and sustainability risks becoming the utopia of the past. Despite the grandiose promises of smart urban technologies, the road to a sustainable smart urban mobility is still long and winding.

References

Acheampong, R. A. and Cugurullo, F. 2019. Capturing the behavioural determinants behind the adoption of autonomous vehicles: Conceptual frameworks and measurement models to predict public transport, sharing and ownership trends of self-driving cars. *Transportation Research Part F: Traffic Psychology and Behaviour.* 62, 349–75.

Acheampong, R., Cugurullo, F., Dusparic, I., & Guériau, M. (2019). An examination of user adoption behavior of autonomous vehicles and urban sustainability implications. *Transportation Research Procedia, 41*, 187–190.

Acheampong, R. A., Thomoupolos, N., Marten, K., Beyazıt, E., Cugurullo, F., and Dusparic, I. 2018. *Literature review on the social challenges of autonomous transport.* STSM Report for COST Action CA16222 "Wider Impacts and Scenario Evaluation of Autonomous and Connected Transport (WISE-ACT).

Alibaba. 2018. *City Brain.* Online [accessed 15/04/2018] available at: www.alibabacloud.com/et/city

Angelidou, M. 2015. Smart cities: A conjuncture of four forces. *Cities.* 47, 95–106.

Barnister, D. 2018. *Inequality in Transport.* Marcham, Oxfordshire: Alexandrine Press.

Battarra, R., Gargiulo, C., Tremiterra, M. R., and Zucaro, F. 2018. Smart mobility in Italian metropolitan cities: A comparative analysis through indicators and actions. *Sustainable Cities and Society.* 41, 556–67.

Byoghavn. 2018. *Nordhavn.* Online [accessed 15/04/2018] available at: www.byoghavn.dk/presse/nyheder/2018/containerterminal+bliver+til+bredygtigt+bykvarter.aspx.

Colding, J., Colding, M., and Barthel, S. 2018. The smart city model: A new panacea for urban sustainability or unmanageable complexity?. *Environment and Planning B: Urban Analytics and City Science.* DOI: https://doi.org/10.1177/2399808318763164.

Coletta, C., Evans, L., Heaphy, L., and Kitchin, R., eds. 2018. *Creating Smart Cities.* London: Routledge.

Contreras, G. and Platania, F. 2018. Economic and policy uncertainty in climate change mitigation: The London smart city case scenario. *Technological Forecasting and Social Change.* DOI: https://doi.org/10.1016/j.techfore.2018.07.018.

Cowley, R., Joss, S., and Dayot, Y. 2018. The smart city and its publics: Insights from across six UK cities. *Urban Research & Practice.* 11(1), 53–77.

Cresswell, T. 2010. Towards a politics of mobility. *Environment and Planning D: Society and Space.* 28(1), 17–31.

Cugurullo, F. 2013a. How to build a sandcastle: An analysis of the genesis and development of Masdar city. *Journal of Urban Technology.* 20(1), 23–37.

Cugurullo, F. 2013b. The business of Utopia: Estidama and the road to the sustainable city. *Utopian Studies.* 24(1), 66–88.

Cugurullo, F. 2016a. Urban eco-modernisation and the policy context of new eco-city projects: where Masdar city fails and why. *Urban Studies.* 53(11), 2417–33.

Cugurullo, F. 2016b. Speed kills: Fast urbanism and endangered sustainability in the Masdar city project. In A. Datta and A. Shaban, eds., *Mega-Urbanization in the Global South: Fast Cities and New Urban Utopias of the Postcolonial State* (pp. 66–80). London: Routledge.

Cugurullo, F. 2016c. Frankenstein cities. In J. Evans, A. Karvonen and R. Raven, eds., *The Experimental City* (pp. 195–204). London: Routledge.

Cugurullo, F. 2018a. The origin of the Smart city imaginary: From the dawn of modernity to the eclipse of reason. In C. Lindner and M. Meissner, eds., *The Routledge Companion to Urban Imaginaries* (pp. 113–124). London: Routledge.

Cugurullo, F. 2018b. Exposing smart cities and eco-cities: Frankenstein urbanism and the sustainability challenges of the experimental city. *Environment and Planning A: Economy and Space.* 50(1), 73–92.

Cugurullo, F. and Ponzini, D. 2018. The transnational smart city as urban eco-modernisation. In A. Karvonen, F. Cugurullo and F. Caprotti, eds., *Inside Smart Cities: Place, Politics and Urban Innovation* (pp. 149–162). London: Routledge.

Dameri, R. P., Benevolo, C., Veglianti, E., and Li, Y. 2018. Understanding smart cities as a glocal strategy: A comparison between Italy and China. *Technological Forecasting and Social Change.* https://doi.org/10.1016/j.techfore.2018.07.025.

Datta, A., 2018. The digital turn in postcolonial urbanism: Smart citizenship in the making of India's 100 smart cities. *Transactions of the Institute of British Geographers.*

De Jong, M., Joss, S., Schraven, D., Zhan, C., and Weijnen, M. 2015. Sustainable–smart–resilient–low carbon–eco–knowledge cities: Making sense of a multitude of concepts promoting sustainable urbanization. *Journal of Cleaner Production*. 109, 25–38.

Duarte, F. and Ratti, C. 2018. The impact of autonomous vehicles on cities: A review. *Journal of Urban Technology*. 25(4), 3–18.

Fagnant, D. and Kockelman, K., 2014. Environmental implications for autonomous shared vehicles, using agent-based model scenarios. *Transportation Research Part C 40: 1-20 13*.

Garau, C. and Pavan, V. M. 2018. Evaluating urban quality: Indicators and assessment tools for smart sustainable cities. *Sustainability*. 10(3), 575.

Glasmeier, A. K. and Nebiolo, M. 2016. Thinking about smart cities: The travels of a policy idea that promises a great deal, but so far has delivered modest results. *Sustainability*. 8(11), 1122.

Grossi, G. and Pianezzi, D. 2017. Smart cities: Utopia or neoliberal ideology?. *Cities*. 69, 79–85.

Guedes, A. L. A., Alvarenga, J. C., Goulart, M. D. S. S., Rodriguez, M. V. R., and Soares, C. A. P. 2018. Smart cities: The main drivers for increasing the intelligence of cities. *Sustainability*. 10(9), 1–18.

Guériau, M., Cugurullo, F., Acheampong, R., and Dusparic, I. (forthcoming). Shared autonomous mobility-on-demand: Learning-based approach and its performance in the presence of traffic congestion. *IEEE Intelligent Transportation Systems Magazine*.

Hashimoto, Y., Gu, Y., Hsu, L. T., Iryo-Asano, M., and Kamijo, S. 2016. A probabilistic model of pedestrian crossing behavior at signalized intersections for connected vehicles. *Transportation Research Part C: Emerging Technologies*. 71, 164–81.

Hawkins, J. and Nurul Habib, K. 2019. Integrated models of land use and transportation for the autonomous vehicle revolution. *Transport Reviews*. 39(1), 66–83.

Hopkins, J. L. and McKay, J. 2018. Investigating 'anywhere working' as a mechanism for alleviating traffic congestion in smart cities. *Technological Forecasting and Social Change*. https://doi.org/10.1016/j.techfore.2018.07.032.

Jones, P. and Lucas, K. 2012. Social impacts and equity issues in transport: An introduction. *Journal of Transport Geography*. 21, 1–3.

Kaika, M. 2017. 'Don't call me Resilient Again!' The New Urban Agenda as Immunology … or what happens when communities refuse to be vaccinated with 'smart cities' and indicators. *Environment and Urbanization*. 29(1), 89–102.

Karvonen, A., Cugurullo, F., and Caprotti, F. 2018a. Introduction: Situating smart cities. In A. Karvonen, F. Cugurullo and F. Caprotti, eds., *Inside Smart Cities: Place, politics and urban innovation* (pp. 1–9). London: Routledge.

Karvonen, A., Cugurullo, F., and Caprotti, F., eds. 2018b. *Inside Smart Cities: Place, Politics and Urban Innovation*. London: Routledge.

Karvonen, A., Cugurullo, F., and Caprotti, F. 2018c. Conclusions: The long and unsettled future of smart cities. In A. Karvonen, F. Cugurullo and F. Caprotti, eds., *Inside Smart Cities: Place, politics and urban innovation* (pp. 291–298). London: Routledge.

Khan, M. S., Woo, M., Nam, K., and Chathoth, P. K. 2017. Smart city and smart tourism: A case of Dubai. *Sustainability*. 9(12), 2279.

Kong, L. and Woods, O. 2018. The ideological alignment of smart urbanism in Singapore: Critical reflections on a political paradox. *Urban Studies*. 55(4), 679–701.

Litman, T., 2017. Autonomous vehicle implementation predictions. Implications for Transport Planning. Presented at the 2015 Transportation Research Board Annual Meeting. Online [accessed 15/04/2018] available at: http://leempo.com/wp-content/uploads/2017/03/M09.pdf

Luque-Ayala, A. and Marvin, S. 2015. Developing a critical understanding of smart urbanism?. *Urban Studies*. 52(12), 2105–16.

March, H. and Ribera-Fumaz, R. 2016. Smart contradictions: The politics of making Barcelona a self-sufficient city. *European Urban and Regional Studies*. 23(4), 816–30.

Martens, K., Golub, A., and Robinson, G. 2012. A justice-theoretic approach to the distribution of transportation benefits: Implications for transportation planning practice in the United States. *Transportation Research Part A: Policy and Practice*. 46(4), 684–95.

Martin, C. J., Evans, J., and Karvonen, A. 2018. Smart and sustainable? Five tensions in the visions and practices of the smart-sustainable city in Europe and North America. *Technological Forecasting and Social Change*. 133, 269–78.

McLean, A., Bulkeley, H., and Crang, M. 2016. Negotiating the urban smart grid: Socio-technical experimentation in the city of Austin. *Urban Studies*. 53(15), 3246–63.

Mohamed, N., Al-Jaroodi, J., Jawhar, I., Idries, A., and Mohammed, F. 2018. Unmanned aerial vehicles applications in future smart cities. *Technological Forecasting and Social Change*. https://doi.org/10.1016/j.techfore.2018.05.004.

Odendaal, N. 2016. Getting smart about smart cities in Cape Town: Beyond the rhetoric. In S. Marvin, A. Luque-Ayala and C. McFarlane, eds., *Smart Urbanism: Utopian vision or false dawn?* (pp. 71–87). London: Routledge.

Shwayri, S. T. 2013. A model Korean ubiquitous eco-city? The politics of making Songdo. *Journal of Urban Technology*. 20(1), 39–55.

Sivak, M. and Schoettle, B., 2015. *Potential Impact of Self-Driving Vehicles on Household Vehicle Demand and Usage*, Sustainable Worldwide Transportation Program, Transportation Research Institute, The University of Michigan. Report No. UMTRI-20153.

Trencher, G. 2018. Towards the smart city 2.0: Empirical evidence of using smartness as a tool for tackling social challenges. *Technological Forecasting and Social Change*. https://doi.org/10.1016/j.techfore.2018.07.033.

Wadud, Z., MacKenzie, D., & Leiby, P. (2016). Help or hindrance? The travel, energy and carbon impacts of highly automated vehicles. *Transportation Research Part A: Policy and Practice*, *86*, 1–18.

Wiig, A. 2018. Secure the city, revitalize the zone: Smart urbanization in Camden, New Jersey. *Environment and Planning C: Politics and Space*. 36(3), 403–22.

World Health Organization, 2015. *Global status report on road safety 2015*. World Health Organization. Online [accessed 15/04/2018] available at: www.who.int/violence_injury_prevention/road_safety_status/2015/en/

Yan, J., Liu, J., and Tseng, F. M. 2018. An evaluation system based on the self-organizing system framework of smart cities: A case study of smart transportation systems in China. *Technological Forecasting and Social Change*. https://doi.org/10.1016/j.techfore.2018.07.009.

Yigitcanlar, T., Kamruzzaman, M., Buys, L., Ioppolo, G., Sabatini-Marques, J., da Costa, E. M., and Yun, J. J. 2018. Understanding 'smart cities': Intertwining development drivers with desired outcomes in a multidimensional framework. *Cities*. https://doi.org/10.1016/j.cities.2018.04.003.

Zhang, W., Guhathakurta, S., Fang, J., and Zhang, G. 2015. Exploring the impact of shared autonomous vehicles on urban parking demand: An agent-based simulation approach. *Sustainable Cities and Society*. 19, 34–45.

40
SUSTAINABLE MOBILITY

Petter Næss

Introduction

Sustainable mobility is a term frequently used in planning and policy-making, but what is meant by this concept is often unclear. The term combines two concepts that are themselves complex and contested: sustainable development and mobility. Hardly surprising, the literature on sustainable mobility varies widely in its focus. This chapter offers an interpretation of the concept rooted in the Brundtland Commission's (1987) definition of sustainable development and discusses some of its implications for transport and land use planning policies in affluent countries.

The European Union started using the term 'sustainable mobility' already in 1992 in a Green Paper on the impact of transport on the environment (CEC, 1992). Around the same time, the OECD initiated work on 'sustainable transport' and organized a number of conferences on the topic (see, for example, OECD, 1997). Following the increased social science focus on transportation as part of a wider concept of mobility in the late 1990s (Kaufmann, 2002; Urry, 2000), 'sustainable mobility' became the most commonly used term, at least in a European context.

According to the Brundtland Commission (1987), the concept of ***sustainable development*** combines ethical norms of welfare, distribution and democracy while recognizing that nature's ability to absorb human-made encroachments and pollution is limited. During the more than three decades that have passed, the term has come to be used in an increasingly diffuse manner. Instead of declaring openly that a sustainable development as defined by the Brundtland commission is not wanted, those whose interests are threatened by the requirements of such development may attempt to re-define the concept. For example, by twisting and stretching the concept to mean 'sustained growth' instead of development and fulfillment of human needs, and by omitting the concern of a fair distribution spatially (between wealthy and poor countries and between different population groups within a country (see Schade & Rothengatter, 2011 for an example related to EU policies for the transportation sector).

However, if sustainable mobility is at all to be relevant as a distinct agenda for policy and research, sustainable development must be something different from the already existing development. In the following, the concept of sustainable mobility is understood in line with the Brundtland Commission's (1987:43) definition of sustainable development:

Sustainable development is development that meets the needs of the present without compromising the ability of future generations to meet their own needs. It contains within it two key concepts:

- the concept of 'needs', in particular the essential needs of the world's poor, to which overriding priority should be given, and
- the idea of limitations imposed by the state of technology and social organization on the environment's ability to meet present and future needs.

The concept of ***mobility*** is used differently in different disciplines of science, and also often metaphorically as a synonym for change. In economics, it refers to reallocation of factors of production, especially the labor force. In sociology, it has traditionally been understood as the movement of individuals or population groups between social positions or places of residence. In transport research and physical planning it refers to locomotion and transportation in society (Høyer, 1999:12). Transportation researchers traditionally used the term mobility as a term referring to potentials for movement (Kasanen, 1994). In the research field of mobility sociology emerging since the 1990s, the term usually includes both the potential for movement and the actualized movements. It also includes the traditional sociological mobility concepts referring to changes in social position or geographical place of residence, as well as the more recent phenomenon of virtual mobility. Today, the concept of mobility is widely accepted as covering potential as well as actualized movement, also in the field of transportation research.

As part of the specific concept of sustainable mobility, the term mobility will in the following be used about physical movement in the form of transportation, while changes in social position, migration, etc. will not be addressed.

The concept of mobility is linked with, but not identical to, the concept of *accessibility*. Accessibility refers to the possibilities to reach the persons and goods required, and for persons and goods to get to their destinations. A low accessibility implies that the transportation of persons or goods between the relevant spots is associated with long distances, a high consumption of time, high inconvenience, or large costs. Conversely, when accessibility is high, barriers of this kind are small. In transport policy, improving accessibility has usually been the explicit purpose of efforts to increase mobility (e.g. transport infrastructure investments). However, accessibility can be influenced also by other means than changes in physical mobility, notably through changes in the geographical location of facilities, and by replacing physical mobility with virtual contact.

Høyer (1999) defined ***sustainable mobility*** simply as mobility in accordance with the goals and requirement of sustainable development. Although elaborating on various aspects and implications of the concept, Høyer did not provide any more specific definition. Such more elaborate definitions appear still to be scarce. One of the few examples is Morency (2013), who defines mobility as sustainable 'when it is created in a way that respects safety and the environment, ensures the provision of life's material needs and guarantees fairness among individuals'. An early attempt to make a more specific definition of the related concept of sustainable transportation stems from the Canadian Center for Sustainable Transportation (CfST) (2002), where sustainable transportation is defined as a system which

- allows the basic needs of individuals and societies to be met safely and in a manner consistent with human and ecosystem health, and with equity within and between generations

- is affordable, operates efficiently, and offers choice of transport mode, and supports a vibrant economy
- limits emissions and waste within the planet's ability to absorb them, minimizes consumption of non-renewable resources, limits consumption of renewable resources to the sustainable yield level, reuses and recycles its components, and minimizes the use of land and the production of noise.

A definition content-wise close to the one above but somewhat more lengthy was adopted by the European Conference of Ministers of Transport (ECMT) in 2004.

Apart from using the term transportation instead of mobility, the above definition is confined to the transport system, whereas the volume of transport is omitted. This makes the definition less satisfactory, as the sustainability of any transport system cannot be assessed independently of the volume of transport using that system. For example, a completely car-based transport system might be environmentally sustainable if the total volume of transport were radically lower than today, say, 5% of the present level. Conversely, a transport system based on rail and buses might be unsustainable if the volume of transport were 20 times higher than today. The volume dimension should therefore be included in the definition of sustainable mobility. Moreover, the meaning attached to the formulation in the CfST definition about offering 'choice of transport modes' should be clarified in order to avoid the interpretation that the market demand for infrastructure for any mode of transport should necessarily be met. For some environmentally straining forms of transport, notably flights and (urban) car travel, the demand must probably be suppressed in some way (through taxes or through absence of capacity increases) if sustainable results in terms of energy use, CO_2 emissions and land consumption are to be achieved. In light of the Brundtland Commission's above-mentioned definition of sustainable development, the phrase about 'a vibrant economy' should arguably be replaced by the formulation 'an economy meeting the population's essential needs'.

On this background, I propose a definition of sustainable mobility largely based on the CfST definition, but structured in such a way that the economic (to meet human needs), environmental (protection of natural resources and ecosystems) and social (equity within and between generations) dimensions of sustainable development are mentioned under a bullet each:

Sustainable mobility is mobility in accordance with the principles of sustainable development, understood as a volume of physical mobility and a transport system where mobility in society

- allows the basic mobility needs of individuals and societies to be met safely and consistent with human health, offers choice among environmentally sustainable transport modes, operates efficiently and supports an economy meeting the population's essential needs
- takes care of ecosystem integrity and limits emissions and waste within the planet's ability to absorb them, minimizes consumption of non-renewable resources, limits consumption of renewable resources to the sustainable yield level, reuses and recycles its components, and minimizes the use of land and the production of noise
- is affordable and consistent with equity both within and between generations, at a global, regional as well as local scale.

Strategies for obtaining sustainable mobility

Below, different strategies for achieving sustainable mobility will be discussed. While the focus of this book is on urban contexts, the discussion below will not be limited to intra-urban travel, since interactional mechanisms may exist between daily-life travel and long-distance leisure travel.

Current mobility patterns in affluent countries are far from meeting the criteria of sustainable mobility as defined above (Banister, 2008; Holden, 2016). Høyer (1999) discussed three main strategies for promoting environmentally sustainable mobility:

- Making each separate mode of transport more energy-efficient and environmentally friendly (the efficiency increase strategy)
- Modal shift from energy-demanding to more energy-efficient and less polluting modes of transport (the substitution strategy)
- Reducing the movement of persons and goods (the reduction strategy).

In a similar vein, Banister (2008) stated, in a much cited article entitled 'The sustainable mobility paradigm', that sustainable mobility requires measures to

- reduce the need for making trips (e.g. by means of information and communication technology),
- shift travel modes towards more environmentally friendly forms,
- reduce trip lengths through land use planning, and
- encourage greater efficiency in the transport system.

Traveling by more environmentally friendly modes

What would a change to more environmentally friendly modes of transportation mean? CO_2 emissions and land efficiency are two important indicators. Since current transportation systems are predominantly fossil-based, CO_2 emissions are closely related to energy use. Even with a large-scale future transition to electric vehicles, the production of electricity for a growing car fleet would not be without environmental impacts (more about this below). Transport infrastructure also occupies space, very often at the cost of farmland, natural ecosystems or existing built environments. Cleaner vehicles and fuels cannot do away with this.

Table 40.1 shows typical CO_2 emissions per passenger kilometer and land efficiency for different modes of travel. The figures must only be taken as rough indicators. Vehicle technologies are changing, and emissions and land efficiency may be different in other geographical contexts than those from which the figures in the table have been calculated. The table still gives an indication of which modes of travel have the smallest and the largest environmental footprints.

Air travel and passenger transport at sea require relatively little land for airports, harbors and adjacent access roads, but their high emission levels per kilometer make their total environmental loads substantial. Due to the aggravated climate impacts of emissions at high altitudes, the environmental problems associated with air travel are even larger than indicated by the levels of CO_2 emissions. Due to its high speed, airplanes also encourage travel over longer distances, with additional emissions as a result.

Overall, non-motorized modes stand out as the environmentally most favorable ones, followed by train, intercity bus, metro/light rail and city bus. It should be noted that

Table 40.1 Mean CO_2 emissions and land efficiency for different modes of travel

Travel mode	*Emission (gram of CO_2 per person kilometer*	*Land efficiency (max.number of passengers that a 3.5 m urban road lane can carry)*
High-speed passenger boat	803	N.A.
Airplane, short/medium-distance flights[2]	340	N.A.
Airplane, long-distance flights[3]	238	N.A.
Passenger boat, slow	205	N.A.
Private car	172	2000
City bus	102	9000
Metro/light rail	70	22000
Intercity bus	52	N.A.
Train	32	80000
Bike	0	14000
Walking	0	19000

Sources: Emissions (CO_2 equivalents) from from www.gronnhverdag.no/nor/Transport/CO2-utslipp-fra forskjelligetransportmidler, based on www.klimakalkulatoren.no; land efficiency from ADB (2012), quoted from Litman (2015), p. 5.7.6.

emissions from the construction of infrastructure are not included in the table. Taking also such emissions into consideration, the differences between rail-based urban public transportation and city buses will be smaller (Strand et al., 2009). Private cars have larger footprints than the above-mentioned modes, but still lower than airplanes and passenger boats (especially high-speed ones). In an urban context, non-motorized modes come out best, followed by transit and with the private car the least environmentally friendly mode.

Modal shift toward more environmentally friendly modes of transport can be obtained through transport infrastructure development prioritizing such modes at the cost of environmentally less favorable modes; through land use development encouraging public transportation, walking and biking; by various other transport policy measures (e.g. road pricing); and through cultural changes (possibly boosted by awareness campaigns directed toward the public).

Limitation of transport volumes

It is worth noticing that both Høyer's and Banister's proposed strategies include not only efficiency improvements in vehicles and transport systems and a change of modes of transportation toward environmentally more favorable modes, but also reductions in transport volumes. Whether or not a reduction (or at least limitation) of the volume of transportation should be an aim of sustainable mobility policies is perhaps the most controversial issue in the discourse on sustainable mobility. Opponents of setting goals of limitation or reduction often argue that this will be detrimental to economic development and reduce people's opportunities on the labor market and participation in leisure activities. However, if environmental sustainability actually requires a halt to (or reversal of) the hitherto ever-increasing transport volumes, any setbacks to international trade and people's ability to go to distant destinations would be a necessary sacrifice to prevent present mobility trajectories from eroding the environmental quality and opportunities of future generations. Moreover, the underlying purpose of transport activity is most often not mobility per se, but accessibility:

to provide good access for the inhabitants to workplaces, schools, service facilities and leisure activities, and good access for companies to a large number of potential employees, suppliers and customers. Apart from tourism, where much of the purpose of trips is to explore new and exotic places, accessibility can be promoted through proximity as well as through mobility.

The idea that there is a need to reduce transportation volumes in addition to developing cleaner, more energy-efficient vehicles and increasing the use of environmentally friendly means of transportation was also expressed in the above-mentioned very first European policy document on sustainable mobility (CEC, 1992) which stated:

> Public and private investment should be guided towards collective transport, whereas urban, industrial and commercial as well as regional development planning should be geared towards reducing the need for mobility. At the same time infrastructure planning should be made subject to restrictions on land intrusion as well as to strict environmental impact assessment procedures at both the strategic and project stages, including evaluation of alternative options.
>
> *(CEC 1992, paragraph 119 (p. 51))*

The recommended restrictions on land use intrusions also clearly go beyond a transition to cleaner and more efficient vehicles, since vehicles occupy space when they are moving as well as when parked, regardless of how clean and energy-efficient they are. Accordingly, the CEC document stated that it would be necessary to 'limit the increase transport demand, particularly in encumbered sectors', in order to limit land intrusion.

Clearly, a dramatic change has taken place in the international policy-making discourse during the 26 years since the 1992 European Green paper, following a strong neoliberal turn over these decades and the increasingly dominant belief in 'decoupling' between consumption growth and negative environmental impacts. According to Holden (2016), there has been a change in focus from one recommending reductions in *overall transport volumes* to curb global environmental impacts to one recommending reduced transport *intensity* to combat local pollution and congestion to increase competitiveness and inhabitants' well-being. How, then are the prospects of obtaining environmental sustainability while putting no limits to growth in the volumes of transportation?

Mobility growth counteracts the effects of technological improvements

A number of studies show that direct emissions from vehicles of some pollutants (such as NO_X and lead) have been reduced at a faster rate than traffic growth, but less so for greenhouse gas emissions (Mock, 2012). Even if some efficiency gain has been achieved, CO_2 emissions from transportation have continued to increase globally as well as in sub-regions of the world (Africa, Europe, Central and South America, North America, Middle East and Asia) (Finel & Tapio, 2012).

Including a life cycle perspective ('well-to-wheel') reveals even greater difficulties in combining mobility growth with environmental sustainability. According to Holden (2016), even very optimistic levels of vehicle technology improvement will be insufficient to obtain sustainable mobility if transport volumes and the use of cars and airplanes continue to grow at present rates. For one thing, growth in passenger transport mileage has so far counteracted reductions in fuel consumption from increased engine efficiency.

Moreover, vehicles tend to become heavier, which counteracts any gains in energy efficiency. Actual fuel consumption and emission levels are also higher than the low values boasted from laboratory tests (and some of the latter figures have turned out to be false, as revealed in the scandal in 2015 about Volkswagen emission levels[1]). Moreover, in a life cycle perspective, the use of alternative fuels tends to change environmental impacts geographically (from the places of driving to the places of fuel or vehicle production) or thematically (other types of negative environmental impacts), rather than reducing the total environmental impacts (Holden, 2016). For example, production of biofuels represents a threat to biodiversity in rain forests and other ecosystems and comes in direct competition with food production on arable land.

The technological possibilities for replacing today's polluting and climate-destructive air travel with environmentally friendly alternatives appear much smaller than what is the case for travel within cities and urban regions. Although a few test flights over long distances based on solar energy have been conducted with small planes carrying one or two persons, doing the same with large aircrafts able to support large-scale international aviation is something completely different. The idea of sustainable growth in international mobility appears highly unrealistic.

Providing accessibility through proximity instead of through mobility

Leaving international tourism aside, most of our mobility takes place for daily-life purposes. Since the large majority of the population in affluent countries live in urban areas, transportation in cities and urban areas is important both in terms of environmental impacts and for people's possibilities to engage in activities in society. A large number of studies over the last 40 years have investigated how different land use characteristics of cities and metropolitan areas influence travel behavior. The insight into the causal mechanisms underlying correlations between land use and transportation has gradually improved. Today, there is overwhelming evidence in support of the 'compact city' as a preferred urban form in order to reduce intra-urban traveling distances and promote environmentally friendly modes of transportation. Compared to sprawling urban structures, a dense and concentrated city is a city of short distances. More specifically, high overall density and a high proportion of dwellings as well as workplaces located close to the city center has been found to reduce car-driving, increase the share of non-motorized travel and facilitate high-quality public transportation services (see, for example, Elldér, 2014; Engebretsen et al., 2018; Ewing & Cervero, 2010; Næss et al., 2019). At an intra-metropolitan scale, compact urban development around the main city center seems clearly preferable to more polycentric structures if the aim is to reduce urban motoring and its emissions. Decentralizing workplaces to suburban nodes will, for example, most often result in longer rather than shorter average commuting distances and increase the proportion of commuters traveling by car (Engebretsen et al., 2018). At a national/provincial scale, 'decentralized concentration' still appears more favorable from a sustainable mobility perspective than concentrating a high proportion of the population in the largest urban regions.

For facilities such as schools, kindergartens, grocery and other non-specialized stores, libraries, facilities for physical exercise, etc., a location offering short travel distances for users and customers is more important than minimizing the commuting distances of the employees. Such facilities should therefore be available in local centers all over the city or metropolitan area. Workplaces that generate much truck transport and/or need much space per employee should also not be located to the inner city. For such workplaces, a suburban location close to intercity road and rail connections would be more preferable (Verroen et al., 1990).

Some have argued that land use planning is a long-term measure compared to measures that could potentially influence travel behavior within a shorter time horizon (such as road pricing). However, the latter measures will affect people's welfare as well as the functionality of cities negatively if the land use development increases the distances between the various facilities of a city, as is the case with urban sprawl. Land use changes take place anyway; the question is whether we want them to support sustainable mobility or pull in the opposite direction. Moreover, compact urban development is favorable not only in terms of sustainable mobility, but also for the protection of farmland and biodiversity (EEA, 2006; Næss, 2001).

The existence of so-called rebound effects adds complexity to the picture. People who do not need to spend money on owning and driving a car can spend their money on other things instead, for example leisure flights. Several studies have indeed shown that residents of dense inner-city districts tend to make more flights than suburbanites do (Holden, 2016; Næss, 2016). Avoiding such effects seems impossible unless the purchasing power decreases, although specific effects can be countered through policy measures such as environmental taxes on leisure flights.

The construction of transport infrastructure is of course also crucial. Mobility-increasing transport investments tend to increase the spatial separation between activities and create more mobility-demanding built environments and location patterns. Such spatial structures also contribute to shape lifestyles, contracts and mobility cultures incompatible with sustainable mobility. This makes it politically more difficult to implement policies aiming to promote a transition toward a sustainable mobility paradigm (Isaksson & Richardson, 2009).

Should accessibility always be increased?

Modernity has increased accessibility and availability of options and commodities tremendously compared to pre-modern societies (although not equally distributed between population groups). Most people probably appreciate that we can choose between more jobs, leisure opportunities and commodities today than two hundred years ago. However, this does not logically imply that a further increase in the future will also be beneficial. The 'acceleration of society' (Rosa, 2015) arguably entails not only steadily increased mobility but also the restless chase for more opportunities. However, a too hectic life where too many choices have to be made can be stressful. Not only mobility, but also accessibility should arguably aim at sufficiency rather than maximizing.

For those who wish so, living a life without being very mobile should also be an option (Ferreira et al., 2017). In an urban context, this is often possible for those who live centrally with short distances to relevant destinations. However, in many cities, such residential locations have become increasingly unaffordable for large population groups over the last decades. Stronger governmental housing policies would be required to counteract this. Better opportunities for people to promote low-mobility, locally oriented lifestyles in rural areas should also be promoted.

Fair distribution of mobility opportunities

Today, mobility levels are very unevenly distributed not only between rich and poor countries, but also between population groups within each country. This is especially the case for long-distance leisure travel. For intra-urban travel, travel distances and modes vary much less with income, at least in affluent countries such as Norway. As discussed above, distances traveled for such purposes can be curbed through sustainability-oriented land use policies. However, people should also be allowed – within ecologically sustainable limits – to travel for leisure purposes beyond what is necessary for maintaining daily-life routines. If everybody is to be granted a certain amount of leisure mobility while at the same time the average mobility level among the population is to be kept within sustainable limits, those population groups who now travel the longest distances for leisure purposes cannot continue to do so. A just distribution of mobility levels therefore requires minimum entitlement norms as well as maximum norms for acceptable mobility. Holden (2016) illustrates this by suggesting a 'sustainable mobility area' for EU residents where the daily distance available by public transportation is at least 11 km per capita while the energy use for passenger transportation does not exceed 8 kWh per capita. In practice, the latter would set a ceiling for the daily number of kilometers, depending on vehicle technology and the shares of different modes of travel.

While Holden's suggested numbers are obviously contestable and depend on several assumptions, his idea about a 'sustainable mobility area' highlights the relevance of setting both maximum and minimum norms for mobility levels. Since much daily-life mobility is, at least within a short or medium term, tied up in existing urban structures and location patterns, it might be relevant to apply maximum norms for mobility levels only for leisure travel, for example by introducing heavy taxes for extra-metropolitan travel exceeding the established maximum norm. Ensuring everybody access to the minimum level of mobility could be promoted by offering free public transport for travel up to this level (and perhaps somewhat above), funded, for example, by revenues from taxation of excessively high mobility.

Concluding remarks

Sustainable mobility is mobility in accordance with the goals and requirements of sustainable development. Sustainable mobility must ensure ecological sustainability, satisfy basic needs and be socially just (across and within generations). Three main strategies should be combined to obtain ecological sustainability: making each separate mode of transport more energy-efficient and environmentally friendly, promoting a modal shift from environmentally unfriendly to more environmentally friendly modes of transport, and reducing the movement of persons and goods. The third strategy may benefit from the fact that the underlying good to be reached through mobility is most often accessibility, not mobility per se. Land use planning can play an important role – positive or negative – in providing proximity or distance to the facilities that people need or wish to visit.

Social sustainability requires that ecologically sustainable mobility must be obtained while securing everyone access to the facilities and people they need to visit in their daily life. The basic mobility level to which everyone should be entitled should also allow for leisure travel within limits set by ecological sustainability requirements. Since the average level of mobility cannot be too high in order to stay within limits of ecological sustainability, the satisfaction of basic mobility needs for all as well as concerns of social justice require that norms for maximum levels of mobility as well as a minimum level to which everyone is entitled be maintained.

Note

1 See https://en.wikipedia.org/wiki/Volkswagen_emissions_scandal.

References

ADB (2012). Solutions for Urban Transport. Asian Development Bank (www.adb.org); at http://farm8.staticflickr.com/7228/7399658942_267b1ba9fc_b.jpg.

Banister, D. (2008). The sustainable mobility paradigm. *Transport Policy*, 15, pp. 73–80.

Brundtland Commission (World Commission on Environment and Development) (1987). *Our Common Future*. Oxford/New York: Oxford University Press.

CEC (Commission of European Communities) (1992). *The Impact of Transport on the Environment: A Community strategy for sustainable mobility. Green Paper*. Brussels: Commission of European Communities.

Center for sustainable transportation (2002). *Definition and vision of sustainable transportation*. Accessed February 2005 at www.cstctd.org (website no longer available).

EEA (2006). *Urban sprawl in Europe: The ignored challenge*. EEA Report 10/2006. Copenhagen: European Environment Agency.

Elldér, E. (2014). Residential location and daily travel distances: The influence of trip purpose. *Journal of Transport Geography*, 34, pp. 121–130.

Engebretsen, Ø.; Næss, P. & Strand, A. (2018). Residential location, workplace location and car driving in four Norwegian cities. *European Planning Studies*, 26(10), pp. 2036–2057.

Ewing, R. & Cervero, R. (2010). Travel and the built environment. *Journal of the American Planning Association*, 76, pp. 1–30.

Ferreira, A.; Bertolini, L. & Næss, P. (2017). Immotility as resilience? A key consideration for transport policy and research. *Applied Mobilities*, 2(1), pp. 16–31, at http://dx.doi.org/10.1080/23800127.2017.1283121.

Finel, N. & Tapio, P. (2012). *Decoupling transport CO_2 from GDP*. FFRC ebook 1/2012. Turku: Finland Futures Research Centre, retrieved August 2018 from www.utu.fi/fi/yksikot/ffrc/julkaisut/e-tutu/Documents/eBook_2012-1.pdf.

Holden, E. (2016). *Achieving sustainable mobility: Everyday and leisure-time travel in the EU*. London/New York: Routledge.

Høyer, K.G. (1999). *Sustainable mobility: The concept and its implications*. PhD thesis. Roskilde: Roskilde University Center.

Isaksson, K. and Richardson, T. (2009). Building legitimacy for risky policies: The cost of avoiding conflict in Stockholm. *Transportation Research Part A: Policy and Practice*, 43(3), pp. 251–257.

Kasanen, P. (1994). *Demand for Mobility and Transportation*. Working Paper at SUSEM network meeting in Aalborg 24 to 25 May 1994.

Kaufmann, V. (2002). *Re-thinking mobility: Contemporary sociology*. Aldershot: Ashgate.

Litman, T. (2015). *Roadway Land Value*. Victoria, British Columbia: Victoria Transport Policy Institute. Retrieved August 2018 from www.vtpi.org/tca/tca0507.pdf

Mock, P. (2012). *Decoupling emissions from growing traffic volume*. Washington: International Council on Clean Transportation, retrieved August 2018 from www.theicct.org/blogs/staff/decoupling-emissions-growing-traffic-volume.

Morency, C. (2013). *Sustainable Mobility: Definitions, concepts and indicators*. Mobile Lives Forum. Retrieved August 2018 from http://en.forumviesmobiles.org/video/2013/02/12/sustainable-mobility-definitions-concepts-and-indicators-622.

Næss, P. (2001). Urban Planning and Sustainable Development. *European Planning Studies*, 9(4), pp. 503–524.

Næss, P. (2016). Urban planning: Residential location and compensatory behaviour in three Scandinavian cities. In Santarius, T.; Walnum, H. J. & Aall, C. (Eds.). *Rethinking climate and energy policies new perspectives on the rebound phenomenon*, pp. 181–207. Switzerland: Springer.

Næss, P.; Strand, A.; Wolday, F. & Stefansdottir, H. (2019). Residential location, commuting and non-work travel in two urban areas of different size and with different center structures. *Progress in Planning*, 128, pp. 1–36.

OECD (1997). OECD Proceedings: Towards sustainable transportation. The Vancouver conference. www.oecd.org/greengrowth/greeningtransport/2396815.pdf

Rosa, H. (2015). *Social acceleration: A new theory of modernity*. New York: Columbia University Press.
Schade, W. & Rothengatter, W. (2011). *Economic aspects of sustainable mobility*. Report for the European Parliament's Committee on Transport and Tourism. Brussels: European Parliament.
Strand, A., Næss, P., Tennøy, A. and Steinsland, C. (2009). *Gir bedre veger mindre klimagassutslipp?* TØI report 1027/2009. Oslo: Institute of Transport Economics.
Urry, J. (2000). *Sociology beyond societies: Mobilities for the twenty-first century*. London: Routledge.
Verroen, E.J., Jong, M.A., Korver, W. and Og Jansen, B. (1990). *Mobility profiles of businesses and other bodies*. Rapport INRO-VVG 1990-03. Delft: Institute of Spatial Organisation TNO.

41
TERMINAL TOWNS

Ida Sofie Gøtzsche Lange

Introduction

A significant part of urban mobilities is represented by the many different nodes of transportation that connect people and places, cities and nations in a global network. These nodes are physical places such as train stations and airports, equipped with specialized facilities to handle large numbers of people and means of transportation, e.g. cars and ships. Public and private organizations invest huge amounts of money to establish, develop, improve and run transportation nodes as they are often seen as strategic points and subjects to economic developments. In this chapter, I will challenge the way we conceptualize this global trend and how we have seemingly elided a vital element of it. I will do this by introducing the concept of Terminal Towns as a way of talking about towns or cities, which are also acting as transport hubs and thus being highly affected by transit. By suggesting this concept, I am questioning how the transportation of people and goods influences the local environment (physically, socially and economically). As the late John Urry stated:

> Issues of movement, of too little movement for some or too much for others or the wrong sort or at the wrong time, are it seems central to many people's lives and to the operations of many small and large public, private and nongovernmental organizations.
> *(Urry 2007, p. 6)*

In other words, urban mobilities are not neutral; neither are they universally good but might in fact come with some (unintended) costs. They can very often become areas of conflict as I will elaborate on in this chapter. First, I will unfold the term 'terminal' and expand on its meanings in relation to the theme of this book, urban mobilities. In this part I will elaborate on a geographical understanding of a terminal as well as on the notion of Shrinking Cities in relation to the etymology of the word 'terminal', before deriving the concept of Terminal Towns. Secondly, to unfold the concept of Terminal Towns further, I will use the case of Hirtshals, Denmark to exemplify the complex dynamics of being a small town and a growing port and the problems that arise between local and global interests of a transportation node. Thirdly, I will elaborate on the port city as a special type of Terminal Town. Finally, I will attempt to ask what constitutes a terminal town or city and give some theoretical and methodological points into studying terminal towns and cities further.

Terminals and the concept of Terminal Towns

The word 'terminal' holds a double meaning with contrasting connotations: as a noun it means '1a: either end of a carrier line having facilities for the handling of freight and passengers', 'b: a freight or passenger station that is central to a considerable area or serves as a junction at any point with other lines' or 'c: a town or city at the end of a carrier line' – in other words a physical place made for transit. As an adjective 'terminal' comes with a very different and distressing meaning of someone/something dying: 'leading ultimately to death//terminal cancer', 'approaching or close to death: being in the final stages of a fatal disease//a terminal patient' or meaning something 'extremely or hopelessly severe//terminal boredom' (Merriam-Webster 2018).

In the context of urban mobilities though, the noun would undoubtedly be the predominant understanding of the word. From a geographical point of view Rodrigue et al. define a terminal as 'any facility where freight and passengers are assembled or dispersed.' (Rodrigue et al. 2009, p. 164). In such understanding, transport terminals are of great importance as 'essential links in transportation chains' (ibid.), paving the way for global exchange and economic growth. Rodrigue, Comtois and Slack further describe how '[t]erminals play a key role in transport systems' (2009, p. 170) and how transport terminals have particular and strong spatial and functional characters:

> [Transport terminals] occupy specific locations and they exert a strong influence over their surroundings. At the same time they perform specific economic functions and serve as foci clusters of specialized services.
>
> *(Rodrigue et al. 2009, p. 164)*

As the quote suggests, transport terminals are enforced by physical, economic and relational dimensions. In terms of the spatial and physical dimension, Rodrigue, Comtois and Slack make a distinction between passenger and freight, as they explain how freight terminals take up much more space than terminals for passenger transportation. They also describe how '(l)ocation and spatial relations play a significant role in the performance and development of a transport terminal.' (Rodrigue et al. 2009, p. 171). For this, there are two dimensions involved: absolute location and relative location, meaning that there are both local geographical conditions involved for where it is appropriate to situate a terminal and relational geographical conditions for where it is strategically desirable for the location in terms of competition and cooperation with other terminals. Described within cluster theory, the concentration of terminals within a region has been identified as a critical element for terminal operators to join forces and compete for customers with terminals in other regions or nations, and thus '(t)he contribution of transport terminals to regional economic growth can often be substantial' (Rodrigue et al. 2009, p. 171). The understanding of terminals operating more or less intentionally within clusters suggests how terminals take part in larger networks of urban mobilities. In terms of cluster theory studies, ports have been more studied than other types of terminals, and it is recognized that 'port activity, historically at least, generates strong agglomeration economies that produce strong spatially distinct port communities' (Rodrigue et al. 2009, p. 171).

Considering economics, terminals are expensive in terms of constructing and maintaining physical buildings and infrastructures, and of transhipment and administration costs. However, corporate profits exceed the expenses, and the activities will also be of benefit on a regional and local level:

> Activities in transport terminals represent not just exchanges of goods and people, but also constitute an important economic activity. Employment of people in various terminal operations represents an advantage to the local economy. [...] It is no accident that centers that perform major airport, port and rail functions are also important economic locales.
>
> *(Rodrigue et al. 2009, p. 170)*

The universality though of this statement can be questioned in terms of the local benefits, at least in the case of ports. No doubt, terminals represent important economic activities. But in concurrency with containerization and technological improvements, the operation of port activities becomes less dependent on local workforces, and port–city relationships become more and more detached, as I will return to later in this chapter.

Going back to the etymology of the word 'terminal', following the path as an adjective, the meaning has not yet been ascribed to the urban mobilities theme in geographical terms, as has the noun. Nevertheless, the idea of a terminal town/city is highly relevant in the present time as many populated places are in fact shrinking all over the world, despite globalization being dominant and the world population growing. This phenomenon is widely described and addressed in the theoretical concept of Shrinking Cities (see also Beauregard 2015, Laursen 2008, Pallagst, Wiechmann & Martinez-Fernandez 2014, Richardson & Nam 2014, Ryan 2012). The concept was first described in 1980s Germany but has been adapted globally as it describes a worldwide problem. Laursen (2008) describes four different kinds of shrinkage: demographic, economic, socio-cultural and physical. She explains how the demographic situation is the overall indicator of whether a city is shrinking or growing. This can be caused by lower birth rates, an aging population and young people emigrating elsewhere for better education, job opportunities or cultural milieus. Economic shrinkage is at play when house prices are falling and the market is holding back investments in the area, and when e.g. schools are closing as the municipality cannot afford to run all their institutions due to decreasing tax revenues and budgets. The socio-cultural dimension of a shrinking city is related to a high level of unemployment, a weakened health care system, lower service levels, a weakened education system and overall a community without many social and cultural resources. Physical shrinkage is the most visible indicator of a shrinking city, as more and more buildings and infrastructures are worn out, some left empty for decay and some being torn down leaving holes in the urban fabric that no one has either the resources or impetus to redevelop. Depending on the degree of shrinkage, some shrinking cities or territories might experience some or all of the abovementioned kinds of shrinkage, but an important insight is to understand the complexity of the situation on different scales. There seems to be a delicate relationship between growth and decline:

> The contemporary urban fabric can be described as a conglomerate of greater and smaller urban concentrations living in the same organism. In this conglomerate there are build and open spaces as well as urban growth and urban decline.
>
> *(Laursen 2008, p. 94)*

This means that one place can be growing within the boundaries of an overall declining territory, and vice versa. Some places might be conceived as high value sites attracting investments for their strategic roles, while other places are left behind as contained in the concepts of Sticky and Slippery Spaces (Graham 2002, Markusen 1996). Depending on how much you zoom out, or which glasses you put on, a place can therefore be denoted as either sticky or slippery, as either growing or shrinking. This complexity is entailed in the concept of Terminal Towns as will be unfolded in a moment. The shrinkage is definitely related to the etymology of being terminal.

Based on the different perspectives above, derived from the meanings of the word 'terminal', I would here like to suggest a theoretical concept of terminal towns. A Terminal Town can be defined as any town where flows crossing are bigger than flows entering and where shrinkage of one or more kind is recorded. In the following, I will give a concrete example of a terminal town, namely Hirtshals, Denmark, showcasing those complex dynamics that form the bedrock for its conflicting meanings and why I would argue that this case can be perceived as a Terminal Town.

Hirtshals: a Terminal Town in Denmark (Lange 2016)

The complex dynamics of being a declining town and a growing port makes Hirtshals an extreme and paradigmatic case (Flyvbjerg 2001) of a terminal town. Hirtshals is situated at the fringe of North Jutland in the municipality of Hjørring. The port of Hirtshals is a strategic node between countries around the North Sea (Norway, Sweden, Great Britain, Iceland and the Faroe Islands) and continental Europe, with very good connections both land- and water-based. This is both in terms of passenger and freight transfers. Looking at the past decade, numbers are all increasing in terms of ships calling at the port, throughput of goods and passengers traveling via the port of Hirtshals, as seen in Table 41.1.

With connections to five destinations in Norway, on average ten ferries make their arrival to the port of Hirtshals every day, letting more than 2.5 million passengers travel through each year. This number is giddy in relation to the town population of approximately 6,000 people. In addition to the ferry connections, routes for freight transfer were established in 2016 to Zeebrugge in Belgium and Göteborg in Sweden (Hirtshals Havn 2018). In 2004 the Danish motorway network was extended to reach Hirtshals (Vejdirektoratet 2014). Thus, you can travel the 70 km from Aalborg (Denmark's fourth largest city) to Hirtshals in about 40 minutes (Krak 2016, Vejdirektoratet 2014). Hirtshals is also connected to the rail system in Denmark, and with frequent departures of 22 minutes

Table 41.1 Traffic numbers for the Port of Hirtshals

Call of vessels, passengers and throughput of goods in the Port of Hirtshals 2007–2018

Year	*2007*	*2008*	*2009*	*2010*	*2011*	*2012*	*2013*	*2014*	*2015*	*2016*	*2017*	*2018*
Ships calling at port	2 089	1 759	1 797	2 298	2 443	2 412	2 628	2 711	2 569	2 614	2 634	2 322
Throughput of goods, 1000 tons	1 284	1 205	1 230	1 315	1 352	1 505	1 449	1 589	1 769	1 781	1 880	1 896
Passengers, 1000 persons	1 838	1 888	2 063	2 178	2 247	2 246	2 345	2 479	2 567	2 581	2 530	2 491

Source: Statistics Denmark, © www.statbank.dk/SKIB101

between Hjørring and Hirtshals (NT 2013), there are also good opportunities for commuting by public transport. In 2005, new modern Desiro train sets were installed. In 2015, a so-called 'combined terminal' (also referred to as freight terminal) of 24,000 m^2 was established on the basis of a government investment of 33.5 million Danish kr. (approximately 4.5 million euro) (Hirtshals Havn 2015). In 2008 an extensive transport centre was inaugurated in the periphery of Hirtshals, accommodating a number of facilities for truck drivers and other business travellers.[1] In light of all these activities, Hirtshals does indeed function as a terminal. In 2008 more than 90 companies in the port participated in a survey on the economic impact of the Port of Hirtshals on the Region of North Jutland. The survey documents, *inter alia*, how the port creates 2,723 jobs (Hirtshals Havn 2008). Though this is positive, the benefits of the locals are relatively small. Many of the jobs are in fact possessed by people living elsewhere; in particular in Hjørring and Aalborg, which is only just possible due to all the good infrastructural connections to the port.

This leads to another terminal aspect of the case. Hirtshals is a shrinking town in terms of demography, socioeconomic aspects and a noticeable decline in the urban milieu. Since the middle of the 1990s the population has fallen from just above 7,000 inhabitants to 5,857 inhabitants in 2018. The decline is also visible in the urban fabric with empty and neglected houses as well as shops and other commercial leases in the inner city. On average, the citizens of Hirtshals have a higher consumption of health care per citizen (4.73) than the total consumption of the municipality (3.89). Hirtshals is also the sub area in the entire municipality which has the lowest education level, with more than 44% inhabitants with primary school education being the highest completed education and only 8% inhabitants with a middle or higher education (Hjørring Kommune 2014). The disposable income per person measured in 2012 was lower in Hirtshals than in Hjørring, and there are more temporary and permanent care services in Hirtshals than in the municipality in general (Hjørring Kommune 2015). Overall, the average household income is lower in Hirtshals city centre and along the busy roads, while it is highest in the more rural neighbourhood Emmersbæk and on the edge of Hirtshals, among other places in the western end of the city near the sea, where many of the homes have sea views (Hjørring Kommune 2014).

Viewing Hirtshals as a terminal, the physical, economic and relational dimensions are rather complex. Though the area of the town stays the same, the physical material structures within that area are in decay, whereas the spatial dimensions of the port are growing with new buildings and extensive landside usage, now exceeding the area of the town (see Figure 41.1). The economic situation is one of growth when examining the growing port, making Hirtshals an important node for the regional and the national economy. However, within the town, a situation of decline in urban functions as shops, banks, a school etc. are closing, and the average income of the inhabitants is lower for this particular town than for the municipal and regional inhabitants. Relationally, the port is benefitting from the cluster of ports in North Jutland, i.e. supported by a maritime centre for operations called Marcod, being 'an independent maritime center which strengthens and facilitates the maritime companies, network and competences in an international maritime industry' (Marcod 2016) with their goal 'to create growth in the Blue Northern Jutland' (ibid.). Flows and connections are widely understood as the basis on which the port, and thus the town, is built, but they also become barriers in several ways to the local environment. In Hirtshals, it becomes particularly clear as the barrier that the port infrastructure produces. This is a physical barrier in terms of road and embankment separating port and town, but also a physical barrier because the city becomes invisible from the roads as the embankment hides the town. Furthermore, there

Figure 41.1 Hirtshals port and town separated by road.

Source: Author's own production based on ortho photo from Styrelsen for dataforsyning og effektivisering, "GeoDanmark ortofoto", Spring 2019

may be a mental and social barrier in a more figurative way, because the town is faced with a negative reputation, while the port and infrastructure are more closely linked to progress (Lange 2016). In this way, the challenges can be found in the characteristics of the Terminal Town concept: the interlinkages between being a shrinking city as well as a growing terminal in a geographical understanding complicate how the place is to be understood – as a smooth transportation node or a lived town for residents, workers and tourists.

The port as a terminal and the port-city relationship

From the specific case of Hirtshals as a terminal town, also being a port city, I will here elaborate more generally on ports as terminals and the characteristics of port–city relationships. To begin with, I will argue for the port as a particular type of terminal (in the sense of a noun). It is a terminal between land and water, connecting and handling flows and exchange. It is also an urban typology covering specialized terminals (container-terminals, ferry-terminals etc.).

> Being the point of interaction between land and sea, ports traditionally served as economic and cultural centres of cities and surrounding regions. However, the contemporary technological advancement in shipping, increase in international trade and the global division of labour had fundamentally transformed the nature of

> ports. Notably, the 'terminalization' of port operations greatly modified their roles in transport networks and global supply chains, which implied an increased spatial and functional segregation between port, urban and regional activities (Ng, 2012, Olivier and Slack, 2006).
>
> *(Ng & Ducruet 2014, p. 785)*

In recent years, the research theme of Port City Relations has had a significant increased focus within the volume of port geography research papers[2] (Ng & Ducruet 2014). This trend is also reflected in the increased number of activities (conferences, sessions and workshops) organized both by professionals and academics on the theme of port–cities interrelations (Hein 2016). Further, networks and programmes have been formed to explore the theme, e.g. the Association Internationale Ville et Ports (AIVP), the Association for the Collaboration between Ports and Cities (RETE), OECD's Port City Programme and the LDE Centre for Metropolis and Mainport, the latter also doing the Port City Futures initiative 'investigating the evolving spatial use and design of port city regions over time, in particular addressing when port and city activities occur in the same places and sometimes conflict' (LDE Metropolis and Mainport, n.d.).

Within the research of ports there have been drawn several different models, from several different perspectives, to capture and explain how ports develop. The most famous might be the Anyport Model, developed by James Bird and published in 1963, describing six consistent stages of port evolution based on an extensive study of the development of major British ports (Bird 1963).

> Anyport's development has been analysed into six eras terminated by events which can be dated. Each of these eras was called into being by increased trade and technical advance in the design of ships. The ship designer has always been the pacemaker in shipping transport innovations since his creation has merely to float and sail economically per ton mile; whereas the port engineer has to cope not only with the demands of ship designers, but also with the physical difficulties of the port's land and water sites. Despite these extra problems the port engineer must provide his new facilities before the ship that has made them necessary comes into service, or trade may be lost. One era supersedes another, but installations remain. [...] Because of the great capital cost of port installations, it is often cheaper progressively to downgrade a dock or quay in traffic importance rather than scrap it altogether. [...] One of the most fascinating aspects of port study is that the various eras of port development can be seen co-existing in present port lay-outs.
>
> *(Bird 1963, pp. 33–34)*

Zooming in on the port–city interface, Hoyle (1988) describes five stages of evolution from *the primitive cityport* having a close spatial and functional association between city and port, via *the expanding cityport* and *the modern industrial cityport* where port facilities expand and gradually grow further away from the city, to *the retreat from the waterfront* where outdated port industries in the city centre close and leave abandoned areas in the city centre, consolidating a clear separation of port and city; and finally *the redevelopment of the waterfront* with urban renewal within the original portcity core, a redevelopment of the so-called 'abandoned doorstep'. Hoyle describes how port–city relationships/interfaces changed especially in the 1970s and 80s (1988):

> The modern seaport acts as a gateway rather than as a central place [...] weakening the traditionally strong functional ties between ports and cities.
>
> *(Hoyle 1988, p. 3)*

> Ports have become more noticeably national gateways, while cities have retained their regional and local functions. Herein lie the seeds of separation between port and city.
>
> *(Hoyle 1988, p. 6)*

More recently a model on nine different typologies of port cities (Ducruet 2007) has been drawn. It comes as a matrix of port–city relationships, having on one axis low to high centrality, describing the size and impact of the city and its urban functions, and on the other axis low to high intermediacy, describing the size and impact of the port from local to global:

> The opposition between city size and port performance is a recurrent issue [...]. It means that the efficiency of port operation contradicts the importance of the urban centre, because of growing congestion at the port–city interface. Although coastal location has been an advantage in attracting population and industries close to ports for economic reasons, the spatial effects of such concentration have increasingly led to diseconomies of scale and of agglomeration. It also confirms in some way the opposition between the hub port, dominated by maritime shipment, and the coastal city or metropolis, dominated by central place functions.
>
> *(Ducruet 2007, pp. 9–10)*

These three models and many others emphasize on different levels the relationship between port and host city and how these evolve and grow (together or apart). From a more specialized point of view, a 5-stage model on the process of container port terminal development has been suggested by McCalla in 2000, culminating in the formation of so-called 'Super-terminals' through the final stage: *redevelopment*:

> Redevelopment may represent a radical approach to port investment and administration. No longer are old practices acceptable; a new approach must be tried to make the port competitive and expand its container throughput. However, this process proceeds in an environmentally aware society, very different from that encountered in the early stages. Not everyone may share the port's enthusiasm for increasing its scale of operations. Consequently, in the attempt to refocus container-handling operations in the port, opportunity is given to environmentalists and local residents to challenge the redevelopment. Thus, the development–environment interface can become a key issue obstructing the model's progression.
>
> *(McCalla 2000, p. 127)*

As McCalla observes, the growing ports are not universally viewed as a boon. There can be many reasons for such objections or reservations expressed towards expanding terminals. Often it is on a local level that the downsides are experienced:

> Ports are important economic drivers for cities and regions, well beyond their often secluded and fenced-off locations. Their impact on land and sea, on people in cities and rural areas is extensive, particularly if shipping and maritime headquarters are also located in the city. [...]. The economic benefits of the port often spread beyond the port, whereas the negative impacts are local.
>
> *(Hein 2016, p. 314)*

Hirtshals has been showcased as an example of such a situation. More generally, I would stress that such conflict between local and global interests is at the core of Terminal Towns.

Conclusion and perspectives for the study of Terminal Towns

In this chapter, I have presented the concept of Terminal Towns. Such places fall victim to a two-edged sword. On one hand, they are largely dependent on their flows and connections; in the case of Hirtshals these flows and connections are in fact somewhat the foundation of the town's existence. Urban mobilities, in the form of transit to and from the port of Hirtshals are bringing progress and growth, economically and spatially. On the other hand, Terminal Towns are challenged and even compromised by the exact same transit flows and connections, as they become barriers in the local environment, not only physically but also mentally and socially.

Hirtshals has been used as an example of a town that encapsulates the ideas nested in 'a geographical understanding of terminal' and 'a shrinking city', thus coming across as a Terminal Town. In the case of Hirtshals, the 'Terminal Town' aspect is empirically supported by the decisive identity of being a port city. Hirtshals is even a port city of the type outport or hub port, with a high degree of intermediacy and a low degree of centrality (Ducruet 2007) which supports the concept of a Terminal Town even more. Thus, I have exemplified the concept of Terminal Towns through the case of a particular port town. As described, ports have special characteristics as infrastructural nodes with often long urban historic meanings attached. In this context the concept of Terminal Towns could be used to study the port–city relationship much further through cases of everything from coastal towns to global port cities (cf. the different typologies of port–city relationships by César Ducruet (2007)). This does by no means imply that all port cities can be perceived as terminal towns, but as the inherent meaning of a port will very often correspond with the meaning of a terminal in the geographical sense, ports could be obvious places to examine for further aspects of the terminal perspective.

This said, the concept of terminal towns is not at all limited to comprising ports. Railway towns, airport cities, border towns and other cities with influential transport centre functions could be investigated in the light of the Terminal Town perspective. Looking for places that at the same time meet the characteristics of 'a geographical understanding of terminal' and of 'a shrinking city' would be the first step to identify a Terminal Town.

To study a place as a Terminal Town it is necessary to unfold potential areas of conflict by including both local and global perspectives, and to ask both which benefits and downsides that stem from the urban mobilities take part in constituting the place in focus. And maybe even more importantly, ask who are being compromised to the detriments of growth for someone else. Thus, the concept of Terminal Towns comes with an ethical obligation to study the power relations that take place in and around areas that experience such complex problems with interrelated growth and decline. Because, as the introductory quote by John Urry reminds us, urban mobilities are very much affecting the lives of people.

Notes

1 The transport centre includes customs, forwarding agencies, petrol station with diesel for trucks and cars and service station facilities. In addition, the transport centre includes driver's lounge, double rooms, bathroom facilities, offices and conference rooms.
2 This increasing trend is found both in geography and non-geography journals, reviewed in the period 1950–2012.

References

Beauregard, R. (2015) 'Shrinking Cities'. *International Encyclopedia of the Social & Behavioral Sciences*, 21: 917–22.

Bird, J. (1963) *The Major Seaports of the United Kingdom*. 1st edn. London: Hutchinson.

Ducruet, C. (2007) 'A metageography of port-city relationships'. In J. Wang, D. Olivier, T. Notteboom & B. Slack, eds. *Ports, Cities, and Global Supply Chains*. Aldershot: Ashgate. pp. 157–72.

Flyvbjerg, B. (2001) *Making Social Science Matter: Why Social Inquiry Fails and How It Can Succeed again/ Bent Flyvbjerg*. New York, NY: Cambridge University Press.

Graham, S. (2002) 'Urban network architecture and the structuring of future cities'. In H. Thomsen, ed. *Future Cities: The Copenhagen Lectures*. København: Fonden Realdania. pp. 109–22.

Havn, H. (2008) *Den Erhvervsøkonomiske Betydning*. Hirtshals: Hirtshals Havn.

Havn, H. (2015) *Kombiterminal*. [Hjemmeside for 'Hirtshals Havn']. 22. maj 2015. Tilgængelig på: www.hirtshalshavn.dk/Aktuelt/Det-sker/Kombiterminal [tilgået: 22/12 2015].

Havn, H. (2018) *Rutenet fra Hirtshals Havn*. [Hjemmeside for 'Hirtshals Havn']. Tilgængelig på: www.hirtshalshavn.dk/Forretningsomraader/Transport/Rutenet-fra-Hirtshals-Havn [tilgået: 25/10 2018].

Hein, C. (2016) 'Port cityscapes: Conference and research contributions on port cities'. *Planning Perspectives*, 31 (2): 313–26.

Hoyle, B. (1988) 'Development dynamics at the port-city interface'. In B.S. Hoyle, D.A. Pinder & M. S. Husain, eds. *Revitalising the Waterfront: International Dimensions of Dockland Redevelopment*. London and New York: Belhaven Press. pp. 3–19.

Kommune, H. (2014) *Hirtshals*. Folder internt i kommunen. Hjørring: Hjørring Kommune [Ikke publiceret materiale].

Kommune, H. (2015) *Hirtshals Som Vækstområde*. PowerPoint-præsentation ved 'Hirtshals Kontorchefseminar' d. 24. juni 2015. Hjørring: Hjørring Kommune [Ikke publiceret materiale].

Krak (2016) *Krak Kort*. [Hjemmeside for 'Eniro/Krak']. Tilgængelig på: http://map.krak.dk/[tilgået: 06/02 2016].

Lange, I. S. G. (2016) *Transit- Eller Leveby?: Et Casestudie Af Hirtshals Som Et Stærkt Mobilitetspåvirket Sted I Gennemfartsdanmark [Transit Town or Living Town? A Case Study of Hirtshals as A Strongly Mobility Affected Place in 'Thoroughfare Denmark']*, Aalborg: Aalborg Universitetsforlag.

Laursen, L. L. H. (2008) *Shrinking cities or urban transformation*. PhD, Aalborg University.

LDE Metropolis and Mainport (n.d.) *Port City Futures*. [Hjemmeside for 'LDE Metropolis and Mainport']. Tilgængelig på: http://portcityfutures.org/[tilgået: 25/10 2018].

McCalla, R. J. (2000) 'From "Anyport" to "Superterminal"'. In David Pinder & Brian Slack, eds. *Shipping and Ports in the Twenty-first Century: Globalisation, Technological Change and the Environment*. Florence: Routledge. pp. 123–124–142.

Marcod (2016) *About MARCOD*. [Hjemmeside for 'Marcod']. Tilgængelig på: www.marcod.dk/en/about-marcod [tilgået: 31/10 2018].

Markusen, A. (1996) 'Sticky places in slippery space: A typology of industrial districts'. *Economic Geography*, 72 (3): 293–313.

Merriam-Webster (2018) *Terminal*. [Hjemmeside for 'Merriam-Webster, Inc.']. Tilgængelig på: www.merriam-webster.com/dictionary/terminal [tilgået: 4/10 2018].

Ng, A. K. Y. & Ducruet, C. (2014) 'The changing tides of port geography (1950–2012)'. *Progress in Human Geography*, 38 (6): 785–823.

NT (2013) *NT - Nordjyllands Trafikselskab*. [Hjemmeside for 'Nordjyllands Trafikselskab (NT)']. Tilgængelig på: www.nordjyllandstrafikselskab.dk/admin/library/media/ntbilleder/2012_31/hirtshalsbanen-fra-02-04-2013-ver3.pdf [tilgået: 25/06 2013].

Pallagst, K., Wiechmann, T., & Martinez-Fernandez, C. (eds.). (2014) *Shrinking Cities: International Perspectives and Policy Implications*. 2nd edn. Ney York: Routledge.

Richardson, H. W. & Nam, C. W. (eds.). (2014) *Shrinking Cities: A Global Perspective*. New York: Routledge.
Rodrigue, J., Comtois, C., & Slack, B. (2009) *The Geography of Transport Systems*. 2nd edn. Oxon: Routledge.
Ryan, B. D. (2012) *Design after Decline: How America Rebuilds Shrinking Cities*. Philadelphia, Pennsylvania: University of Pennsylvania Press.
Urry, J. (2007) *Mobilities*. Cambridge & Malden: Polity Press.
Vejdirektoratet (2014) *Motorvejen Bjergby – Hirtshals*. København: Vejdirektoratet.

INDEX

For Product Safety Concerns and Information please contact our EU
representative GPSR@taylorandfrancis.com
Taylor & Francis Verlag GmbH, Kaufingerstraße 24, 80331 München, Germany

www.ingramcontent.com/pod-product-compliance
Ingram Content Group UK Ltd.
Pitfield, Milton Keynes, MK11 3LW, UK
UKHW050053200926
475702UK00019B/2810

* 9 7 8 0 3 6 7 4 9 1 5 6 7 *